*This book is dedicated
to the memory of my mother and father,
who provided a beacon for my life
by the qualities they exhibited
and the principles they embraced:
love of family, hard work, and integrity.*

CONTENTS

14   OPTOELECTRONICS   428

APPENDICES

PREFACE

This book is written to serve as a textbook for an undergraduate course, at the junior or senior level, for students majoring in electrical or computer engineering. An appropriate background for those using the book includes courses in mathematics that cover ordinary differential equations, a course in modern physics, and a course in basic circuit theory. So that a course on devices becomes more relevant and interesting to students, it should be accompanied by a course in electronic circuits that relates applications of the devices to circuits.

It is the intention of the author that the major contents of this book can be covered in a one-semester course. It is however, left to the discretion of the individual instructor adopting this book to select the chapters or sections that he or she considers most important while maintaining the continuity and rigor of the discussions.

CONTENTS

This book focuses on three major categories of devices: the junction diode, the bipolar junction transistor, and the field-effect transistor in its various forms. Of the fourteen chapters, the first four are devoted to the study of the physical and electrical conducting properties of semiconductors. Two chapters each, not necessarily in this order, are then used to cover the PN junction diode, the bipolar junction transistor, and the metal-oxide semiconductors FET. Finally the junction FET, the metal-semiconductor FET, optoelectronic devices, and fabrication technology are each the subject of one chapter.

In the various chapters on devices, derivations of analytical relations are described in the middle or at the end of the chapter. The sequence of presentation of material is as follows: The first section of each chapter deals with the general construction and operation of the device, using diagrams and typical characteristics. This is followed by an explanation, using drawings, of the steps in the fabrication of the device. The third section is devoted to the detailed description of the graphical characteristics and their relation to the operation of the device. In the next section, analytical relations are derived which are related to and complement the operation and the graphical characteristics.

STYLE AND PRESENTATION

Teachers who adopt this book will note three characteristics in the style of presentation and in the subject coverage that the author has developed over a long teaching career. First, there is less emphasis, compared to some other textbooks, on some of the difficult (for students) concepts of quantum mechanics and their underlying mathematical complexity. In this attempt the author has been careful to avoid compromising the accuracy and thoroughness of the material for which an understanding of the basic quantum mechanical concepts is necessary.

Second, as a general rule, short sentences are used in short sections or paragraphs. The subsections are brief, but numerous, thus giving the student the opportunity to assimilate the material before proceeding to the next subsection without loss of continuity. Long sentences and long paragraphs that deal with complex material tend to discourage, if not confuse, the student and hence do not enhance the learning process.

Third, and whenever necessary and possible, numerical examples are used. These examples serve both to illustrate the applications of the analytical relations and to acquaint the student with the order of magnitude of certain properties of the device.

NOVEL APPROACHES

To complement the attempt of this author at providing a book that is most helpful to students, two other rather unique approaches have been attempted: first, the inclusion of sections on Review Questions, Highlights, and Exercises in each chapter and second, the manner of introduction of the fabrication techniques.

Review Questions, Highlights, and Exercises

In each of the thirteen chapters of the book, excluding a chapter on fabrication technology, two special sections are introduced. One section is introduced at about the middle and the other one at about the end of the chapter. Each section consists of the following three subsections: Review Questions, Highlights and Exercises. The Review Questions are to be used by the student to test his or her understanding of the material that has been presented. The Highlights summarize the important concepts and relations described in the preceding pages. The Exercises consist of application of analytical relations derived in the test.

These three subsections serve to further enhance the learning process through two mechanisms. First, they identify the basic and important concepts presented. Second, they provide the student, through the Review Questions, the background and opportunity to ask questions in class that will help to clarify and reinforce the concepts that are introduced. Through a reading of the text, the Review Questions, and the Highlights, the student will feel more confident and encouraged to identify the importance of his or her own questions and thus, through these questions, present the feedback that is so necessary to the process of teaching and learning.

Fabrication

The second unique arrangement grew out of the question as to where, in the sequence of chapters, a chapter on the fabrication of devices should best be located. Placing such material in an early chapter that precedes the study of the devices seemed out of order since at that point the students know very little about the different devices. On the other hand, placing it toward the end of the book makes it of marginal interest to the student. Only the very interested student will turn the pages back to the chapter where each device is studied, to try to connect the fabrication to the operation.

In this book a compromise and workable arrangement has been employed, namely to integrate the fabrication of the major devices with the study of their operation. To my knowledge, this has not been attempted in any other textbook. A separate and early chapter is devoted to the description of the various major fabrication processes (mask generation, diffusion, oxidation, epitaxy, etc.). At the beginning of the chapter or section dealing with a particular device except for the PN diode, following a basic description of the operation and schematic layout of the device, the complete fabrication of the device is carried out, illustrated by a sequence of drawings.

Through this arrangement, the illustration of the fabrication of the device is integrated with the discussion of its operation so that the student can relate the operation of the device to its actual construction.

This author believes that this technique of the introduction of fabrication adds an extra dimension to the student's conception of the device. Exposing the student to the delicate processes and to the relevant dimensions involved places the outlook of the student on the devices at a more realistic level.

ACKNOWLEDGMENTS

The impetus to write this book grew out of the author's interest in the teaching of this field. This interest has been nurtured by the large number of students who have attended his courses and participated in the class discussions. Those students helped to highlight the teaching and presentation methods that are most important to the learning process. To those students the author is grateful.

The author acknowledges with appreciation the valuable comments made by Dr. Doug Shire on the chapter Fabrication Technology.

The author gratefully acknowledges the following reviewers: Betty Lise Anderson, Ohio State University; Griff L. Bilbro, North Carolina State University; James J. Coleman, University of Illinois at Urbana-Champaign; Linda Head, SUNY Binghamton; Robert Hunsperger, University of Delaware; Patrick McCann, University of Oklahoma; H. G. Parks, University of Arizona; Sannasi Ramanan, Rochester Intitute of Technology; Gregory Stillman, University of Illinois at Urbana-Champaign; Massood Tabib-Azar, Case Western Reserve University; Tingwei Tang, University of Massachusetts at Amherst; John Uyemura, Georgia Institute of Technology; John Wager, Oregon State University; Mark Weichold, Texas A&M University; Wanda Wosik, University of Houston

The author acknowledges also with appreciation and gratitude the work of Mrs. Maureen Morano in her untiring efforts at the typing of many versions of the manuscript. Her expertise and efficiency and the superb quality of her work have been an invaluable asset.

Last, but not least, the author wants to praise the support, tolerance, and encouragement of his wife during the four years of the development of this manuscript.

chapter 1

ATOMIC STRUCTURE AND QUANTUM MECHANICS

1.0 INTRODUCTION

Major advances in the fields of semiconductor materials and semiconductor devices have taken place within the last four decades. These advances have been largely responsible for the information revolution, both in the processing and transmission of intelligence.

Materials used in the fabrication of large-scale circuits may be classified as: good conductors, insulators, and semiconductors. The basic difference between the three is their resistance to current flow, defined in terms of the resistivity of the material. Good conductors, such as copper and aluminum, have a resistivity of less than 10^{-3}ohm-cm and are used for low-resistance wiring and interconnections in electric circuits. Insulators that have resistivities greater than 10^5ohm-cm are used as isolators of circuits and devices, and in the formation of capacitors.

Between these two limits of resistivities lie the semiconductor materials: the pure elemental semiconductors, silicon and germanium, and some II-VI, IV-VI, and III-V compounds, the most important of which is gallium arsenide.

Their mid-range resistivity, or conductivity, is not the direct reason for the importance of semiconductors. Rather, it is the extent to which their properties are influenced by light, temperature, and more importantly, by the addition of minute amounts of special impurities. Extensive changes take place in the resistivity of silicon when one part of an impurity is added to a million parts of silicon. Silicon is abundant in nature in the form of sand (silica) and clay. However, before silicon can be used in devices, major purification of the material is required. An application that illustrates the use of the three types of materials in the formation of integrated circuits follows.

Integrated circuits are fabricated on 8 to 20cm diameter circular sections of *silicon* known as *wafers*. The major components in integrated circuits are transistors,

1

capacitors, and resistors. In the fabrication process, controlled amounts of particular impurities are added to the silicon in order to form the various regions of a transistor. Certain sections of the wafers are oxidized to form *silicon dioxide*, which serves as an insulator and as an isolator of different circuit parts. After the various devices are formed, a *good conductor*, such as *aluminum*, is deposited on the surface to interconnect the various parts of the circuits. Finally, the complete circuit, with external leads attached, is encapsulated in a plastic or ceramic package.

Our objective in this book is to study, first, some of the properties of pure semiconductors and those semiconductors to which impurities are added. Having studied these properties, we then investigate the operation and the current-voltage characteristics of semiconductor devices, such as diodes and transistors. To do that, we need to consider the internal structure of semiconductors, the types of current carriers, and the modes of transport of these carriers.

1.1 CRYSTALS AND THE UNIT CELL

Based on the internal arrangement of the atoms, a solid is labeled as *amorphous, crystalline,* or *polycrystalline.* In crystalline solids, the atoms are arranged in an orderly three-dimensional array that is repeated throughout the structure. Amorphous solids have their atoms arranged in a very random manner with no repeated pattern. The atoms in polycrystalline solids are so arranged that, within certain sections, some sort of a pattern of the atoms exists but the various sections are randomly arranged with respect to each other.

Most semiconductors are crystalline in nature. Let us look into the internal arrangement of the atoms of semiconductors in the basic building block known as the *unit cell.* The arrangement inside the unit cells of silicon and germanium is known as the *diamond lattice* because this arrangement is a characteristic of diamond. Diamond is a form of carbon, which is an element in Column IV of the periodic table. In the diamond lattice, shown in Fig. 1.1, the atoms are arranged within a cube having dimension L, where L is known as the lattice constant.

The unit cell for the diamond lattice has an atom in each corner of the cube (8), one at the center of each of the six faces (6), and four (4) internal to the cube located along the diagonals. From this total of 18 atoms, the atom at each corner is shared by eight cells and each face atom is shared by two cells.

We will use the above information in the following examples to calculate the density of atoms and the mass density of silicon.

EXAMPLE 1.1

a) Determine the number of atoms in each cell of the diamond lattice.
b) Determine the density of atoms in silicon, given the lattice constant $L = 5.43\text{Å}$.

Solution

a) Eight atoms are shared by eight cells, six atoms are shared by two cells, and four atoms are internal to the cell so that:

$$\text{the number of atoms in each cell} = 8/8 + 6/2 + 4 = 8.$$

b) Volume of cell = $(5.43 \times 10^{-8})^3$ cm^3.

$$\text{Density} = 8/\text{volume} = 5 \times 10^{22} \text{ atoms/cm}^3.$$

EXAMPLE 1.2

Determine the density of silicon given Avogadro's number is 6.023×10^{23} atoms/mole.

Solution   From the periodic table of elements, the atomic weight of silicon is 28.09. The density of silicon atoms was found in Example 1.1 to be 5×10^{22} cm^{-3}.

$$\text{Density} = \frac{5 \times 10^{22} \text{ atoms/cm}^3 \times 28.09 \text{ g/mole}}{6.023 \times 10^{23} \text{ atoms/mole}} = \frac{2.33 \text{ g}}{\text{cm}^3}.$$

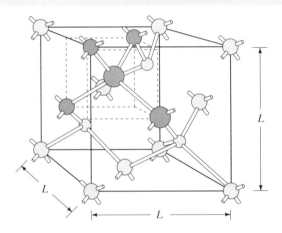

Figure 1.1   Unit cell of the diamond lattice. *Source: Electronic Materials Science* by Mayer/Lau, © 1990. Reprinted by permission of Prentice-Hall, Inc., Upper Saddle River, NJ.

Gallium Arsenide, a III-V compound formed from Gallium (III) and Arsenic (V), crystallizes in the form of a slightly different unit cell, known as the *Zincblende Structure*. In this structure, Ga and As atoms are found at alternate locations inside the unit cell.

On close study of the unit cell lattice of Fig. 1.1, we note that each atom is attached to the four nearest neighboring atoms. This attachment of atoms, in fact, represents the force that holds the lattice atoms together and is known as *covalent bonding*.

Each atom of Column IV elements has four electrons in its outermost shell. The covalent bonding results from the sharing of electrons between atoms. When each atom, say A, is shown bonded to four neighboring atoms, each of the four neighboring atoms contributes one electron to the bond with A. Atom A, therefore, contributes one electron to each bond, so that two electrons are shared by atom A with each of the four atoms. This sort of bonding accounts for some physical properties of these solids.

1.2 WHAT ARE WE LOOKING FOR NOW?

Eventually, we are interested in determining the current in a semiconductor device in response to the application of a source of energy, such as an electric source or light source. To do that, we need to know the types and densities of the current carriers and their masses.

In a semiconductor, there are bands of energy in which electrons can exist and other bands in which they cannot. We need to establish the bases for the formation of these bands. The existence of bands results from the allocation of specific energy levels to an electron in an atom and the consequent displacement of these energy levels by introducing the effect of the forces that atoms exert on each other. As a result, each electron in the solid has a specific energy level; the combinations of these levels form bands.

Before we study the formation of bands, we will present a perspective of the theories that have evolved and have attempted to explain experimentally observed properties of heated materials. Classical mechanics' concepts fail to explain these properties as well as all other phenomena that take place at the atomic levels. Such explanations required the evolution of the science of Wave Mechanics, which treats electrons as particles that have wave-like properties.

1.3 FAILURE OF CLASSICAL MECHANICS AT ATOMIC LEVELS

Two experimentally observed phenomena evolved that could not be explained by the theories of Classical or Newtonian mechanics. These were: blackbody radiation and the sharp discrete spectral lines emitted by heated gases. First, we will consider blackbody radiations.

A blackbody is a heated solid labeled as an ideal radiator of electromagnetic waves. When a solid is heated, it emits radiation over a certain band of frequencies. The heated atoms vibrate so that the amplitudes and frequencies of the vibrations seem to resemble the radiating antenna of a broadcast station. Classical mechanics predict that this heated solid emits radiation over a continuous band of frequencies. Measured responses, however, indicate that this is not true. As a matter of fact, radiation takes place only over a certain band of frequencies. Furthermore, the frequency spectrum of the radiation changes as the temperature of the solid is varied, as shown in Fig. 1.2.

A similar phenomenon occurs when certain gases are heated. It was experimentally observed that heated gases emit radiation in small discrete quantities at certain discrete spectral wavelengths. Again, Classical mechanics had no explanation for this.

The conclusion that was reached confirmed that Classical mechanics could not predict phenomena that occur at microscopic scales or at atomic levels.

1.4 PLANCK'S HYPOTHESIS

The first breakthrough came about with the work of Max Planck. Planck's hypothesis, presented in 1901, was that light is emitted or absorbed in discrete units of energy called *photons*. The vibrating atoms of a heated body emit radiation at a frequency v and the energy is restricted to certain discrete, or quantized values, given by

$$E_n = nhv \text{ for } n = 0,1,2,3,\ldots \tag{1.1}$$

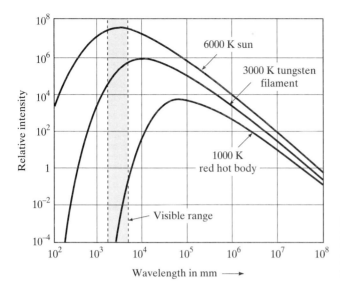

Figure 1.2   Relative intensity of radiation from a blackbody versus wavelength at various temperatures. From E. Uiga, *Optoelectronics*, Prentice Hall (1995)

where v is the frequency of radiation and h is a factor that Planck obtained by a theoretical fit to the experimental results. This is labeled Planck's constant and its value is 6.62×10^{-34}J-s.

Planck's hypothesis confirmed that, in addition to the quantizing of electromagnetic waves, light may also be viewed as consisting of particles labeled *wave packets*, or *photons*. This laid the foundation for the theory of the dual nature of light. It also paved the way in explaining the energy levels of electrons in a solid, as it considered electrons to be particles of matter while at the same time possessing wave-like properties.

At this time, Rutherford explained electron behavior as resulting from a circular motion around the nucleus.

Following upon the conclusions of Rutherford and Planck's hypothesis, Niels Bohr, in 1913, put forward a model confirming the planetary-like motion of electrons around an atom and included the quantum-like theory of Planck.

We will now consider Bohr's theories beginning with his classical model of the atom.

1.5 BOHR'S CLASSICAL MODEL OF THE ATOM

The work of Rutherford compared the nucleus of an atom to the sun and the electrons that revolve around the nucleus to the planets that revolve around the sun. Under the influence of gravitational forces, the planets move around the sun in a quasi-circular orbit.

Bohr suggested that within the atom, gravitational forces (due to masses) become negligible when compared to the electrostatic forces (due to charges) that control the orbit of the electron. The electrical forces are determined by Coulomb's Law, which relates the force exerted by one charged object on another.

In accordance with Bohr's model, we will use Coulomb's Law for the one-electron hydrogen atom and determine an expression for the total energy that the electron possesses, assuming that it is rotating in a fixed circular orbit about the nucleus.

Energy of Hydrogen Electron

In a hydrogen atom, the electrostatic attractive force between the nucleus, consisting of the proton, having a charge $+q$ and the electron having a charge of $-q$ is given by

$$F = \frac{-q^2}{4\pi\varepsilon_0 r^2} \tag{1.2}$$

where ε_0 is the permittivity of free space in Farads/meter, r is the separation between the electron and the nucleus in meters, and F is in newtons. We assume that the mass of the proton is much greater than that of the electron so that the proton is fixed at the center of the system. To maintain motion in a fixed circular orbit having radius r, the electrostatic force is balanced by an equal and opposite centripetal force. We assume that the positive reference direction for the force is outward so that the attraction force in Eq. (1.2) is negative.

Because of the presence of the positively charged nucleus, an electric field is created around it. Electric field intensity, $\mathscr{E}$, defined as the force on a unit positive charge, is directed outward and is thus positive. From Eq. (1.2), we obtain the expression for $\mathscr{E}$ at the location of the electron as

$$\mathscr{E} = \frac{q}{4\pi\varepsilon_0 r^2} \tag{1.3}$$

where, for the units used in Eq. (1.2), $\mathscr{E}$ is in volts/meter. The electric field intensity approaches zero as r approaches infinity. The electrostatic potential, V, due to the electric field, is defined by

$$V = -\int_0^x \mathscr{E}\, dx \tag{1.4}$$

Assuming that the reference for zero potential is at infinity, and since the system is assumed to possess spherical symmetry, we determine the expression for the potential of a point at r as

$$V = \int_\infty^r -\frac{q\,dr}{4\pi\varepsilon_0 r^2} = \frac{q}{4\pi\varepsilon_0 r} \tag{1.5}$$

The potential energy of the electron at a distance r, defined as the product of the potential at that point and the charge, becomes

$$W(\text{P.E.}) = V(-q) = \frac{-q^2}{4\pi\varepsilon_0 r} \tag{1.6}$$

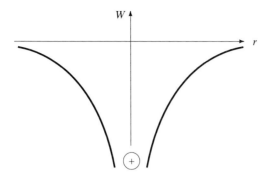

Figure 1.3 Potential energy of the electron in the field established by the nucleus.

The negative sign implies that in moving an electron from infinity to r, negative work is done in contrast to the positive work that needs to be expended in moving a positive charge towards the nucleus. This implies that an electron at point r has to be restrained to prevent it from moving towards the nucleus. A sketch of the potential energy is shown in Fig. 1.3. We note that the actual drawing has spherical symmetry so that it has the shape of a funnel.

We are interested in the total energy that the hydrogen electron possesses, which, in addition to potential energy, includes the kinetic energy resulting from the velocity of the electron. As it moves in the circular orbit, the electron is subjected to a centripetal force (mv^2/r), which exactly balances the attraction force of the nucleus, causing an angular acceleration v^2/r.

By setting the magnitude of the force in Eq. (1.2) equal to the magnitude of the centripetal force, we have

$$\frac{mv^2}{r} = \frac{q^2}{4\pi\varepsilon_0 r^2} \tag{1.7}$$

After solving for mv^2, the expression for the kinetic energy becomes

$$K.E. = \frac{mv^2}{2} = \frac{q^2}{8\pi\varepsilon_0 r} \tag{1.8}$$

The total energy, E, of the electron is

$$E = \text{K.E.} + W(\text{P.E.}) = \frac{-q^2}{4\pi\varepsilon_0 r} + \frac{q^2}{8\pi\varepsilon_0 r} = \frac{-q^2}{8\pi\varepsilon_0 r} \tag{1.9}$$

In accordance with the references established earlier, the electron has the highest energy (least negative) as its distance from the nucleus approaches infinity.

The Hydrogen Atom as a Radiating Antenna

The relations we just derived indicate that the electron moves in a circular orbit having a radius r, velocity v, and total energy E. If a hydrogen atom is heated, the electron absorbs energy and moves to a higher level, corresponding to a new total energy E. It is thus possible for the electron to occupy any orbit, depending upon the energy it gains and, the orbital radius is determined from Eq. (1.8.) Hence, according to this classical theory, the electron can take on continuous values of energies E, depending upon the radii of the orbit. This, however, contradicts the results of *diffraction experiments*. Diffraction is defined as the scattering that light rays undergo when they pass through slits of a solid. The light rays emerge deflected, forming fringes of parallel light, dark, and colored bands. The diffraction experiments, using electrons, indicated that electrons can only occupy certain discrete energy levels as they emit radiation only at certain discrete frequencies.

The question is then: How does the electron emit radiation?

Electron motion at constant speed results in a constant electric current. Since the electron that is orbiting about a proton is subjected to centripetal acceleration, its velocity increases with time, resulting in a time-dependent current. From the laws of electromagnetic theory, a time-varying current generates a time-varying magnetic field. By Maxwell's equation, a time-varying magnetic field induces a time-dependent electric field. Electromagnetic fields and electromagnetic waves are generated, leading to radiation that is similar to that emanating from a television or radio transmitting antenna. Such radiation of energy may be possible over a wide band of frequencies.

If the hydrogen electron is to emit radiation, it must continuously lose energy, revolve in smaller helical orbits with decreasing velocity, and get closer and closer to the nucleus. Hence, if a charge is accelerated, it radiates energy and therefore experiences a loss of energy. This dilemma, faced by Bohr's classical model, led him to qualify it by his quantized model.

1.6 BOHR'S QUANTIZED MODEL

The classical theory proposed by Bohr obviously failed when it was applied to the hydrogen atom. In trying to explain the dilemma he faced, Bohr proposed a model that confirmed the concept of the quantization of energy. The model was without proof, as embodied in his two proposals known as *postulates*.

His first postulate was based on his classical theory, with the exception that an electron could remain in a circular orbit without radiating any energy. His second postulate stated that a quantum of radiation is emitted or absorbed when an electron moves from one energy level to another. We summarize the substance of the model as follows:

1. Electrons revolve around the nucleus only in certain definite circular orbits, every orbit corresponding to a certain level of energy.
2. These orbits are labeled *stationary orbits*. Electrons could stay in these stationary orbits without radiating any energy.
3. The transfer of an electron from an orbit of lower energy to an orbit of higher energy requires absorption of radiation by the atom, while a fall from a higher energy level to a lower energy level orbit results in the emission of radiation. This energy difference, whether absorbed or emitted, is

$$E_2 - E_1 = h\nu \tag{1.10}$$

4. The only stationary orbits of the electrons are those for which the angular momentum is also quantized.

Applying conclusion number 4 above to the single hydrogen atom, we have

$$mvr_n = nh/2\pi \qquad n = 1,2,3,\ldots \tag{1.11}$$

where m is the electron mass, v is the linear electron velocity, r_n is the orbital radius for a given value of n, and n is known as a *quantum number*. Since we are still referring to circular orbits of the electron, Eq. (1.7) applies. When solved for the velocity it gives

$$v = \left(\frac{q^2}{4\pi\varepsilon_0 rm}\right)^{1/2}$$

Substituting the above expression for v in Eq. (1.11) and solving for r_n, we get

$$r_n = \frac{n^2 h^2 \varepsilon_0}{m\pi q^2} \tag{1.12}$$

The expression for the velocity becomes

$$v = \frac{q^2}{2\varepsilon_0 nh}. \tag{1.13}$$

The electronic orbits assumed in the Bohr theory are circular so that the expressions derived earlier for the potential energy and total energy, Eqs. (1.6) and (1.9), are valid.

By substituting the relation for r from Eq. (1.12), as developed from Bohr's theory, in the expression for total energy, E, of an electron, as given by Eq. (1.9), we obtain an expression for the quantized energy of the hydrogen electron as

$$E_n = \frac{-q^4 m}{8n^2 h^2 \varepsilon_0^2} \tag{1.14}$$

If we use rationalized MKS units, as found in Appendix A, for the constants in Eq. (1.14), the unit of energy will be the joule. We now define a unit of energy known as the *electron-volt*, eV, where one electron-volt is the energy acquired by an electron when elevated through a potential difference of one volt. One $eV = 1.6 \times 10^{-19}$ joules. Replacing all the constants by their values, Eq. (1.14) becomes

$$E_n(eV) = -13.6/n^2, n = 1,2,3,\ldots \tag{1.15}$$

where E is the energy of the allowed energy level of the discrete orbit, which is obtained by replacing n with the relevant integer. It is obvious that the electron can exist only at certain discrete energy levels. It is thus possible for an electron to move between any two such levels with the consequent release of radiation, going from a higher (higher n) to a lower (lower n) energy level. The corresponding frequency of radiation is determined from Eq. (1.10). The electron can also be moved to a higher

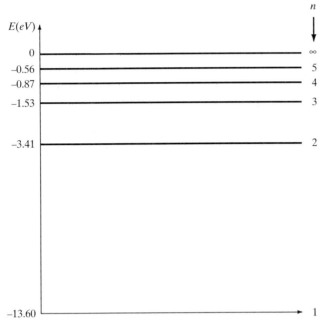

Figure 1.4 Energy level diagram for the hydrogen atom. Six of the infinite number of possible transitions are indicated.

level by the absorption of radiation, again in accordance with Eq. (1.10). In addition to the energy, both the orbital radius and velocity are quantized, as expressed by Eqs. (1.12) and (1.13) respectively. The allowed energy levels in the hydrogen atom, as determined from Eq. (1.15), are usually represented by an energy level diagram, as shown in Fig. 1.4.

Later, Sommerfeld, following upon Bohr's conclusions, verified that the electron orbits were not circular but elliptical. This was later confirmed by others.

Planck's hypothesis provided an explanation for the discreteness of radiation and the particle-like theory of light. Bohr confirmed the quantization of the energy, the orbital radius, and the orbital velocity of the hydrogen electron. Thus, the quantum concept was extended to phenomena taking place at atomic dimensions.

The Bohr theory constituted a big step forward in the explanation of the hydrogen spectra but it did not provide complete answers. While the emission frequencies predicted by Bohr, for the hydrogen atom, are quite close to those that were experimentally observed, a major weakness of Bohr's theory is that it cannot be extended to the modeling of atoms that are more complex than the hydrogen atom. More complete answers were provided by de Broglie and Schrodinger, who provided the basis of the new science of *Wave Mechanics*.

REVIEW QUESTIONS

Q1-1 Give examples of physical systems to which the theories of classical mechanics apply.

Q1-2 On what was Planck's hypothesis based?

Q1-3 What was Bohr's contribution to the science of Wave Mechanics and what were the weaknesses of his postulates?

Q1-4 What prompted Bohr to quantize the energy of an electron?

Q1-5 Why was the hydrogen atom selected for consideration?

HIGHLIGHTS

- Classical mechanics could not provide explanations of the behavior of solids, such as electromagnetic radiation from certain bodies.

- Max Planck hypothesized that it is possible for bodies to radiate energy at certain frequencies, depending upon the energy that is imparted to them. He demonstrated that light has not only wave-like properties but particle-like properties, as well.

- Neils Bohr first hypothesized that electrons can take on any value of energy in their orbit around the nucleus. He later modified this theory.

- In his attempt to provide explanations of experimental results, Bohr further postulated that electrons can exist in certain orbits only, each orbit corresponding to a certain energy. He further stated that the transfer of an electron from one orbit to another was accompanied by a release or a gain of energy. Bohr's explanation served to model the hydrogen atom but could not be extended beyond the one electron model.

EXERCISES

E1-1 a) Determine the binding energy for an electron in a hydrogen atom at a quantum number 3 level. Give the answer in eV and in joules.

b) Repeat the above determination for an electron in a silicon coulombic potential, using the results of the hydrogen atom. The relative dielectric constant for silicon is 11.8 and the effective mass of a silicon electron is assumed to be 1.18 times that of the hydrogen electron.

$$\text{Ans:} \quad \text{a) } E = -1.5\text{eV} \qquad \text{b) } E = 0.02 \times 10^{-19}\text{J}$$

E1-2 a) Use Planck's hypothesis to calculate the wavelength associated with a 1eV photon.

b) Repeat (a) for an electron.

$$\text{Ans:} \quad \text{b) } \lambda = 1.22 \times 10^{-9}\text{m}$$

1.7 WAVE MECHANICS

Attempts to explain electron motion in terms of wave motion were prompted by the experimental observations of a number of scientists. They observed that diffraction patterns could be observed and recorded photographically when beams of electrons were passed through crystals. These results indicated that electrons had characteristics usually associated with waves, and thus should obey wave-motion equations. It was de Broglie, in 1924, who laid the foundation of Wave Mechanics. He confirmed Bohr's conclusion that the ratio of the energy of a wave to the frequency was a constant, *h*, and postulated that the product of the *momentum* of the electron and the *wavelength* was also equal to this constant. This constant is, again, *Planck's constant*. This relationship between dynamic properties of the electron, such as momentum, and wave properties, such as wavelength, forms the basis of wave mechanics. Thus, particle and wave properties were related so that de Broglie's two conclusions, stated in equation form, are

$$\frac{E}{v} = h, \quad \text{(a)}$$

$$\lambda mv = h \quad \text{(b)} \tag{1.16}$$

where h is Planck's constant and λ is the de Broglie wavelength. Wavelength, λ, in meters, is the ratio of the velocity of light in meters/second to the frequency in hertz and $1/v$ is the spatial period of the wave.

The experimental observations of other scientists provided conclusive proof of the relationships formulated by de Broglie.

It might be useful to emphasize the generality of de Broglie's hypothesis, which he predicted from relativity theory. According to his theory, a hypothetical wave is associated with any particle such that Eqs. (1.16a) and (1.16b) hold. The magnitude of the intensity of the de Broglie wave, at any point, is a measure of the probability of finding the particle at that point. In essence, the motion of the particle is describable in probability terms rather than in terms of the deterministic laws of classical mechanics. This probability interpretation and the attendant uncertainty tie in immediately with Heisenberg's Uncertainty Principle, which will be discussed shortly.

The "plausibility" of Eqs. (1.16) may be established as follows:

1. The relation $E = hv$ is an expression of Planck's quantum hypothesis that energy is absorbed or emitted in quanta, the size of a quantum being hv.
2. A light quantum is labeled a photon, so that the energy of a photon of light of frequency v is hv.
3. According to Einstein's theory, matter and energy are related by $E = mc^2$: Hence, a photon would have a mass of hv/c^2 and momentum hv/c or $\mathbf{p} = h/\lambda$ as given by Eq. (1.16a), where $\mathbf{p}$ is the momentum of the photon and c is the velocity of light.

We want to emphasize, at this point, that we are not implying the physical existence of electron waves. Rather, we are thinking of a moving particle as having wave-like properties associated with it. That a particle has wave-like properties follows from electron diffraction experiments, which show that electron beams behave like light beams, and thus must obey the same wave relations as light. Before carrying the analogy any further, let us summarize the accepted dual theories of light:

a) Light has both particle-like and wave-like properties.
b) Light is composed of photons. Thus, a light ray consists of particles (i.e., photons) each of which possesses an energy, hv, where v is the frequency of the light.
c) The intensity of light at a location is the density of the photons at that location.

The analogy between light and electron beams is extended so that the travel and diffraction of a light wave in a medium of varying refractive index is analogous to the case of electron beams traveling through a varying force field. Both phenomena are mathematically analogous. It is found that when, in the optical case, the refractive index varies over dimensions of the same order of magnitude as the wavelengths, classical theories cannot explain the phenomena that occur. Similarly, classical mechanics fails when a force field in a material varies over a distance of the order of a de Broglie wavelength. The reason we have to look to the wave theory for an explanation of the phenomena occurring at atomic distances is partly a result of the Heisenberg Uncertainty Principle.

1.8 HEISENBERG'S UNCERTAINTY PRINCIPLE

Heisenberg's Uncertainty Principle, which was derived from rigorous mathematical physics, states that the minimum value of the product of the uncertainties in the values of two quantities, whose product has the dimension of action, $ML^2/T(\text{kg m}^2/\text{sec})$, is given by Planck's constant, h. Two such sets of quantities are momentum and position, and energy and time. The Principle states that the product of the uncertainties in momentum p and position x and those of energy E and time t are

$$\Delta p\, \Delta x \geq h/2\pi \quad \text{(a)}$$
$$\Delta E\, \Delta t \geq h/2\pi \quad \text{(b)}$$

$$(1.17)$$

A numerical example will clarify this result.

EXAMPLE 1.3

a) Determine the uncertainty in the position of a bullet of mass 20g travelling with an uncertainty in its velocity of 10cm/s. Planck's constant is $h = 6.62 \times 10^{-34}$J-s.

b) Repeat part (a) for an electron that has an uncertainty in its velocity of 10^{-3}cm/s.

Solution

a) At constant mass m, $\Delta v \Delta x \approx h/2\pi$

$$\Delta x = 6.62 \times 10^{-34}/(2\pi \times 20 \times 10^{-3} \times 10 \times 10^{-2}) = 5.26 \times 10^{-32}\text{m}$$

b) $\Delta x = 6.62 \times 10^{-34}/(2\pi \times 9.1 \times 10^{-31} \times 10^{-3} \times 10^{-2}) = 11.57\text{m}$

We note, while Δx for the bullet is negligible compared to its dimensions, the Δx for the electron is many orders of magnitude larger than the diameter of the atom.

In accordance with the Uncertainty Principle, it is thus meaningless to speak of the motion of an electron in a circular or elliptic orbit because such a statement implies that we can trace the path of the electron by making at least two successive observations of its position. In fact, for this example, if the position of the electron is determined very accurately, so that Δx is small, there is a lack of precision in the uncertainty of the momentum and Δp increases. If the limit of Δx is zero, then the uncertainty in the momentum is infinite. The best that one can hope for is to set up a probability pattern around the nucleus, showing the region in which there is an appreciable probability of finding the electron.

Thus, the Uncertainty Principle highlighted another difficulty with the Bohr theory, as it is not possible to assign a certain orbit radius or momentum to a particle.

As was shown in the example, for large-scale phenomena where Δx for the bullet is negligible compared to the dimensions of the bullet, the Uncertainty Principle does not pose any serious difficulties, whereas for the electron, the position is indeterminate. Because the position, velocity, and momentum of an electron cannot be determined using classical mechanics, one has to revert to wave mechanics to study the behavior of the electron.

1.9 SCHRODINGER'S EQUATION

It was Schrodinger who, in building on de Broglie's work, first proposed the wave equation known by his name. He incorporated the quantization theory proposed by Planck and the wave-like nature of matter as proposed by de Broglie. *This equation is as basic to wave mechanics as Newton's laws are to classical mechanics and as Maxwell's equations are to electromagnetic theory.* Schrodinger's equation in the steady-state expresses the probability of locating a particle at a point in space where the *wave function* $\Psi(x,y,z)$ is a measure of that probability and is given by

$$\nabla^2\Psi + \frac{8\pi^2 m}{h^2}(E - W)\Psi = 0 \tag{1.18}$$

where m is the mass of the particle, W is the potential energy of the particle, and E is its total energy. This is the equation of a wave and when applied to electron waves, where m is the mass of the electron, it gives a measure of the probability of finding a given electron at a certain location. Indeed, the quantity $|\Psi|^2 dV$ is the probability that a particle with potential energy W and total energy E will be located in the spatial volume dV at the point resulting from expressing Ψ in terms of x, y and z. Stated in another manner, $|\Psi|^2 dV$ is the probability that a particle having potential energy W will be located within dV and, if it is there, that it will then have total energy E. The wave function is allowed to be a complex quantity and will, in general, for the steady-state, be a function of x, y and z.

Wave Mechanics is, in effect, the fundamental branch of mechanics, and ordinary Newtonian mechanics is derived from it. Thus, when a particle moves in a force field, such that the change of potential occurs over distances that are very large compared with the wavelength associated with the particle, the laws of Newtonian mechanics are applicable. However, if the change in potential occurs over distances comparable to the wavelength, as in the case of the periodic potential of a crystal, then Newtonian mechanics does not apply and the wave nature of the particle gives rise to completely new phenomena unaccounted for.

We summarize the preceding discussion as:

1. The electron is not to be regarded as a wave but as a particle, with an associated hypothetical de Broglie wave.

2. Over a small region where the electron is known to exist, a probability *wave packet* could be set up. The wave packet is a quantum mechanical measure that can be assumed to be localized at a given point in space. It can be said to be analogous to a classical particle that can be located in a given point in space. The wave packet consists of the summation of constant energy wave functions, Ψ, assembled about a center frequency so that this packet can be localized in a given point in space. A sketch of a wave packet is shown in Fig. 1.5. Over a distance that is largely compared with the wavelength, the motion of the electron can be described by the motion of the wave packet, such motion being governed by the laws of classical mechanics, although there is some uncertainty in the simultaneous values of position, momentum, velocity, etc.

3. When considering the motion of the electron over distances that are small compared with the wavelengths, such as inside the wave packet or around the nucleus, no attempt should be made to locate the position of the electron along its path at two successive instants of time. Any attempt to determine the first position will disturb the electron and cause it to trace out a different path.

Under such circumstances, it is appropriate to speak only of the probability of locating the electron at a certain point. This probability can be determined by a solution of Schrodinger's equation.

Calculations in quantum mechanics use the probability function to determine the position, velocity, and momentum of a system. As we shall see later, finding Ψ is not the ultimate result we are after. This quantity is a hypothetical one; a mathematical tool used to obtain, mainly, the allowed system energies. In finding Ψ, two postulates are of considerable importance. These are:

Postulate 1.   The function Ψ and its first derivative are finite, continuous, and single-valued.

Postulate 2.   The probability per unit length (per unit volume in the three-dimensional case) of finding a particle at a particular position and at a certain instant of time is $\Psi\Psi*$, where $\Psi*$ is the complex conjugate of Ψ. The reason for multiplying the function by its complex conjugate is that the probability must be a positive real quantity, whereas Ψ is complex for time-dependent cases. If we integrate $\Psi\Psi*$ over the entire system (entire volume), the result must be unity. Hence,

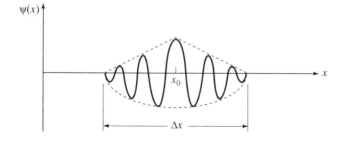

Figure 1.5   The composition of a wave packet at a certain instant. Particle is most likely to be at x_0 but a smaller probability is that it could be found somewhere in the range Δx.

$$\int_{\text{space}} \mathbf{\Psi}\mathbf{\Psi} * \, dx \, dy \, dz = \int_{\text{space}} |\mathbf{\Psi}|^2 \, dx \, dy \, dz = 1 \qquad (1.19)$$

where the product $dxdydz$ is the incremental volume. This is equivalent to saying that it is certain that the particle is somewhere in space.

In the next section, we apply Schrodinger's equation to a simple structure in order to illustrate the concepts of Wave Mechanics. We then apply the solution to the study of the energy levels in the hydrogen atom.

Application to the Potential Well

We shall assume, as shown in Fig. 1.6, that an electron is located in a potential well such that the potential energy W is zero between $x = 0$ and $x = a$, and infinite elsewhere. The infinitely deep one-dimensional well is, in a way, typical of the Wave Mechanics problem that one encounters in the study of the electron energies in a hydrogen atom.

The total energy of the electron is kinetic, since $W = 0$. Therefore, the electron will bounce between the walls, and, if perfectly elastic, will assume a stable state. We will determine an expression for Ψ in the potential well.

Applying Schrodinger's equation in the steady-state ($t\rightarrow\infty$) inside the one-dimensional box where the potential energy W is zero, we have

$$\frac{d^2\Psi}{dx^2} + \mathbf{k}^2\Psi = 0 \qquad (1.20)$$

where $\mathbf{k}$, labeled the *wave vector,* or wave number, is given by

$$\mathbf{k} = \sqrt{8\pi^2 mE/h^2}$$

The solution to Eq. (1.20) is

$$\Psi(x,k) = A\sin \mathbf{k}x + B\cos \mathbf{k}x \qquad (1.21)$$

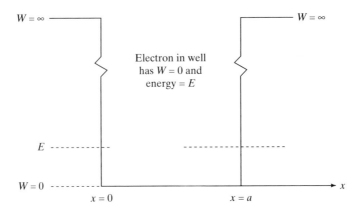

Figure 1.6 Electron in a potential well.

The electron cannot penetrate outside the potential well, so that the wavefunction is zero at $x = 0$ and at $x = a$. Thus,

$$\Psi = 0 \text{ at } x = 0 \quad \text{and} \quad \text{at } x = a$$

from which we obtain,

$$B = 0, \mathbf{k}a = \mathbf{n}\pi \quad \text{and} \quad \Psi = A\sin \mathbf{k}x \tag{1.22}$$

where $\mathbf{n} = \pm 1, \pm 2, \pm 3, \ldots$. Then, setting the expression for $\mathbf{k}$, which follows Eq. (1.20), equal to $n\pi/a$, we have

$$\frac{a^2 8\pi^2 mE}{h^2} = \mathbf{n}^2\pi^2 \tag{1.23}$$

where $\mathbf{n}$ is an integer. The energy corresponding to the integer $\mathbf{n}$ in Eq. (1.23) becomes

$$E_n = \frac{\mathbf{n}^2 h^2}{8ma^2} \tag{1.24}$$

The probability of locating the electron at any x is $(A \sin \mathbf{k}x)^2$ or $(A \sin \mathbf{n}\pi x/a)^2$. If a particle is located at a certain point, or the instant it is there, Eq. (1.24) expresses its energy for that particular integer or quantum number, $\mathbf{n}$. The variations of Ψ and $|\Psi|^2$ in the well are shown in Fig. 1.7. Energy is therefore quantized (i.e., the particle can take on only certain discrete values of energy). The particle spends most of its time at the point at which $|\Psi|^2$ is a maximum, as illustrated in the figure.

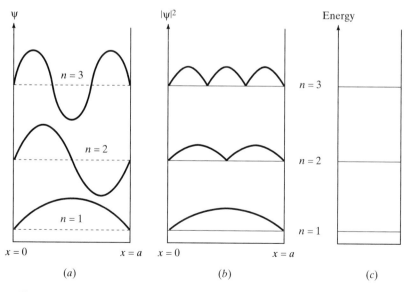

Figure 1.7 Variations of (a) Ψ and (b) $|\Psi|^2$ along the well for different values of $\mathbf{n}$, showing the standing wave patterns. (c) Allowed energy levels for $\mathbf{n} = 1, 2,$ and 3.

From Eq. (1.22) and Fig. 1.7, we observe that, at the points at which $\Psi = 0$, for the first cycle of Ψ, $\mathbf{k}x = \pi$. Since $\mathbf{k}$ is $\mathbf{n}\pi/a$, we write

$$\frac{\mathbf{n}\pi x}{a} = \pi. \tag{1.25}$$

Thus, at these points, $x = a/\mathbf{n}$ and since the distance between zeros is a half-wave-length (one wavelength corresponds to the spatial distance over one cycle), then

$$a/\mathbf{n} = \lambda/2$$

and

$$a = \mathbf{n}\lambda/2. \tag{1.26}$$

From the above, we conclude that to satisfy the boundary conditions for a stationary state, the wavelength is related to the dimensions of the box, as shown by Eq. (1.26). In other words, the wavefunction represents a standing wave similar to that which is observed on a vibrating string. The equation for the energy is also written by using the equality following Eq. (1.20), as

$$E_n = \frac{h^2\mathbf{k}^2}{8m\pi^2} \tag{1.27}$$

We can relate the energy to the momentum by using the expression in Eq. (1.16b) and replacing the wavelength by its equivalence from Eq. (1.26) so that the momentum, $\mathbf{p}$, becomes

$$\mathbf{p} = h/\lambda = \mathbf{n}h/2a \tag{1.28}$$

Since $\mathbf{k}$ is $\mathbf{n}\pi/a$, the expression for the energy can be expressed as

$$E = \mathbf{p}^2/2m \tag{1.29}$$

By using Eq. (1.29) in Eq. (1.27), the momentum is expressed in terms of $\mathbf{k}$ as $\mathbf{p} = h\mathbf{k}/2\pi$.

A sketch of energy versus $\mathbf{k}$ and $\mathbf{p}$ is shown in Fig. 1.8, where the dots illustrate, pictorially only, the various allowed energy levels.

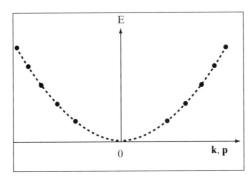

Figure 1.8 Energy versus $\mathbf{k}$ and $\mathbf{p}$ for $\mathbf{k} = \mathbf{n}\pi/a$ and $\mathbf{p} = h\mathbf{k}/2\pi$.

In accordance with Eq. (1.22) and as indicated earlier, n can take on both positive and negative integer values. The wavefunction $\mathbf{\Psi}$ takes on positive and negative values so that

$$\mathbf{\Psi}_{-n} = -\mathbf{\Psi}_{+n},\text{ where they both refer to the same state.}$$

For a particle in a three-dimensional box, assuming the same constraints in three dimensions as for a one-dimensional case, we expect a standing wave of the form

$$\mathbf{\Psi} = \sin\frac{\mathbf{n}_x\pi x}{a}\sin\frac{\mathbf{n}_y\pi y}{b}\sin\frac{\mathbf{n}_z\pi z}{c} = \sin\frac{2\pi x}{\lambda_x}\sin\frac{2\pi y}{\lambda_y}\sin\frac{2\pi z}{\lambda_z}$$

where x, y, and z have separate quantizations with quantum numbers n_x, n_y, and n_z corresponding to the dimensions a, b, and c, respectively. When this wavefunction is substituted in Schrodinger's equation, an expression for E is obtained.

Application to a Potential Wall

This application, which is illustrated by the solution of the following example, is used to determine the probability of locating an electron having energy E in a region of potential energy, W, where W is higher than E.

EXAMPLE 1.4

An electron with a total energy E moves, as shown in Fig. 1.9, in the one-dimensional Region 1 in which the potential energy may be taken as zero, so that $W = 0$ for $x < 0$. At $x = 0$, there is a potential energy barrier of height $W > E$, as shown in the accompanying figure.

a) Verify that the solution of the Schrodinger equation in Region 1 is
$\Psi_1 = C\sin ax + D\cos ax$ and in Region 2 is $\Psi_2 = Ae^{-x/d_0} + Be^{x/d_0}$, where a and d_0 are real numbers.

b) Determine the constants B, C, and D in terms of A.

Solution

a) Schrodinger's equation in Region 1 becomes

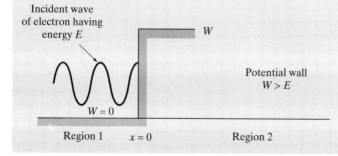

Figure 1.9 An electron in Region 1; incident on Region 2.

$$\frac{d^2\Psi_1}{dx^2} + \frac{8\pi^2 m}{h^2} E\Psi_1 = 0$$

or

$$\frac{d^2\Psi_1}{dx^2} + \mathbf{k}_1^2\Psi_1 = 0, \text{ where } \mathbf{k}_1^2 = \frac{8\pi^2 mE}{h^2}$$

The expression, $\Psi_1 = C \sin ax + D \cos ax$, is a solution to the differential equation where $a = k_1$. In Region 2, Schrodinger's equation becomes

$$\frac{d^2\Psi_2}{dx^2} + \frac{8\pi^2 m}{h^2} (E - W)\, \Psi_2 = 0 \text{ for } E < W$$

or

$$\frac{d^2\Psi_2}{dx^2} - \mathbf{k}_2^2\Psi_2 = 0 \text{ where } \mathbf{k}_2^2 = \frac{8\pi^2 m}{h^2}(W - E)$$

The expression, $\Psi_2 = Ae^{-x/d_0} + Be^{x/d_0}$, is a solution to the differential equation in Region 2 where $d_0 = 1/\mathbf{k}_2$.

b) Applying Postulate 1 on page 15, at $x = 0$, gives

$$\Psi_1 = \Psi_2 \text{ and } \frac{d\Psi_1}{dx} = \frac{d\Psi_2}{dx}$$

In order for Ψ to remain finite as $x \to \infty$, B must equal zero. The solutions at $x = 0$ must match so that

$$C \sin ax + D \cos ax\big|_{x=0} = Ae^{-x/d_0}\big|_{x=0}$$

therefore,

$$D = A$$

By matching the derivatives at $x = 0$, we have

$$aC = -\frac{A}{d_0} \quad \text{and} \quad C = -\frac{A}{ad_0}$$

Hence,

$$B = 0, D = A, \text{ and } C = -\frac{A}{ad_0}$$

The results of this example are:

$$\Psi_1 = -\frac{A}{ad_0} \sin ax + A \cos ax$$

$$\Psi_2 = Ae^{-x/d_0}$$

The solutions are shown sketched in Fig. 1.10.

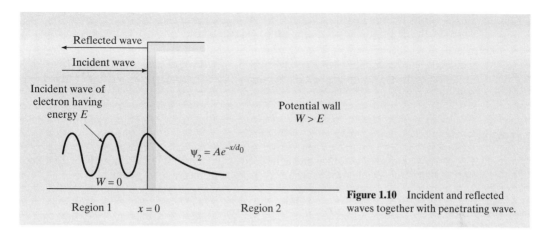

Figure 1.10   Incident and reflected waves together with penetrating wave.

The fact that Ψ_2 has a nonzero value implies that it is possible for an electron having energy E to exist in a region where the barrier energy, W, is greater than the energy of the electron.

This example can be modified to include a third region, labeled Region 3, identical to Region 1 but placed to the right of Region 2. It can be shown that it is possible for a particle having energy E in Region 1 to cross into Region 3 while going through a potential barrier of energy W, where $W > E$. The probability of this occurring is higher if the thickness of Region 2 is reduced. Hence, the particle is said to tunnel through the barrier of Region 2. This concept is applied in the study of the Esaki diode, also known as the tunnel diode, and in the formation of ohmic contacts.

Having established the significance and plausibility of Schrodinger's equation, we will use it in the next chapter to establish the quantum numbers and, hence, the energy levels that a hydrogen electron can have. We will then extend the results to the many-electron sample, such as silicon.

REVIEW QUESTIONS

Q1-6 What is the significance of Heisenberg's principle?

Q1-7 What was de Broglie's contribution to Wave Mechanics?

Q1-8 Briefly state the significance of Ψ in Schrodinger's equation.

Q1-9 Interpret the solution of Schrodinger's equation to the deep potential well and how it relates to the science of Wave Mechanics.

HIGHLIGHTS

- The properties of semiconductors are determined by phenomena on an atomic scale and, hence, require the explanations provided by quantum mechanics' concepts.
- Wave Mechanics established the intimate link between the dynamic and the wave properties of the atom. Herein lies the analogy between a beam of electrons and a light beam.
- De Broglie suggested that quantum-mechanical concepts in their duality applied not only to electromagnetic waves but to electrons as well.

- Another link between particle-like properties and wave-like properties of matter was provided by the de Broglie relationship, which expresses wavelength in terms of momentum.
- The most far-reaching step was taken when Schrodinger developed the extremely important and complex equations that describe the properties of electrons in a physical system. It describes the large number of energies that an electron can have in a region where the energy is confined.

EXERCISES

E1-3 A tennis player serves a 75g ball at a speed of 200 km/h. If the uncertainty in the velocity of the ball is 10cm/s, determine the uncertainty in its position.

$$\text{Ans:}\quad \Delta x = 1.4 \times 10^{-32} \text{m}$$

E1-4 Use Schrodinger's second postulate to determine the expression for A in the solution of Schrodinger's equation, in one dimension, for the infinitely deep potential well.

$$\text{Ans: } A = \sqrt{2/a}$$

E1-5 Determine the energy in eV of an electron for $n = 1$ in the infinitely deep well when $a = 50\text{Å}$.

$$\text{Ans: } E = 14.92 \times 10^{-3} \text{ eV}$$

PROBLEMS

1.1 Determine the density of GaAs given the lattice constant $L = 5.65\text{Å}$, and the molecular weight is 144.63/mole. Avogadro's number is 6.02×10^{23} atoms/mole.

1.2 Use the Bohr model for the hydrogen atom to plot potential energy and total energy as the radius r from the proton increases. Clearly identify the magnitudes of the kinetic energy at two separate radii.

1.3 Determine the velocity of an electron in the ground state of the hydrogen atom.

1.4 The laws of classical physics apply to the motion of a particle provided the dimensions of the system are much larger than the deBroglie wavelength. For the following electrons, determine whether the laws of classical physics apply:

   a) An electron is accelerated in the beam of a cathode-ray tube that has an accelerating voltage of 30KV.

   b) An electron that is accelerated by a potential of 100V in a device whose dimensions are of the order of 2cm/s.

   c) An electron in a hydrogen atom.

1.5 Determine the energy of a photon having wavelengths $\lambda = 10,000\text{Å}$ and $\lambda = 10\text{Å}$. Express the energy in eV and J.

1.6 The antenna of an AM radio station transmitter radiates 100KW of power at 1000KHz.

   a) Calculate the energy of each radiated photon.

   b) Calculate the number of photons radiated per second.

1.7 An oscillator is operating at a frequency of 10MHz.

   a) Calculate the energy of the quantum of radiation of the oscillator.

    b) Calculate the number of quanta in 10^{-6}J.

1.8   **a)** Derive an expression for the wavelength of spectral lines emitted by the transitions from the excited states to the ground state using the Bohr relation for the electron energy in the hydrogen atom.

    b) Calculate the wavelength of the first four spectral lines.

1.9 The velocity of a certain free particle is 5×10^5m/sec. The mass of the particle is 10^{-30}kg. Determine:

    a) the particle energy.

    b) the de Broglie wavelength.

1.10 In accordance with physics statistics, the average energy of an electron in a medium of free electrons at thermal equilibrium is $3kT/2$, where k is Boltzmann's constant and T is in degrees kelvin. Determine, for the electron,

    a) its velocity,

    b) its momentum,

    c) the de Broglie wavelength at $T = 300K$.

1.11 An electron is moving with a velocity of 10^5m/s. Determine, for the electron:

    a) its momentum,

    b) its de Broglie wavelength in m and Å,

    c) its energy in J and eV.

1.12 An electron has a de Broglie wavelength of 100Å. Determine:

    a) electron momentum,

    b) electron velocity.

1.13 For an infrared radiation of 1 μm, determine:

    a) the frequency of the radiation,

    b) the energy of the photon in eV.

1.14 The uncertainty in the position of a particle having mass 10^{-30}Kg is 10Å. Determine the uncertainty in:

    a) the momentum of the particle,

    b) kinetic energy of the particle.

1.15 For the electron in Problem 1.11, determine the wave vector **k**.

1.16 For an infinitely deep potential well having $a = 100$Å, determine for an electron the energy levels for $n = 1, 2$ and 3. Calculate the energy in eV and Joules.

1.17 An electron is located in a one-dimensional potential energy well having width of 3Å. Determine

    a) the kinetic energy of the electron in the ground state.

    b) the frequency of the spectral radiation of an electron that drops from the next higher state to the ground state.

1.18 For a particle that has a mass of 2 grams and energy $1.5kT$, determine the de Broglie wavelength at $T = 300K$.

chapter 2

ENERGY BANDS AND CURRENT CARRIERS IN SEMICONDUCTORS

2.0 INTRODUCTION

We concluded in Chapter 1 that in accordance with the Uncertainty Principle it is not possible to specify, at the same time, the location or the momentum of an electron in a solid. This Principle points to one of the weaknesses of the Bohr hypothesis, which assumed that electrons could be assigned to certain orbits, which in essence implied that their position was known. We then decided that Wave Mechanics' concepts are needed to explain the behavior of electrons in solids.

In this chapter, we will apply Schrodinger's equation to obtain information on the hydrogen atom. We then project the results, with the necessary modifications, to the many-electron solid. We establish the existence of discrete energy levels from which, because of their large number and the closeness of these levels, energy bands result. The highest valence band and the lowest conduction band are of paramount interest. These bands are separated by a region in which no electrons of the semiconductor can exist: *This is the forbidden band*.

At each of these two top bands, a different carrier is said to exist. These carriers are the electron and a vacant space known as the *hole*.

We will use Schrodinger's equation to formulate the conditions existing in the hydrogen atom.

2.1 APPLICATION OF SCHRODINGER'S EQUATION TO THE HYDROGEN ATOM

The Bohr theory of the hydrogen atom is no longer satisfactory for two major reasons. First, it assumes that the electron rotates around the nucleus in a given orbit.

This is not acceptable as it implies that the position of the electron at a given time is known and this violates Heisenberg's Uncertainty Principle. Second, it is not possible to apply Bohr's theory to an atom that has more than one outer electron. We will therefore consider a quantum-mechanical solution since it is not constrained by these limitations. We will apply the results of the solution of the Schrodinger equation to determine the possible energy states of the hydrogen electron and to study the many-electron problem.

Quantum Numbers

The hydrogen atom is readily approximated by a nucleus of charge $+q$, fixed in position, in whose field the electron orbits. The electron, when located at distance r from the nucleus has potential energy, given by Eq. (1.6), as $-q^2/4\pi\varepsilon_0 r$. The potential energy distribution of the hydrogen atom is shown in the sketch of Fig. 2.1(a).

The potential distribution is, in a way, similar to the potential well we considered in Chapter 1. The potential energy (negative) is highest at infinite r and lowest at the nucleus. A major difference between the two is the more complex boundary

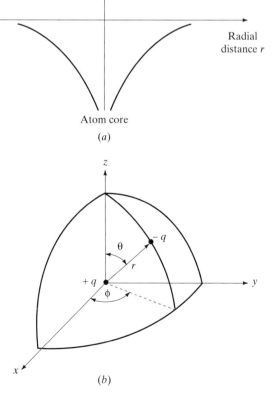

Figure 2.1   (a) Illustration of potential well of hydrogen atom. (b) Spherical polar coordinates for Schrodinger's equation.

profile of the hydrogen atom potential distribution. Furthermore, the potential well is a one-dimensional problem, whereas we wish to solve for the hydrogen atom in three dimensions.

We will proceed with an outline of the general steps in the solution of Schrodinger's equation for the hydrogen atom and then present the results of this solution. We begin by substituting the expression for the potential energy in Schrodinger's equation, Eq. (1.18), in three dimensions, so that we have,

$$\left(\frac{-h^2}{8\pi^2 m}\nabla^2 - \frac{Zq^2}{4\pi\varepsilon_0 r}\right)\Psi = E\Psi \tag{2.1}$$

We shall not solve this equation but we will outline the steps involved in the solution. First, the system, or the hydrogen atom, consists of a proton of charge q and an electron of charge $-q$ and mass m. The position of the proton is fixed at the origin, as shown in Fig. 2.1(b), and the electron is at a point whose spherical coordinates are (r, θ, and ϕ). Second, Eq. (1.18), written in three dimensions, is transformed into the new coordinate system. Third, the method of separation of variables is used to solve the Schrodinger equation. By this method, we assume a solution of the form

$$\Psi\,(r,\theta,\phi) = R(r)\,H(\theta)\,I(\phi) \tag{2.2}$$

When the assumed solution is substituted in the spherical-polar form of Schrodinger's equation, three separate differential equations result; one in r only, one in θ only, and one in ϕ only. In the solution of the three equations, and to obtain physically acceptable results, three quantum numbers emerge; *n*, ℓ and *m$_\ell$*, known as the *principal, azimuthal,* and *magnetic quantum* numbers respectively. This is analogous to the single quantum number *n*, obtained for the electron in the potential well example discussed in the previous chapter.

Let us digress briefly to explain how a solution is obtained so as to identify the sources of these quantum numbers and their relation to the physical properties of the electron and to the electronic orbits.

We will now consider a simplified Schrodinger equation and examine the results. This equation is independent of θ and ϕ but is a function of r, hence labeled the radial equation. The solutions to this equation have spherical symmetry. Before proceeding any further, we note that just as in the example of the potential well discussed in Chapter 1, our eventual interest is not in the expressions for the wavefunctions. For this work, we are interested in the conditions for which the wavefunctions satisfy the Schrodinger equation and the boundary conditions.

The radial form of the differential equation in Ψ has a series of solutions. The simplest and most basic solution, Ψ, expresses Ψ_1 as a function of r and a factor identified as r_0. By substituting this solution in Schrodinger's radial equation, we obtain expressions for E and r_0 (see prob. 2.6). When higher order solutions for Ψ, such as Ψ_2 and Ψ_3, are included, we obtain energy as a function of the *quantum number, n*. This number is a measure of the total energy of the electron in a particular state as well as being a measure of the diameter of the major axis of the elliptical orbit, which is twice the radius of the circular orbit determined by Bohr.

The quantum number, **n**, is an integer that takes on the value 1, 2, and 3. Because of the association of **n** with the energy of the electron, the energy levels in **n** are also designated by the electronic shells *K, L*, and *M*, corresponding to **n** = 1, 2, and 3.

A quantum-mechanical interpretation of the orbital radius of the hydrogen atom is that it is no longer a fixed number (and the electron having a fixed orbit) but that there is a high probability of locating the electron at a distance identified by a quantum number.

Two other quantum numbers result from the complete solution of the equations in each of θ and φ of the Schrodinger equations.

The *azimuthal number,* ℓ, results from the solution of the equation in θ such that its value determines both the quantized angular momentum as $\ell h/2\pi$ and the diameter of the minor axis of the quantized elliptic orbit of the electron as

$$\frac{\text{minor axis diameter}}{\text{major axis diameter}} = \frac{\ell}{\boldsymbol{n}}$$

The values of the set of azimuthal quantum numbers cover the range

$$\ell = 0 \text{ to } \ell = \boldsymbol{n} - 1$$

In the solution of the equation for φ, the *magnetic orbital quantum number,* $\boldsymbol{m}_\ell$ appears, determines the direction of the orbital angular momentum, and has integral values that vary from $-\ell$ to $+\ell$, including $\boldsymbol{m}_\ell = 0$. The word *magnetic* results from the fact that an electron in an orbit represents an accelerating charge and, hence an electric current that has a magnetic field.

The existence of these three quantum numbers was predicted earlier by Sommerfeld. Later, Goudsmit and Uhlenbeck, guided by their knowledge of spectra more complicated than that of hydrogen, predicted that the electron itself possesses a magnetic moment and an angular momentum quite independent of its rotation in an orbit around the nucleus.

Furthermore, it was observed that in a many-electron atom, the actual number of quantum states is twice that predicted by the combination of **n**, ℓ and $\boldsymbol{m}_\ell$. It was then postulated that the electron, in its orbit around the nucleus, spins on its own axis in two unique and opposite directions with respect to its orbital momentum. Thus, each electron can exist in two spin states. As a result, a fourth quantum number, labeled the *spin quantum number,* $\boldsymbol{m}_s$ was included, which has a value of $\pm 1/2$. This intrinsic spin with respect to the orbital angular momentum results in the doubling of the number of quantum states. Therefore, every quantum state is identified by four quantum numbers, designated by **n**, ℓ, $\boldsymbol{m}_\ell$, and $\boldsymbol{m}_s$.

The Pauli Exclusion Principle

In 1925, Pauli formulated the principle that there can be only two electrons of opposite spin in a quantum state. He concluded this from a study of atomic spectra at about the same time that the extra quantum number, $\boldsymbol{m}_s$, was identified. He, in effect, concluded that no more than two electrons can have the same distribution in space. Thus, any quantum state, identified by the four quantum numbers, can accommodate no more than one electron. Actually, the quantum numbers represent another way of defining the wavefunction of a given electron.

To comply with the Pauli Exclusion Principle, it follows that each electron in the solid, regardless of the number of electrons, is identified by four quantum numbers.

In Table 2.1 are listed the various quantum numbers together with the maximum number of electrons that can be accommodated in an isolated atom.

The reader will question our reference to many electrons, whereas the origin of three of the four quantum numbers resulted from a solution of Schrodinger's equation for the one-electron model. In fact, the levels shown in Table 2.1 can be interpreted as the *possible* states that the hydrogen electron can have. Furthermore, if the results for the many-electron problem are to be derived from the single electron model, then the various levels defined by the quantum numbers n and ℓ can accommodate the number of electrons shown in the table.

TABLE 2.1 Quantum Numbers and Number of Electrons.

Shell	$n > 0$	$0 \leq \ell \leq n-1$	$-\ell \leq m_\ell \leq \ell$	$m_s = \pm 1/2$	Levels	Maximum Number of Electrons
	n	ℓ	m_ℓ	m_s		
K	1	0	0	$\pm 1/2$	$1s$	2
		0	0	$\pm 1/2$	$2s$	2
L	2	1	0	$\pm 1/2$		
		1	-1	$\pm 1/2$	$2p$	6
		1	$+1$	$\pm 1/2$		
					$3s$	2
M	3				$3p$	6
					$3d$	10

It is easily verified from Table 2.1 that the maximum number of electrons in a shell is $2n^2$. It is also to be expected that the higher n shells are farther from the nucleus and have the higher energies.

Thus, the first n level contains two electrons, the second contains eight, the third 18, and so on. Since the levels were originally assigned by spectral studies, the levels are denoted by $s, p, d, f, g,$ etc., where $\ell = 0$ corresponds to the s level, $\ell = 1$ to the p level, and so on. (The letter s refers to sharp, the p to principal, the d to diffuse, the f to fundamental, and so on.) Carbon, for example, with an atomic number of 6 (six electrons), has $1s^2\ 2s^2\ 2p^2$, with the first number referring to the shell and the superscript number referring to the number of electrons in that particular level. Silicon has $1s^2\ 2s^2\ 2p^6\ 3s^2\ 3p^2$ for an atomic number of 14. The $3p$ level can accommodate a maximum of six electrons but only two are there, leaving four empty spaces in that level. A schematic representation of an isolated silicon atom is shown in Fig. 2.2, illustrating the location of the electrons in the shells.

We note that the lower levels are filled first, with the uppermost energy levels only partially filled with electrons.

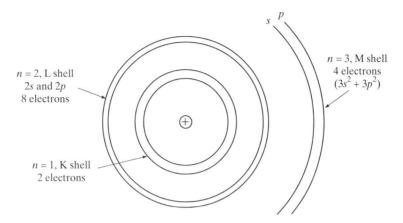

Figure 2.2   Representation of a silicon atom.

The Many-Electron Solid

The transition from the one-electron problem to the many-electron atom does not require new solutions. All that must be done is to assign to the electrons the energy levels determined by the one-electron model. In addition to this, other effects appear as a consequence of having many electrons around the nucleus.

Naturally, we expect that these electrons will exert forces on one another. If the attractive force of the nucleus is assumed to be zero at infinity and negative elsewhere, then the repulsive forces exerted by the other electrons are positive in nature. This, therefore, raises the values of the energy levels that are calculated by using the one-electron model. The forces referred to above are due to the electrons of the same atom. The forces mentioned above, in conjunction with the Exclusion Principle, have the following effects:

1. Energy levels calculated using the one-electron model and a certain quantum number n are lower than the actual levels. The number n also denotes the shell identification as shown in Table 2.1.

2. Electrons having the same quantum number n but different l numbers do not all have the same energy levels. Within a given shell, with a given value of n, the s electron ($l = 0$), has the lowest energy, the p electron ($l = l$) the next higher, the d higher, and so on. The smaller the value of l, the closer the wave function penetrates to the nucleus. The energy levels are actually depressed. Those with high l do not penetrate into the interior of the atom at all. They find themselves in a field of charge, in which the nucleus is shielded by all other electrons and they have energies that are almost hydrogen-like. Figure 2.3 shows the distribution of these levels schematically for a single atom.

3. Electrons having the same n and l quantum numbers but different m_ℓ and m_s quantum numbers have slightly different energies corresponding to different energy levels.

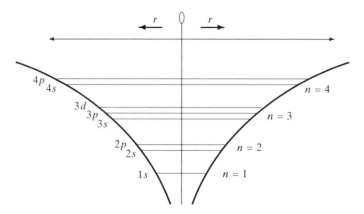

Figure 2.3 Locations of energy levels for a single atom.

4. Theoretically, each electron in a solid represents a quantum state defined by four quantum numbers and has energy levels different from every other electron.

A more complete model will be included later.

2.2 ENERGY LEVEL SPLITTING

We saw in the last section that while the major subdivision of the energies of electrons are determined by the quantum number **n**, every electron in an atom or in a complete crystal occupies a quantum state defined by the four quantum numbers and has its own energy level.

As a mental exercise, let us isolate a single atom of a solid and consider the effects on the energy levels as another atom of the same solid is moved from a further distance away and closer to the first atom. Originally, the electrons of these two atoms had energy levels that were identical. As the second atom is moved closer to the first atom, the outermost electronic orbits tend to overlap. The energy levels of the two atoms are slightly modified so as to accommodate the Exclusion Principle.

When a solid is formed from N atoms, theoretically, N energy levels are formed for each energy level that exists in one atom. This phenomenon is known as *energy level splitting,* and at the normal spacing of the atoms in a solid, takes place mainly for the higher energy levels. Obviously, the degree to which splitting takes place increases with the extent of interactions of the atoms.

In Fig. 2.4, we illustrate energy splitting by showing the effects of bringing five atoms of the same solid close to one another. We note that the higher energy levels associated with the higher **n** quantum numbers and farthest away from the nucleus are affected first as the orbits begin to overlap. The electrons in the lower energy levels are initially shielded, but as the atomic spacing is decreased, lower levels begin to split.

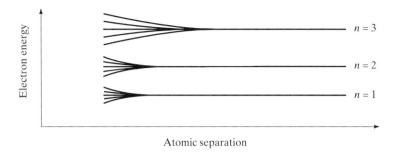

Figure 2.4   The splitting of an energy level as five atoms of the same solid are brought together.

2.3 ENERGY BAND FORMATION

Realizing that each atom of silicon has 14 energy levels, corresponding to 14 electrons, and that there are approximately 10^{22} atoms in a cubic centimeter, it is easy to imagine the consequences of the splitting of energy levels when the atoms come together to form a cubic centimeter of silicon. Granted, the splitting (of major consequence) in silicon takes place mainly at the $3p$ and $3s$ levels. The result of the splitting is the formation of bands of energy at the normal atomic spacing R_0. This effect is illustrated in Fig. 2.5 for a metal, an insulator, and for silicon. The band gap energy refers to the width of the forbidden band separating the conduction and valence bands.

In Fig. 2.5, we have shown the top two or three bands. It is worthwhile noting that, at the normal atomic spacing in both the insulator and silicon, the top band is separated from the next lower band by a region where no energy levels of the solid exist. This region is known as the *forbidden band* or *band gap*. No such band exists in copper.

The top band is known as the *conduction band,* followed at a lower level by the *forbidden band* and the *valence band.* At the temperature of absolute zero, electrons occupy the lowest energy levels so that there are no electrons in the conduction band, although the levels are there. It is apparent from the electron-volt width of the gap that it requires less energy, thermal or otherwise, to move an electron from the valence band to the conduction band of silicon than is needed to accomplish the same in the insulator.

The specific characteristics of the bands, including the width of the forbidden band, the band gap energy, fundamentally determine the properties of a material. We observe that for copper, as shown in Fig. 2.5 (a), the energy bands overlap. Thus, it is fairly easy for an electron to move around in copper; this electron can move from filled levels to higher empty levels. One can supply the energy to move the electron by raising the temperature or by subjecting the metal to a source of light. (We are assuming that the lower energy levels are occupied first and that conduction can occur only in the partially filled bands.)

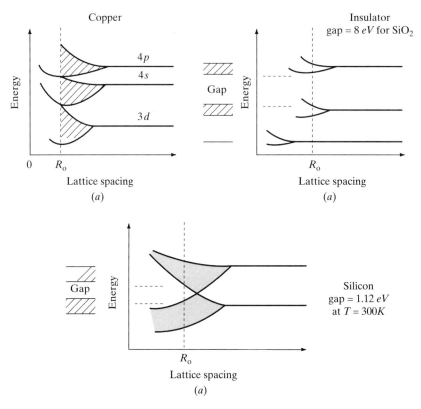

Figure 2.5   The splitting of energy levels as atomic spacing decreases for (a) a metal such as copper, (b) an insulator such as SiO_2, and (c) silicon. R_0 is the normal atomic spacing.

We also note, for an insulator in Fig. 2.5 (b), the existence of a wide gap separating the filled bands from the empty bands. This forbidden gap, in which no electrons can exist, is the reason for the poor conductivity of insulators. A great amount of energy is required to excite the electrons from the filled levels to the higher empty levels.

The band structure of silicon indicates the existence of a forbidden gap, which is relatively narrow compared to that of an insulator, making silicon a better conductor than the insulators because the electrons require a much smaller amount of energy to move from the filled lower bands to the unfilled upper bands. Electrons at absolute zero temperature occupy the lowest levels of energy in accordance with the Pauli Exclusion Principle so that each is theoretically stacked, in energy, over another until all electrons are accommodated.

Let us consider silicon in more detail. The atom has $1s^2$, $2s^2$, $2p^6$, $3s^2$, and $3p^2$, where all states up to those in the $3p$ level ($\ell = 1$) are filled to the maximum number of electrons that they can accommodate. The $3p$ level has six states, which require six electrons to fill all the states. Only two of the six states are filled with electrons. At higher temperatures, some electrons will be raised from the $3s$ level to empty states at $3p$.

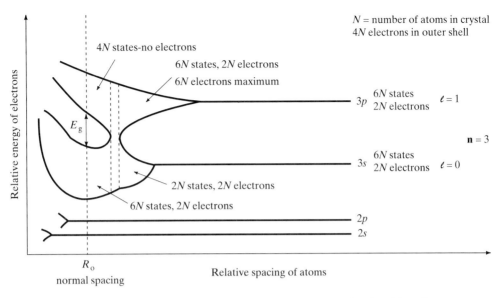

Figure 2.6   The formation of energy bands in silicon as atomic spacing is reduced ($T = 0$).

The process of energy band formation in silicon is interesting and we will study the splitting in more detail by considering the representation shown in Fig. 2.6.

Varying the spacing of the atoms, as shown in Fig. 2.6, is obviously a theoretical exercise. However, we assume that N silicon atoms are brought near each other. At large atomic spacing, the $3p$ levels in the M shell and all lower energy levels are shown as straight lines, as the atoms have not begun to influence each other. Of course, the straight line $3p$ includes six N quantum states and the $3s$ lines include two N states. The energies of the states are very close to each other.

As the spacing is reduced, the $3p$ levels split first followed by the $3s$ levels. At still smaller spacing, the two bands formed by the $3p$ and $3s$ levels merge into a single band containing eight N states. Following this, the new band splits into two distinct bands and, at the normal atomic spacing, two bands (conduction and valence) are formed separated by the gap. There are four N levels in each of the bands.

At very low temperatures and since electrons tend to occupy the lower levels first, there will be four N electrons that completely fill the valence band. The conduction band contains no electrons. At higher temperatures, electrons will be excited from the valence band to the conduction band. From studies of energy level splitting, we present the following results:

- A *forbidden band* is formed between the top band and the next lower band. Ideally, states that accommodate electrons of silicon cannot exist in the forbidden band.
- The top band is the *conduction band* and the next lower band is the *valence band*. There are many other bands below the valence band.

- The width of the forbidden band, the band gap energy E_g, depends on the lattice constant and the energy levels around this band. The lattice constant represents the normal atomic separation.
- Low energy levels produce narrow bands with a wide gap and high levels produce wide bands with a narrow gap.
- At the temperature of absolute zero, electrons occupy the lowest possible energy levels.
- The width of the band gap, E_g, decreases with temperature in accordance with the equation in Table 2.2, in which we have shown the values of E_g and the dependence of E_g on temperature for GaAs, Si and Ge.

TABLE 2.2 Band Gap Variation with Temperature

$$E_g(T) = E_g(0) - \frac{\alpha T^2}{(T + \beta)} \text{ (eV), } T \text{ in K}$$

Material	$E_g(0)$	$\alpha(\times 10^4)$	β	$E_g(300)$
GaAs	1.519	5.405	204	1.422
Si	1.17	4.73	636	1.125
Ge	0.7437	4.774	235	0.663

Source: C. D. Thurmond, "The Standard Thermodynamic Function of the Formation of Electrons and Holes in Ge, Si, GaAs and GaP," *Journal of Electrochem. Society* 122, 1133 (1975)

It should be noted that band structures are not exact models of electronic energies. In order to obtain them, many approximations and simplifying assumptions have to be made. Some experimental proof of the model is obtainable from studies of x-ray spectra. Properties such as conductivity, magnetic effects, and optical effects depend, to a certain extent, on the presence of foreign atoms, lattice defects, or imperfections, whereas the band theory is based on a hypothetically perfect lattice.

A band model that results from Fig. 2.6 is shown in Fig. 2.7.

REVIEW QUESTIONS

Q2-1 Where do the quantum numbers come from?

Q2-2 If each electron in a solid is permitted to occupy only one energy level, estimate the separation in eV between two adjacent levels in silicon, assuming that an energy band is about 1eV.

Q2-3 What phenomena cause energy levels to split?

Q2-4 When atoms are brought together to form a solid, which energy levels split first: the high levels farther from, or the one closer to, the nucleus and why?

Q2-5 What particular aspects completely specify the electrical conducting properties of a solid?

Q2-6 If Z is the atomic number of a solid, what is the charge of the nucleus?

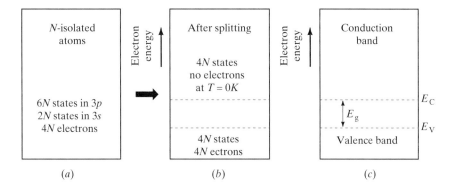

Figure 2.7 Development of the Band Model from Fig. 2.6: (a) available states and electrons in N atoms, (b) splitting causes redistribution of states and electrons, (c) final band formation for $T > 0$. The number of electrons available in the conduction band depends on temperature. At $T = 0K$, there are none.

HIGHLIGHTS

- The solutions of the time-independent Schrodinger's equation in r, θ, and ϕ provide the three quantum number n, ℓ, and m_ℓ respectively, where n defines the total energy of an electron in a particular quantum state.

- The Pauli exclusion principle requires that in an atom that contains many electrons, no more than one electron can exist in any one quantum state. This resulted in a fourth quantum number m_s for the spin of two electrons. Thus, each electron is specified by four quantum numbers.

- As atoms of the same solid are, theoretically, brought together, the energies of the out-ermost electrons are modified. Each level, such as the $3p$ level, which accommodates six electrons, is split into closely spaced levels.

- Where N interacting atoms of a solid are brought to the normal atomic spacing, R_0, the old levels are split into very closely spaced levels. These levels are so close to each other that bands are effectively formed. The top two bands of semiconductors in which electrons can exist are the conduction band and the valence band. These two bands are separated by a band in which states are available but no electrons can exist. This band is the forbidden band.

EXERCISES

E2-1 A certain atom has 12 electrons. Identify, in increasing order of energy, the set of quantum numbers that correspond to each of the 12 electrons.

E2-2 How many of the electrons in an atom of silicon, at thermal equilibrium, form each of the conduction band and the valence band?

E2-3 Determine the electron configuration, in ascending order of energy, for the following elements: Ge, Ga, As, and P.

2.4 MATHEMATICAL MODEL OF BAND FORMATION

In the last section, we determined that an electron in a hydrogen atom is allowed to possess discrete values of energy, E, with each energy level identified by a set of four quantum numbers. No information was provided as to the preclusion of the existence of electrons in certain energy bands, such as the forbidden band separating the conduction and valence bands. Nor was information available as to the shape of the energy distribution as a function of the wave number $\mathbf{k}(n\pi/a)$.

In this section, we will consider a mathematical model for the potential energy distribution in an array of atoms and present the results of this approximate model.

The electronic potential energy of the hydrogen atom is given by Eq. (1.6) and plotted in Fig. 2.8(a) for one atom and plotted in Fig. 2.8(b) for two atoms. Assuming a one-dimensional hydrogen crystal, the potential energy sketched as a function of distance, would appear as shown in Fig. 2.8(c) with a lattice constant, L. The precise

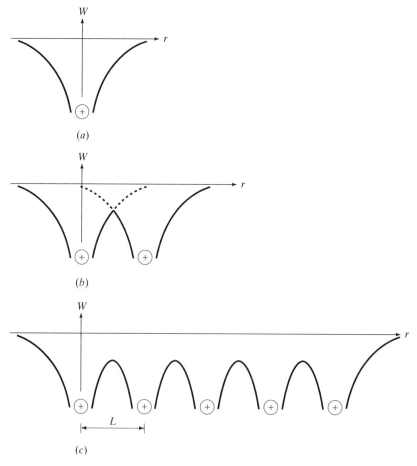

Figure 2.8 (a) Potential energy of an electron in the vicinity of one hydrogen atom; (b) identical to instance (a) except two atomic cores are included; and (c) identical to instance (a) except an array of atoms are shown.

nature of the potential energy distribution for an array of atoms is very complex so that the solution of Schrodinger's equation for such a distribution is very difficult.

Kronig–Penney Model

In an effort to obtain a mathematical solution that will confirm energy band formation, including the existence of forbidden bands, Kronig and Penney used a one-dimensional model of the potential energy distribution in a solid. The model shown in Fig. 2.9 exhibits some similarity to the one-electron potential distribution of Fig. 2.8(c). It consists of a regular array of square-well potentials. In spite of its crude similarity to Fig. 2.8(c), the solution of Schrodinger's equation using this model describes, fairly accurately, the effects that are observed in real solids.

It is important to be aware of the assumptions that are made in the model and its solution. These are:

1. Electron interaction with the core is purely coulombic in nature.
2. Electron to electron interaction is precluded.
3. Non-ideal effects, such as collisions with the lattice and the presence of impurities, are neglected.
4. Atoms are fixed in position, although, in fact, the atoms may be vibrating.

In the Kronig–Penney model, the zero level of potential energy is assumed to be in the vicinity of the core and the peak value W_0 occurs midway between the cores.

An electron having a certain mass, m, and energy, E, is assumed to exist in this potential energy distribution, which is established by the atom cores. The problem is to determine the distribution of the energy levels as a function of the wave number **k**. A separate Schrodinger equation is set up for each of the regions $x = 0$ to $x = a$, where $W = 0$, and $x = -a$ to $x = L$, where $W = W_0$. The distance $(a + b)$ is labeled the lattice constant L.

Subject to the boundary conditions, two solutions for the wavefunctions are obtained, one for each of the regions $0 < E < W_0$ and $E > W_0$. It is necessary to repeat here that our information is not obtained from the expressions for the wavefunctions but in the expressions that result from applying the boundary conditions

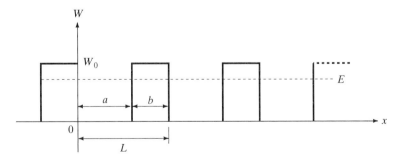

Figure 2.9 One dimensional Kronig–Penney model of the potential energy distribution.

and from the equations resulting from substitution of the expressions for the wave functions in Schrodinger's two equations. The two important solutions consist of an expression on the left-hand side of each of the equations, which include E and W_0, and one term on the right-hand side of each equation, given by cos $\boldsymbol{k}$ $(a + b)$, whose value must lie between -1 and $+1$. Certain values of energy, E, make the left-hand side of these equations greater than 1 or less than -1. Those ranges of energies result in the forbidden bands. The other energy ranges are those that are either occupied by electrons or that could be occupied by electrons. A plot of these two equations is exhibited in Fig. 2.10(a). For the sake of comparison, we show in Fig. 2.10(b) the distribution of energy, E, for the potential well discussed in Chapter 1.

The following observations can be made from the plots:

- The Kronig–Penney model plot displays discontinuities and perturbations when compared to the parabolic shape of the free particle solution. These modifications are greatest at the lower values of energy. At the higher values of energy, the solution for this model is similar to that of the free particle. The implication is that the greater the electron energy, the less is the importance of the periodic potential in the crystal.

- The K–P model plot has discontinuities at $\boldsymbol{k}L = \pm n\pi$. One can conclude that at the values of energies corresponding to these discontinuities, electron waves cannot propagate in the solid. These values of $\boldsymbol{k}$ mark the boundaries of the energy zones that make up the forbidden bands.

- At higher values of E, the width of the permitted bands increases while the width of the forbidden bands is reduced.

- The left-hand sides of the equations that are used to sketch Fig. 2.10 are not changed if $\boldsymbol{k}L$ changes by $\pm 2\pi$. For reasons related to the masses of the particles and that will become more clear later, it is convenient to shift the curves of the second and third allowed bands by $\pm 2\pi/L$ along the x-axis. The values

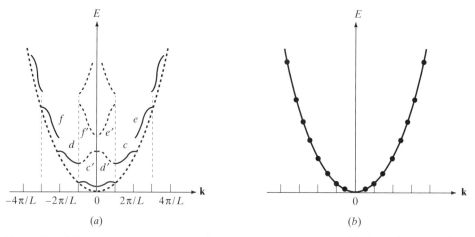

Figure 2.10 (a) Energy versus wave number $\boldsymbol{k}$ using the Kronig–Penney model. The dashed curve that is superimposed is the free particle solution. (b) Energy versus wave number for the potential well.

of the second and third allowed band for positive k are shifted by $-2\pi/L$ and those for negative k are shifted by $2\pi/L$. In this manner, the sketch of Fig. 2.10 is limited to the region $-\pi/L$ to π/L, as shown by the dotted curves marked c', d', e', and f', replacing the curves c, d, e, and f. The new curves are shown sketched in Fig. 2.11 and they represent the energy profile for the valence and conduction bands corresponding to $\boldsymbol{n} = 1, 2$, and 3.

We have shown in Fig. 2.11 energy curves for three values of $\boldsymbol{n}$. Obviously, by use of the two equations referred to earlier, it is quite possible to extend the sketches to higher values of $\boldsymbol{n}$. However, the ones shown in the figures are adequate to illustrate the relevance and benefits of the Kronig–Penney model.

It is important to realize that the one-dimensional Kronig–Penney model bears a very general resemblance to the actual conditions existing in a crystal. In a real three-dimensional crystal, the E-**k** relationships are much more complicated than those obtained in Fig. 2.10. However, the Kronig–Penney model results have exhibited two properties that are extremely important. First, bands exist in which electrons cannot exist and, second, the shape of the E-**k** curves at the locations of the forbidden bands indicate that an opposite concavity exists between the shapes of the allowed bands above and below a certain forbidden band. We will refer to this property in relation to the effective mass later in this chapter.

Direct and Indirect Semiconductor

The actual band structures of semiconductors are much more complex than those shown in Fig. 2.11. One distinguishing feature of semiconductors is the location of the conduction band energy minimum with respect to the valence band maximum on the E-**k** diagrams. In silicon and germanium, and as shown in Fig. 2.12(b), the valence band maximum does not occur at the conduction band minimum. The valence band maximum in all semiconductors occurs at **k** = 0, whereas the conduction band minimum for Si and Ge occurs at a different **k**, indicating a difference in momentum between these two points. In gallium arsenide, the conduction band minimum and the valence band maximum occur at **k** = 0, as shown in Fig. 2.12(a).

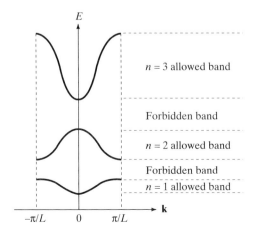

Figure 2.11 Energy distribution from the Kronig–Penney model for the lower three allowed bands.

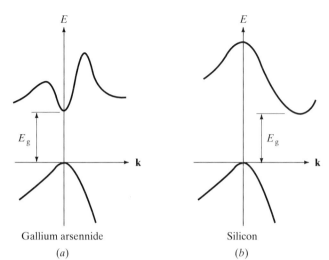

Gallium arsennide
(*a*)

Silicon
(*b*)

Figure 2.12   *E*-**k** sketches for (a) direct band gap and (b) indirect band gap semiconductors.

Hence, GaAs is known as a *direct band gap* semiconductor, and Si and Ge are known as *indirect band gap* semiconductors.

In a direct band gap semiconductor, a photon of light energy, $h\nu$, can excite an electron from the top of the valence band to the bottom of the conduction band. Similarly, an electron in the conduction band of a direct semiconductor can fall directly into an empty state in the valence band and emit photons of light having energy E_g. In indirect semiconductors, an electron in the conduction band cannot fall directly into the valence band because it must undergo a change in energy and a change in momentum. A photon by itself cannot excite an electron from the top of the valence band of an indirect semiconductor to the bottom of the conduction band because the photon has sufficient energy to cause the transition but does not possess the necessary momentum for this transition.

An electron moving between the valence band and the conduction band of an indirect semiconductor can occur through a defect in the semiconductor or by the action of phonons, which can provide sufficient momentum to assist indirect transitions.

The application to which the direct semiconductors become important is the optical device. In this case, GaAs is a principal semiconductor used in semiconductor lasers and light-emitting diodes.

2.5 COVALENT BOND MODEL

A representation that complements the energy band diagram known as the *covalent bond model* is shown in Fig. 2.13. This diagram is a two-dimensional form of the diamond lattice structure shown in Fig. 1.1 in which each atom is bonded to its four nearest neighbors.

The bonding is a result of the fact that each atom shares four outermost-orbit electrons with four adjacent atoms. These four electrons occupy the 3*s* and 3*p* levels in the energy band representation, shown in Fig. 2.6. At $T = 0K$, there will be 4*N*

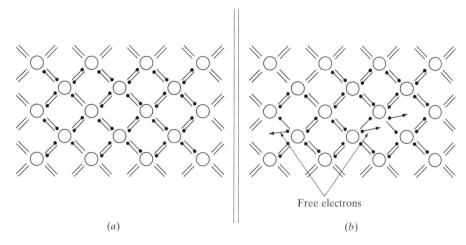

(a)   (b)

Free electrons

Figure 2.13   Covalent bond model of a semiconductor (a) at low temperature and
(b) at room temperature.

lines in Fig. 2.13(a), N being the number of atoms in the crystal, which normally fill
the lower band (the valence band) shown in Fig. 2.6 and Fig. 2.7(b).

The lines shown in the figure represent one electron each and the circles represent the atom core, which includes the nucleus and all other electrons of the atom
except the ones in the outermost orbits, such as the $3s$ and the $3p$ levels in silicon. It
is of interest to realize that it is the covalent bonding that imparts hardness to
Group IV semiconductors, such as Ge and Si. Of course, carbon is the hardest material. In III-V semiconductors, such as GaAs, the binding forces represent the covalent bonding in addition to the ionic forces so that the strength of the bonding is
high. This translates into a higher band gap energy and a higher melting point when
compared to Group IV semiconductors having the same atomic number and lattice
constant.

At the normal atomic spacing, R_0, and for a semiconductor at 0K, all electrons
are attached to their cores and are said to occupy their lowest possible levels of
energy in the band diagram. An increase of temperature imparts energy to the electrons, and at 300K (room temperature), many electrons are moved from the top of
the valence band to the bottom of the conduction band. If the imparted energy is
exactly equal to the band gap energy, the electrons at the top of the valence band
are transferred to the bottom of the conduction band and possess potential energy
E_C. For energy greater than the band gap E_g, the electrons at the top of the valence
band move up to the conduction band and acquire kinetic energy; their total energy
places them above E_C.

At $T = 300$K, approximately one in 10^{12} silicon electrons escapes a covalent
bond and becomes what is known as a *free electron* able to travel throughout the
structure. In the energy band model, an electron becomes free when it has been
moved from the top of the valence band to the bottom of the conduction band.

With this in mind, we will note one distinction between metals and semiconductors. The conductivity of metals decreases with increasing temperature, whereas,

as we shall see later, the conductivity of semiconductors, under certain conditions, may increase with increasing temperature.

In metals, the conduction (free) electrons are distant from their nucleus, move freely among atoms, and are bonded to different atoms at different times. A metal is thus said to consist of an array of positive ions surrounded by closed-shell electronic orbits immersed in a gas of free electrons. There is no forbidden gap that the electrons have to overcome. While there are approximately 10^{10} free electrons per cm^3 in silicon at room temperature, a metal will have approximately 10^{20} electrons per cm^3.

2.6 CURRENT CARRIERS—ELECTRONS AND HOLES

A unique feature of semiconductors that results from their particular energy band diagram is that two types of carriers exist: electrons in the conduction band and holes in the valence band. In metals, conduction consists of the controlled motion of electrons only. So, where do the holes come from and how do they behave as current carriers?

The Hole

When a pure semiconductor, such as silicon, which is initially at $T = 0K$, acquires thermal energy equal to or greater than the band gap energy, E_g, electrons are excited from the top of the valence band and into the conduction band. They become free electrons. For every electron that leaves the valence band, a vacancy in the covalent bonding is left behind into which another electron in that valence band may move. When an electric field is applied to the silicon, the electrons in the conduction band acquire velocity in a direction opposite to that of the field. Similarly, the electrons in the valence band, which moved to fill the vacancies, gain velocity. By having the vacancy occupied by another electron, the vacancy moves in the direction of the field. *This vacancy is the hole.* Thus, both the electron and the hole cause electric current in the same direction with the hole moving in the direction of the field and the negatively charged electron moving opposite to the direction of the electric field.

One crude analogy is that of a two-level parking garage where, initially, all the cars are arranged in a single row in the ground floor and the upper floor is completely empty. Until a car is moved to the first floor from the ground floor, there can be no motion of cars in either floor since the ground floor is full and the first floor is empty. When the front car is moved upstairs, then all the cars behind the space that the front car occupied can now move one car-length forward, thus causing the empty space to move towards the back. The car that was moved upstairs is available for motion and it is analogous to our electron in the conduction band. The vacancy that was created on the ground floor is analogous to the hole and it moves in a direction opposite to the motion of the cars.

Another analogy is represented by the cylinders shown in Fig. 2.14.

The bottom cylinder in Fig. 2.14(a) is completely filled with a liquid and the top cylinder is empty. These cylinders represent the valence and conduction bands

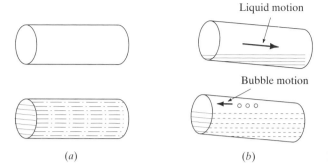

Figure 2.14   Analogy to bands in silicon.

respectively of silicon at absolute zero. In this state, there can be no motion of liquid in the bottom cylinder and, obviously, none in the top cylinder. When a small volume of liquid is transferred from the bottom cylinder to the top cylinder, a possibility exists for motion of liquid in both cylinders: the small volume of liquid in the top and the bubble left behind in the bottom cylinder.

By tipping both cylinders to the right side in Fig. 2.14(b), gravity forces the liquid in the top cylinder to move right and the bubble in the bottom cylinder to move left. The small volume of liquid and the bubble move in opposite directions. The force of gravity is analogous to an applied electric field, the small volume is analogous to the free electron and the bubble is analogous to the hole.

Analytical Description of the Hole

A more analytical description of the hole is determined as follows:

Electrons are thermally excited from the valence band to the conduction band, leaving empty states in the valence band. When an electric field is applied, electrons in the conduction band are accelerated, and so are the electrons in the valence band as they move into the empty states. The current density (amps/m^2) of electrons in the valence band, J_{vb}, can be determined by a summation of the motion of all the electrons in the valence band as:

$$J_{vb} = \sum - qv_d \tag{2.3}$$

where $(-q)$ is the charge of an electron, v_d is its drift velocity, and the negative sign indicates that the direction of current is opposite to the direction of motion (v_d) of the electrons.

Mathematically, one can also state that the current density in the valence band is made up of two components: the summation of the motion of all the electrons in a completely filled band (no vacancies) minus that current associated with the missing electrons as,

$$J_{vb} = \underset{\text{filled band}}{\sum - qv_d} - \underset{\text{empty states}}{\sum - qv_d} \tag{2.4}$$

Since the current in a filled band is zero, the current resulting from the availability of empty states is the current in the valence band given by

$$J_{vb} = \sum qv_d \qquad (2.5)$$

Because the direction of motion of the carriers is in response to an electric field, we conclude that the hole has a positive charge and the free electron has a negative charge since the electric field causes the holes and the electrons to move in opposite directions.

Electron and Hole Energies

The generation of an electron-hole pair in the bands is shown in Fig. 2.15(a) and in the covalent picture is shown in Fig. 2.15(b). In the band picture, the electron has moved up in energy, and in the covalent bond illustration, the electron is free to roam with its place becoming available for occupancy.

In the energy band diagram, the word "energy" refers to the energy of the electrons. When the energy of the electron is increased, that electron occupies a higher energy level in the band. This applies equally well to conduction band and valence band electrons.

An increase in the energy of a hole is caused by the raising of the energy level of the electron. As electrons move up within the valence band, holes become available and move down the band occupying lower levels of the electron energies. *Thus, an increase in the energy of a hole is associated with downward motion in the valence band.*

An electron located at the level of the bottom of the conduction band at rest has potential energy E_C, and zero kinetic energy. Similarly, a hole located at the top of the valence band E_V at rest, has potential energy E_V, and zero kinetic energy.

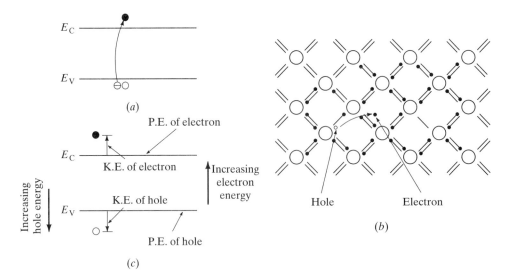

Figure 2.15   Generation of electron/hole pair; (a) energy band model, (b) covalent bond model, and (c) directions of energy increases of electrons and holes.

Electrons located above E_C and holes below E_V, as shown in Fig. 2.15(c), are said to have kinetic energy given by the difference between their energy location and the respective band edge energy.

2.7 EFFECTIVE MASS

We have just concluded that a hole possesses a positive charge, q, and the electron's charge is $-q$ where $q = 1.6 \times 10^{-19}$ coulombs. However, the mass of the electron in the conduction band and the mass of the hole in the valence band are quite different and both differ from the mass of an electron in a vacuum.

The mass of a particle will be defined as the ratio of the net force on the particle to the acceleration that the particle experiences. In using the mass of an electron in vacuum or free space, one completely ignores the effect of the crystal in which the electron is immersed. In fact, the forces within the crystal are much greater than the force exerted by an electric field of the strength normally applied to an electron. The forces in the crystal are known as *lattice forces.*

In order to account for the lattice forces in the equation of motion, we introduce the following two equations:

$$\boldsymbol{F} + \text{lattice forces} = m_0 \frac{dv}{dt}$$

$$\boldsymbol{F} = m_n^* \frac{dv}{dt} \tag{2.6}$$

where $\boldsymbol{F}$ is the externally applied force, as by the electric field, m_0 is the mass of an electron in a space free of lattice forces, and m_n^* is the mass of the electron that includes the effect of the forces in the crystal. Such a mass, m_n^*, is labeled the effective mass of an electron and m_p^* is the effective mass of a hole. We will now determine expressions and definitions for the effective masses.

In Chapter 1, we determined that for both a free particle and for the particle in the well, the energy is given by

$$E = \frac{p^2}{2m_0} \tag{2.7}$$

where $\boldsymbol{p}$ is the particle momentum and m_0 is the free-electron mass. The motion of an electron in the conduction band of a semiconductor is analogous to that of a free particle, except that the conduction band electron is subjected to lattice forces. In this case, we label this mass as the *effective mass* of an electron and we define the energy as

$$E = \frac{p^2}{2m_n^*} \tag{2.8}$$

where m_n^* is the effective mass of an electron and $\overline{p}$ is labeled the crystal momentum in reference to the forces of the crystal that act on the electron. In an operation similar to Eqs. (1.27) and (1.29), momentum is also written as

$$p = h\mathbf{k}/2\pi \tag{2.9}$$

where h is Planck's constant and $\mathbf{k}$ is the wave number.

The expression for the energy of an electron in the lattice is obtained by using Eq. (2.9) in Eq. (2.8),

$$E = h^2\mathbf{k}^2/8\pi^2 m_n^* \tag{2.10}$$

The important conclusion is that the energy is directly proportional to $\mathbf{k}^2$ and inversely proportional to the effective mass of an electron. It is obvious from Fig. 2.11 that the effective mass is not necessarily constant, it is a constant only if the relation of E to $\mathbf{k}$ is purely parabolic.

Based on Eq. (2.10), the effective mass of an electron is defined as

$$m_n^* \propto \left(\frac{d^2E}{d\mathbf{k}^2}\right)^{-1} \tag{2.11}$$

On close study of the E-$\mathbf{k}$ plot of Fig. 2.11 for an electron in a varying field, it is apparent that the second derivative of E with respect to $\mathbf{k}$ is negative when the E-$\mathbf{k}$ curve is concave downwards and it is positive when the curve is concave upwards. The curves near the minimum of the conduction band and near the maximum of the valence band are nearly parabolic so one can safely assume that the mass of the electron is constant in those two regions. However, near the minimum of the conduction band, the E-$\mathbf{k}$ curve is concave upwards so that the mass of an electron is positive, whereas near the maximum of the valence band, the curve is concave downwards so that the mass of an electron is negative.

The direction that a particle will be accelerated when an electric field is applied is determined by the sign of the mass and the charge as

$$a = q\,\mathscr{E}/m \tag{2.12}$$

A positive mass electron (one near the bottom of the conduction band) having a negative charge will be accelerated in a direction opposite to that of the field, in accordance with Eq. (2.12). The motion of an electron opposite to the direction of the electric field constitutes a current in the direction of the field since current is defined as being opposite to the direction of the motion of an electron.

An electron (negative charge) near the top of the valence band, which has a negative mass, may be regarded, as far as its motion is concerned and in accordance with Eq. (2.12), as a positive mass-positive charge particle as the two negative signs in Eq. (2.12) are replaced by a positive sign. Therefore, the electron near the top of the valence band is accelerated and causes current in the direction of the electric field.

In conclusion, an electric field applied to a semiconductor accelerates electrons in the conduction band in an opposite direction to the electric field and accelerates positive particles in the valence band in the same direction as the electric field. The result is a current in the direction of the electric field consisting of electrons (positive mass, negative charge) moving in the conduction band and the particles (positive mass, positive charge) moving in the valence band. These particles are the holes.

Approximate values of the effective masses of electrons m_n^* and holes m_p^* relative to the masses in a vacuum, are listed in Table 2.3. We note from the table that the effective masses of electrons and holes in gallium arsenide are smaller than those in silicon. We refer to the E-**k** diagrams of Fig. 2.12 and note that the curvature of the E vs. **k** diagram, in particular near the bottom of the conduction band in GaAs, is considerably greater than that in Si. Consequently, and in accordance with Eq. (2.11), electrons in GaAs have a much smaller effective mass than in Si.

TABLE 2.3 Density of states effective masses of electrons and holes in Si, Ge, and GaAs and 300K

	Si	Ge	GaAs
m_n^*/m_0	1.18	0.55	0.065
m_p^*/m_0	0.81	0.37	0.52

These masses are labeled as *density of states effective masses*. A different definition for masses with different values is used for calculations of the mobilities of electrons and holes. They are labeled *conductivity effective masses*. We will define mobility of a particle as the ratio of the velocity of the particle to the electric field. In this book, all references for effective mass will be to that of the density of states.

2.8 CONDUCTORS, SEMICONDUCTORS, AND INSULATORS

The feature that distinguishes these three types of materials is the extent of their ability to conduct electrical current. For conduction of current to take place, the following requirements must be met:

1. There must be energy bands that are partially filled with electrons. Since electrons occupy the lowest bands first, these partially filled bands are located at or near the top of an energy band.

2. An electric field must be applied to accelerate the electrons in the partially filled bands. In being accelerated, electrons gain energy, but compared to the energies separating the bands, if a forbidden band exists, this energy is very small.

For energy bands to become partially filled with electrons, either electrons are lifted, by acquiring energy, from a completely filled band to a completely empty band, or there are bands of energy that are empty and overlap filled bands without the existence of forbidden bands.

The structural difference between insulators and semiconductors is that insulators have a very wide band separating the valence band from the conduction band, whereas semiconductors have a much smaller band gap energy. In conductors, the top occupied band is only partially filled with electrons. We show in Fig. 2.16 the relationship between conduction and valence bands in the three materials.

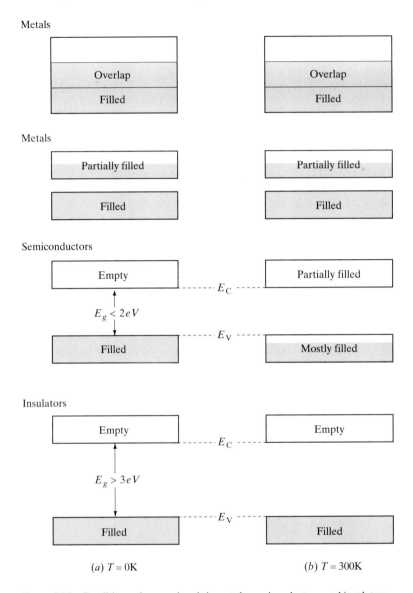

Figure 2.16   Conditions of energy bands in metals, semiconductors, and insulators at (a) very low temperatures and (b) at $T = 300K$.

At low temperatures, the conduction band of a semiconductor is practically empty. Applying an electric field to the solid accelerates the electrons in the valence band and increases their kinetic energy. This increase in energy is nowhere sufficient to move the electron across the gap into the conduction band. Therefore, no current flows. By raising the temperature of the semiconductor, some electrons at the top of the valence band gain enough thermal energy to move into the conduction band. Both the electrons in the conduction band and the holes in the valence can now be accelerated, gain kinetic energy, and therefore carry a current when an electric field is applied. The conductivity at 300K is small, compared to metals, hence the name semiconductor.

The band gap energy of an insulator, such as diamond or silicon dioxide, is several times greater than that of semiconductors, such as silicon or gallium arsenide. Even at very high temperatures (high enough to approach the melting point), very few electrons acquire enough thermal energy to be raised to the conduction band. Therefore, the resistivity of insulators is very high and their conductivity is very low.

In metals, the top band is only partially filled with electrons. Bands become partially filled if either the number of electrons is not sufficient to fill the band or, more commonly, a completely filled valence band overlaps an empty conduction band. An electric field applied to the metal at a temperature of 300K easily moves electrons to higher empty levels and causes a current to flow. Thus, the conductivity of metals is very high.

REVIEW QUESTIONS

Q2-7 Distinguish between a direct and an indirect semiconductor. Give an example of each.

Q2-8 Suggest an application for indirect semiconductors in relation to their absorption and emission of photons.

Q2-9 Trace on Fig. 2.12(b) the motion of a hole in a semiconductor to which an electric field is applied.

Q2-10 Why is the mass of an electron negative for an electron located at the top of the valence band?

Q2-11 Why are the effective masses of holes and electrons smaller in GaAs than in Si?

Q2-12 Why is the ability to conduct electricity, or conductivity, higher in metals than in semiconductors at room temperature?

Q2-13 Why is the conductivity of insulators, when compared to that of a semiconductor, negligible?

HIGHLIGHTS

- No forbidden band exists in a metal; a wide gap (several eV) exists in insulators. The gap is 1.12eV in silicon and 1.41eV in gallium arsenide.
- The effective mass of an electron is inversely proportional to the second derivative of the expression for energy as a function of the wave vector. Thus, at the top of the valence band, the mass is negative; it is positive at the bottom of the conduction band.
- A negative-mass, negative-charge electron that is subjected to an electric field is accelerated in the direction of the electric field, opposite to that of a negative-charge, positive-mass electron. The negative-mass, negative-charge electron is interpreted to be a hole, which has positive charge and positive mass.

EXERCISES

E2-4 The energy gap of GaAs is 1.42V. Determine:

a) the minimum frequency of light that will cause the transition of an electron from the valence band to the conduction band.

    b)   The wavelength of this light.

Ans: a) $f = 3.43 \times 10^{14}$Hz   b) $\lambda = 8.74 \times 10^{-7}$m

E2-5   At room temperature, it is calculated that there are 10^{10} electrons per cm^3 that have moved from the valence band to the conduction band of silicon.

    a)   What is the density of holes?

    b)   What fraction of the electrons have moved into the conduction band?

Ans: a) $p = 10^{10}$ cm^{-3} b) 1 in 10^{13}

PROBLEMS

2.1 A certain atom has the following subshells:

$$2f, 3d, 3f, 3g, 5g$$

Determine:

    a)   The values of the quantum numbers n and ℓ that correspond to each of the subshells.

    b)   Which of the subshells are allowable?

2.2 For the subshells of Problem 2.1, determine the number of electrons in each.

2.3 The expression for the potential energy of an electron in a one-dimensional crystal lattice is given by Eq. (1.6). Plot the potential energy of the electron inside the lattice considering three atomic cores (Zq) located at $x = 0, x = a$ and $x = 2a$. Choose a suitable scale for potential energy and cover the region from $x = 0$ to $x = 3a$.

2.4 The E-**k** diagram for a particular energy band of a certain material is shown in the figure below. An electric field is applied to the material in such a direction as to cause a force in the negative **k**-direction. Determine:

    a)   The signs of the effective masses of the wavepackets made up of groups of states near A, B, C and D.

    b)   The direction of the velocity of each of the wavepackets.

    c)   The direction of acceleration of each wavepacket.

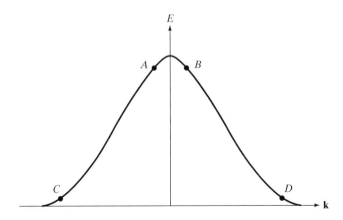

2.5 The *E*-**k** diagram for a certain material is shown in the figure below. The diagram for a free electron is shown dotted. Sketch as accurately as possible:

    a) dE/dk versus **k**.

    b) d^2E/dk^2 versus **k**.

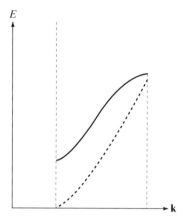

2.6 The wavefunction represented by the time-independent equation for the hydrogen atom, in spherical coordinates, is given by:

$$\frac{1}{r^2\partial r}\left(r^2\frac{\partial\Psi}{\partial r}\right) + \frac{1}{r^2\sin\theta}\frac{\partial}{\partial\theta}\left(\sin\theta\,\frac{\partial\Psi}{\partial\theta}\right) + \frac{1}{r^2\sin^2\theta}\frac{\partial^2\Psi}{\partial\phi^2}$$

$$+ \frac{8\pi^2m}{h^2}\left(E + \frac{q^2}{4\pi\varepsilon_0 r}\right)\Psi = 0$$

    a)   Reduce the equation to one in *r* only, independent of θ and ϕ.

    b)   For the equation determined in (a), first, assume a solution of the form $\Psi_1 = C\exp(-r/r_0)$ and, second, use Postulate 2 in Chapter 1 for the volume of a spherical shell of radius *r* and thickness *dr* in order to determine an expression for *C*.

    c)   Determine the expression for r_0 by substituting the solution for Ψ_1 in the differential equation found in (a).

    d)   Determine the expression for the total energy of the electron in the lowest energy state. Given:

$$\int_0^\infty x^n e^{-ax}\,dx = n!/a^{n+1}$$

chapter 3

INTRINSIC AND EXTRINSIC SEMICONDUCTORS

3.0 INTRODUCTION

We learned in the last chapter that each electron in a solid is assigned a specific level of energy and that these levels are so close to each other that they merge into bands of energy. We also found that these bands of energy that electrons may occupy are separated by forbidden energy gaps where no electron of the solid can exist. In semiconductors, the three top bands are the valence band, the forbidden gap, and the conduction band.

In this chapter, we will determine expressions for the densities of electrons and holes in the conduction band and valence band respectively, for a semiconductor into which no impurities have been deliberately introduced. We will then determine the densities where small traces of impurities are added to the semiconductor.

So that we can determine current-voltage relationships, we need to explain the processes that cause electrons and holes to move. These we will study in Chapter 4.

3.1 DENSITY OF STATES

The densities of electrons and holes in the conduction and valence bands respectively are dependent on two factors. First, we need to know the density of states available for occupancy. Then, we determine the probability of occupancy of the various states at their respective energy levels.

The density of states refers to the number of available electron states for unit volume per unit energy at a certain energy level. The expression for the density of states in a metal, derived in Appendix C, is given by

$$N(E) = \frac{\pi}{2}\left(\frac{8m}{h^2}\right)^{3/2} E^{1/2} \tag{3.1}$$

where $N(E)$ is the density of states function for free electrons in a metal measured per unit volume, per unit energy, located about the energy level E. Its significance is clearer if we state the following: To obtain the total number of energy states per unit volume in a given energy range, dE, about E, we multiply $N(E)$ by dE. In fact, it becomes an integration process.

We will assume that the function given by Eq. (3.1) is valid for semiconductors, provided we use the effective mass for the electron, since the electron moves in the periodic potential of the crystal lattice and it may be assumed to be a free electron. The densities of electrons and holes will be determined by using this function together with a probability of occupancy of a state function.

The energy E corresponds to a set of quantum numbers and can therefore take on only certain discrete values. A sketch of the distribution of states is shown in Fig. 3.1. This curve is not continuous but is made up of a set of discrete points with the adjacent states so close to each other that, for all practical purposes, one can safely consider it continuous.

An analogy should help clarify the significance of the variation of $N(E)$ and E. Let us consider an oval-shaped stadium, with seats at all elevations around the field, and label its volume as a unit of volume. At the higher elevations, one can count more seats around the stadium for a given level than at the lower elevations. Thus, we can say that the number of seats per unit volume (the whole stadium) per meter of elevation corresponds to $N(E)$. The number of seats per meter of elevation at the higher elevations is larger than the number of seats per meter of elevation at the lower elevations. Similarly, the density of states $N(E)$ is larger at higher energy levels E.

Classical mechanics states that at absolute zero all electrons have zero energy, and that when a material is heated from absolute zero, each particle absorbs an amount of energy equal to kT, where k is Boltzmann's constant and T is the absolute temperature. However, Wave Mechanics theories obviously do not agree with those of classical mechanics. First, only very few electrons have zero energy at zero temperature. Second, when the material is heated, only those electrons close to a certain energy level, known as the *Fermi level*, can be excited to higher unoccupied levels. The significance and relevance of the Fermi energy level will be clarified in

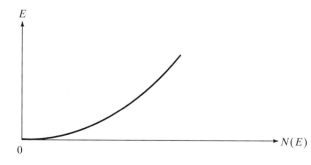

Figure 3.1 Distribution of energy states; E versus $N(E)$.

the following section. Therefore, when a material is heated from absolute zero, not every electron gains energy kT, as classically happens. According to Wave Mechanics, only those electrons within an energy range of the order of kT of the Fermi level can be thermally excited. Furthermore, in the study of energy band formation, we found that the solid has states of varying energy, even at absolute zero, and that is because of the *Exclusion Principle.*

3.2 FERMI-DIRAC DISTRIBUTION FUNCTION

The relations derived in Appendix C describe conditions occurring at absolute zero. As we raise the temperature, we expect some electrons to move to higher energy levels and thus to energy levels previously unoccupied. The question then is: What is the distribution of the electrons with respect to energy as the temperature is raised above zero, subject to the assumption that the density of states is independent of temperature?

We also noted that the Pauli Exclusion Principle states that a quantum state can be occupied by only two electrons that have opposite spin. At absolute zero, the electrons therefore occupy the lowest energy levels possible. Because of the Pauli Exclusion Principle, the distribution of electrons is governed by Fermi statistics. The Fermi statistics describe what occurs as the temperature is raised above zero. It describes the probability that a state of energy E is filled by two electrons of opposite spin.

The Fermi-Dirac distribution predicts that as the temperature increases, an energy state corresponding to energy E will have a higher probability of being occupied than at a lower temperature. The Fermi function is given by

$$f(E) = \frac{1}{\exp[(E - E_F)/kT] + 1} \tag{3.2}$$

where $f(E)$ is the probability that a state with energy E is occupied, E_F is the Fermi energy, k is Boltzmann's constant, and T is the absolute temperature. We will discuss shortly the physical significance of E_F.

The above expression can also be interpreted to mean that at a certain temperature T, the probability of occupancy of a state is lower, the higher the energy level E of that state. Thus, the states at higher energy levels are less likely to be occupied than states at lower energy levels. Since we observe from Eq. (3.1) that the density of states is higher at the higher energy level, we can conclude that at the higher energy levels, where the states are more numerous, the probability of occupancy of a state is much lower than at the lower-level, less-numerous states. The reason for this seemingly abnormal behavior results from the fact that electrons initially occupy the lowest-level states where the Pauli Exclusion Principle allocates only two electrons of opposite spin to each state.

Although E_F is defined in Appendix C, we shall relate its significance in this context as well. At temperatures approaching absolute zero and for $E < E_F$, Eq. (3.2) indicates that $f(E)$ will be equal to 1 and, for $E > E_F$, $f(E)$ will be zero. Thus, *at absolute zero, E_F is the dividing energy below which all states are occupied with*

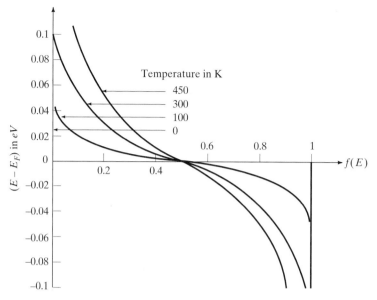

Figure 3.2   Fermi-Dirac distribution function for three temperatures. Temperature is measured in K.

electrons and above which all states are vacant. This is the same definition for E_F we arrived at in Appendix C.

As the temperature increases, some electrons acquire enough energy to move into states above E_F. Figure 3.2 shows the Fermi-Dirac distribution function for three temperatures. Note the rounding-off of the curve at the higher energy levels. Also note that at $E = E_F$, $f(E)$ is always 1/2. We can therefore define E_F as being that energy level for which the probability of occupancy is 1/2; or we can say that over a long period of time, the probability is that half the states are filled at the level E_F.

We have seen from Eq. (3.1) that $N(E)$ gives the density of the states at energy E. At thermal equilibrium, the density of electrons, dn, having energy between E and $E + dE$ is then

$$dn = f(E) \times N(E) \, dE \qquad (3.3)$$

By substituting for $f(E)$ from Eq. (3.2) and for $N(E)$ from Eq. (3.1) and using the effective mass for a semiconductor, we can then write Eq. (3.3) as

$$dn = \frac{(\pi/2)(8m^*/h^2)^{3/2} \, E^{1/2} \, dE}{1 + \exp[(E - E_F)/kT]} \qquad (3.4)$$

From Eq. (3.4), we can then calculate the incremental density of free electrons corresponding to a specific interval of energy dE, for a fixed Fermi level, and as a function of the temperature T.

A word of caution is in order here concerning the location of E_F in semiconductors. In a later chapter, we shall note that the level E_F lies in the forbidden band. You may then question our statements here concerning occupancy of E_F, even though it is in the forbidden gap. The answer is that energy states obviously exist in the forbidden band but these states do not accommodate electrons of the semiconductor.

Since the probability of occupancy of a state is $f(E)$, the probability of vacancy of a state becomes $[1-f(E)]$.

In the following example, we illustrate the application of the Fermi distribution function to a metal.

EXAMPLE 3.1

(a) Determine the probability of occupancy of a state that is located at 0.259eV above E_F at:

$$\text{i) } T = 0\text{K} \qquad \text{ii) } T = 300\text{K} \qquad \text{iii) } T = 600\text{K}$$

(b) Determine the probability of vacancy of a state that is located at 0.4eV below E_F at $T = 300$K.

(c) Repeat part (b) if the state is at 0.01eV above E_F at $T = 300$K.

Solution

(a) i) At $T = 0$K

$$f(E) = \frac{1}{1 + \exp\dfrac{0.259}{0}} = 0$$

ii) At $T = 300$K

$$f(E) = \frac{1}{1 + \exp\dfrac{0.259}{0.0259}} = 4.54 \times 10^{-5}$$

iii) At $T = 600$K

$$f(E) = \frac{1}{1 + \exp\dfrac{0.259}{0.0518}} = 6.69 \times 10^{-3}$$

(b) The probability of vacancy is $[1 - f(E)]$ so that at 0.4V it becomes

$$1 - f(E) = 1 - \frac{1}{1 + \exp\dfrac{-0.4}{0.0259}} \cong 0$$

(c) The probability of vacancy at 0.01V becomes

$$1 - \frac{1}{1 + \exp\dfrac{0.01}{0.0259}} = 0.595$$

3.3 CARRIER DENSITIES

Semiconductors become useful only after special impurities are added to them. In its almost pure form, and when no impurities are added, the semiconductor is labeled as *intrinsic*. It is *extrinsic* when selected impurities are added and the semiconductor is said to be *doped* with impurities.

Densities of States in the Conduction and Valence Bands

Initially, we will determine expressions for the electron and hole densities in an intrinsic semiconductor. We will assume that the expression for the density of states given by Eq. (3.1) applies for all conduction band and valence band energies with the appropriate effective masses substituted.

There are about 5×10^{22} atoms/cm^3 in silicon. At $T = 0$K, the $4N$ quantum states in the valence band are filled with electrons and the $4N$ states in the conduction band are empty. These states are distributed throughout the bands. The states of importance in semiconductor devices, as we shall see later, are the ones near the top of the valence band and the states near the bottom of the conduction band.

To obtain the densities of states near the band edges, we let E_C represent the minimum electron energy in the conduction band and E_V represent the maximum hole energy in the valence band. We recall that an increase of the energy of an electron in the conduction band corresponds to the electron moving up on the energy scale in the conduction band. An increase of the energy of a hole in the valence band, then, corresponds to the hole moving down into the valence band. Thus, we change the variable for integration so that Eq. (3.1) becomes $(E - E_C)$ for the conduction band and it becomes $(E_V - E)$ for the valence band states. Therefore, the functions for the densities of states in the conduction and valence bands, as shown in Fig. 3.3, become

$$N_n(E) = \frac{\pi}{2}\left(\frac{8m_n^*}{h^2}\right)^{3/2} (E - E_C)^{1/2} \text{ for } E > E_C \tag{3.5}$$

where $N_n(E)$ is the number of states per unit volume per unit energy at E in the conduction band. The corresponding density of hole states in the valence band at E is

$$N_p(E) = \frac{\pi}{2}\left(\frac{8m_p^*}{h^2}\right)^{3/2} (E_V - E)^{1/2} \text{ for } E < E_V \tag{3.6}$$

where $N_p(E)$ is the density of states in the valence band and is assumed to be located at E. Using the above expressions, we determine the electron density in the conduction band from

$$n = \int_{E_C}^{\infty(\text{top})} N_n(E) f_c(E) \, dE \tag{3.7}$$

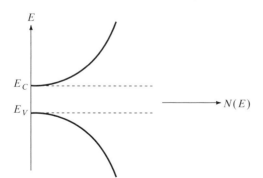

Figure 3.3   Distribution of the density of states in the conduction and valence bands.

where $f_c(E)$ is given by Eq. (3.2). The hole density, which represents the density of vacant states, is given by

$$p = \int_{-\infty(\text{bottom})}^{E_V} N_p(E) f_v(E) \, dE \tag{3.8}$$

where $f_v(E)$ is the probability of vacancy of a state in the valence band and is $[1 - f_c(E)]$.

The Boltzmann Approximation

To simplify the integration in Eqs. (3.7) and (3.8), we make an important assumption that is related to the location of the Fermi level. This assumption, known as the *Boltzmann approximation*, consists in dropping the unity term in the denominator of the expression for the Fermi function given by Eq. (3.2). Based on this approximation, Eq. (3.2), for the probability of the occupancy of a state by an electron, becomes

$$f_c(E) = \exp\left[-(E - E_F)/kT\right] \text{ for } E > E_F \tag{3.9}$$

The reason for restricting the energy level to values greater than E_F is that an energy level $E < E_F$ in Eq. (3.9) makes the probability greater than unity, which is meaningless.

Let us briefly digress to investigate the condition for which the approximation of Eq. (3.9) is fairly valid. Upon comparing $f(E)$ from Eqs. (3.2) and (3.9) at various values of $(E - E_F)$, we note that for $(E - E_F) = 3kT$, $f(E)$ calculated from Eq. (3.9) differs from that of Eq. (3.2) by about 5 percent and at $(E - E_F) = 4kT$ the difference is about 1.8 percent. We will arbitrarily state that Eq. (3.9) is valid provided the Fermi level is at least $3kT$ below the bottom of the conduction band where all free electrons reside.

For energy levels below E_F, the probability of a vacancy (or hole occupancy) can be written from Eq. (3.2) as

$$1 - f_c(E) = 1 - \frac{1}{\exp\left[(E - E_F)/kT\right] + 1} = \frac{\exp\left[(E - E_F)/kT\right]}{1 + \exp\left[(E - E_F)/kT\right]} \tag{3.10}$$

At values of $(E_F - E) = 3kT$, the exponential term in the denominator of Eq. (3.10) becomes very small so that the validity of this equation is restricted to values of $(E_F - E) > 3kT$. Equation (3.10) becomes

$$1 - f_c(E) \cong \exp - \left[(E_F - E)/kT\right] \tag{3.11}$$

The restriction to Eq. (3.11) implies that the Fermi level is at least 3kT above the top of the valence band.

Since the electrons occupy states in the conduction band and holes represent unoccupied states in the valence band, we conclude that the validity of the Boltzmann approximation is restricted to the range of Fermi energies extending from $3kT$ above the top of the valence band to $3kT$ below the bottom of the conduction band.

It is possible that in a highly doped semiconductor, as we shall see later, the Fermi level moves to within less than $3kT$ from the band edges and even into the bands. Such semiconductors are said to be *degenerate*.

We conclude that for a non-degenerate semiconductor, the Fermi-Dirac equation represented by f_c for electrons and f_v for holes respectively becomes

$$f_c(E) = \exp - (E - E_F/kT) \quad \text{(a)}$$

$$f_v(E) = \exp - (E_F - E)/kT) \quad \text{(b)} \tag{3.12}$$

Expressions for Electron and Hole Densities

Using the expressions in Eqs. (3.5) and (3.6) together with Eqs. (3.11) and (3.12) in Eqs. (3.7) and (3.8), we have for electrons

$$n = \frac{\pi}{2}\left(\frac{8m_n^*}{h^2}\right)^{3/2} \int_{E_C}^{\infty} (E - E_C)^{1/2} \exp-\left(\frac{E - E_F}{kT}\right)dE \tag{3.13}$$

For holes, the expression is

$$p = \frac{\pi}{2}\left(\frac{8m_p^*}{h^2}\right)^{3/2} \int_{-\infty}^{E_v} (E_V - E)^{1/2} \exp-\left(\frac{E_F - E}{kT}\right)dE \tag{3.14}$$

Sketches graphically describing the operations in Eqs. (3.7) and (3.8) are shown in Figs. 3.4 and 3.5 for the electron and hole distributions in the conduction and valence bands respectively, for different locations of E_F.

Before we proceed with evaluating the integral, we will clarify two questions that an interested reader may raise. The first concerns changing the limits of integration. We have replaced the top of the conduction band by plus infinity and the bottom of the valence band by minus infinity. The reason is to simplify the integration. However, the change is quite justifiable since the exponential functions $f_c(E)$ and $f_v(E)$ decrease so rapidly as we move away from the band edges that the densities of the carriers become negligible as we move a few kT's away from the edges of the conduction and valence band edges. The second question concerns the effective masses. We confirmed at the end of Chapter 2 that the effective mass, m^*, is a function

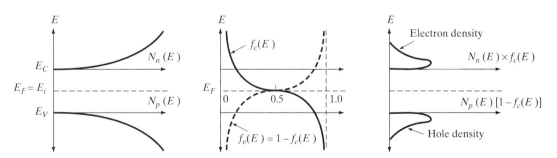

Figure 3.4   The product of the distribution of energy states and the Fermi function gives the distribution of electrons and holes for E_F midway between E_C and E_V.

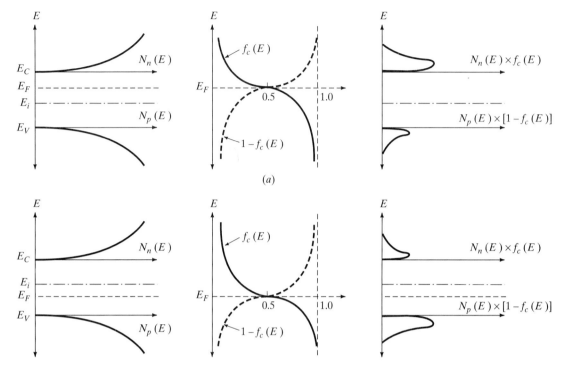

Figure 3.5 This is Fig. 3.4 repeated except that (a) the Fermi level is assumed to be above the middle of the gap and (b) the Fermi level is assumed to be below the middle of the gap. Note that at $E = E_F$, $f(E) = 0.5$.

of the energy level. In Eqs. (3.13) and (3.14) we have, however, assumed that this mass is constant and we placed it outside the integrand. The justification is that since the exponential in the expressions decays very rapidly as we move away from the conduction and valence band edges, E_C and E_V, our interest is in the carrier densities over the narrow portions above E_C and below E_V. In these small regions, the effective mass is relatively constant as we indicated earlier.

To integrate Eq. (3.13), we let $x = (E - E_C)/kT$ so that the integral becomes

$$n = \frac{\pi}{2}\left(\frac{8m_n^*}{h^2}\right)^{3/2} \int_0^\infty (xkT)^{1/2}\, kT \exp-\left(\frac{xkT + E_C - E_F}{kT}\right)dx$$

$$= \frac{\pi}{2}\left(\frac{8m_n^*}{h^2}\right)^{3/2} \exp-\left(\frac{E_C - E_F}{kT}\right)(kT)^{3/2}\int_0^\infty x^{1/2}\, e^{-x}\, dx \qquad (3.15)$$

The integral has a value of $\dfrac{\pi^{1/2}}{2}$ and n becomes

$$n = \frac{\pi}{2}\left(\frac{8m_n^*}{h^2}\right)^{3/2} \exp-\left(\frac{E_C - E_F}{kT}\right)(kT)^{3/2}\frac{\pi^{1/2}}{2}$$

$$= 2\left(\frac{2m_n^* kT\pi}{h^2}\right)^{3/2} \exp-\left(\frac{E_C - E_F}{kT}\right) \qquad (3.16)$$

A similar expression is obtained for p as

$$p = 2\left(\frac{2m_p^* kT\pi}{h^2}\right)^{3/2} \exp-\left(\frac{E_F - E_V}{kT}\right) \tag{3.17}$$

The expressions for n and p in Eqs. (3.16) and (3.17) can be written as

$$n = N_c \exp-\left(\frac{E_C - E_F}{kT}\right) \tag{3.18}$$

$$p = N_v \exp-\left(\frac{E_F - E_V}{kT}\right) \tag{3.19}$$

where N_c and N_v are given by

$$N_c = 2\left(\frac{2\pi m_n^* kT}{h^2}\right)^{\frac{3}{2}} \tag{3.20}$$

$$N_v = 2\left(\frac{2\pi k m_p^* T}{h^2}\right)^{\frac{3}{2}} \tag{3.21}$$

For m in kg., k in J/K and h in J-s, N is in m^{-3}.

The terms N_c and N_v are referred to as the effective density of states of the conduction and valence bands respectively. The values for N_c and N_v are shown in Table 3.1 where the difference between N_c and N_v result from the different values of the effective masses of electrons and holes. Relative masses of electrons and holes are shown in Table 2.3.

We remind the reader that the above relations were derived subject to the approximation that E_F was no closer than $3kT$ units of energy to either the conduction band or valence band edges.

TABLE 3.1   Effective density of states and band gap energy at $T = 300K$.

Semiconductor	$N_c(cm^{-3})$	$N_v(cm^{-3})$	$E_g(eV)$
Si	3.22×10^{19}	1.83×10^{19}	1.12
Ge	1.03×10^{19}	5.35×10^{18}	0.66
GaAs	4.21×10^{17}	9.52×10^{18}	1.42

By taking the product of n and p from Eqs. (3.18) and (3.19), we obtain the interesting result

$$np = N_c N_v \exp-\left(\frac{E_C - E_V}{kT}\right) \tag{3.22}$$

Intrinsic Carrier Density

In an intrinsic semiconductor, one to which no impurities have been added, at $T > 0K$, the density of electrons in the conduction band must equal the density of holes in the valence band because for every electron that is excited to the conduction band, a hole is created in the valence band. We define this density as n_i and n_i^2, from Eq. (3.22) becomes

$$np = n_i^2 = N_c N_v \exp-\left(\frac{E_C - E_V}{kT}\right) \qquad (3.23)$$

The intrinsic carrier density, n_i, assuming that the effective masses do not change with temperature and using Eqs. (3.20) and (3.21) in Eq. (3.23), becomes

$$n_i = K_1 T^{3/2} \exp\frac{-E_g}{2kT} \qquad (3.24)$$

where K_1 is a constant independent of temperature and E_g, equal to $(E_C - E_V)$, is the band gap energy of the semiconductor. High temperatures and smaller band gaps favor large values of intrinsic carrier density. It is important to note that the T in the exponent has a much greater effect than the factor $T^{3/2}$. We will illustrate these effects in Example 3.2.

By substituting for the constants in Eq. (3.24) and at $T = 300K$, we find that the intrinsic carrier density in silicon is approximately $1 \times 10^{10} cm^{-3}$, signifying that this is the density of electrons in the conduction band and the density of holes in the valence band. Mainly because of the larger band gap, the intrinsic carrier density in gallium arsenide at $T = 300K$ is $2.49 \times 10^6 cm^{-3}$.

There are 5×10^{22} atoms per cubic centimeter in silicon and each atom has four electrons in the covalent bond representing the 3s and 3p levels. These same levels form the conduction and valence bands in silicon. This results in a density of 2×10^{23} electrons per cm³ available for conduction, all residing in the valence band at $T = 0K$. At $T = 300K$ the density of electrons (in the conduction band) is $1 \times 10^{10} cm^{-3}$. Approximately, therefore, one in 10^{13} of the electrons available at $T = 0K$ at the top of the valence band is elevated at room temperature from the top of the valence band to the conduction band. This represents a very small proportion of the available electrons.

By substituting for the constant K_1 in Eq. (3.24) from the expressions in Eqs. (3.20) and (3.21), we have

$$n_i = 2\left(\frac{2\pi kT}{h^z}\right)^{3/2} (m_n^* m_p^*)^{3/4} \exp\left(\frac{-E_g}{2kT}\right) \qquad (3.25)$$

We observe that the intrinsic carrier density is independent of the location of the Fermi level and depends strongly on the *band gap energy*, E_g, and on the temperature. The strong dependence on temperature is a result of the rapidly varying term in the exponent since the number of carriers that acquire thermal energy is increased as more of the electrons that are deeper in the valence band are able to move up to the conduction band. This dependence is illustrated in Fig. 3.6. The intrinsic carrier density decreases exponentially with an increase of E_g, accounting for the very low conductivity of insulators.

Calculations of the intrinsic carrier density are illustrated by Example 3.2.

EXAMPLE 3.2

Calculate the intrinsic carrier density of silicon at:

a) $T = 300K$ b) $T = 600K$

Include effect of variation of E_g with temperature.

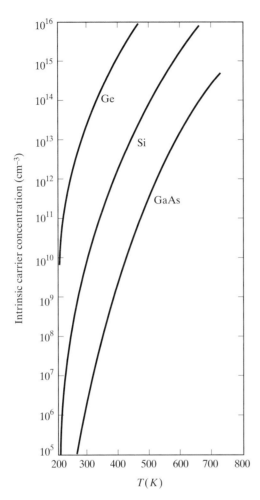

Figure 3.6 Dependence of the intrinsic carrier concentration on temperature.

Solution

a) $n_i^2 = N_c N_v e^{-E_g/kT}$

Using the values of N_c and N_v from Table I and using an E_g of $1.12 eV$, we determine n_i^2 as

$$n_i^2 = 0.9772 \times 10^{20} \text{cm}^{-6}$$

$$n_i = 9.885 \times 10^9 \text{cm}^{-3}$$

b) At $T = 600$K, we determine E_g, from the relation in Table 2-2, to be $1.032 eV$. Using Eq. (3.24) and $n_i(300\text{K}) = 10^{10}\text{cm}^{-3}$, we have

$$\frac{n_i^2(600\text{K})}{n_i^2(300\text{K})} = \left(\frac{600}{300}\right)\left[\frac{\exp(-1.032/(8.62 \times 10^{-5} \times 600))}{\exp(-1.12/(8.62 \times 10^{-5} \times 300))}\right] = 1.137 \times 10^{11}$$

$$n_i(600\text{K}) = 3.337 \times 10^{15}\text{cm}^{-3}$$

We highlight the influence of T in the exponential by (a) assuming that E_g does not change with temperature and (b) comparing the factors by which each of the temperature terms, in the expression for n_i^2, affects n_i. For an increase of T from 300K to 600K, the T^3 term increases n_i by a factor of 2.8, whereas the same increase in T causes the exponential term to increase n_i by about a factor of 50,000.

It is important to indicate that there are discrepancies among various sources in the values of the intrinsic carrier density. The reason for this uncertainty in the calculated value is a result of the very complex relations from which the values of the effective masses are determined.

In this book and for silicon at $T = 300$K, we will use a value of 1×10^{10} carriers per cubic centimeter. This value agrees fairly well with the calculation using the values of N_c and N_v listed earlier. There is no sacrifice of accuracy in this approximation. In fact, the advantage is that it is a number that can easily be remembered.

Location of Fermi Level

One major assumption that we made in deriving the expressions for the electron and hole densities for the intrinsic semiconductor is that the Fermi level is at least 3 kT eV distant from the edges of the conduction and valence bands. We investigate the strength of the validity of this assumption for an intrinsic semiconductor by equating the expressions for the intrinsic electron and hole densities in Eqs. (3.18) and (3.19) and solving for E_F, so that

$$E_F = \frac{E_C + E_V}{2} + \frac{kT}{2} \ell n \left(\frac{N_v}{N_c} \right) \tag{3.26}$$

We use the expressions for N_c and N_v from Eq. (3.20) and (3.21) to obtain

$$\frac{N_v}{N_c} = \left(\frac{m_p^*}{m_n^*} \right)^{3/2} = \left(\frac{m_n^*}{m_p^*} \right)^{-3/2} \tag{3.27}$$

The expression for E_F becomes

$$E_F = \frac{E_C + E_V}{2} + \frac{3}{4} kT \ell n \frac{m_p^*}{m_n^*} = \frac{E_C + E_V}{2} - \frac{3}{4} kT \ell n \left(\frac{m_n^*}{m_p^*} \right) \tag{3.28}$$

where $(E_C + E_V)/2$ represents the middle of the band gap. The Fermi level for an intrinsic semiconductor is labeled E_i so that

$$E_i = E_F = E_V + \frac{E_g}{2} - \frac{3}{4} kT \ell n \left(\frac{m_n^*}{m_p^*} \right) \tag{3.29}$$

We observe that E_i is temperature-dependent so long as the effective masses of electrons and holes are unequal. The second term in the right-hand side of Eqs. (3.26) and (3.28) is of the order -0.01eV for silicon at room temperature. For all practical purposes, one can state that the Fermi level in an intrinsic semiconductor, E_i, is approximately halfway between E_C and E_V, since E_C in Eq. (3.28) is given by $E_V + E_g$ so that $(E_C + E_V)/2$ becomes $(E_V + [E_g/2])$.

In the following example, we obtain order of magnitude values for the location of E_i.

EXAMPLE 3.3

Determine the location of the Fermi level with respect to the middle of the band gap in intrinsic silicon and intrinsic gallium arsenide at $T = 300\text{K}$. The value of k is given as $8.61 \times 10^{-5}\text{eV/K}$.

Solution   By using the values of N_v and N_c from Table 3.1, we calculate

$$E_i = \frac{E_C + E_V}{2} + \frac{kT}{2}\ell n\left(\frac{N_v}{N_c}\right)$$

For intrinsic silicon,

$$E_i = (E_C + E_V)/2 + \frac{8.61 \times 10^{-5} \times 300}{2}\ell n\frac{1.83 \times 10^{19}}{3.22 \times 10^{19}}$$

$$E_i = (E_C + E_V)/2 - 0.0073\text{eV}$$

For intrinsic gallium arsenide,

$$E_i = (E_C + E_V)/2 + 0.012915\,\ell n\frac{9.52 \times 10^{18}}{4.21 \times 10^{17}}$$

$$E_i = (E_C + E_V)/2 + 0.0403\text{eV}$$

We note that the Fermi level at 300K in intrinsic silicon is 0.0073eV below the midgap, while in gallium arsenide, E_i is 0.0403eV above the midgap.

REVIEW QUESTIONS

Q3-1 Determine the number of allowed states in a crystal that has N atoms.

Q3-2 Why is the Boltzmann approximation used?

Q3-3 For what values of $(E_C - E_F)$ is Eq. (3.9) valid?

Q3-4 In the equations derived so far, the effective mass has been assumed constant. Is this true? Why?

Q3-5 In Eq. (3.15), the upper limit of integration was set at infinity. Why?

Q3-6 Briefly explain why the narrower the band gap, the higher is the intrinsic carrier density in a semiconductor.

Q3-7 Briefly explain why the intrinsic carrier density increases with an increase of temperature.

HIGHLIGHTS

- The density of states function $N(E)$ represents the number of free electrons in a metal per unit volume per unit energy at energy E.
- The Fermi-Dirac function $f(E)$ represents the probability that a state located at energy E is occupied by an electron. Therefore, $(1 - f(E))$ represents the probability that a vacancy exists at energy E.
- The energy level E_F is that energy at which the probability of occupancy is one-half.
- The product of $f_C(E)$ and $N_n(E)$ given by Eqs. (3.6) and (3.5) represents the number of electrons per unit volume per unit energy at E.

- The intrinsic carrier density represents the density of each of the electrons in the conduction band and holes in the valence band.

EXERCISES

E3-1 Determine the number of electronic states in the conduction band of silicon located between energies of 1.0 and 1.1eV. Assume a volume of $10^{-16}m^3$.

E3-2 Calculate the intrinsic carrier density of Ge.

Ans: $n_i = 2.17 \times 10^{13} cm^{-3}$

E3-3 a) Determine the location of the Fermi level in intrinsic Ge, with respect to the bottom of the conduction band.

b) Determine the location of E_F when measured from the top of the valence band.

Ans: a) $E_C - E_F = 0.338eV$

3.4 EXTRINSIC SEMICONDUCTORS

Extrinsic semiconductors are formed by the addition of small amounts of selected impurities to pure semiconductors. These impurities are elements in Column III or Column V of the periodic table. The addition of as little as one impurity atom to a million semiconductor atoms has considerable effect on the conducting properties of the semiconductor. The atomic dimensions and the electronic structures of these impurities are similar to those of the semiconductor. The effect of adding the impurity is to increase the density of one of the two carriers and, hence considerably alter the electrical conductivity. This process cannot be effective in good conductors, such as copper, since the density of electrons is so large that it can hardly be changed by the addition of impurities.

These impurities are also known as *dopants* and the process of adding them is known as *doping the semiconductor*. The result of the doping generates either of two types of extrinsic semiconductors, identified by the type of the carrier whose density is increased. They may be N-type or P-type, where the electron density is increased in the N-type and the hole density is increased in the P-type.

It is important to indicate here, a fact which will be explained later, that the semiconductor is charge neutral whether in the intrinsic condition or after impurities are added and the semiconductor becomes extrinsic.

N-type Semiconductor

When elements from Column V of the periodic table are added to silicon, an *N*-type semiconductor results. Typical dopants from Column V elements are phosphorous, arsenic, and antimony. These have five valence electrons in their outermost orbit and are known as *donor impurities* because they have one electron in excess of what is needed for the covalent bonding.

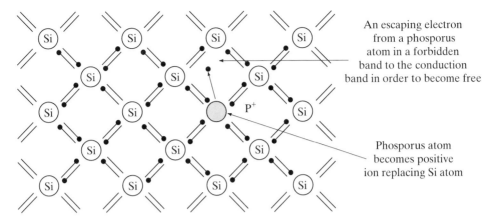

Figure 3.7 The addition of a 5-electron donor phosphorus atom to the covalent bond structure generates a free electron since only four are required to complete the bond.

Each impurity atom occupies the space of one silicon atom among, say, a million atoms of the semiconductor. For covalent bonding with the four nearest semiconductor atoms, four electrons are needed and the fifth electron of the donor atom is not required in the crystal binding. This electron becomes a free electron influenced only by the crystal structure and it is said to be "loosely" bound. A sketch of the resulting structure is shown in Fig. 3.7.

In the band picture, the donor atoms occupy energy levels in the forbidden band that are slightly below the bottom of the conduction band.

The free electrons that are generated *intrinsically* escape from a semiconductor covalent bond when that bond absorbs thermal energy. Using the band picture, the intrinsically generated semiconductor electron gains enough energy to jump across the band gap and becomes free. The band gap energy is about 1.1eV in silicon at $T = 300\text{K}$. The question is then, how much energy is required to release one electron from a donor atom?

We will use the binding of energy of the hydrogen electron to obtain an estimate of the binding energy of the donor electron. The hydrogen electron in its ground state, $n = 1$, has energy given by Bohr's relationship, Eq. (1.14), to be

$$E_h = \frac{-m_0 q^4}{8\varepsilon_0^2 h^2} = -13.6\text{eV}$$

An energy of 13.6eV is required to free the hydrogen electron.

To obtain a similar quantity of energy for our donor electron, we have to correct the expression for the energy of the hydrogen electron by replacing the relative value of the mass of the donor electron and the relative permittivity of the semiconductor.

Because the hydrogen electron is in the field of the hydrogen nucleus, whereas the donor electron is under the influence of the potential field of the semiconductor lattice, we expect the masses to be different. The donor electron is moving in a semi-conductor lattice that has permittivity $\varepsilon_r\varepsilon_0$ where ε_r, the relative permittivity, has a value of 11.8 for silicon.

Using the above information, we obtain the approximate expression for the binding energy of a donor electron, E_D, in silicon to be,

$$\frac{E_D}{E_h} = \frac{-m_n^*}{m_0\,(11.8)^2}$$

where m_n^*/m_0 is the relative effective mass in silicon and is equal to 1.18.

An *order to magnitude value* for E_D is obtained by assuming the masses to be equal so that E_D becomes,

$$E_D = \frac{-13.6}{(11.8)^2} \cong -0.1\text{eV}.$$

The above indicates that an energy of 0.1eV is sufficient to excite the fifth donor electron into the conduction band so that it becomes a free electron leaving behind an ionized donor atom. Thus, we can state that the donor energy level in silicon is very close to the conduction band edge.

More accurate values for ionization energies are listed in Table 3.2.

An energy band representation of the condition of donor atoms in silicon is shown in Fig. 3.8, clearly illustrating the proximity of the donor energy level to the conduction band. Since electrons in the donor impurity occupy different energy levels, the higher the temperature of the sample (semiconductor + impurity), the greater is the thermal energy and the larger is the number of electrons that can be elevated to the conduction band.

The important difference between this ionization mechanism to produce donor electrons and the intrinsic process to produce electrons is that the ionized impurities are fixed charges in the lattice and no holes are produced.

Thus, one can state that at $T = 300$K, each donor atom contributes a free electron to the conduction band leaving behind the ionized donor atom. The question

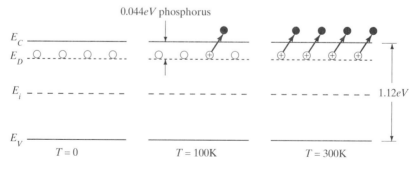

Figure 3.8 Energy band representation of donor impurity in silicon and the effect of temperature. The phosphorus donor atoms have energy about 0.044eV below E_C and are completely ionized at $T = 300$K.

becomes: Is the total resulting density of electrons, intrinsic and donor, equal to the sum of the donor density and the intrinsic carrier density? The answer is no and we shall investigate this in a later section. Suffice it to recall here that the semiconductor is charge-neutral.

P-Type Semiconductor

A semiconductor is said to be *P*-type when an element from Column III of the periodic table is added to it. The common dopants are: boron, aluminum, gallium, and indium. The atoms of these dopants have three electrons in their outermost orbit and when a dopant atom replaces a semiconductor atom, in the lattice structure, a space in the covalent bond is available into which an electron of the semiconductor can move. Hence, the dopant atom accepts an electron. This type of dopant is known as an *acceptor impurity*.

The electron that occupies the available space in the acceptor bonding comes from the electrons that are in the valence band of the semiconductor and, once captured, this electron causes the acceptor atom to become negatively ionized. A sketch of the resulting covalent bond arrangement is shown in Fig. 3.9.

The condition for ionization of acceptor atoms is analogous to that of donor atoms. The acceptors' atoms are located at energy levels slightly above the valence band. At room temperature, there is sufficient thermal energy to excite electrons from the valence band into the acceptor level. The absence of an electron from the top of the valence band generates a hole. Thus, for every acceptor atom, a hole is generated at room temperature and the acceptor atom becomes negatively ionized. This condition is illustrated in Fig. 3.10.

Thus, at room temperature, each donor atom provides a free electron to the conduction band and each acceptor atom generates a hole in the valence band. At

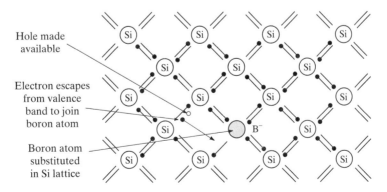

Hole made available

Electron escapes from valence band to join boron atom

Boron atom substituted in Si lattice

Figure 3.9   A boron atom accepts an electron to complete the covalent bond and hence creates a vacancy, a hole, into which another electron can move.

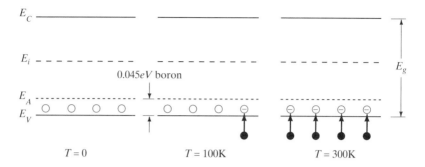

Figure 3.10 The acceptor energy level in the energy band diagram of silicon and the effect of temperature on the ionization of the acceptor atoms. At $T = 300K$ all acceptor atoms are ionized by the addition of electrons from the valence band.

room temperature, all donor atoms are positively ionized, each atom having lost an electron to the conduction band. Similarly, each acceptor atom is negatively ionized at room temperature. Listed in Table 3.2 are the results of more accurate calculations of the ionization energies of donors and acceptors in silicon. *The ionization energy is defined as the energy required to ionize a donor or an acceptor atom.*

TABLE 3.2 Ionization Energies of Impurities in Silicon

Element	Function	Ionization energy (eV)
Boron	Acceptor	0.045
Aluminum	Acceptor	0.057
Gallium	Acceptor	0.065
Phosphorus	Donor	0.044
Arsenic	Donor	0.049
Antimony	Donor	0.039

Source: Reprinted from S. M. Sze and J. C. Irvin, "Resistivity, Mobility and Impurity Levels in GaAs, Ge and Si at 300K," *Solid State Electronics* 11, 599 (1968), with kind permission from Elsevier Science Ltd., The Boulevard, Langford Lane, Kidlington 0X5#1GB, UK.

In addition to the controlled addition of donor and acceptor impurities, other impurities may exist at energy levels more distant from the band edges than the controlled impurities. These impurities are randomly available and usually act as centers of recombination generation or traps. Because they exist around the middle of the energy gap, they can act as energy levels at which electrons and holes meet to recombine. More will be said later about recombinations.

3.5 THERMAL EQUILIBRIUM

In an earlier section, we derived an expression for the intrinsic carrier density, the electron density, and the hole density, as a function of the temperature and the band energy gap. We concluded that at room temperature the intrinsic carrier density in silicon is approximately $1.0 \times 10^{10} \text{cm}^{-3}$. This carrier density results from the transfer of electrons to the conduction band as they receive thermal energy. The question that comes to mind is: As long as the thermal energy is available, why is it that the density of the carrier does not continuously increase, but rather reaches a fixed value?

The answer is that as carriers are continuously generated, and when there is no other source than thermal energy, electrons and holes continuously recombine and pairs of carriers disappear. In moving through the lattice, they encounter obstacles causing them to lose energy and disappear. Thus, the rate of generation is accompanied by the equal and opposite rate of recombination. This condition exists at thermal equilibrium. *Thermal equilibrium is therefore defined as the state in which a process is accompanied by an equal and opposite process, while the system is held at constant temperature, and no external source of energy acts on it.*

In the next chapter, we will present analytical relationships for the processes of carrier generation and recombinations.

3.6 DENSITIES OF CARRIERS IN EXTRINSIC SEMICONDUCTORS

When dealing with doped semiconductors, we refer to electrons in N-type semiconductors, the more numerous carriers, as the majority carriers. Holes in N-type semiconductors are minority carriers. In P-type semiconductors, holes are the majority carriers and electrons are the minority carriers.

We stated earlier that the density of electrons, when donors are added to silicon, is *not* equal to the sum of the intrinsic electron density and the density of the ionized donor atoms. What, then, is the resulting density?

In an earlier section, and prior to determining the intrinsic carrier density, we found that the product of n and p, n_i^2, is a constant for a certain semiconductor at a certain temperature and is given by $np = n_i^2 = N_c N_v \exp\left(-E_g/kT\right)$. There is no condition in the expression that restricts it to intrinsic semiconductors because E_g does not change with impurity concentration and N_c and N_v are constants. This product is therefore a constant equally valid for intrinsic as well as for doped semiconductors, provided that we refer to the carrier densities at thermal equilibrium (a condition when there are no sources of energy other than thermal energy). We label the extrinsic values of the electron and hole density at thermal equilibrium as n_0 and p_0 respectively so that we can write,

$$n_0 p_0 = n_i^2 \qquad (3.30)$$

In order to determine the values of n_0 and p_0 for a doped semiconductor, we need another relationship that relates them.

Charge Neutrality

We consider the general case when both donors and acceptors are added to a semiconductor. Assuming thermal equilibrium, the material includes particles that have positive charges and others that have negative charges. The negative charges are made up of the electrons in the conduction band and the acceptor atoms which are ionized. The holes in the valence band and the ionized donor atoms constitute the positive charges. Since the intrinsic silicon is charge neutral and the added impurities are neutral, the resulting mix is also neutral so that, charge neutrality requires,

$$q(n_0 + N_A^-) = (p_0 + N_D^+)q \tag{3.31}$$

where N_A^- denotes the density of acceptor atoms and N_D^+ refers to the density of donor atoms, assuming they are all ionized and n_0 and p_0 are the thermal equilibrium values of electron and hole densities respectively. By replacing p_0 in Eq. (3.30) by its equivalence from Eq. (3.31), a quadratic equation in n_0 is obtained. The solution to the quadratic equation is given by Eq. (3.32) where the positive sign has been selected for the term under the square root because that term has a larger magnitude than the first term and n_0 cannot be negative.

$$n_0 = \frac{N_D - N_A}{2} + \left[\left(\frac{N_D - N_A}{2} \right)^2 + n_i^2 \right]^{1/2} \tag{3.32}$$

We have dropped the positive and negative superscripts from the symbols for impurity densities since we have already assumed that all impurity atoms are ionized at room temperature.

For an N-type semiconductor, either $N_D >> N_A$ or N_A is zero. From Eq. (3.32), we find n_0 for an N-type semiconductor to be,

$$n_0 = \frac{N_D}{2} + \left[n_i^2 + \left(\frac{N_D}{2} \right)^2 \right]^{1/2} \tag{3.33}$$

The density of the holes is then calculated from Eq. (3.30) to be,

$$p_0 = \frac{n_i^2}{n_0} \tag{3.34}$$

In solving Eqs. (3.30) and (3.31) simultaneously and when numerical values are given for N_D and N_A, it is always advisable to obtain a quadratic equation for the larger expected carrier density, the expected majority carrier density. The reader should try solving for the minority carrier density to appreciate this caution.

If the donor density N_D is much greater than n_i, which is quite common in most extrinsic semiconductors, Eq. (3.33) can be simplified so that

$$n_0 \cong N_D \text{ and } p_0 \cong \frac{n_i^2}{N_D} \tag{3.35}$$

For a donor density of 10^{16}cm^{-3} in silicon at $T = 300\text{K}$, where $n_i = 1 \times 10^{10} \text{cm}^{-3}$, we find,

$$n_0 \cong 10^{16} \text{cm}^{-3} \text{ and } p_0 = 1 \times 10^4 \text{cm}^{-3}.$$

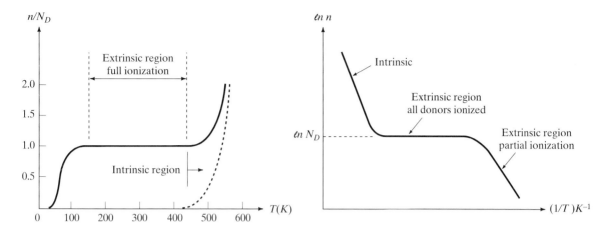

Figure 3.11   Variation of majority carrier density with temperature. In the logarithmic sketch, the slope of the line in the intrinsic region is a measure of the band gap energy E_g.

We note, therefore, that the density of majority carriers is much greater than the density of minority carriers.

An interesting supplement to the results of this section is obtained by plotting the ratio of the majority carrier density to the donor density as a function of temperature, as shown in Fig. 3.11.

We assume that a donor impurity having density N_D of 10^{16}cm^{-3} is added to a silicon sample. At temperatures near zero K, the electron density in the conduction band is zero since the thermal energy at that temperature is not sufficient to ionize any donor atoms and certainly not enough to excite electrons from the valence band into the conduction band. As the temperature is increased, some donor atoms become ionized, losing their electrons to the conduction band, while there is still not sufficient energy to move electrons out of the valence band.

We notice that up to about 150K, the electron density is smaller than the donor density because not all donor atoms are ionized and very few electrons have been excited from the valence band to the conduction band. The region 0–150K is known as the *freeze-out region*. At temperatures beyond 150K, all donor atoms are assumed to become ionized so that $n \cong N_D$. This concentration of electrons remains constant at the value of N_D up to room temperature, at which $N_D >> n_i$. The range from $n = N_D^+ \cong N_D$ up to where n_i begins to increase beyond its room temperature value is known as the *extrinsic region*. This range extends from about 150 to 400K. Above 400K the intrinsic carrier density begins to increase rapidly and eventually reaches a value greater than N_D. The material begins to become intrinsic again and the electron density becomes much larger than N_D because of the large number of electrons that are thermally excited from the valence band to the conduction band. For temperatures exceeding 400K, the region is labeled the *intrinsic region*. In the intrinsic region of Fig. 3.11, the densities of both holes and electrons increase, however, their thermal equilibrium values are still determined by Eqs. (3.30) and (3.31).

Calculations of electron and hole densities in extrinsic silicon are shown in Examples 3.4 and 3.5.

EXAMPLE 3.4

To a sample of intrinsic silicon we add 10^{17} atoms per cc of phosphorus and 9×10^{16} atoms per cc of boron.

Calculate the density of the majority and of the minority carriers at:

a) $T = 300\text{K}$, **b)** $T = 600\text{K}$

Solution

a) at $T = 300\text{K}$, we assume that all impurities are ionized so that,

$$np = n_i^2 = 1 \times 10^{20} \tag{1}$$

Neutrality required that

$$n + N_A = p + N_D$$

$$n + 9 \times 10^{16} = p + 10^{17} \tag{2}$$

Solving equations 1 and 2 simultaneously, we have

$$n \cong 10^{16}\text{cm}^{-3}$$

$$p = \frac{n_i^2}{n} = 10^4\text{cm}^{-3}$$

We observe that the electrons supplied by the donor atoms dominate the resulting electron density.

b) at $T = 600\text{K}$, $n_i^2 = 11.137 \times 10^{30}\text{cm}^{-6}$ so that

$$np = 11.137 \times 10^{30} \tag{3}$$

$$n + 9 \times 10^{16} = p + 10^{17} \tag{4}$$

Solving equations 3 and 4 simultaneously, we have

$$n \cong 1.1 \times 10^{16}\text{cm}^{-3}$$

$$p = \frac{n_i^2}{n} = 1.014 \times 10^{15}\text{cm}^{-3}$$

Comparing the results to those of part (a), we note the effect of the thermally generated electrons and holes.

EXAMPLE 3.5

A sample of intrinsic silicon is doped with 10^{17} atoms/cm^{-3} of phosphorus. Calculate the density of the electrons and the holes at:

a) $T = 300\text{K}$ **b)** $T = 600\text{K}$

Solution

a) At $T = 300\text{K}$, we find that

$$n_i = 1 \times 10^{10} \text{cm}^{-3}$$

$$np = n_i^2 = 1 \times 10^{20} \text{cm}^{-6} \tag{1}$$

Neutrality requires that

$$n = p + N_D = p + 10^{17} \tag{2}$$

Solving equations 1 and 2 simultaneously, we have

$$n \cong 10^{17} \text{cm}^{-3}, p = \frac{n_i^2}{n} = 1 \times 10^3 \text{cm}^{-3}$$

b) At $T = 600\text{K}$,

$$n_i^2 = 11.137 \times 10^{30} \text{cm}^{-6}$$

$$np = n_i^2 \tag{3}$$

$$n = p + 10^{17} \tag{4}$$

Solving equations 3 and 4 simultaneously,

$$n \cong 1.0011 \times 10^{17} \text{cm}^{-3}$$

The hole density is calculated to be

$$p = \frac{n_i^2}{10^{17}} = 1.11 \times 10^{14} \text{cm}^{-3}$$

Note the large increase in the hole density while the electron density remained practically constant.

Additional expressions for n_0 and p_0

We have determined in Section 3.3 that, for an intrinsic semiconductor, the Fermi level E_F lies approximately halfway between the bottom of the conduction band and the top of the valence band at $(E_C + E_V)/2$. We have denoted this intrinsic Fermi level by the symbol E_i. Since n_i is useful in calculating carrier densities in extrinsic semiconductors, we will find E_i useful as a reference level when dealing with extrinsic semiconductors.

Equation (3.18) can therefore be modified by replacing E_F by E_i and n by n_i, so that

$$n_i = N_c \exp\left(\frac{E_i - E_C}{kT}\right) \tag{3.36}$$

Similarly, and since we are referring to an intrinsic semiconductor, Eq. (3.19) becomes

$$n_i = N_v \exp\left(\frac{E_V - E_i}{kT}\right) \tag{3.37}$$

Equations (3.36) and (3.37) give

$$N_c \exp\frac{-E_C}{kT} = n_i \exp\frac{-E_i}{kT} \tag{3.38}$$

and

$$N_v \exp\frac{E_V}{kT} = n_i \exp\frac{E_i}{kT} \tag{3.39}$$

Replacing the terms on the left-hand side of Eqs. (3.38) and (3.39) by their equivalence from Eqs. (3.18) and (3.19), and solving for n and p, we have

$$n = n_i \exp\left(\frac{E_F - E_i}{kT}\right) = n_0 \tag{3.40}$$

$$p = n_i \exp\left(\frac{E_i - E_F}{kT}\right) = p_0 \tag{3.41}$$

For an extrinsic semiconductor in thermal equilibrium, the n and p in Eqs. (3.40) and (3.41) have become n_0 and p_0. In each of these two equations, the variables are the thermal equilibrium carrier densities and the energy distance between E_i and E_F. Thus, for an intrinsic semiconductor, E_F is at E_i, whereas for an extrinsic semiconductor and as n_0 and p_0 change, the distance between E_i and E_F changes. The location of E_i with respect to the conduction and valence band edges has not changed. Therefore, E_i provides a useful reference for extrinsic semiconductors.

3.7 FERMI LEVEL IN EXTRINSIC SEMICONDUCTORS

From Eq. (3.40), we find the location of the Fermi level in an N-type semiconductor, where $n = n_0 \cong N_D$, as

$$E_F - E_i = kT \ell n \frac{n_0}{n_i} \cong kT \ell n \frac{N_D}{n_i} \tag{3.42}$$

For a P-type semiconductor, $p = p_0 \cong N_A$, and from Eq. (3.41) we have

$$E_i - E_F = kT \ell n \frac{p_0}{n_i} \cong kT \ell n \frac{N_A}{n_i} \tag{3.43}$$

We observe from Eqs. (3.42) and (3.43) that for an N-type semiconductor the Fermi level is higher than E_i and moves up closer to the conduction band. For a P-type semiconductor, the Fermi level moves down towards the valence band since E_F is less than E_i. We recall that E_i, the Fermi level for an intrinsic semiconductor, is approximately midway between the conduction and valence bands. Calculations for the locations of the Fermi level in extrinsic silicon are carried out in the following example.

EXAMPLE 3.6

(a) Determine the location of the Fermi level, with respect to E_C and E_V, when 10^{16} phosphorus atoms per cc are added to a sample of intrinsic silicon at $T = 300K$.

(b) Repeat part (a) if 10^{15} boron atoms per cc replace the phosphorus atoms.

(c) Repeat part (a) if $T = 600K$.

(d) Repeat part (a) if 10^{18} atoms replace the 10^{16} atoms of phosphorus.

Solution

From Example 3.3, we have E_i at 300K to be 0.0073eV below the middle of the band gap.

(a) By using Eq. (3.42) for phosphorus donor atoms, where $n_0 = N_D$, we have

$$E_F - E_i = kT \ell n \frac{N_D}{n_i} = 0.02583 \ell n \frac{10^{16}}{1 \times 10^{10}} = 0.358eV$$

$$E_i = \frac{E_C + E_V}{2} - 0.0073eV \text{ and } E_F = E_i + 0.358eV = \frac{E_C + E_V}{2} + 0.3507eV$$

From Table 2.2, E_g (at 300K) $= 1.12eV = E_C - E_V$, by using this in the expression for E_F above, we obtain $E_F = E_C - 0.209eV = E_V + 0.9107$.

(b) We use Eq. (3.43) for boron acceptor atoms, where $p_0 \cong N_A$, as

$$E_i - E_F = 0.02583 \ell n \frac{10^{15}}{1 \times 10^{10}} = 0.298eV$$

and from Example 3.3 we have

$$E_i = \frac{E_C + E_V}{2} - 0.0073, \text{ also } E_C = E_V + 1.12, \text{ we find}$$

$$E_F = E_V + 0.255eV \text{ and } E_F = E_C - 0.865eV.$$

(c) At $T = 600K$, n_i is determined in Example 3.2 as 3.337×10^{15} cm^{-3}, $E_g = 1.032eV$ and

$$n_0 = 1.2 \times 10^{16} \text{ cm}^{-3}, \text{ so that } E_F - E_i = 0.05166 \ell n \frac{1.2 \times 10^{16}}{3.337 \times 10^{15}}, E_F = E_i + 0.065eV.$$

Assuming that E_i is approximately at $(E_C + E_V)/2$, E_F is located approximately in the middle of the band gap.

(d) $E_F - E_i = 0.02583 \ell n (10^{18}/10^{10}) = 0.475eV$, $E_F = E_i + 0.475$.

Using $E_i = \frac{E_C + E_V}{2}$, we have $E_F = E_V + 0.56 + 0.475 = E_V + 1.035 = E_C - 0.085eV$

This example shows that the Fermi level moves toward the conduction band level, E_C, when donor atoms are added and moves towards the valence band when acceptors are added. As the temperature is increased, for both donors and acceptors, the Fermi levels move back towards the middle of the gap as the semiconductor approaches the intrinsic state.

Keeping in mind that the Fermi level for an intrinsic semiconductor is located approximately midway between the bottom of the conduction band and the top of the valence band, the following conclusions can be drawn from the example:

- The Fermi level moves toward E_C when donor impurities are added. It moves toward E_V when acceptors are added.

- Increasing the temperature beyond a certain value causes the semiconductor to become intrinsic and the Fermi level to move toward the middle of the band gap.

- Increasing the doping density in part (d) caused E_F to move yet closer to E_C. Further increase of the doping to 10^{19} cm^{-3} causes the conductor to become degenerate, since E_F is less than $3kT$ from E_C.

  In conclusion, we can state that, at a fixed temperature, the location of E_F is a measure of the doping of a semi-conductor.

REVIEW QUESTIONS

Q3-8 In an intrinsic semiconductor, explain why E_F is not located exactly midway between the valence and conduction bands.

Q3-9 As the temperature of a sample of semiconductor that is doped with donor impurities is increased, in what direction does E_F move?

Q3-10 Briefly define the phrase "thermal equilibrium"?

Q3-11 What is a degenerate semiconductor?

Q3-12 The configuration of electrons in a certain element is $4s^2 4p^3$. Is it a donor or an acceptor element?

Q3-13 By referring to the periodic table, determine which element is represented by the configuration $4s^2 4p^3$.

Q3-14 The density of atoms in silicon is 5×10^{22} cm^{-3}. Since each atom has four valence electrons, the density of states becomes 2×10^{23}. Why are N_v and N_c so much smaller than this number?

HIGHLIGHTS

- By adding controlled amounts of impurities to semiconductors, the densities of holes and electrons can be markedly changed. These impurities may be elements either from Column V of the periodic table, known as donors, or they may be from Column III, known as acceptors. Phosphorus and arsenic are donors; boron and gallium are acceptors.

- When donor impurities are added to an intrinsic semiconductor, the more numerous carriers, electrons, are called majority carriers and the holes are minority carriers. When both donors and acceptor impurities are added, the one with the larger density determines which carrier becomes the majority.

- A unique property of semiconductors is the large changes that are made in the densities of the carriers by the addition of impurities, also known as dopants. The changes in the carrier densities are translated into changes in conductivity.

- A semiconductor that has been doped is known as an extrinsic semiconductor.

- Donor and acceptor atoms occupy energy states in the forbidden band, donor atoms are close to E_C, and acceptor atoms are close to E_V. In these locations it takes very little energy to excite a donor electron to go into the conduction band and also to excite an electron from the valence band to an acceptor level, thus generating a hole. A donor atom that loses an electron and an acceptor atom that gains an electron are said to be ionized.

- In intrinsic semiconductors, the Fermi level E_F is located close to the middle of the band gap. The Fermi level moves up towards the conduction band when donors are added and moves closer to the valence band when acceptors are added.

EXERCISES

E3-4 A sample of semiconductor is doped with $N_D = 10^{13} cm^{-3}$. The intrinsic carrier density is $10^{13} cm^{-3}$. Determine n and p.

Ans: $n = 1.61 \times 10^{13} cm^{-3}$

E3-5 If $(N_c N_v)^{0.5}$ is theoretically the largest possible value of intrinsic carrier density that can be generated, determine as a fraction of this value, the number of electrons that can be thermally excited into the conduction band at 300K in germanium.

Neglect changes of E_g with temperatures given $E_g = 0.66eV$.

Ans: 3×10^{-6}

E3-6 Repeat Ex. 3-6 for silicon at 300K

Ans: 4×10^{-10}

E3-7 Determine the density in cm^3 and m^3 of free electrons in silicon if the Fermi level is 0.2eV below E_C at 300K.

Ans: $n = 1.08 \times 10^{16} cm^{-3}$

E3-8 A sample of silicon is doped with $N_D = 5 \times 10^{16} cm^{-3}$ at room temperature. Determine the location of the Fermi level relative to the conduction band edge.

Ans: $E_C - E_F = 0.16eV$

3.8 WHICH SEMICONDUCTOR?

Three semiconductors have been used in the fabrication of devices since the early days of the semiconductor industry. They are germanium, silicon, and gallium arsenide. Germanium was used early on but since the 1960s silicon has been the dominant semiconductor. Gallium arsenide has just recently acquired great importance for certain devices. The comparison among the three throws a light on the particular properties that make a semiconductor attractive for device fabrication.

Semiconductors can be classified into two major categories: elemental and compound semiconductors. Elemental semiconductors, such as germanium and silicon, are classified in Group IV of the Periodic Table of elements and have four valence electrons. Compound semiconductors are composed of a combination of Group III and Group V elements or Group II and Group VI elements. Examples of important compound semiconductors are gallium arsenide, gallium phosphide, indium phosphide, and indium arsenide.

Germanium has two distinct attractive qualities: In the first place, it can be refined and processed more easily than the others. In the second place, both electrons and holes have higher mobilities than the corresponding carriers in silicon, as can be seen in Table 3.3. The higher mobility translates into faster switching and higher operating frequency limits.

A disadvantage of germanium is its high sensitivity to temperature because of its relatively narrow band gap, which may cause instability in a device. Instability results because a higher temperature increases the density of electrons that are excited to the conduction band, thus increasing the current that results in higher heat dissipation, which increases the temperature. The more serious problem is the difficulty of introducing controlled amounts of impurities into small selected areas. Because of this, one has to work with large areas. This results in carriers having to take longer times in covering distances within the device. This longer travel time means slower switching speed.

Silicon has important advantages that are unmatched by germanium. It is abundant in nature in the form of sand and quartz. Thus, the cost of the starting material is negligible. Because silicon has a wider energy gap (forbidden band) than germanium, it can be used at higher temperatures, thus greatly reducing a cause of instability. Silicon devices can be operated safely at temperatures of about 200°C, whereas germanium devices are limited to a safe operating temperature of 80°C.

Silicon has a major processing advantage in that it forms a stable oxide, silicon dioxide. The silicon dioxide offers a top quality insulator and when used in device processing, it provides a good barrier for the diffusion of impurities in certain selected areas. Furthermore, it becomes possible to operate with very small dimensions of the order of one micron or less. Faster switching and higher frequency limits are the benefits.

Thus, silicon is a top quality semiconductor while providing an insulator with excellent properties.

Gallium arsenide has the major advantage of a mobility that is approximately five times that of silicon, since a higher mobility is translated into a higher velocity of the carriers for a fixed electric field. Of course, this property offers possibilities of faster switching. It is, however, more difficult to process and is more expensive than silicon. One of its major applications, as we shall see later, is in optical devices.

We have listed in the accompanying table some properties of the three major semiconductors.

TABLE 3.3   Properties of Silicon, Germanium and Gallium Arsenide at $T = 300K$

Property	Unit	Si	Ge	GaAs
Density of atoms	cm^{-3}	5×10^{22}	4.4×10^{22}	2.2×10^{22}
Energy gap	eV	1.12	0.66	1.42
Effective mass m^*/m_O				
electron		1.182	0.553	0.0655
hole		0.81	0.357	0.524
Effective density of states	cm^{-3}			
conduction band		3.22×10^{19}	1.03×10^{19}	4.21×10^{17}
valence band		1.83×10^{19}	5.35×10^{18}	9.52×10^{18}
Intrinsic carrier density	cm^{-3}	1×10^{10}	2.17×10^{13}	2.49×10^{6}
Mobility at low doping	cm$^2(V-s)^{-1}$			
electron		1350	3900	8800
hole		480	1900	400
Breakdown field	V/cm	3×10^5	10^5	4×10^5
Relative permittivity	dimensionless	11.8	15.8	13.1
Melting point	°C	1410	940	1240

PROBLEMS

3.1 Draw the energy band diagrams, showing E_C, E_V, E_F, and E_i, for the following; assuming all impurities are ionized,

    a)   Intrinsic silicon at 300K.

    b)   Silicon doped with 10^{17} boron atoms cm^{-3} at 300K.

3.2 A silicon sample is doped with 10^{16} cm^{-3} of phosphorus atoms. Assume that all phosphorus atoms are ionized at 300K and determine:

    i) The electron density n_0.

    ii) The hole density p_0.

    iii) The location of the Fermi level with respect to E_i.

3.3 Repeat Problem 3.2 if the phosphorus doping is at a level of $10^{18}cm^{-3}$.

3.4 Repeat Problem 3.2 for a boron doping of $10^{16}cm^{-3}$.

3.5 The Fermi level in a silicon sample at equilibrium is located $0.40eV$ below the middle of the band gap. At $T = 300K$,

    a)   Determine the probability of occupancy of a state located at the middle of the band gap.

    b)   Determine the probability of occupancy of the acceptor states if the acceptor states are located at $0.04eV$ above the top of the valence band.

    c)   Check if the assumption of complete ionization in part (b) is valid.

3.6 The effective conduction band density of states is $N_c = 3.22 \times 10^{19}cm^{-3}$ and the effective valence band density of states is $N_v = 1.83 \times 10^{19}cm^{-3}$. Assume that N_C and N_v are located at the conduction band and valence band edges respectively. Let $T = 300K$.

    a)   For Problem 3.5, determine the thermal equilibrium electron density.

    b)   Determine the thermal equilibrium hole density.

    c)   Calculate the $p_0 n_0$ product.

3.7 In a silicon sample at equilibrium the Fermi level is located above the middle of the band gap by $0.38eV$. The phosphorus donor states are located $0.04eV$ below the conduction band. Determine the percentage ionization of the phosphorus atoms at $T = 300K$.

3.8 A silicon sample at 300K has an acceptor density of $10^{15}cm^{-3}$. Determine:

    a)   The hole density.

    b)   The electron density.

3.9 A certain silicon sample at 300K has $p_0 = 4 \times 10^{10}cm^{-3}$. Determine:

    a)   The electron density.

    b)   The acceptor density if the donor density is 1.25×10^{10} cm^{-3}.

3.10 A certain silicon sample at 300K has $N_D = 1 \times 10^{16}cm^{-3}$ and $N_A = 0.8 \times 10^{16}cm^{-3}$. Determine:

    a)   The majority carrier density.

    b)   The minority carrier density.

3.11 Determine the electron and hole thermal equilibrium densities and the location of the Fermi level, with respect to E_C, for a silicon sample at 300K that is doped with

 a) $1 \times 10^{15} \text{cm}^{-3}$ of boron.

 b) $3 \times 10^{16} \text{cm}^{-3}$ of boron and $2.9 \times 10^{16} \text{cm}^{-3}$ of phosphorus.

3.12 Determine the approximate donor binding energy for silicon given $\epsilon_r = 11.8$ and $m_n^*/m_0 = 0.26$.

3.13 An N-type silicon sample has an arsenic dopant density of 10^{17}cm^{-3}. Determine:

 a)   The temperature at which half the impurity atoms are ionized.

 b)   The temperature at which the intrinsic density exceeds the dopant density by a factor of 10. Assume E_g does not change with T.

 c)   Assuming complete ionization, calculate the minority carrier density at 300K and the location of the Fermi level referred to E_C.

3.14 A silicon sample has the energy band diagram shown in the figure below. Given $E_g = 1.12eV$, and $n_i = 10^{10} \text{cm}^{-3}$.

 a)   Sketch the potential V as a function of x.

 b)   Sketch the electric field $\mathscr{E}$ as a function of x.

 c)   Determine the values of n_0 and p_0 at $x = x_1$ and $x = x_2$.

3.15 A silicon sample is doped with 10^{16} donor atoms/cm³. Draw an energy level diagram showing the location of the Fermi level with respect to the middle of the band gap for:

 a)   77K (liquid nitrogen).

 b)   300K.

 c)   600K.

Neglect the change of E_g with temperature.

3.16 **a)**   At $T = 300K$, what percentage of the electrons in a cm³ of silicon are located in the conduction band?

 b)   Repeat for $T = 500K$.

3.17 On the energy band diagram, electron energy is measured upwards. Explain why hole energy is measured downwards.

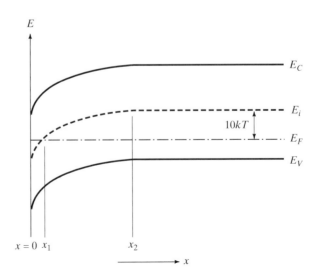

3.18 In the band diagram shown below, which carrier has the larger kinetic energy? Explain the reason for your answer.

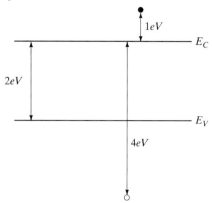

chapter 4

CARRIER PROCESSES: DRIFT, DIFFUSION AND GENERATION-RECOMBINATION

4.0 INTRODUCTION

In the last chapter, we identified electrons and holes as the current carriers in a semiconductor. The electrons exist in the conduction band and the holes are in the valence band. In an intrinsic semiconductor, at thermal equilibrium, the density of electrons is equal to the density of holes because the rate of generation of electron-hole pairs is balanced by the rate of recombination of electrons and holes.

The addition of small traces of controlled impurities increases the density of one carrier while decreasing the density of the other. The density of electrons is enhanced by the addition of donor impurities while the addition of acceptor impurities increases the density of holes.

In order to determine expressions for the carrier currents in terms of an applied voltage and other factors, it is necessary to discuss and define the processes by which carriers are caused to move.

Carrier motion is caused by two conditions; the application of an electric field whose force accelerates the carriers and, second, a difference of carrier concentration between two points causes carriers to move by diffusion from regions of high concentration to regions of low concentration. Furthermore, the processes of carrier generation and recombination affect the resulting densities of the carriers.

It is by considering the combined effects of carrier motion, generation, and recombination, on the distribution and motion of the carriers, that we can proceed to set up the expressions from which we determine the relations describing the operation of semiconductor devices.

4.1 VELOCITY LIMITATIONS

In the last chapter, we derived relationships for the densities of carriers in an extrinsic semiconductor. When we consider a bar of semiconductor, to which an electric field is applied, the current in the bar will depend on the density of the carriers, the cross-sectional area of the bar and the velocity with which the carriers move. For an N-type semiconductor, where $n_0 >> p_0$, the current is given by

$$I\left(\text{amps} = \frac{\text{couls}}{\text{sec}}\right) = n_0 \left(\frac{\text{electro}}{\text{cm}^3}\right) q \left(\frac{\text{couls}}{\text{electron}}\right) v \left(\frac{\text{cm}}{\text{sec}}\right) \times A \ (\text{cm}^2)$$

The question we need to answer is: How do we determine the mean velocity of the carriers? Obviously, the velocity depends on the acceleration, which in turn depends on the electric field intensity. For a constant electric field, what is the limitation to the velocity that the carriers can acquire?

In our discussions in earlier chapters, we assumed that the solid possesses a perfectly periodic lattice and disregarded any irregularities in it. In a metal at $T > 0K$, electrons are continuously in motion due to the thermal energy they acquire. If a constant force is applied to the electrons, as by an electric field, the electrons in the perfectly periodic crystal would be accelerated, gain energy, move to empty states at higher energy levels within the same band, and continue to gain in velocity and energy.

Electric current is proportional to the number and velocity of the electrons and, therefore, the current in a material to which an electric field is applied, in the absence of other phenomena that might limit its value, will tend to increase without limit as the velocity increases. However, we know from Ohm's law that the current reaches a fixed value indicating a constant velocity. The velocity we are referring to is a *mean velocity,* averaged over the velocities of all the free electrons.

The mean velocity is constant because of a resistive force that prevents further acceleration. This resistive force is a direct result of the collision of the electrons with atoms or with the regions in the lattice of the material that disturb the motion of the electrons. The collison could also be with foreign atoms.

Before further discussion of the nature or effect of the collisions, let us present the condition of the electrons in a semiconductor when they are not under the influence of an electric field.

4.2 THERMAL VELOCITY

Electrons and holes in semiconductors are in constant motion because of the thermal energy they receive. Since they are in motion, they are not associated with any particular lattice position. At any one time, the electrons and holes move in random directions with a mean random velocity. This velocity, known as the *thermal velocity,* is of the order of 10^7cm/s for electrons in silicon. Because of the motion in random directions, the current resulting from the motion of all the carriers in any one direction is zero.

As a measure of the energy of the electron, we establish the level of the energy of an electron at rest to be at the bottom of the conduction band E_C. The kinetic energy of an electron is measured by the energy separation above E_C, as $E - E_C$.

It has been established that, at thermal equilibrium, the mean-square thermal velocity of the electron is related to temperature by the relation*

$$\frac{1}{2} m_n^* v_{\text{th}}^2 = \left(\frac{3}{2}\right) kT \qquad (4.1)$$

where m_n^* is the conductivity effective mass of the free electron, v_{th} its mean thermal velocity, k is Boltzman's constant, and T is the temperature in degrees Kelvin. For silicon at $T = 300K$, this velocity has been calculated to be approximately 2.3×10^7cm/s. The mean kinetic energy for all the electrons at thermal equilibrium is $E - E_C = (3/2)kT$, and this translates into 0.04eV at 300K, which is slightly above the conduction band edge E_C, and for silicon represents approximately 1/25 of the band gap energy E_g.

Electrons traveling in a solid under the influence of a small applied electric field collide with the lattice, exchange energy with the lattice, and start all over. Depending upon the magnitude of the field, the electron gives up a certain amount of heat to the lattice.

4.3 COLLISIONS AND SCATTERING

When a relatively low-intensity electric field is applied to a metal or a semiconductor, electrons acquire a mean velocity in accordance with Ohm's law. This new velocity is superimposed on the thermal velocity and its magnitude is much smaller than the mean thermal velocity. This new velocity component refers to the average rate of motion of the electron population in the direction of the force of the electric field. This velocity is known as the *drift velocity, v_d*.

Sketches comparing the random field-free motion of electrons with the field-directed motion are shown in Fig. 4.1.

It is important to point out that since the drift velocity is, in most devices, much smaller than the thermal velocity, the drift velocity can be considered a perturbation of the thermal velocity.

The electron that is moving in a solid, subjected to an electric field, collides with the atoms and with the nonideal lattice structure, loses most of its velocity, and then repeats the process. This process of collision is more accurately known as *scattering*.

It is important to point out that quantum-mechanical calculations indicate that scattering does not take place, and hence no energy exchange occurs, when an electron is traveling in a perfectly periodic and stationary lattice. A non-stationary lattice results when the atoms vibrate by acquiring heat.

Scattering does not refer only to a direct collision between the moving elec-

---

*From Halliday, Resnick, and Walker, *Fundamentals of Physics*, p. 582; copyright © Wiley (1993). Reprinted by permission of John Wiley & Sons, Inc.

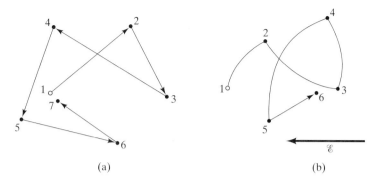

Figure 4.1   (a) Mean random thermal motion of electrons at $T > 0$K and with zero electric field. (b) Field-directed motion, at $T > 0$K, added to the thermal motion.

trons and the atomic core. It is caused more by the variation in the potential distributions in the solid when the crystal structure is not perfectly periodic. Such a solid is labeled *aperiodic*. Aperiodicity is caused by three factors, one of which is a temperature of the solid above zero. Aperiodicity may also result from imperfections in the solid and from the presence of impurities. There are therefore three phenomena that individually or jointly cause departures from the periodic potential.

At normal operating temperatures, the lattice atoms vibrate about their equilibrium positions. These vibrations disturb the periodicity and change the periodic potential, which generates an electric field that scatters the carriers. An electron travelling through the lattice undergoes scattering, which causes changes in the magnitude and direction of its velocity. The scattering results both from collisions with the vibrating atoms themselves and with the varying potential field that results from the displacement of the vibrating atoms. The higher the temperature, the larger is the amplitude of the atomic vibrations, the larger is the scattering cross section, the larger is the disturbance of the potential field, and therefore the higher the probability of the scattering taking place.

A second mechanism that causes scattering results from the presence of ionized impurity atoms. These atoms produce an electric field that changes the potential distribution in the crystal. This type of scattering is more important at lower temperatures when the thermal scattering is weaker. The higher the impurity concentration, the higher is the probability of this type of scattering.

A third cause of scattering, although less important than thermal or impurity scattering, results from imperfections such as vacancies in the crystal structure and crystallographic defects.

Crystal imperfections cause a change in the periodic potential and thus create an electric field. An electron passing near an imperfection interacts with the field resulting in a change of the direction of motion.

We conclude that at lower temperatures, impurity scattering is more dominant, while at the higher temperatures, thermal (lattice) scatterings is more effective.

4.4 COLLISIONS EFFECTS

Drift Velocity

The effect of the scattering, or the collisions, on the motion of a carrier is equivalent to a frictional resistive force that controls the acceleration of the particle and limits its velocity to the drift velocity.

The motion of an electron that is subjected to an electric field, in the x-direction, obeys the equation,

$$m_n^* \frac{d^2x}{dt^2} = -q\,\mathcal{E}_x - w\frac{dx}{dt} \tag{4.2}$$

where $\mathcal{E}_x$ is the electric field intensity, m_n^* is the electron effective mass, q is the electronic charge, and the second term on the right hand side of Eq. (4.2) is the frictional resistive force where w is a constant dependent on the solid.

The electric field does not accelerate the electron continuously because the resistive force increases as the velocity increases until such time that the resistive force balances the force of the electric field until the steady-state, when the velocity becomes the drift velocity. The drift velocity can be found by setting the impulse (force × time), applied to an electron during the time τ_c between collisions, equal to the momentum gained by the electron in that period. In the steady state, the drift velocity is determined from

$$-q\mathcal{E}_x\tau_c = m_n^* v_d$$

where τ_c is the mean time between collisions, known as the mean scattering time, and the drift velocity is given by

$$v_d = -q\mathcal{E}_x \frac{\tau_c}{m_n^*} \tag{4.3}$$

We digress briefly to consider the transient behavior of the electron velocity.

For an electron starting from rest at $t = 0$, the solution to Eq. (4.2) is

$$v(t) = \frac{dx}{dt} = -\frac{q\mathcal{E}_x}{w}\left[1 - \exp\left(\frac{-wt}{m_n^*}\right)\right] \tag{4.4}$$

On comparing Eqs. (4.3) and (4.4), we find that $\tau_c = m_n^*/w$ so that Eq. (4.4) becomes

$$v(t) = \frac{-q\mathcal{E}_x\tau_c}{m_n^*}\left[1 - \exp\left(\frac{-t}{\tau_c}\right)\right] \tag{4.5}$$

Thus, after a certain time $t > \tau_c$, the particle velocity will have a steady-state value given by the drift velocity, the negative sign indicating that the velocity is in a direction opposite to that of the electric field.

Suppose that, as the electron is being accelerated, the electric field is removed and at this time assume its velocity is v_0. For this condition, the velocity in accordance with the solution to Eq. (4.2) becomes

$$v(t) = v_0 \exp(-t/\tau_c) \tag{4.6}$$

The time constant τ_c may also be interpreted as the factor that controls the rate at which the velocity and the current decay to zero after the field is removed. This time is known as the *relaxation time* and it is of the order of 10^{-15}s.

The steady-state value of the drift velocity in the presence of the electric field is also found by setting $t >> \tau_c$ in Eq. (4.4). This time has also been represented to refer to a measure of the time it takes for carriers in a semiconductor to move and neutralize charges.

The reader may have concluded that where we speak of the velocity of an electron, we are implying that it is possible to isolate the particle. In fact, the velocity here refers to the mean of the velocities of all the electrons that are free.

We defined earlier scattering in the presence of an electric field, as the process of collision between the field-directed electron and the lattice structure. We will now illustrate this operation on an energy band diagram.

Collisions and Energy Exchanges

In Fig. 4.2(a), we show an N-type semiconductor with metal contacts attached to both ends. At point A, a voltage $+V$ is applied with respect to contact B, which is at ground potential. We will study the effect on the energy of an electron released from rest at point B. Before we do that, we will review the energy relations when the bar is replaced by a vacuum enclosure. At terminal B of the vacuum enclosure, an electron is released from rest, and since there are no obstacles and hence no collisions along the way, the electron accelerates and strikes plate A. At point B, the electron lost potential energy qV; at impact with terminal A, all the potential energy is converted to kinetic energy and hence heat at point A.

In the energy band diagram of the semiconductor bar of Fig. 4.2(b), an electron held at point B has potential energy qV and its potential energy when it arrives

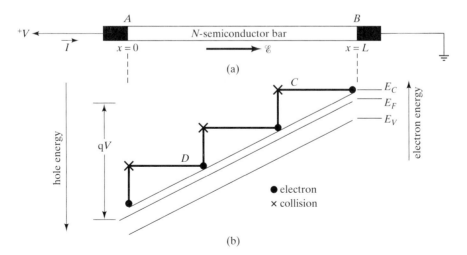

Figure 4.2   Illustrating conduction in a bar and by use of the energy band diagram (a) semiconductor bar with a voltage applied; (b) motion of electron on energy band diagram.

at A is zero. Since the bottom of the conduction band at E_C is the level of the potential energy of an electron at rest, then the energy band diagram for an electron in this illustration shows a decreasing potential energy and hence a decreasing level of E_C as we proceed from B to A in Fig. 4.2(b). This explains the slope of the energy band diagram.

An electron released from rest at point B in the bar is accelerated by the electric field in Fig. 4.2(a), acquires kinetic energy, and just before point C, its total energy is unchanged as it replaces the potential energy it lost by kinetic energy. When it collides with the lattice structure at point C, it loses all the kinetic energy it acquired, its total energy is potential, and the electron drops to the level E_C corresponding to point C. It is accelerated again, collides with the lattice, and drops back to the lower E_C level at point D. This process continues until the electron reaches point A, where it has lost all the qV potential energy that it had at point B. This potential energy has been converted, along the way, into heat in the lattice.

4.5 MOBILITY

Referring to Eq. (4.3), we observe that the drift velocity of electrons (or holes) is given by the product of the electric field intensity and a factor that depends on the electronic charge, the effective mass, and the relaxation time of the semiconductor. We label that factor the *mobility of a carrier,* μ, so that for an electron we have

$$v_d = -\mu_n \mathscr{E} \tag{4.7}$$

where v_d is the drift velocity of an electron in cm/s, μ_n is its mobility in cm²/volt-sec and $\mathscr{E}$ is the electric field intensity in V/cm. The negative sign denotes that, for an electron, the direction of the velocity is opposite to the direction of the electric field intensity. By using Eq. (4.3), in Eq. (4.7), the mobility of the electron becomes

$$\mu_n = \frac{q\tau_c}{m_n^*} \tag{4.8}$$

The analogous expressions for hole motion are

$$v_d = \mu_p \mathscr{E} \tag{4.9}$$

and the mobility of the hole is

$$\mu_p = \frac{q\tau_c}{m_p^*} \tag{4.10}$$

Because the time interval between collisions depends mainly on temperature and impurity concentration, the mobility depends on those two parameters as well as on the material through the effective mass. With reference to Fig. 2.12, which showed a much greater curvature near the bottom of the conduction band of GaAs than that of Si, the effective mass of an electron in GaAs is much smaller than in Si. As a result, the electron mobility in GaAs is four to five times greater than in silicon.

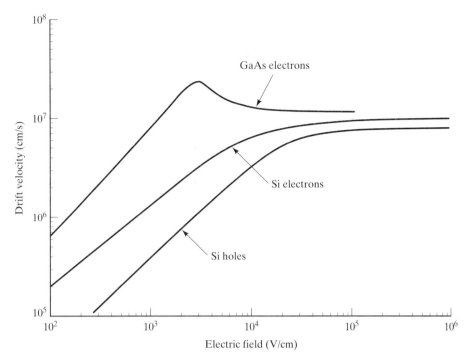

Figure 4.3 Variation of drift velocity with electric field intensity for gallium arsenide and silicon. *Source:* Reprinted with permission from Ruch and Kino, "Measurement of the Velocity-Field Characteristics of Gallium Arsenide," *Applied Physics Letters* 10, 40 (1967) copyright American Institute of Physics; and Caughey and Thomas "Carrier Mobilities in Silicon Empirically Repated to Doping and Field," *Proc. IEEE* 55, 2192 (1967) © IEEE.

The above treatment assumes that the time interval between collisions is independent of the magnitude of the applied electric field intensity. This is valid provided the drift velocity is much smaller than the thermal saturation velocity, v_{th}, which is approximately 10^7 cms/s for electrons in silicon.

Experimental results of the measurements of the drift velocity of electrons and holes are shown in Fig. 4.3. It is quite evident that at very large values of electric field intensity, the drift velocity approaches the saturation velocity and hence the mobility decreases with increasing electric field intensity.

We also note, in Fig. 4.3, the dependence of the drift velocity, and hence the mobility in the linear region, on the material through its dependence on the effective masses of the electrons and the holes.

Effects of Impurity Concentration and Temperature on Mobility

The dependence of the mobility on the impurity concentration, through the mean scattering time τ_c, results from the scattering of the carriers by the coulombic effect of the ionized impurities as the carriers travel in the solid. High doping density

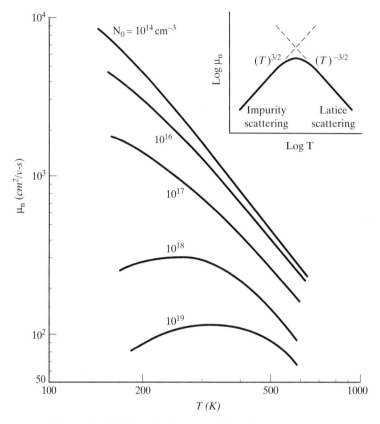

Figure 4.4   Variation of electron mobility with temperature and doping in silicon.
Source: S.M. Sze, *Semiconductors: Devices, Physics and Technology,* p. 33; copyright
© Wiley (1985). Reproduced by permission of John Wiley & Sons, Inc.

means more scattering, lower drift velocities, and mobilities. The variation of mobility with scattering through its dependence on impurity concentration and temperature is shown in Fig. 4.4 and highlighted in the inset.

The question then is: Why is it that, in the inset of Fig. 4.4, the mobility is seen to increase with an increase of temperature in a certain temperature range, yet decreases with an increase of temperature in another temperature range?

At low temperatures, lattice vibrations are so small so that as the temperature increases, the carriers receive thermal energy that increases their velocity, thus reducing the time that the electron spends in the vicinity of the lattice and impurity atoms. This reduces the scattering cross section, which causes the mobility to increase with temperature.

The increase of mobility with temperature continues until, at a certain temperature, the lattice vibrations dominate and the scattering cross section increases, thus increasing the possibility of collision with the lattice and decreasing the mobility with an increase of temperature.

Expressions for the Mobility

When two scattering processes occur at the same time, the combined effect on the mobility is determined from the probabilities of the two processes. If in an incremental time, dt, the probability of scattering by process 1 is dt/τ_1, where τ_1 is the time intervals between two collisions and the probability of scattering by process 2 is dt/τ_2, then the total probability dt/τ_c is the sum of the two probabilities so that we can write

$$dt/\tau_c = dt/\tau_1 + dt/\tau_2$$

$$\text{or } \frac{1}{\tau_c} = \frac{1}{\tau_i} + \frac{1}{\tau_t} \tag{4.11}$$

Where τ_i and τ_t refer to the collision times due to impurities and temperature respectively. The mobility μ can be expressed in terms of the mobilities due to impurities and due to temperature μ_i and μ_t respectively as

$$\frac{1}{\mu} = \frac{1}{\mu_i} + \frac{1}{\mu_t} \tag{4.12}$$

It has been shown that the mobility due to lattice scattering (temperature), μ_t, is proportional to $1/T^{3/2}$. The mobility caused by impurity scattering, μ_i, has been determined to be proportional to $T^{3/2}/N$, where N is the total impurity density $(N_D + N_A)$.

Empirical expressions for the mobilities of electrons and holes in silicon in cm^2/V-s, as a function of doping density and temperature have been derived,[*] and shown below, where T is in K, $T_n = T/300$, and N is in cm^{-3}.

$$\mu_n = 88T_n^{-0.57} + \frac{7.4 \times 10^8 T^{-2.33}}{1 + \left[\dfrac{N}{(1.26 \times 10^{17} T_n^{2.4})}\right] 0.88 T_n^{-0.146}} \tag{4.13}$$

$$\mu_p = 54.3T_n^{-0.57} + \frac{1.36 \times 10^8 T^{-2.23}}{1 + \left[\dfrac{N}{(2.35 \times 10^{17} T_n^{2.4})}\right] 0.88 T_n^{-0.146}} \tag{4.14}$$

At $T = 300$K these expressions reduce to:

$$\mu_n = 88 + \frac{1252}{1 + 0.698 \times 10^{-17} N} \tag{4.15}$$

$$\mu_p = 54.3 + \frac{407}{1 + 0.374 \times 10^{-17} N} \tag{4.16}$$

These relations agree fairly well with experimental results.

---

[*]From Arora et al., "Electron and Hole Mobilities in Silicon as a Function of Concentration and Temperature," *IEEE Transactions on Electron Devices*, Vol. 29, No. 2, p. 2192, February 1982. © 1982, IEEE.

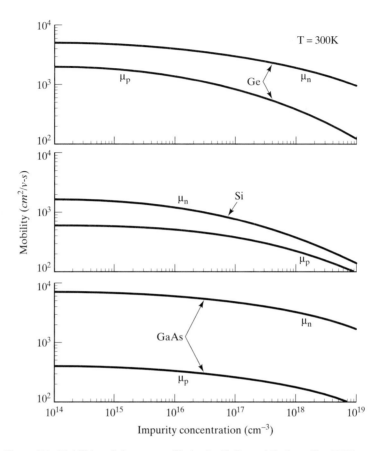

Figure 4.5   Mobilities of electrons and holes for Si, Ge and GaAs at $T = 300K$, as a function of total impurity density ($N_D + N_A$). *Source:* Reprinted from S. M. Sze and J. C. Irvin, "Resistivity, Mobility and Impurity Levels in GaAs, Ge and Si at 300K," *Solid State Electronics, 11,* 599 (1968) with kind permission from Elsevier Science Ltd., The Boulevard, Langford Lane, Kidlington 0X5 1GB, UK.

Plots of measurements of the mobility versus total impurity concentration ($N_D + N_A$), at $T = 300K$, are shown in Fig. 4.5. We observe that at low concentration levels the mobility is independent of the concentration.

We conclude this section by summarizing the factors that influence the mobility:

1. The mobility depends very strongly on the temperature and on irregularities in the crystal, which may be due to the growing of the crystal or the impurities.

2. So long as the drift velocity is less than the thermal velocity, the mobility varies directly with, and is proportional to, the time between collisions.

3. If the drift velocity approaches the saturation velocity through an increase of the electric field intensity, the mobility decreases with an increase of the electric field.

4. The mobility, therefore, is affected to varying degrees by: irregularities in the lattice structure, temperature, concentration of added impurities, and the electric field intensity.

4.6 DRIFT CURRENT AND CONDUCTIVITY

When an electric field is applied to a semiconductor, the electrons and holes acquire drift velocities and the consequent motion of carriers results in an electric current. This current is known as *drift current*. The electron drift current is given by:

$$I_n = -A\,q\,n\,v_d \qquad (4.17)$$

where A is the cross sectional area in cm^2, normal to the direction of current flow, q is the charge in coulombs, n is the electron density in cm^{-3}, v_d is the electron drift velocity in cm/sec, and I is in amperes. The direction of the current is opposite to that of the motion of the electrons. By replacing for the velocity from Eq. (4.7) into Eq. (4.17), we have

$$I_n = A\,q\,n\,\mu_n\,\mathscr{E} \qquad (4.18)$$

The current due to electron motion is thus in the same direction as the electric field intensity. The electron current density in amps/cm^2 is

$$J_n = \frac{I_n}{A} = q\mu_n n\mathscr{E}$$

The factor multiplying $\mathscr{E}$ is known as the *conductivity* σ_n so that

$$J_n = \sigma_n\mathscr{E} \qquad (4.19)$$

where the unit of the conductivity is (ohm-cm)$^{-1}$. Analogously the current density for holes becomes

$$J_p = q\mu_p p\mathscr{E} = \sigma_p\mathscr{E} \qquad (4.20)$$

The total drift current density, J_t, is

$$J_t = J_n + J_p = \sigma_n\mathscr{E} + \sigma_p\mathscr{E} \qquad (4.21)$$

The total conductivity is

$$\sigma_t = \sigma_n + \sigma_p = q\mu_n n + q\mu_p p \qquad (4.22)$$

The addition of small traces of impurities has dramatic effects on the conductivity of semiconductors. We illustrate this by an example.

EXAMPLE 4.1

a) Determine the conductivity of intrinsic silicon at 300K.

b) Repeat part (a) for a sample of silicon doped with 10^{17}cm^{-3} of phosphorus.

Solution

a) For intrinsic silicon at 300K, $n = p = n_i = 10^{10} cm^{-3}$, the mobilities are calculated using Eqs. (4.15) and (4.16) to be $\mu_n = 1340 cm^{-3}$ and $\mu_p = 461.3 cm^{-3}$. The intrinsic conductivity becomes

$$\sigma_i = q\, n_i(\mu_n + \mu_p) = 2.88 \times 10^{-6}(\text{ohm-cm})^{-1}$$

b) The densities of electrons and holes are calculated to be $10^{17} cm^{-3}$ and $10^3 cm^{-3}$ respectively and the mobilities are calculated using Eqs. (4.15) and (4.16) to be $\mu_n = 825 cm^2/V\text{-}s$ and $\mu_p = 350 cm^2/V\text{-}s$. The extrinsic conductivity becomes

$$\sigma \cong 1.6 \times 10^{-19} (825 \times 10^{17} + 350 \times 10^3) = 13.2(\text{ohm-cm})^{-1}$$

Since the density of silicon atoms is approximately $10^{23} cm^{-3}$, the addition of one part of phosphorus to a million parts of silicon increased the conductivity of the sample by a factor of about one million.

4.7 RESISTIVITY AND RESISTANCE

The reciprocal of the conductivity, the resistivity ρ, has units of ohm-cm. In general, the semiconductor is either N-type or P-type. For a semiconductor in thermal equilibrium, the resistivity becomes

$$\rho = 1/(q\mu_n n_0 + q\mu_p p_0) \tag{4.23}$$

where n_0 and p_0 are the electron and hole densities, in thermal equilibrium, respectively in cm^{-3}. For an N-type semiconductor $n_0 \gg p_0$ and $n_0 \cong N_D$, Eq. (4.23) becomes

$$\rho \cong 1/(q\mu_n N_D) \tag{4.24a}$$

For a P-type semiconductor, the expression for the resistivity is given approximately by

$$\rho \cong 1/(q\mu_p N_A) \tag{4.24b}$$

By using values of the mobility calculated from the linear region of Fig. 4.3, and calculating resistivity using Eqs. (4.24(a)) and (4.24(b)), one obtains a plot of the resistivity versus impurity density. The results are shown in Fig. 4.6.

REVIEW QUESTIONS

Q4-1 Identify the processes that cause scattering.

Q4-2 Upon what physical factors does mobility depend?

Q4-3 Why is the electron mobility in GaAs about three times that in silicon?

Q4-4 Why and how does the mobility depend on doping? Explain.

Q4-5 An electric field is applied to a sample of semiconductor in thermal equilibrium. If the field is in the positive x-direction, in what directions are the motions of holes and electrons?

Q4-6 In the above question, in what directions are the electron and hole currents?

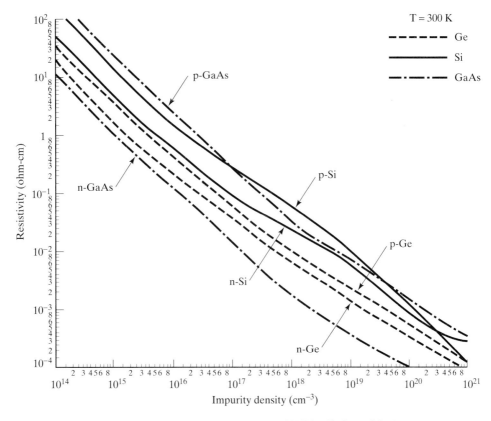

Figure 4.6   Resistivity versus impurity density at 300K for Si, Ge, and GaAs. *Source:* Reprinted from S. M. Sze and J. C. Irvin, "Resistivity, Mobility and Impurity Levels in GaAs, Ge and Si at 300K," *Solid State Electronics, 11,* 599 (1968) with kind permission from Elsevier Science Ltd., The Boulevard, Langford Lane, Kidlington 0X5 1GB, UK.

HIGHLIGHTS

- Electrons in a semiconductor at $T > 0$ are continuously moving at a very high velocity in all directions. This velocity is known as the mean thermal velocity. At thermal equilibrium, the net velocity of all electrons in any one direction is zero, therefore the current is zero.

- When an electric field is applied to a semiconductor, electrons acquire energy and a velocity that is much smaller than the mean thermal velocity. Now there is a net mean velocity opposite to the direction of the electric field. This velocity is known as the drift velocity.

- The reason for the limited drift velocity is the electrical interaction of the electron with the imperfect solid.

- The factor that relates the velocity to the electric field intensity is labeled the mobility. At low values of electric field the velocity increases linearly with an increase of the field resulting in a relatively constant mobility. Just as electrons in the conduction band acquire drift velocity and mobility, so do the electrons in the valence band.

- The mobility is also influenced by the temperature and the doping of the semiconductor. The higher the doping the lower the mobility.
- The current resulting from an applied electric field, known as the drift current, is the product of the electric field intensity, and the sum of the conductivities of holes and electrons. Conductivity, the reciprocal of resistivity, is the product of the carrier density, the charge of the carrier, and the mobility.

EXERCISES

E4-1   A sample of silicon is doped with 10^{15} cm^{-3} of phosphorus and 5×10^{15}cm^{-3} of boron. Determine:

    a)   the density of holes and electrons

    b)   the mobility of holes and electrons

        Ans: a) $p = 4 \times 10^{15}$cm^{-3}, $n = 2.5 \times 10^{4}$cm^{-3}

                b) $\mu_p = 350$cm^2/V-s, $\mu_n = 1500$cm^2/V-s

E4-2   An electric field is applied to a sample of Ge that is doped with 10^{15}cm^{-3} of boron. The field has intensity of 100 V/cm. Calculate:

    a)   the velocity of electrons and holes

    b)   the conductivity of the sample

        Ans: a) $v_d(n) = 5 \times 10^{5}$cm/s, $v_d(p) = 2 \times 10^{5}$cm/s

                b) $\sigma = 0.16$ (ohm-cm)$^{-1}$

E4-3   For the sample in E4-2 calculate a) the resistivity b) the drift current density J_t

        Ans: a) $\rho = 3.12$ ohm-cm,   b) $J_t = 32$ A/cm^2

E4-4   Determine the resistance of the sample in E4-2 if its length is 10 cm and its area is 1 cm^2.

        Ans: $R = 31.2\Omega$

EXAMPLE 4.2

A sample of silicon at 300K has resistivity of 5 ohm-cm. Assume that arsenic is the only dopant. Determine the impurity concentration.

Solution   Arsenic is a donor impurity; hence the resistivity is given by,

$$\rho = \frac{1}{q\mu_n n}$$

Both μ_n and n are unknown.

One method to determine n is to use Fig. 4.6 and read N_D to be approximately 9×10^{14}cm^{-3}.

To confirm this value, we use Fig. 4.5 to read the mobility for $N_D = 9 \times 10^{14}$ and find $\mu_n \cong$ 1400 cm^2/s from which we calculate the resistivity to be 4.96 ohm-cm.

One can, therefore, assume a starting value for N_D, determine the mobility from Fig. 4.5, calculate the resistivity, and then refer to Fig. 4.6 to obtain a new value for N_D.

The resultant impurity concentration is 9×10^{14}cm^{-3}.

EXAMPLE 4.3

A sample of silicon at 300K of length 2.5cm and a cross-sectional area of 2mm² is doped with 10^{17}cm^{-3} of phosphorus and 9×10^{16} cm^{-3} of boron. Determine:

a) The conductivities of the sample due to electrons and that due to holes.

b) The resistance of the sample.

Solution

a) The conductivities of electrons and holes are given respectively by,

$$\sigma_n = q\mu_n n \quad \text{and} \quad \sigma_p = q\mu_p p$$

We need to determine n and p from the following relations

$$p + N_D = n + N_A \quad \text{and} \quad pn = n_i^2$$

The results were determined in Example 3.4 to be

$$n = 10^{16}\text{cm}^{-3} \quad \text{and} \quad p = 1 \times 10^4 \text{cm}^{-3}$$

The values of the mobilities are found from Eqs. (4.15) and (4.16), using $N = N_A + N_D$, to be

$$\mu_n = 626\text{cm}^2/\text{V-s} \quad \text{and} \quad \mu_p = 292\text{cm}^2/\text{V-s so that,}$$

$$\sigma_n = 1.6 \times 10^{-19} \times 626 \times 10^{16} = 1.001 \text{ (ohm-cm)}^{-1}$$

$$\sigma_p = 1.6 \times 10^{-19} \times 292 \times 1 \times 10^4 = 4.67 \times 10^{-13} \text{ (ohm-cm)}^{-1}$$

b) The total conductivity becomes

$$\sigma = \sigma_n + \sigma_p \cong 1 \text{ (ohm-cm)}^{-1} \text{ and the resistivity is } \rho = 1/\sigma$$

$$R = \rho l/A = 1/\sigma A = 2.5/(1 \times 2 \times 10^{-2}) = 125 \text{ ohms}$$

We will now briefly discuss and explain the ohmic resistance of metals and semiconductors in the light of our discussion of the scattering phenomena.

Resistance of Metals and Semiconductors

Equation (4.6) indicates that the electron motion undergoes damping as it travels throughout the lattice structure. Eventually, its velocity becomes zero. The electrons are constantly traveling with high velocities, and when an electron collides with the lattice or with an impurity atom, it loses some of its energy and thus starts out with a new velocity.

This damping of the electron wave is actually caused by any disturbance in the periodic lattice. These irregularities in the structure scatter the electron waves, just as light is scattered in the atmosphere. In getting scattered, the electron wave is converted from a plane wave traveling in a certain direction with a certain momentum into waves traveling in all directions and corresponding to no momentum at all. The electron has then lost its momentum.

When we consider a specific material, we find that its electrical resistance is due to both perturbations in the solid lattice and due to the temperature. Thus, the resistivity has two components: one independent of temperature and the other temperature-dependent.

Since resistance can be defined as the ratio of applied voltage to current, an increase of current with temperature signifies decreased resistance. The current is proportional to the number of carriers and to their net mean velocity in the direction of the applied electric field. In a metal, when the temperature is increased, the density of free electrons hardly changes because of the already large density, but the average drift velocity decreases because of the increased vibration of the atoms and molecules. Thus, with an increase of temperature, the resistance of a metal increases.

Semiconductors, on the other hand, have a negative temperature coefficient of resistance. Their resistance decreases with temperature. In some ways, this property is an advantage of semiconductors; on the other hand, it is a source of instability. When the temperature increases, the incremental velocity of the carriers due to the electric field decreases, but the density of the carriers (free electrons and holes) increases appreciably. The increase in the density of carriers more than offsets the decrease of their velocity. Electrons from the valence band acquire thermal energy and move into the conduction band to become free electrons, leaving vacancies in the valence band that account for the holes. Thus, the current increases with an increase of temperature, which decreases the resistance, thus increasing the current, which then increases the temperature. Instability results and a runaway situation arises, which causes the destruction of the semiconductor.

Thus, we observe that irregularities in a material cause it to resist the flow of current, and that, when the temperature is increased, the resistances of metals and semiconductors vary in an opposite manner.

4.8 PARTICLE DIFFUSION AND DIFFUSION CURRENT

Diffusion

Carrier motion in semiconductors, which results in currents, is either caused by the application of an electric field, which causes a drift current, or by the process of diffusion of carriers.

The process of diffusion consists of the motion of carriers away from regions of high concentrations of carriers to regions of low concentration of carriers. This process is well illustrated when a light is flashed at a small region near the center of a long thin bar of semiconductor material. Electron-hole pairs, in addition to those available at thermal equilibrium, are then generated at the center because of the light energy that is absorbed by the semiconductor. These excess carriers will move randomly to the right and to the left of the generation zone, resulting in an outwards flux of particles away from the center. These carriers are said to diffuse (just as a drop of ink in a glass of water quickly mixes with water).

We illustrate this process by first considering Fig. 4.7(a), which shows a distribution of P particles per unit volume. We are interested in the rate of motion of these particles per unit area and unit time across a plane at x_1. The particles are in constant thermal motion from left to right, and from right to left at x_1. Since the density of the particles is greater to the left of x_1 than it is to the right of x_1, more particles

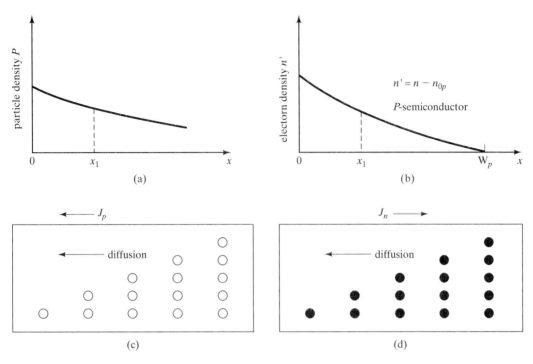

Figure 4.7   Illustrating diffusion: (a) diffusion of particles at x_1; (b) electrons injected at $x = 0$ cause electron density profile in P-semiconductor; (c) hole diffusion and hole diffusion current; and (d) electron diffusion and electron diffusion current.

will cross in the direction of decreasing density. The particle flow across x_1, in the positive direction, is given by

$$F = -D \frac{dP}{dx}\bigg|_{x=x_1} \qquad (4.25)$$

where F is the flux per unit area, per unit time, P is the volume density of particles, dP/dx is the gradient at x_1 and D is labeled the *diffusion constant*. For F to have units of particles per unit area per unit time, D has to have the dimension of distance squared per unit time. According to Eq. (4.25), and since the derivative is negative at x_1, particle flow is in the positive x-direction. In three dimensions, Eq. (4.25) becomes

$$F = -D\nabla P \qquad (4.26)$$

In the process of particle diffusion in semiconductor devices, the flux consists of the motion of electrons or holes across a surface. We will translate the rate of motion of carriers into an electric current since the motion involves charged electrons or holes. We consider in Fig. 4.7(b) a P-semiconductor region into which electrons are either injected at $x = 0$ from an N-semiconductor or are generated by a strong source of light focused at $x = 0$. We assume no electric field exists in the sample and that the density of excess electrons (excess over those available in the P semiconductor) decreases with increasing x.

Because the gradient is negative in the direction of electron diffusion, the flux of electrons (motion per unit area per unit time) passing through the plane at x_1, as determined from Eq. (4.26), is given by

$$F_n = -D_n \frac{dn'}{dx}\bigg|_{x=x_1} \tag{4.27}$$

where the gradient dn/dx has unit of electrons per cm^3 per cm of length, D_n is in cm^{2}s^{-1} so that the unit of F_n is electrons cm^{-2}s^{-1}. Because the gradient is negative, F_n represents a positive flux of electrons moving from left to right (increasing x) at x_1.

Diffusion Current

Up to this point, the charge of the carrier has not been introduced. To convert the expression to current density, we introduce the charge $-q$ coulombs for an electron so that the expression in Eq. (4.27) is written in terms of C cm^{-2}s^{-1}, which has units of current density as

$$J_n = (-q)(-D_n)\frac{dn}{dx} = qD_n \frac{dn}{dx} \tag{4.28}$$

For a negative gradient, indicating a decrease of n with x, the current density is negative. This is to be expected since a motion of electrons in the positive direction results in a current in the negative direction. If holes are injected into an N region, Eq. (4.27) applies as well and is multiplied by the charge $+q$ to obtain an expression for the hole current density. The direction of hole motion is in the direction of positive current flow causing J_p to be positive since dp/dx is negative.

In general, the current densities due to the diffusion of electrons and holes are given by

$$\begin{aligned} J_n &= qD_n\left(\frac{dn}{dx}\right) &\text{(a)} \\ J_p &= -qD_p\left(\frac{dp}{dx}\right) &\text{(b)} \end{aligned} \tag{4.29}$$

When J_n and J_p are in amps/cm^2, q is in coulombs, the diffusion constants are in cm^{2}s^{-1} and the derivatives represent the gradients in one dimension. Diffusion of electrons and holes and the directions of the diffusion currents are shown in Figs. 4.7(c) and 4.7(d).

The diffusion constant for either holes or electrons is related to the mobility by Einstein's relation in Eq. (4.30) (see Appendix D). At a fixed temperature, D is directly proportional to μ. This is to be expected since both mobility and diffusion constants are factors in the expressions for drift and diffusion currents respectively. Therefore,

$$D = \left(\frac{kT}{q}\right)\mu \tag{4.30}$$

At 300K, $kT/q = 0.0259$ V and for μ in cm^2/V-sec, D is in cm^2/s. Thus, D has the same dependence that μ has on doping density. However, because of kT/q, D is proportional to $T^{-0.5}$, where lattice scattering is dominant, and to $T^{-2.5}$ when impurity scattering is the major scattering phenomenon.

Application of the expression for the diffusion current is shown in Example 4.4.

EXAMPLE 4.4

The region shown in Fig. 4.7(b) is a section of a silicon device that is at room temperature ($T = 300$K). This region is doped with 10^{15}cm^{-3} acceptor atoms. A stream of minority carriers is injected at $x = 0$ and the distribution of minority carriers in the sample is assumed to be linear, decreasing from a value of 10^{11}cm^{-3} at $x = 0$ to the equilibrium value at $x = W$, where W is 10 microns.

Determine the diffusion current density of electrons.

Solution   The thermal equilibrium densities of majority and minority carriers are calculated to be:

$$p_0 \cong 10^{15}\text{cm}^{-3} \text{ and } n_0 = 1 \times 10^5\text{cm}^{-3}$$

Using the density of impurity atoms, we determine the mobility of electrons in the P region from Eq. (4.15) to be:

$$\mu_n = 1331 \text{ cm}^2/(\text{V-s})$$

The electron diffusion current density is given by:

$$J_n = q\, D_n \frac{dn}{dx}$$

The diffusion constant is obtained from Einstein's relationship at $T = 300$K,

$$D_n = \frac{kT}{q}\mu_n = .0259 \times 1331 = 34.48 \text{ cm}^2/\text{s}.$$

The slope of the electron density $\dfrac{dn}{dx}$, is $-\left(\dfrac{10^{11} - 1 \times 10^5}{10 \times 10^{-4}}\right) = -10^{14}\text{cm}^{-4}$

so that $J_n = 1.6 \times 10^{-19} \times 34.48 \times (-10^{14}) = -0.55\text{mA/cm}^2$

The motion of electrons is in the positive x-direction, causing a current in the negative x-direction.

4.9 CARRIER CURRENTS

In this chapter, we have examined the processes that cause carrier currents—namely drift and diffusion. In the next section we will study the mechanisms of carrier recombination and generation.

By combining the expressions for the current densities of holes and electrons due to drift and diffusion, we have the following expressions for J_n, J_p, and J_t, representing the electron current density, the hole current density, and the total current density respectively as:

$$J_n = q\mu_n n \mathscr{E} + qD_n\left(\frac{dn}{dx}\right) \qquad \text{(a)}$$

$$J_p = q\mu_p p \mathscr{E} - qD_p\left(\frac{dp}{dx}\right) \qquad \text{(b)} \qquad \qquad (4.31)$$

$$J_t = J_n + J_p \qquad \text{(c)}$$

The units of the symbols are: J in amps/cm², q in coulombs, μ in cm²/V-s, n and p are in cm⁻³, D_n and D_p in cm²/s and dn/dx and dp/dx are in cm⁻⁴.

It is evident from Eqs. (4.31) that in order to derive relations for the currents in a device that is subjected to a certain applied voltage in the steady-state, one needs to derive expressions for both the electric field intensity distribution as a function of distance, and the spatial distributions of the electron and hole densities. Fortunately, at the current levels at which most of the devices normally operate, the drift current of minority carriers is neglected. Our emphasis will be on the diffusion current of minority carriers. In the following section, we will set up the continuity equation from which we obtain expressions for the distribution of minority carriers.

4.10 RECOMBINATION AND GENERATION

Rates of R-G

So far, we have discussed the two processes of carrier motion—drift and diffusion. A third category of carrier actions that involve a variation in the densities of electrons and holes are the processes of generation and recombination. In fact, they are two separate processes that take place simultaneously and their rates are equal only at thermal equilibrium. These two processes indirectly affect the currents by changing the carrier densities involved in drift and diffusion.

Generation is the process of creating new carriers, holes, and electrons.

Recombination is the inverse of generation, whereby an electron and hole disappear simultaneously.

In an intrinsic semiconductor at thermal equilibrium, electrons and holes are continuously generated and continuously recombine. Because the intrinsic carrier density is determined solely by the energy gap and temperature, thus constant under given conditions, the rate of generation of electron-hole pairs is balanced by the equal rate of recombination. In extrinsic semiconductors, the carrier densities are determined by both the impurity doping density and the intrinsic carrier density, hence one can also state that in thermal equilibrium, the rates of generation and recombination are equal. We remind the reader that thermal equilibrium is a condition in which no external forces (such as light) or an electric field are applied to the semiconductor. For intrinsic semiconductors and extrinsic nondegenerate semiconductors, the relation $np = n_i^2$ is valid at thermal equilibrium.

The question is then, when are the generation and recombination rates not equal? To answer that question, we point out that semiconductor devices operate normally under nonequilibrium condition. This translates into $np \neq n_i^2$ indicating that external effects are influencing the carrier densities and carrier distributions.

Of course, the tendency for systems under nonequilibrium conditions is to return to equilibrium. For example, when excess minority carriers are injected into one end of a semiconductor bar, electron and hole densities will tend to obtain equilibrium values by causing the rate of recombination to exceed the rate of generation. Thus, the rates of generation and recombination are not balanced when the semiconductor is perturbed, causing the concentration of one type of carrier to exceed its equilibrium value. The forces of nature tend to restore conditions back to where they stood prior to the perturbation.

An N-type semiconductor has equilibrium majority and minority carrier densities of n_0 and p_0 respectively. If this semiconductor is exposed to a source of energy that increases the density of the minority carriers, then the process of generation-recombination acts to reduce the concentration of these carriers by causing the recombination rate to exceed the generation rate. If, on the other hand, due to extraction of carriers, the carrier density were to decrease, then the generation rate will exceed the rate of recombination until equilibrium is established.

Direct Generation-Recombination

Electrons and holes in semiconductors are generated when an electron is displaced directly from the valence band to the conduction band due to absorption of thermal energy or exposure to light energy. This process is labeled *direct generation* when the electron is excited directly from the valence band to the conduction band, and it results in an additional electron-hole pair. *Direct recombination* occurs when an electron falls from the conduction band directly into the valence band, eliminating both the electron and a hole. Such direct generation-recombination processes are very unlikely in silicon and germanium, as discussed in Section 2.4, because of the shape of the *E-k* diagram of Fig. 2.12. Direct recombination generation is the common process in III-V compounds such as gallium arsenide, gallium phosphide and indium antimonide. This direct recombination results in light emission. Sketches illustrating direct recombination-generation are shown in Fig. 4.8a.

Direct generation of electron-hole pairs may be caused by one or more of the following processes: exposure to thermal energy, exposure to light energy, or impact ionization. Impact ionization occurs when the collision of an accelerated electron with an atom transfers enough energy to the atom to cause the generation of new carriers. Under favorable conditions an avalanche process may ensue that results in excessive heat being generated, which destroys the device.

To generate an electron-hole pair, the amount of energy imparted to the electron in the valence band must be at least of the order of the energy gap E_g. Similarly, the recombination process is accompanied by a release of light or heat energy of the order of the energy gap. In a semiconductor at thermal equilibrium, the processes of generation and recombination take place continuously and their rates are equal.

Indirect Generation-Recombination

The principal recombination-generation (R-G) process in germanium and silicon is of the indirect type where a third party, acting as a catalytic agent, is involved. This third party is a localized state in the forbidden band, which serves as a "stepping stone" between the conduction and valence bands. Such states are at energy levels

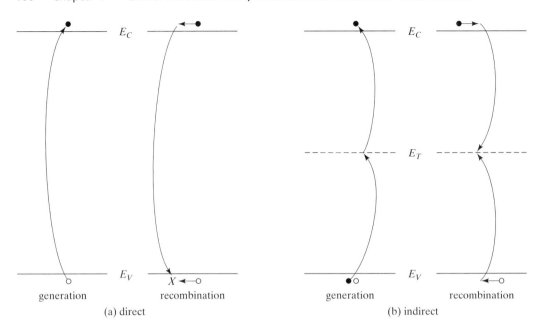

(a) direct   (b) indirect

Figure 4.8 Direct and indirect processes of R-G.

in which silicon and germanium atoms cannot exist. They are, however, available only for foreign atoms. Sketches showing indirect R-G processes are shown in Fig. 4.8(b). Each of the indirect generation and the recombination processes occur in two steps, where the "stepping stone" is a localized state having an energy level E_T near the middle of the forbidden band.

The frequency and location of recombination are probabilistic in nature. The energy level of an impurity recombination center is determined as follows: An electron transition from E_C to the recombination center E_T has a probability of occurrence of $P_n(E_T)$ whereas the probability of a hole transition from E_V to E_T is $P_p(E_T)$. Since a recombination is completed by the transition of both particles, the probability of the occurrence of such an event is proportional to the product, $P_n(E_T)P_p(E_T)$, of the two probabilities. The maximum of such a product is found to occur midway in the band gap at $(E_C + E_V)/2$.

There are four processes that make up the generation and recombination in indirect semiconductors. These are shown in Fig. 4.9, where an acceptor localized state exists slightly above the middle of the forbidden band. The first process is the *capture of an electron* by the localized state, while the inverse, second process, whereby an electron is moved from the localized state to the conduction band, is known as *emission of an electron.*

The third process, known as the *hole capture*, is described by the transfer of an electron from the localized state to the valence band. Finally, the fourth process consists in the transition of an electron from the valence band to the localized state, leaving a hole behind and called *hole emission.*

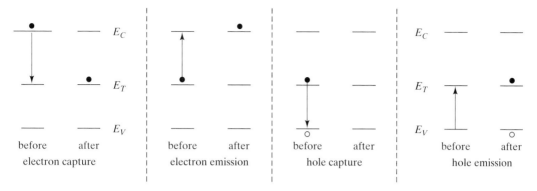

Figure 4.9   Interaction of free carriers with localized states in indirect generation-recombination. The R-G center is an acceptor type.

The localized states are also known as recombination-generation *centers* or *traps*.

The obvious question becomes: Where do these centers or traps come from?

These traps occur because of *lattice imperfections* (such as a missing atom in the covalent band structure), which result from *dislocations* or defects in the lattice or result from the *presence of impurities*, which may be deliberately introduced.

The lattice imperfections normally result from the preparation of the semiconductor sample and hence from the fabrication process. Such defects may be greatly reduced by very high-quality growth or by the production of very pure silicon.

The impurities that cause traps may be inherent to the solid but are commonly introduced so as to control the lifetimes of minority carriers. For the impurities in silicon to introduce efficient recombination-generation centers, the energy levels of these impurities must lie in the immediate vicinity of the middle of the forbidden band, in contrast to Column III (boron) and Column V (phosphorus) impurities, where energy levels are close to the valence and conduction bands respectively.

Gold and iron in silicon and copper in germanium introduce such recombination centers, some above (acceptors) and some below (donors) the middle of the forbidden band, but in both cases, no more than 0.1eV from the middle of the band.

Low-Level Injection and Recombination

Consider an N-type semiconductor in thermal equilibrium where $n_0 >> p_0$. This is now perturbed by some mechanism that causes p_0 to increase by Δp. To preserve neutrality, n_0 must also increase by Δn and $\Delta n = \Delta p$.

We will assume that $\Delta p << n_0$. Such a perturbation, which caused the excess minority carrier density (excess over equilibrium) to be much smaller than the majority carrier density, is said to have caused *low-level injection*. This condition of low-level injection is that in which most devices are normally operated when excess minority carriers are injected. Because the change in n_0, Δn, is so small compared to n_0, we assume for all practical purposes that n_0 is unchanged.

To illustrate the concept of low-level injection, assume $n_0 = 10^{15}\text{cm}^{-3}$ in silicon, so that $p_0 = n_i^2/n_0 = 10^5\text{cm}^{-3}$. An increase in p_0, Δp_0, of 10^{10}cm^{-3} will cause an increase in n_0, Δn_0, of 10^{10}cm^{-3}, which is negligible compared to n_0.

Generation-recombination is a continuous process, but since the hole density has been increased to a value greater than its thermal equilibrium value, the rate of recombination must exceed the generation rate. We are assuming a trap density N_T, located at a trap energy level E_T near the middle of the forbidden band. Since the material is N-type, the Fermi level E_F is above the middle of the energy gap and E_F is greater than E_T. Because $E_T < E_F$, the Fermi-Dirac relationship predicts that the E_T level is practically full of electrons. The recombination process consists of the capture of holes from the valence band. Once the holes are captured, some of the much more numerous electrons are also captured.

Analytical Relations

Expressions are needed for the time-variation of carrier densities under perturbed nonequilibrium conditions for the typically more common indirect recombination-generation processes. Before determining the expressions, and partly as a reminder, we define some terms as follows:

$$n, p = \text{electron, hole density, cm}^{-3}$$
$$n_0, p_0 = \text{thermal equilibrium electron, hole density cm}^{-3}$$
$$\Delta n = n - n_0 = \text{deviation of electron density from the thermal equilibrium value}$$
$$\Delta p = p - p_0 = \text{deviation of hole density from the thermal equilibrium value.}$$

Δn and Δp can be both positive or negative. A positive deviation implies an excess of carriers while a negative deviation corresponds to a deficit of carriers.

$$N_T = \text{number of recombination-generation centers (traps) per cm}^3.$$

When the source of excitation that caused the perturbation is removed, it will take some time for the density of holes to return to its equilibrium value via the indirect recombination-generation process. The question becomes: On what factors does the time rate of change of hole density depend?

First, to discard an excess hole, that hole must move from its location in the valence band to E_T, where the recombination traps are. These traps, of density N_T, are filled with electrons because $E_F > E_T$. Therefore, the rate of decay of hole density, $\partial p/\partial t$, is proportional to N_T since the larger the density of traps, the faster the recombination will occur. Furthermore, the number of holes scrambling to rejoin electrons increases with the number of excess holes since the greater the number of holes available for recombination, the more of them will recombine per unit time and the faster the recombination rate. Hence, the rate of hole density decay is proportional to Δp. Both factors can be included in an equation that states

$$\partial p / \partial t = -K_1 N_T \Delta p \qquad (4.32)$$

where K_1 is a constant of proportionality (positive) and the negative sign indicates a decrease in the hole density because of recombination. Similarly, the analogous expression for minority carrier electrons in a perturbed P semiconductor is given by

$$\partial n / \partial t = -K_2 N_1 \Delta n \qquad (4.33)$$

The units of $K_1 N_T$ and $K_2 N_T$ are $(sec)^{-1}$ so that one can define the two time constants as

$$\tau_p = 1/K_1 N_T \qquad (a)$$

$$\tau_n = 1/K_2 N_T \qquad (b) \qquad (4.34)$$

By using the relations of Eq. (4.34) in Eq. (4.33), we write

$$\partial p / \partial t = -\Delta p / \tau_p \qquad (a)$$

$$\partial n / \partial t = -\Delta n / \tau_n \qquad (b) \qquad (4.35)$$

It is to be emphasized here that the relations in Eqs. (4.35) are for indirect thermal generation-recombination process and are:

- for minority carriers
- for low-level injection.

Minority Carrier Lifetime

Since Eqs. (4.35) represent the rate of decay of the excess minority carriers, Δp in N-type and Δn in P-type semiconductors, it stands to reason that the time constants τ_p and τ_n represent a mean time interval during which all excess minority carriers have recombined. Some will recombine at $t = 0$ after the perturbation is removed and others will take a longer time. In that case, we define τ_p and τ_n as the average time an excess minority carrier will survive in an environment of a great number of majority carriers. These are labeled the *lifetimes of minority carriers.*

The lifetimes cannot be accurately predicted as they depend on the density of traps N_T, which itself is not a predictable quantity since it varies from one sample of semiconductor to another and is also subject to changes during fabrication of devices.

However, it is possible to increase the density of traps and hence reduce the lifetime by the addition of certain metals to the semiconductor such as gold in silicon.

At certain gold concentrations, an increase of two orders of magnitude of gold density may serve to reduce the lifetime by three orders of magnitude. The lifetimes in typical samples may vary from a microsecond to a nanosecond. The main purpose of decreasing the lifetime of minority carriers, as we shall explain in a later chapter, is to increase the switching speed of semiconductors or improve the high-frequency response of devices.

REVIEW QUESTIONS

Q4-7 Explain, in your own words, what is meant by diffusion in general and, as relating to semiconductors, in particular.

Q4-8 If the electron density at one end of a semiconductor bar is increased relative to the other end, does the hole density take on a similar distribution? Why?

Q4-9 Explain why the rate of generation of electrons in a semiconductor is given by n_0/τ_n where n_0 is the equilibrium electron density and τ_n is the lifetime of the minority carrier electrons.

Q4-10 What is the rate of recombination and what does it depend on?

Q4-11 Explain the following statement: "If the minority carrier density at one point in a semiconductor exceeds the thermal equilibrium value, the recombination rate exceeds the generation rate."

HIGHLIGHTS

- The current in some electronic devices is strictly a drift current. In others, it is a combination of drift and diffusion. Diffusion of carriers results when the carrier density in a semiconductor varies with distance.

- Diffusion current is directly proportional to the slope of the carrier density distribution. Diffusion current is the product of the slope, the charge q, the cross-sectional area, and the diffusion constant.

- Diffusion constant is a measure of the ease of diffusion of the carriers and is directly proportional to the mobility.

- While carriers are accelerated in a semiconductor, two other processes are also active; they are generation and recombination. Electrons and holes disappear as carriers by recombining and are continuously replenished by generation. Generation is the process of the transfer of electrons from the valence band to the conduction band.

- To determine drift and diffusion currents, it is necessary to know the distribution, with distance, of the carriers. The continuity equation, which is derived in the next section, is used, subject to appropriate boundary conditions, to determine the distribution of the minority carriers in a semiconductor.

EXERCISES

E4-5 The electron density in a silicon sample decreases linearly from 10^{16}cm^{-3} to 10^{15}cm^{-3} over a distance of 10cm. The area of the sample is 6cm^2 and the diffusion constant is 15cm^2/s. Determine the diffusion current of electrons.

$$\text{Ans: } I = -12.96\text{mA}$$

E4-6 The distribution of electrons in a semiconductor sample is given by $n(x) = 10^{15}\exp(-x/L)$ cm^{-3}. The length $L = 10$cms and the diffusion constant is 25cm^2/s. Determine the diffusion current density at $x = 0$.

$$\text{Ans: } J = -4 \times 10^{-4}\text{A/cm}^2$$

4.11 THE CONTINUITY EQUATION

We will now derive the continuity equation, which will be used to determine expressions for the distribution of minority carriers in an extrinsic semiconductor. After having determined these expressions, then we can deal with the diffusion currents of minority carriers.

We consider in Fig. 4.10, a section of an N-type semiconductor into which excess minority carrier holes have been injected and are moving within the section, and we focus our attention on the hole density in a very small slice, Δx, of the section. Since hole current is flowing, a certain number of holes are brought in by the hole current at x and a certain number is carried out by the current at $(x + \Delta x)$. In addition and within the slice, generation and recombination take place. We assume low-level injection ($\Delta p \ll n_0$) as was discussed in the preceding section. The hole density in the slice is p and the equilibrium hole density is p_0.

The number of *coulombs of holes* per square centimeter per second that flow into the slice is $J_p(x)$. The number of *coulombs* entering per second is $J_p(x)A$ and the number of *holes* that enter the slice per second is $J_p(x)A/q$. Similarly, the number of holes that leave the slice per second is $J_p(x + \Delta x)A/q$. Therefore, the rate of change (per second) of holes within the slice caused by the hole currents is the difference between the two

$$[J_p(x) - J_p(x + \Delta x)]A/q \quad \text{holes/s}$$

Because of recombination within the slice, the number of holes entering the slice at x is larger than the number of holes leaving at $x + \Delta x$ as shown in Fig. 4.11.

Since the number of holes within the slice ($pA\Delta x$) is greater than the equilibrium value ($p_0A\Delta x$), the number recombining, per second, within the slice is $pA\Delta x/\tau_p$ and the number generated is $p_0A\Delta x/\tau_p$. Thus, the rate of change of the number of holes due to recombination and generation is

$$-(p - p_0)A\Delta x/\tau_p$$

where τ_p is the lifetime of holes in the N material and the negative sign is inserted to indicate a decrease in p as long as $p > p_0$.

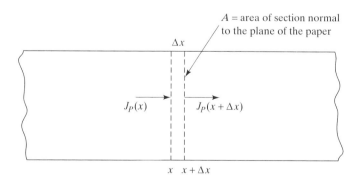

Figure 4.10   Section of N-type semiconductor for the formulation of the continuity equation.

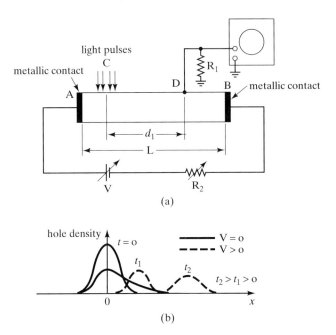

Figure 4.11   (a) Haynes–Shockley experimental setup; (b) the effects of drift, diffusion, and recombination on the minority carrier hole pulse.

Because of the assumed one-dimensional nature of the flow of minority carriers, holes enter and leave the slice only across the faces x and $(x + \Delta x)$. The total number of minority carriers within the slice is $pA\Delta x$. Thus, we can write

$$\frac{\partial}{\partial t}(pA\Delta x) = \frac{-(p - p_0)}{\tau_p}A\Delta x + \frac{[J_p(x) - J_p(x + \Delta x)]A}{q}$$

In effect, the statement of the equation is: Rate of change of holes in Δx = rate of (generation-recombination) + the rate at which holes enter the slice less the rate at which they leave. The partial derivative has been used to indicate a variation of p with time and distance. By dividing the previous equation by $A\Delta x$, we obtain the continuity equation in three dimensions for holes in N-type semiconductors

$$\frac{\partial p}{\partial t} = -\frac{p - p_0}{\tau_p} - \frac{1}{q}\operatorname{div} J_p \tag{4.36}$$

The analogous expression for electrons in P-type semiconductor is given as

$$\frac{\partial n}{\partial t} = -\frac{n - n_0}{\tau_n} + \frac{1}{q}\operatorname{div} J_n \tag{4.37}$$

In the chapter on diodes, we will use the continuity equation, after replacing the current density by its diffusion component equivalence from Eq. (4.29), to determine the steady-state distributions of minority carriers in the regions subject to the relevant boundary conditions. We will assume that the drift currents of minority carriers are negligibly small compared to the diffusion currents, and determine expressions for the currents as a function of distance. The applied voltage will appear in the boundary conditions.

The results referred to above include three properties of the minority carriers. They are: mobility, diffusion constant, and lifetime. In the following section, we briefly describe an experimental procedure that was first used to calculate these properties. Our interest in this experiment is more in the illustrations it provides of the carrier processes rather than in the detailed determination of the three properties.

4.12 HAYNES–SHOCKLEY EXPERIMENT

In 1951, two researchers at Bell Labs, Haynes and Shockley, reported an experiment they performed that enabled them to measure mobility, diffusion constant, and lifetime. It is important to emphasize again that, in determining current-voltage relations of semiconductor devices, it is the carrier processes of minority carriers that will serve as the mechanisms of the derivations.

We will briefly explain the experiment without entering into the details of the calculations that follow the results. The main purpose of describing the experiment in this book is to illustrate the combined three carrier processes that minority carriers undergo.

As shown in Fig. 4.11(a), three contacts are made to an N semiconductor bar. The two contacts at A and B are used to connect a battery in series with a resistance R_2. The function of the battery is to establish an electric field that will cause carriers to drift in the bar with holes moving from left to right. At point C, we focus pulses of a beam of light on the bar and the light energy moves more electrons from the conduction to the valence band, thus generating, at C, additional electrons and holes. At point D, a special contact is made to the bar, which we will study in the following chapter, that permits only minority carriers to be picked up.

The light pulse beam generates a pulse of electrons and holes at C. The number of electron-hole pairs generated at C, because of low-injection, is so small that the number of electrons is negligible compared to those in the bar. However, the number of holes is much greater than those (minority carriers) in the bar. The pulse of holes generated drifts in the bar, is picked up at D, and the current of holes causes a voltage drop across R_1 to ground. This voltage across R_1 is connected to the Y terminal of the oscilloscope and appears as a pulse on the scope. The X-time scale of the scope is synchronized with the timing of the light pulse generator. In brief, pulses of holes are light-generated at C and displayed on the scope after travelling in the bar.

Our interest here is in the shape of the hole pulse as it travels from C to D. The hole pulse is subjected to the three carrier processes as shown in Fig. 4.11(b). The pulse moves along the bar as a result of *drift*, its amplitude decreases and the pulse becomes wider because of *diffusion* of holes both to the right and to the left of the center of the pulse. The area of the pulse is reduced because of *recombination*.

The electric field intensity in the bar is determined from the voltage source and the distance d. The time it takes the pulse to travel from C to D is measured on the scope so that the average drift velocity is calculated and hence the mobility is determined. Other results of measurements enabled Haynes and Shockley to calculate the diffusion constant and the lifetime.

Although nowadays there are more precise methods of measuring minority carrier properties, the experiment explicitly illustrates the effects of the three carrier processes that minority carriers are subjected to.

PROBLEMS

4.1 The intrinsic resistivity of a semiconductor at 300K is 3×10^5 ohm-cm. Given $\mu_n = 1700$ and $\mu_p = 350$, cm^2/V-s, determine the intrinsic carrier density.

4.2 An intrinsic semiconductor has resistivity of 2×10^5 ohm-cm at $T = 300$K and has $\mu_n = 4000$ and $\mu_p = 1000$cm^2/V-s. Calculate the resistivity for

    a)   An acceptor doping of 10^{15}cm^{-3}.

    b)   A donor doping of 10^{17}cm^{-3}.

    Assume that the mobilities are constant with doping.

4.3 A semiconductor (*N*-type) bar is injected at one end by minority carrier holes and an electric field of 100V/cm is uniformly applied along the length of the bar that moves the holes a distance of 2cm in 10µs. Determine

    a)   The drift velocity of holes.

    b)   The diffusion constant of holes at $T = 300$K.

4.4 An *N*-type semiconductor bar is 2cm long, has a cross-sectional area of 0.1cm^2, an electron density of 5×10^{14}cm^{-3} and a resistivity of 10 ohm-cm. A 10V battery is connected across the ends of the bar. Determine

    a)   The time it takes an electron to travel the length of the bar.

    b)   The energy in eV and in Joules delivered to the bar.

4.5 At $T = 300$K, the intrinsic carrier density of a GaAs sample is 1.8×10^6cm^{-3}, $\mu_n = 8500$, and $\mu_p = 400$cm^2/V-s. An electric field applied to a bar of extrinsic GaAs causes equal electron and hole current densities. Determine

    a)   The equilibrium electron and hole densities.

    b)   The net doping density.

4.6 A semiconductor has an intrinsic resistivity of 3×10^4 ohm-cm. Donor and acceptor atoms are added with densities of 10^{14}cm^{-3} and 5×10^{12}cm^{-3} respectively. Given $\mu_n = 1600$ and $\mu_p = 600$cm^2/V-s, determine the current density if the applied electric field is 100mV/cm.

4.7 A bar of intrinsic semiconductor has a resistance of 5 ohms at 360K and 50 ohms at 330K. Assume that the change in resistance is a result of the change in n_i only. Calculate the band gap energy of the semiconductor.

4.8 A certain silicon sample has $N_A = 5.01 \times 10^{16}$cm^{-3} and $N_D = 5.02 \times 10^{16}$cm^{-3}. Given $\mu_n = 1200$ and $\mu_p = 400$cm^2/V-s, determine

a) The conductivity.

b) Repeat part (a) if $N_A = 10^{17}$ and $N_D = 1.1 \times 10^{17}\text{cm}^{-3}$. Use the mobilities of part (a).

4.9 Determine the resistance of a bar of silicon that has the following properties: length L = 0.8cm, area A = 1mm^2, $N_D = 3 \times 10^{16}\text{cm}^{-3}$, $N_A = 10^{14}\text{cm}^{-3}$, μ_n = 1000 and μ_p = 500cm^2/V-s.

4.10 A silicon bar has a length of 1cm, a height of 0.01cm, and a depth of 0.2cm. At T = 300K, determine the resistance of the bar for the following conditions:

a) Intrinsic.

b) A donor doping of 10^{15}cm^{-3}.

c) An acceptor doping of 10^{15}cm^{-3}.

4.11 a) For *minimum conductivity* of a semiconductor sample, determine an expression for the electron and hole density, at a given temperature, in terms of μ_n, μ_p, and the intrinsic carrier density.

b) Use μ_n = 3900, μ_p = 1800cm^2/V-s, and $n_i = 10^{10}\text{cm}^{-3}$ to calculate the hole density and the maximum resistivity.

4.12 Given a Ge sample that has, at a given temperature, $n_i = 2.5 \times 10^{13}\text{cm}^{-3}$, μ_n = 3900 and, μ_p = 1900cm^2/V-s, determine:

a) The intrinsic conductivity.

b) The minimum conductivity.

4.13 Given a semiconductor that has a mobility ratio that is independent of impurity density given by $K = \mu_n/\mu_p$ and $K > 1$, determine an expression for the maximum resistivity in terms of K and the intrinsic resistivity.

4.14 Determine all the possible values of hole and electron densities that cause the conductivity of a semiconductor to be equal to the intrinsic conductivity.

4.15 An electric field of 10V/cm is applied to an intrinsic silicon sample. If the carriers drift 1cm in 100μs, determine, at T = 300K:

a) The drift velocity.

b) The diffusion constant.

c) The conductivity.

4.16 A bar of heavily doped P-silicon, to which an electric field is applied, has a drift current density of 50A/cm^2. The hole drift velocity is 50cm/s. Determine the hole density.

4.17 An N-type Ge bar has a resistivity of 5 ohm-cm. Determine the time it takes an electron to travel 5×10^{-3}cm if the current density is 0.1A/cm^2. Use the plots in the text.

4.18 A voltage is applied between two contacts that establishes an electric field of 100V/cm in the space separating the contacts. For an electron starting from rest at the first contact, determine:

a) The drift velocity of an electron if the space between the contacts is occupied by an N-semiconductor having μ_n = 3900 cm^2/V-s

b) The velocity of an electron at a distance of 1cm if the space between the contacts is a vacuum.

4.19 A bar of N-type silicon 1cm long is doped with 10^{15}cm^{-3} of donor atoms at T = 300K and a voltage of 5V is applied to the ends of the bar. Determine:

a) The hole drift current density.

    b)   The total drift current density.

4.20 A hole current of 10^{-4}A/cm^2 is injected into the side ($x = 0$) of a long N-silicon bar. Assuming that the holes flow only by diffusion, and that at very large values of x, the distribution of excess holes decays to zero. Determine:

    a)   The steady-state excess hole density at $x = 0$.

    b)   Repeat part (a) at $x = 100\mu$m.

Given $\mu_p = 400$cm^2/V-s, $\mu_n = 1600$cm^2/V-s, and the lifetime of holes is 25μs.

4.21 The bar of Prob. 4.20 has $N_D = 10^{15}$cm^{-3}, determine:

    a)   The rate of generation of electron-hole pairs at $x = 100\mu$m.

    b)   The rate of recombination of electron-hole pairs at $x = 100\mu$m.

chapter 5

THE PN JUNCTION DIODE

5.0 INTRODUCTION

In earlier chapters, we have studied the properties of semiconductors under equilibrium and non-equilibrium conditions. It was indicated that pure intrinsic semiconductors are of very limited use. Semiconductors that are doped with impurities form the basis of the devices we are about to study.

A semiconductor that has been doped with acceptor impurities and into the surface of which donor atoms are diffused forms an extremely interesting junction known as the PN junction diode. The current-voltage characteristic of a typical diode is shown in Fig. 5.1.

In closely studying this characteristic, we observe the two most important properties of diodes. In the first place, where the voltage is positive, very small voltages, less than one volt, cause large currents, whereas when the voltage is negative the current is extremely small. Second, the slope of the characteristic in the first quadrant is extremely large and the slope in the third quadrant, preceding the sudden drop, is very small.

The first property points to the use of the diode as a rectifier in converting alternating (AC) voltages to undirectional voltages and eventually through filtering to direct voltages (DC). The second important application of the diode is its use as a switch from an almost short circuit (low V_a/I) in the first quadrant to an open circuit (high V_a/I) in the third quadrant.

The PN junction diode provides an essential background to the study of the bipolar and junction field-effect transistors.

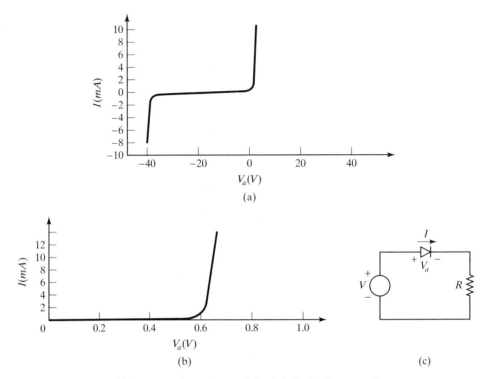

(a)

(b)   (c)

Figure 5.1 (a) Current-voltage characteristic of diode; (b) Current-voltage characteristic in the positive I and V regions; (c) Circuit from which (b) is obtained showing diode symbol.

5.1 SPACE-CHARGE REGION

Formation of Region

Let us assume that a slab of N material and a slab of P material are brought together in a manner in which their structures line up and a single crystal is formed, as shown in Fig. 5.2. For the sake of simplicity, we assume that the (metallurgical) boundary between the P and N regions represents a *step junction* or an *abrupt junction*. It is assumed, in an abrupt junction, that the transition from the P to the N region takes

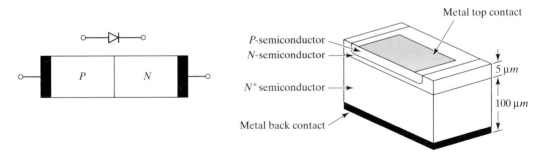

Figure 5.2 (a) Two-dimensional diode section and symbol. (b) Typical construction.

place within an extremely small distance. There is thus a sudden change in doping in going from P to N.

In general, diodes are fabricated with one region more highly doped than the other, the more highly doped region identified by a superscript plus, such as P^+N and N^+P. The majority carrier density in the more highly doped region may be about three orders of magnitude greater than the majority carrier density of the other region.

In our illustration, we will use a P^+N diode, where in Fig. 5.3(a) we have two sketches, one to illustrate the hole densities in both regions and one to illustrate the electron densities in the two regions. The equilibrium hole density in the P-region is p_{0p} and in the N region is p_{0n}, whereas n_{0p} and n_{0n} represent the equilibrium electron densities in the P and N regions respectively.

Because of the large gradients of charges (electrons and holes) that exist near the surface of contact, holes diffuse from P to N and electrons diffuse from N to P. The diffusions are shown in Fig. 5.3(b) for each type of carrier.

By assuming that N_A and N_D, in the P and N regions respectively, are each much greater than the intrinsic carrier density, then $p_{0p} \approx N_A$ and $n_{0n} \approx N_D$.

For every electron that leaves the N region, because of diffusion across the surface, an uncompensated positively ionized donor is left behind, and for every hole that leaves the P region, an uncompensated negatively ionized acceptor is created. These carriers leave from points near the surface of contact. As a result, the P and N regions are separated by what is known as a *depletion region*, a region that is depleted of holes and electrons but contains positively ionized donor atoms on one side and negatively ionized acceptors atoms on the other side. Because this region consists mainly of ionized charged impurities, it is also known as the *space-charge region*. The PN junction diode in equilibrium consists of charge-neutral P and N regions separated by a space-charge region. Because of the high electrostatic field produced in the depletion region, the concentrations of mobile carriers, holes and

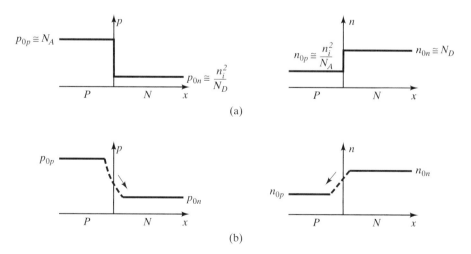

Figure 5.3   (a) Dopant and carrier densities before contact; (b) Carrier densities after contact showing diffusion of carriers.

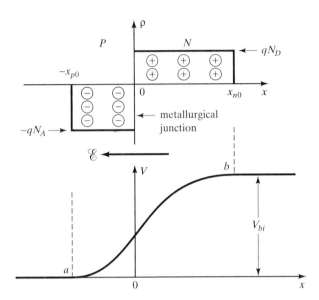

Figure 5.4 Space-charge region, direction of electric field, and built-in voltage at thermal equilibrium.

electrons, are very small in comparison to the impurity concentrations over most of the space-charge region. Conditions after contact are shown in Fig. 5.4.

We note in Fig. 5.4 that the space-charge region has been assumed to contain ionized impurity atoms only and we also assume that outside of the depletion region the materials are neutral. The N side of the depletion region is positively charged and the P side is negatively charged, and since the electric field intensity is defined to be the force on a unit positive charge, it is shown in Fig. 5.4 to be directed from right to left.

We observe in the figure that the N-side of the depletion layer extends from $x = 0$ to $x = x_{n0}$, whereas the P side extends from $x = 0$ to $x = -x_{p0}$. The line separating the two regions is at $x = 0$, which is the metallurgical junction of the P and N regions. The subscripts 0 refer to thermal equilibrium.

In a later section of this chapter, analytical relations will be developed and expressions will be derived for V_{bi} and for the amount of bending of the bands.

Barrier Voltage and Energy Bands

The charges in the depletion region cause an electric field, which results in a voltage across the depletion region labeled the built-in voltage, V_{bi}, and shown in Fig. 5.4.

The built-in voltage, V_{bi}, is accompanied by bending of the energy bands, as shown in Fig. 5.5. The bending is explained as follows: Voltage between two points a and b is defined as the energy expended or acquired in moving a unit positive charge from a to b. If the positive charge is at a and b is at a higher voltage V, then work is expended in moving the charge. After arriving at b, the positive charge has gained potential energy equal to the energy expended that is given by (qV), where q is the charge in coulombs of the electron.

If an electron is moved from a to b, with b at a higher voltage, V, the electron loses potential energy so that at b the electron is said to have lower potential energy than at a. This energy at b is equal to $(-qV)$.

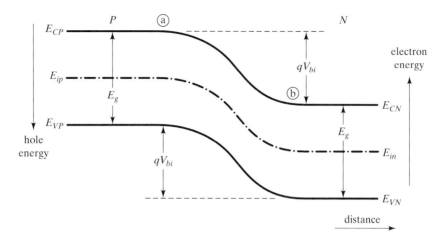

Figure 5.5   Bending of energy bands.

Since in Fig. 5.4, the voltage at x_{n0} is higher than the voltage at $-x_{p0}$ by V_{bi}, an electron at $x \geq x_{n0}$ has less potential energy than an electron at $x \leq x_{p0}$. We note in Fig. 5.5 that the respective energy levels in the neutral N region are lower than those in the P region by qV_{bi}. Hence, the energy bands in the N region are said to bend down relative to those in the P region.

We recall that electrons reside in the conduction band and holes reside in the valence band. Furthermore, electron energy is measured upwards, whereas hole energy is measured downwards.

The presence of charges on either side of the junction create an electric field directed from N to P. The positive and negative charges (impurities ions), shown in Fig. 5.4, are generated because the electrons and holes that are associated with them left by diffusing to the other side. We label the *built-in* voltage as the *contact* potential and also as the *barrier voltage*.

Drift and Diffusion Currents

Holes from P to N and electrons from N to P continue to diffuse because of the gradients of carrier densities resulting in a total diffusion current from P to N. Simultaneously, the electric field at the junction forces the holes to drift from N to P and the electrons to drift from P to N resulting in a total drift current from N to P. We have two currents, drift and diffusion, for both electrons and holes in opposite directions.

The total current, the sum of the two drift and two diffusion currents, must be zero. A non-zero current causes energy dissipation, which is not possible since there is no source in the diode to provide this energy.

The student may wonder if the total electron current and the total hole current are each zero. The current of each carrier has to be zero because of the following reasoning: Since the diode is at thermal equilibrium, the product np is equal to n_i^2. If we assume that the electron current is not zero, then the hole current is equal in magnitude and opposite in direction to the electron current. There is thus a net transfer of electrons and holes in the same direction causing an increase in the densities of both in one of the regions and violating the condition of thermal equilibrium.

5.2 ANALYTICAL RELATIONS AT EQUILIBRIUM

In this section, we will establish analytical relations for the conditions in the diode at thermal equilibrium. After reviewing some basic concepts in electrostatics, we will determine the location of the Fermi level throughout the diode. We will then derive an expression for the built-in voltage at the junction. This is followed by the derivation of analytical relations for the electric field, the potential distributions, and the concomitant bending of the energy bands in the diode.

Electrostatics of the Space Charge Region

The electric field is the force exerted on a unit positive charge. The force on an electron is $-q\mathscr{E}$, where q is the magnitude of the electronic charge.

We also know that force is the negative of the gradient of potential energy so that the force acting on an electron becomes

$$-q\mathscr{E} = -(\text{gradient of potential energy of the electron})$$

The expression for the electric field becomes

$$\mathscr{E} = \frac{1}{q}\,\text{grad}\,E \tag{5.1}$$

where E is the potential energy of the electron. An electron in the conduction band has potential energy of E_C, the bottom of the band. The gradient of E could be replaced by the gradient E_C. Since the gradients of E_C, E_i, and E_V are the same, we choose to use E_i in Eq. (5.1) for reasons that will become apparent shortly. Using E_i, Eq. (5.1) in one dimension becomes

$$\mathscr{E} = \frac{1}{q}\frac{dE_i}{dx} \tag{5.2}$$

The electrostatic potential, ϕ, is defined in one dimension by

$$\mathscr{E} = -\frac{d\phi}{dx} \tag{5.3}$$

Upon comparing Eqs. (5.2) and (5.3), we write

$$\phi = \frac{-E_i}{q} \tag{5.4}$$

Finally, we use Poisson's relation in Eq. (5.5) to relate the potential, the potential energy, and the electric field to the diode constants. Poisson's equation in one dimension is

$$\frac{d^2\phi}{dx^2} = -\rho/\varepsilon \tag{5.5}$$

where ρ is the volume charge density, $(\varepsilon = \varepsilon_r\varepsilon_o)$ is the permittivity, ε_r is the relative dielectric constant, and ε_o is the permittivity of free space.

By using Eqs. (5.3) and (5.4) in Eq. (5.5) we have

$$\frac{d^2E_i}{dx^2} = \frac{q\rho}{\varepsilon} \tag{5.6}$$

$$\frac{d\mathscr{E}}{dx} = \frac{\rho}{\varepsilon} \tag{5.7}$$

Constancy of the Fermi Level

In the previous section, we concluded that, at thermal equilibrium, the net current of both electrons and holes across the diode junction is zero. We refer to Eq. (4.31b), and repeated here, for the hole current density in one dimension

$$J_p = q\mu_p p\mathscr{E} - q\,D_p\frac{dp}{dx} \tag{5.8}$$

The expressions for the hole density, given by Eq. (3.41), and its derivative are

$$p = n_i \exp\left[(E_i - E_F)/kT\right] \qquad \text{(a)}$$

$$\frac{dp}{dx} = \frac{n_i}{kT}\{\exp\left[(E_i - E_F)/kT\right]\}\left[\frac{dE_i}{dx} - \frac{dE_F}{dx}\right] \qquad \text{(b)} \tag{5.9}$$

By using the expressions in Eqs. (5.9) in Eq. (5.8), replacing D_p by $\mu_p\,kT/q$, $\mathscr{E}$ by its equivalence from Eq. (5.2), and setting $J_p = 0$, we have

$$\left(\frac{dE_i}{dx}\right)n_i\exp\left(E_i - E_F\right)/kT = [n_i\exp\left(E_i - E_F\right)/kT]\left[\frac{dE_i}{dx} - \frac{dE_F}{dx}\right] \tag{5.10}$$

Equation (5.10) is simplified so that $\dfrac{dE_F}{dx} = 0$.

This result indicates that the Fermi level is constant as we move from the N region to the P region. The identical result is obtained by using expressions for the electron density.

We conclude that in equilibrium, the Fermi level must be constant throughout the semiconductor, as shown in Fig. 5.6.

Built-In Voltage in Terms of Fermi Potential

In our earlier discussion, we concluded that in a diode in equilibrium, an electric field and a potential are established across the junction, and that the energy bands are bent in both the N and P regions. We also just established that, whereas $E_C, E_i,$

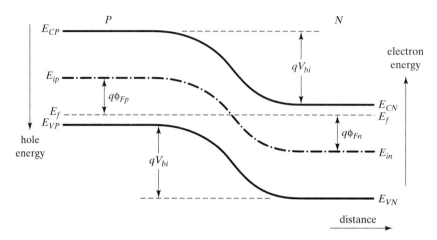

Figure 5.6   Energy distributions, the Fermi level, E_F, and the Fermi potentials at equilibrium.

and E_V are bent, the Fermi level E_F is flat throughout and not subject to any bending. The question is then: By how much energy are the bands bent or what is the magnitude of the potential across the junction?

The potential is developed across the space charge region, which we have assumed consists entirely of ionized donors and acceptors. We explained this assumption by acknowledging that there are some carriers that are continuously crossing the depletion layer. The exception is that their density is small in comparison to the densities of the ionized impurities.

Since the intrinsic energy levels, E_{ip} in the P region and E_{in} in the N region, are bent and the Fermi level E_F is constant across the space-charge region, we define *Fermi potentials* in each of the charge neutral P and N regions, ϕ_{Fp} and ϕ_{Fn}, as

$$\phi_{Fp} = -\frac{E_F - E_{ip}}{q}, \ \phi_{Fn} = -\frac{E_F - E_{in}}{q} \tag{5.11}$$

We remind the reader that in the P region, E_{ip} is higher than E_F so that ϕ_{Fp} is positive. In the N region, E_F is above E_{in} so that ϕ_{Fn} is negative. Since E_F is constant throughout, the total potential separation between E_{ip} and E_{in} is given by $(\phi_{Fp} + |\phi_{Fn}|)$, which we label the built-in voltage V_{bi}. The bands of E_C, E_i, and E_V are each bent by qV_{bi}.

A sketch of the energy bands at equilibrium showing the Fermi potentials is shown in Fig. 5.6.

Built-In Voltage in Terms of Doping Densities

Since the electron and hole currents across the junction are each zero, we use Eq. (4.31(b)) for the hole current in one dimension as

$$J_p = q\mu_p p\mathscr{E} - qD_p \frac{dp}{dx} = 0 \tag{5.12}$$

By using the Einstein relation to replace D_p, we have

$$q\mu_p p \mathscr{E} = q \left(\frac{kT}{q} \mu_p \right) \frac{dp}{dx}$$

or

$$\mathscr{E} dx = \frac{kT}{q} \frac{dp}{p} \qquad (5.13)$$

By integrating across the depletion layer from $-x_{p0}$, where the hole density is p_{0p}, to x_{n0}, where the hole density is p_{0n}, we have

$$\frac{kT}{q} \int_{p_{0p}}^{p_{0n}} \frac{dp}{p} = \int_{-x_{p0}}^{x_{n0}} \mathscr{E} dx \qquad (5.14)$$

where the first subscript in $p, 0$, refers to the equilibrium value and the second subscript refers to the relevant P or N regions.

The term on the right-hand side of Eq. (5.14) is the negative of the built-in voltage, V_{bi}, across the depletion layer so that after integrating we have

$$V_{bi} = \frac{kT}{q} \ell n \frac{p_{0p}}{p_{0n}} \qquad (5.15)$$

where p_{0p} is the equilibrium hole density in the P region and p_{0n} is the equilibrium hole density in the N region. Since N_A in P and N_D in N are each much greater than n_i, and $p_{0p} \cong N_A$ and $n_{0n} \cong N_D$, so that $p_{0n} = n_i^2/N_D$, we have

$$p_{0p} \cong N_A \text{ and } p_{0n} \cong \frac{n_i^2}{N_D} \qquad (5.16)$$

Using the relations of Eqs. (5.16) in Eq. (5.15), we have

$$V_{bi} = \frac{kT}{q} \ell n \frac{N_A N_D}{n_i^2} \qquad (5.17)$$

The built-in voltage V_{bi}, also known as the barrier voltage, depends on the doping of the P and N regions and on temperature and band gap, through n_i^2. The barrier voltage for an abrupt silicon junction, which has energy band gap of 1.2eV, varies between 0.7 and 1.0 volt while that for an abrupt gallium arsenide junction, whose band gap energy is 1.4eV, varies between 1.0 and 1.4 volts.

The appearance of a voltage when two materials are in contact raises the possibility of associating this voltage with voltage sources. This contact voltage is not a source of voltage, hence it is not a source of energy. Any attempt to measure this voltage by connecting a voltmeter across the external P and N contacts will provide an indication of zero volts. So what happened to the barrier voltage V_{bi}? The metallic contacts between the voltmeter leads and the P and N regions provide a contact potential equal to and opposite to the diode barrier voltage. Furthermore, if the diode voltage were a source, connecting a resistance across the diode terminals would cause a current and energy will be dissipated. This is absurd since there is no source of energy in the circuit.

Electric Field and Potential in the Space Charge Region

In this section, we will derive expressions for the electric field in the space charge region, also labeled the depletion region, and for the potentials of the region. We start with Poisson's equation as defined by Eq. (5.7), written in one dimension for the general case

$$\frac{d\mathcal{E}}{dx} = \frac{\rho}{\varepsilon} = \frac{q}{\varepsilon}(p - n + N_D - {}_{NA})$$ (5.18)

We identify the limits of the depletion region as extending from $x = -x_{p0}$ to $x = x_{n0}$, as shown in Fig. 5.7. From $x = 0$ to $x = -x_{p0}$, the portion of the depletion region extracted from the P region is included, and from $x = 0$ to $x = x_{n0}$ the balance of the depletion region is in the N region.

To obtain expressions in closed form, some simplifying assumptions, classified under the *depletion approximation*, are made.

The assumptions of the depletion approximation are:

1. In the space charge region from $x = -x_{p0}$ to $x = x_{n0}$, only ionized impurities exist. The acceptor density covers the distance $x = -x_{p0}$ to $x = 0$, and from $x = 0$ to $x = x_{n0}$ only donor impurities are available. Throughout the depletion region, the densities of electrons and holes are much smaller than N_A and N_D and are neglected.

2. Outside the space charge region, two neutral regions, one on each side of the depletion region, exist. They include electrons and holes. The sum of all the charges is zero so that these regions are labeled as neutral and the electric field is zero.

The whole diode, taken as one unit, is charge neutral and the total charge per unit area on one side of $x = 0$ is equal in magnitude and opposite in sign to the charge on the other side of $x = 0$ as shown in Eq. (5.19).

$$-q\,N_A^-\,x_{p0} = q\,N_D^+ x_{n0}$$ (5.19)

where N_A^- and N_D^+ are the densities of the ionized impurities on the P side and N side of $x = 0$ in the depletion region respectively.

The doping is not symmetrical, and ratios of N_A to N_D or N_D to N_A in practical diodes have values of about 100 so that we can conclude from Eq. (5.19) that the depletion layer extends deeper into the lightly doped region. The expression for $\mathcal{E}(x)$ in the N side of the space-charge region, subject to the assumptions of the depletion approximation, is obtained by integrating Eq. (5.18) as

$$\mathcal{E}(x) = \frac{qN_D}{\varepsilon}x + C_1 \text{ for } 0 \leqslant x \leqslant x_{n0}$$

At $x = x_{n0}$, $\mathcal{E}(x) = 0$, since we are at the boundary of the neutral N region. So,

$$C_1 = -\frac{qN_D}{\varepsilon}x_{n0}$$

and the expression for the electric field becomes

$$\mathscr{E}(x) = \frac{qN_D}{\varepsilon}(x - x_{n0}) \text{ for } 0 \leqslant x \leqslant x_{n0} \tag{5.20}$$

For $-x_{p0} \leqslant x \leqslant 0$, where $\rho = -qN_A$, we derive an analogous expression as

$$\mathscr{E}(x) = -\frac{qN_A}{\varepsilon}(x + x_{p0}) \text{ for } -x_{p0} \leqslant x \leqslant 0 \tag{5.21}$$

At $x = 0$, the electric field must be continuous since there is no layer of charge at that point. We obtain the expression for the largest value of $\mathscr{E}$, $\mathscr{E}_{max}$, from Eqs. (5.20) and (5.21). Since both of the two expressions have a maximum at $x = 0$ we have

$$\mathscr{E}_{max} = -\frac{qN_D}{\varepsilon}x_{n0} = -\frac{qN_A}{\varepsilon}x_{p0} \quad \text{at} \quad x = 0 \tag{5.22}$$

A section of the diode, the charge distribution, the electric field intensity, the potential and the potential energy distribution are shown in Fig. 5.7 for $N_A \gg N_D$. The expression for $\mathscr{E}_{max}$ from Eq. (5.22) is used in Eqs. (5.20) and (5.21) so that they are rewritten as

$$\mathscr{E}(x) = \mathscr{E}_{max}\left(1 - \frac{x}{x_{n0}}\right) \text{ for } 0 \leqslant x \leqslant x_{n0} \tag{a}$$

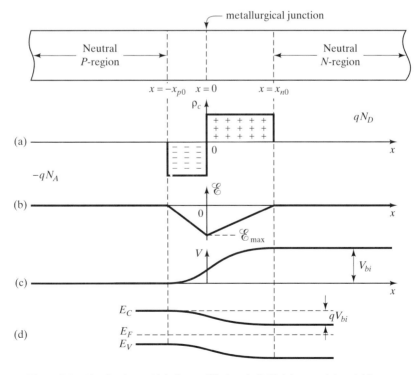

Figure 5.7 Distributions of (a) charge, (b) electric field, (c) potential, and (d) energy at equilibrium.

$$\mathscr{E}(x) = \mathscr{E}_{\max}\left(1 + \frac{x}{x_{p0}}\right) \text{ for } -x_{p0} \leq x \leq 0 \qquad \text{(b)} \qquad (5.23)$$

We determine an expression for the potential ϕ_p across the P side of the depletion layer by starting with Eq. (5.3) in integral form.

$$\phi_p(x) = -\int \mathscr{E}(x)dx \text{ for } -x_{p0} \leq x \leq 0$$

By using Eq. (5.23b), integrating and arbitrarily assuming the zero of potential to be at $x = -x_{p0}$ in order to determine the constant of integration, we have

$$\phi_p(x) = -\mathscr{E}_{\max}\left(x + \frac{x_{p0}}{2} + \frac{x^2}{2x_{p0}}\right) \text{ for } -x_{p0} \leq 0 \leq x \qquad (5.24a)$$

Similarly, ϕ_n is determined by using Eq. (5.23(a))

$$\phi_n(x) = -\int \mathscr{E}(x)dx \text{ for } 0 \leq x \leq x_{n0}$$

After integrating, and since the potential is continuous at $x = 0$, we set $\phi_n = \phi_p$ at $x = 0$ so that

$$\phi_n(x) = -\mathscr{E}_{\max}\left(x + \frac{x_{p0}}{2} - \frac{x^2}{2x_{n0}}\right) \text{ for } 0 \leq x \leq x_{n0} \qquad (5.24b)$$

Although the above equation is only valid for the limits shown, the zero level for $\phi_n(x)$ is at $x = -x_{p0}$. The total voltage across the space-charge region is given by setting $x = x_{n0}$ in the expression for ϕ_n so that

$$V_{bi} = -\frac{\mathscr{E}_{\max}}{2}(x_{p0} + x_{n0}) = -\frac{\mathscr{E}_{\max}W}{2} \qquad (5.25)$$

where V_{bi} is the total voltage from P to N and W is the total width of the depletion layer. Since $\mathscr{E}_{\max}$ is a negative quantity, V_{bi} is a positive quantity. The distribution of total potential energy is determined from $q\phi_p$ and $q\phi_n$.

Width of the Space Charge Region

We will now obtain an expression for the total depletion width in terms of the physical properties of the diode and the built-in voltage.

By using the expression for $\mathscr{E}_{\max}$ from Eq. (5.22) in Eq. (5.25), and realizing that $(x_{p0} + x_{n0})$ is the width of the depletion region, we have

$$V_{bi} = \frac{qN_Ax_{p0}}{2\varepsilon}(x_{p0} + x_{n0}) \qquad (5.26)$$

where $\varepsilon = \varepsilon_r\varepsilon_0$.

The expression for x_{n0} from Eq. (5.19) is used in Eq. (5.26) and the resulting equation is solved for x_{p0} as

$$x_{p0} = \left(\frac{2\varepsilon V_{bi} N_D}{q N_A (N_A + N_D)} \right)^{\frac{1}{2}} \qquad (5.27)$$

The width of the depletion layer, W, becomes, by using Eqs. (5.19) and (5.27),

$$W = x_{p0} + x_{n0} = x_{p0} + x_{p0} \frac{N_A}{N_D} = x_{p0} \left(1 + \frac{N_A}{N_D} \right)$$

$$W = \left(\frac{2\varepsilon}{q} \frac{(N_A + N_D)}{N_A N_D} V_{bi} \right)^{\frac{1}{2}}. \qquad (5.28)$$

We will use an example to illustrate the application of the expressions and we will calculate typical orders of magnitudes for the various quantities.

EXAMPLE 5.1

An abrupt (step) junction diode made of silicon has $N_A = 10^{17} \text{cm}^{-3}$ and $N_D = 10^{15} \text{cm}^{-3}$. The diode is at 300K and has area $= 10^{-5} \text{cm}^2$. The relative permittivity of silicon is 11.8.
    Calculate:

a) The built-in voltage, V_{bi}.
b) The depletion widths x_{n0} and x_{p0}.
c) The maximum value of the electric field intensity.
d) The charge stored in each of the depletion regions.

Solution
From Eq. (5.17),

$$\textbf{a)} \quad V_{bi} = 0.0259 \ \ell n \frac{10^{15} \times 10^{17}}{1 \times 10^{20}} = 0.715V$$

We determine x_{p0} from Eq. (5.27) and then x_{n0} from Eq. (5.19)

$$\textbf{b)} \quad x_{p0} = \left(\frac{2 \times 11.8 \times 8.854 \times 10^{-14} \times 0.715 \times 10^{15}}{1.6 \times 10^{-19} \times 10^{17}(1.01 \times 10^{17})} \right)^{\frac{1}{2}}$$

$$x_{p0} = 0.0961 \text{ microns}$$

$$x_{n0} = x_{p0} \frac{N_A}{N_D} = 0.961 \text{ microns}$$

$$W = x_{p0} + x_{n0} = 0.970 \text{ microns}$$

The maximum value of the electric field is found from Eq. (5.22)

$$\textbf{c)} \quad \mathscr{E}_{max} = -q \frac{N_D x_{n0}}{\varepsilon} = \frac{-1.6 \times 10^{-19} \times 10^{15} \times 96.1 \times 10^{-6}}{11.8 \times 8.854 \times 10^{-14}}$$

$$\mathscr{E}_{max} = -1.48 \times 10^4 \text{V/cm}$$

$$Q^+(\text{depletion layer}) = q N_D x_{n0} A$$

$$\textbf{d)} \quad Q = 1.6 \times 10^{-19} \times 10^{15} \times 96.1 \times 10^{-6} \times 10^{-5}$$

$$Q = 153.3 \times 10^{-15}C$$

It is worth noting that the depletion layer extends much deeper into the region with the lighter doping as shown by Eq. (5.19) and as calculated in the preceding example.

5.3 CONDITIONS IN THE DIODE WITH VOLTAGE APPLIED

Biasing a diode is the process of connecting a voltage source (or a current source) between the metallic contacts of the N and P regions. The biasing voltage, depending upon its direction, causes the diode to conduct either in the *forward direction*, as in the first quadrant of the characteristics of Fig. 5.1, or in the *reverse direction*. In the forward direction, a small voltage causes a large current, whereas in the reverse direction, the current is negligibly small unless the voltage is so high that breakdown occurs, as shown in the large increase in the current at a certain fixed voltage in the third quadrant.

Forward bias is achieved by connecting the positive lead of the voltage source to the P region contact and the negative lead to the N region contact. Obviously, a diode is said to be reverse-biased when the opposite connections are made. We will determine the effect of the biasing on the characteristics of the space-charge region of Fig. 5.4.

When a forward bias is applied to the diode by making P positive with respect to N, as shown in Fig. 5.8, the question is: How is this voltage distributed across the diode? There are five places that can share the voltage as determined by the ohmic drops. The two aluminum contacts to each of the P and N regions have very low resistivity and hence low resistance so that there are negligible voltage drops across them. On comparing the conductivities of the three remaining regions, we find that the neutral regions have much higher conductivities, hence lower resistance, compared to the depletion region. The neutral regions have an abundance of carriers

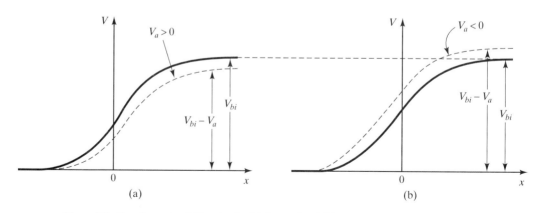

Figure 5.8   Junction potential barrier for (a) forward and (b) reverse bias.

and the space-charge region is depleted of carriers, causing it to have a high resistance. Thus, the applied voltage, at normal current levels, may be assumed to appear totally across the depletion region.

The applied voltage V_a, opposes the built-in voltage, V_{bi}, when V_a is connected, to make P positive with respect to N and V_a aids the built-in voltage when it is connected to make N positive with respect to P. This is shown in Fig. 5.8(a), where the former connection is labeled *forward bias*, and shown in Fig. 5.8(b), where the latter connection is labeled reverse bias.

Because the applied voltage adds to or subtracts from the built-in voltage, the electric field intensity must change. This necessitates a change in the amount of charge in each side of depletion layer. The dopings N_A and N_D cannot change, thus a change in the width of the depletion layer must accompany the application of a voltage to the diode. Applying a forward bias reduces the electric field intensity, thus reducing the charge and reducing the width of the depletion layer, whereas a reverse bias increases that width. The effects of biasing on charge distributions, on the electric field, on the potential distribution, and on the energy bands are shown in Figs. 5.9 and 5.10.

Since we concluded that all the applied voltage, at least at the normal current levels, appears across the depletion layer, the voltage, V_j, across this layer becomes

$$V_j = V_{bi} - V_a \tag{5.29}$$

where V_{bi} is the barrier voltage and V_a is the applied voltage (with V_a being positive for forward bias and negative for reverse bias).

The expression for the depletion layer width in Eq. (5.28) is modified by replacing V_{bi} by V_j as

$$W = \left[\frac{2\varepsilon}{q}\left(\frac{N_A + N_D}{N_A N_D}\right)V_j\right]^{\frac{1}{2}} \tag{5.30}$$

It is important to remind the reader that the derivations in this chapter are based on the assumption that the P and N regions form a step junction, or an *abrupt junction*, at the metallurgical boundary.

REVIEW QUESTIONS

Q5-1 For a PN junction diode in thermal equilibrium, explain why each of the electron and hole currents are zero.

Q5-2 Briefly explain why the section separating the P and N regions is labeled (a) a space charge region (b) a depletion region.

Q5-3 A voltage appears across the depletion region of a PN junction diode in thermal equilibrium. Can this voltage be measured by connecting a voltmeter to the metal terminals of the diode?

Q5-4 Since a voltage appears across the depletion region, which region has the higher voltage, the P or the N region?

Q5-5 The depletion region extends deeper into the lightly doped (N) region of the PN diode. Explain why.

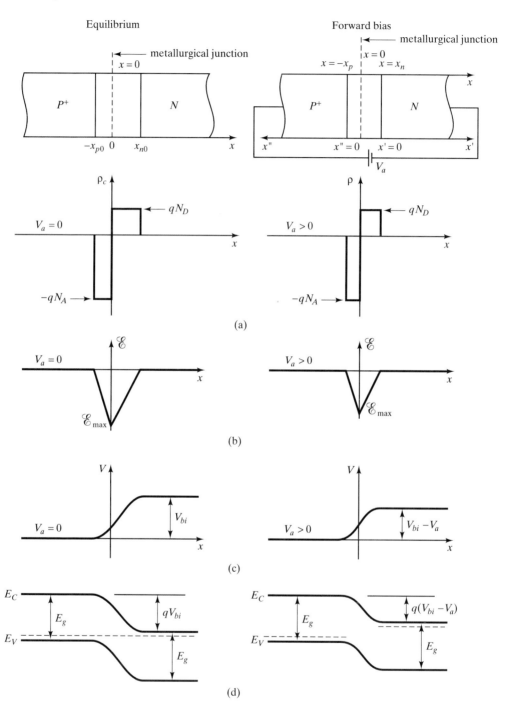

Figure 5.9 Distributions of (a) charge density, (b) electric field, (c) potential, and (d) energy at equilibrium and at forward bias ($V_a > 0$) for $N_A \gg N_D$.

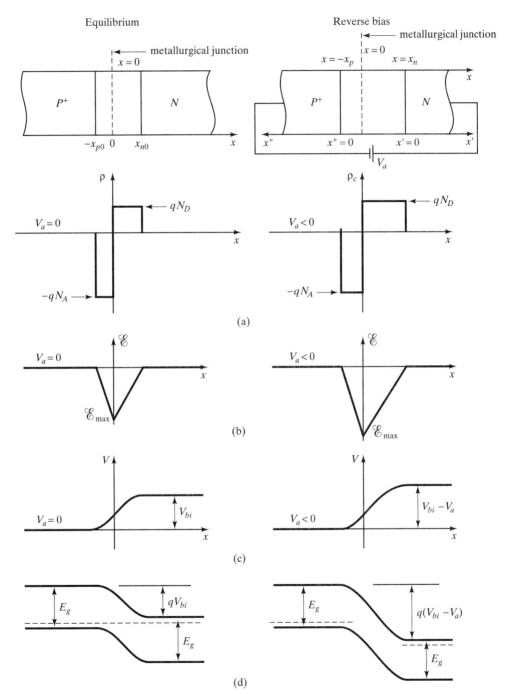

Figure 5.10 Distributions of (a) charge density, (b) electric field, (c) potential, and (d) energy at equilibrium and at reverse bias ($V_a < 0$) for $N_A \gg N_D$.

Q5-6 In Fig. 5.3(a), the change from the P side of the depletion layer to the N side looks perfectly abrupt. This is not realistic. Why?

HIGHLIGHTS

- When a P semiconductor and an N semiconductor are placed in intimate contact, holes diffuse (rush) from P to N and electrons rush from N to P.
- As a result of the transfer of carriers, positively charged donor ions are left behind on the N side of the junctions and negatively charged acceptor ions accumulate on the P side. The whole region that consists of ions is known as the space charge region.
- The existence of positive charges on one side and negative charges on the other side causes an electric field directed from N to P.
- The electric field acts to force electrons to drift from P to N and holes from N to P so that a balance is established at which each of the hole and electron currents is zero.
- The existence of an electric field between $-x_{p0}$ and x_{n0} requires a voltage difference between the P and N regions. This voltage is known as the built-in voltage.
- Because a voltage difference is formed between the P and N regions, an energy difference exists between them. Thus, an energy hill or barrier is established across the depletion region. Consequently, the energy bands are bent and the Fermi levels of the two regions are aligned.

EXERCISES

E5-1   A PN$^+$ silicon junction diode is doped with $N_D = 10^{16} \text{cm}^{-3}$ and $N_A = 10^{15} \text{cm}^{-3}$.
   a)   Determine the location of the Fermi levels, with respect to the bottom of the conduction band, in each of P and N.
   b)   Show that the separation between the Fermi levels is equal to the built-in voltage calculated from the doping.
   Ans: **a)** for $N, E_C - E_F = 0.21\text{eV}$
         for $P, E_C - E_F = 0.86\text{eV}$

E5-2   For the diode of the above exercise, determine:
   a)   The width of the portion of the depletion region in N.
   b)   The total width of the depletion layer.
   c)   The electric charge in each of the two segments of the depletion region given the area is $0.2 \times 0.3\text{cm}$.
   Ans: **a)** $0.088\mu\text{m}$
         b) $0.968\mu\text{m}$

5.4 CURRENTS IN DIODE

Motion of Carriers with Bias Applied

In this section, we will derive the expression for the current-voltage characteristic of the PN junction diode. First, we will analyze qualitatively the currents across the junction both at equilibrium and with an applied voltage. Let us establish some rules concerning the location and motion of the carriers:

1. Electrons in the conduction band have potential energy E_C and kinetic energy $(E - E_C)$, where E is the energy level located above E_C of the electron. Energy of electrons is measured upwards in the energy diagram.

2. Holes in the valence band have potential energy E_V and kinetic energy $(E_V - E)$, where E is the energy level of the hole. Energy of the hole is measured downwards in the energy diagram.

3. Electrons in P and holes in N are minority carriers. They move across the junction because of the electric field at the junction with electrons moving from P to N and holes from N to P. These are drift currents.

4. Majority carriers move across the junction, electrons from N to P and holes from P to N, because of the large concentration gradients at the junction. These gradients result from the large concentration of electrons in N and holes in P, and the small density of holes in N and electrons in P. Currents resulting from these gradients are diffusion currents.

5. When electron and holes move from one region to another, across the depletion region, their motion occurs at a fixed energy level E. In this motion, while the total energy is constant, the energy of the carrier changes from potential to kinetic or vice versa.

At thermal equilibrium, each of the electron and hole currents is zero. The electron diffusion current caused by electrons moving from N to P is equal and opposite to the electron drift current caused by electrons from P to N. The same analogous processes occur for holes. A schematic diagram to illustrate the motions is shown in Fig. 5.11.

Two properties of the figures in the following pages require mentioning here. First, we observe in Fig. 5.10 that the magnitude and sign of the slope of the energy band diagram in the depletion layer are a measure of the intensity and direction of the electric field respectively. We note the electric field is directed from right to left. At forward bias, and as shown in Fig. 5.12, the slope has decreased, indicating a decrease in the magnitude of the electric field intensity as compared to the condition at equilibrium.

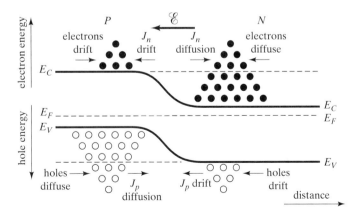

Figure 5.11   Illustration of carrier motion at equilibrium.

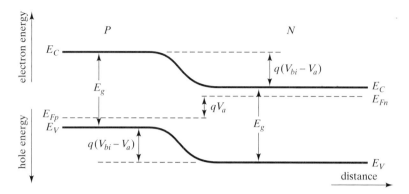

Figure 5.12 Energy band diagram with forward bias. Note the negative slopes of the energy levels in the space-charge region. The slope decreases at forward bias, reflecting a decrease in the electric field.

The second item is the use of filled and hollow dots to represent electrons and holes arranged in pyramid-like structures in the conduction band for electrons and in the valence band for holes. Pictorially, these represent the energy distributions of the carrier densities that we have previously determined analytically from the product of the density of states function and the Fermi-Dirac distribution. The distributions of carrier densities represent an exponentially decreasing function of energy, as measured upwards from the bottom of the conduction band for electrons and as measured downwards from the top of the valence band for holes.

When a bias is applied to the PN junction diode, we assume that the P region is fixed in energy and we allow the N-region energy levels to move up for forward bias and down for reverse bias. Since the Fermi levels in the neutral regions are fixed with respect to E_C and E_V, the relations for the energy distributions of the carriers apply and the number of carriers available for transfer from one region to another varies as the energy levels move up or down.

Conditions with Forward Bias

When a forward bias is applied to the diode, the electric field at the junction is reduced and so is the voltage across the junction. If we assume that the energy levels in the P region remain fixed, then the energy levels in the N region are raised by qV_a, as shown by Fig. 5.12.

The number of electrons, minority carriers in the P region, above E_C, is the same as that at equilibrium. They constitute the electron drift current. The number of electrons in N that have energies above the E_C of the P region is considerably greater than that at equilibrium. These electrons will diffuse to P. Thus, we have an electron diffusion current across the junction that is greater than the drift current caused by electrons moving from P to N. Analogously, holes will diffuse from P to N and this diffusion current is much greater than the drift current due to holes from N to P.

An illustration of the carrier motions on the energy band diagram is shown in Fig. 5.13.

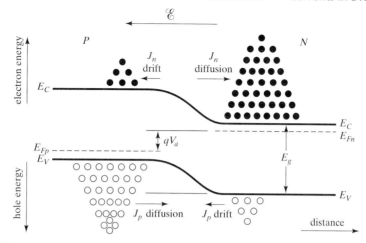

Figure 5.13   Carrier motions with forward bias ($V_a > 0$). Note the larger number of holes in P and electrons in N that are available for diffusion (compared to equilibrium).

Conditions With Reverse Bias

When a reverse bias is applied, the drift currents do not change since the density of minority carriers has not changed. The slope of the majority carrier density distribution across the junction is still very large but the currents due to diffusion are small because of the smaller number of these carriers, electrons in N and holes in P, that have energies that are greater than the E_C of P and E_V of N respectively. This is illustrated in Fig. 5.14. Stated otherwise, electrons in N and holes in P have a higher energy barrier to overcome, compared to the conditions at equilibrium, so that the net current is negative. This current is negative because it is composed of a larger

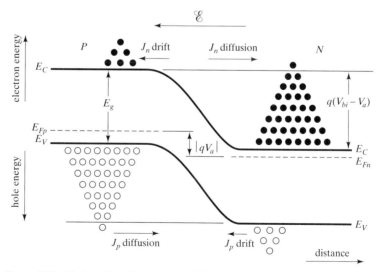

Figure 5.14   Motion of carriers at reverse bias ($V_a < 0$). Note the smaller number of holes in P and electrons in N that are available for diffusion (compared to equilibrium).

number of electrons crossing from P to N than those crossing from N to P and, similarly, a larger number of holes crossing from N to P than from P to N, all compared to equilibrium.

Thus, the current is limited mainly by the availability of thermally generated minority carriers and is, in fact, independent of the applied bias for an applied reverse bias of a few tenths of a volt or larger.

Assumptions for Ideal Diode Equation

The expression we will derive for the current-voltage characteristic of the diode is labeled ideal because of the following assumptions that are made:

1. The space-charge region boundaries represent a step-junction from the bulk P and N regions.
2. No carriers exist in the space-charge region, they just traverse it. This region consists of ionized impurities only.
3. In the bulk of the diode, outside the depletion layer, the semiconductor is neutral. This is valid due to the limited conductivity that causes the ohmic voltage drop to be negligibly small.
4. Operation is at a temperature such that all impurity atoms are ionized.
5. Perfect ohmic contacts are made to the ends of the P and N regions. A perfect ohmic contact is one that is assumed to have zero resistance and allows easy current flow in both directions so that the voltage across the contact is zero. As we shall see later, the minority carrier density at the ohmic contact is the value at equilibrium.
6. When the diode is forward-biased, low injection is assumed. This means that as electrons from N and holes from P are injected into the opposite region, the density of the carriers at the boundary of the new region, where they become minority carriers, is much smaller than the equilibrium density of the majority carriers in that region.
7. No recombination or generation takes place in the depletion region so that both electron and hole current are constant across this region.

We will derive the diode equation by determining expressions for the minority carrier currents in each of the neutral regions. The sum of the minority carrier currents at the junction constitutes the total current.

The reason for using the minority carriers in each region is because of assumption three (the electric field, and hence the voltage across the neutral regions, is zero). This is not quite true, as we will determine. There is an electric field that causes, at low-injection levels, negligible minority carrier drift currents but large majority carrier drift currents.

Hence, the minority carrier currents in the neutral region result from diffusion only. The procedure we will follow consists of determining expressions for the distribution of minority carriers, subject to the boundary conditions, from the continuity equations. Expressions for the minority carrier currents will be obtained as a function of distance. The two minority carrier currents at the edges of the depletion layer will be added to obtain the total current.

Solution of Continuity Equation

We refer to the continuity equation, Eq. (4.36) for holes in the N region, in one dimension, in the steady state and replace J_p by its diffusion component, $-qD_p\, dp/dx$, so that

$$-\frac{p - p_{0n}}{\tau_p} + D_p \frac{d^2 p}{dx^2} = 0 \qquad (5.31)$$

where p is the hole density at x, p_{0n} is the equilibrium hole density in N, and τ_p is the lifetime of holes in the N region.

We let p' represent the excess hole density, $(p - p_{0n})$, and Eq. (5.31) becomes

$$\frac{d^2 p'}{dx^2} = \frac{p'}{D_p \tau_p} \qquad (5.32)$$

In order to make the expressions less cumbersome, we will label the point x_n as our new point of reference, $x' = 0$, so that $x' = x - x_n$, where x_n and $-x_p$ are the limits of the depletion layer with bias applied in contrast to x_{n0} and $-x_{p0}$, which are for thermal equilibrium. The general solution to Eq. (5.32) becomes

$$p' = B_1 \exp\!\left(\frac{-x'}{L_p}\right) + B_2 \exp\!\left(\frac{x'}{L_p}\right) \qquad (5.33)$$

where B_1 and B_2 are constants of integration and L_p, known as the diffusion length for holes in the N region, is given by

$$L_p = \sqrt{D_p \tau_p} \qquad (5.34)$$

The diffusion length represents the average distance that excess minority carriers, holes in this case, diffuse before they recombine. To determine B_1 and B_2, we need two boundary conditions. They are

$$p'(0) = p_{0n}\left[\exp\!\left(\frac{qV_a}{kT}\right) - 1 \right] \text{at } x' = 0$$

$$p'(W_n) = 0 \qquad\qquad \text{at } x' = W_n \qquad (5.35)$$

The voltage V_a is the applied voltage and W_n is the distance in the N region from the edge of the depletion layer to the ohmic contact.

We shall indicate in a subsequent section* how we arrive at the first boundary condition. The second boundary condition indicates that at the end of the N region, the excess carrier density $(p - p_{0n})$ is zero. In other words, the hole density takes on its equilibrium value. This is not due to recombination in the bulk of the N region but rather to the ohmic contact at the end of the N region. Even if the width of the N region is so small that no recombination occurs, the hole density at the ohmic contact will be p_{0n}. The ohmic contact is a surface of very high recombination rate because of the great abundance of electrons.

When we use the conditions of Eqs. (5.35) in Eq. (5.33), we obtain

---

*See subsection titled "Boundary Condition at Junction."

$$p' = \frac{p_{0n}\left[\exp\left(\frac{qV_a}{kT}\right)-1\right]\left[\exp\left(\frac{x'}{L_p}\right)-\exp\left(\frac{2W_n-x'}{L_p}\right)\right]}{1-\exp\left(\frac{2W_n}{L_p}\right)}$$ (5.36)

Before we proceed further, let us distinguish two cases: $W_n \gg L_p$ and $W_n \ll L_p$

$$W_n \cong 10L_p \text{ and } W_n \cong 0.1L_p$$

Replacing p' by $(p - p_{0n})$ and x' by $(x - x_n)$, the excess hole density in the N region for each of the two cases becomes*

$$p' = p - p_{0n} = p_{0n}\left[\exp\left(\frac{qV_a}{kT}\right)-1\right]\exp-\left(\frac{x-x_n}{L_p}\right) \text{ for } W_n \gg L_p \quad \text{(a)}$$

$$p' = p - p_{0n} = p_{0n}\left[\exp\left(\frac{qV_a}{kT}\right)-1\right]\left(1-\frac{x-x_n}{W_n}\right) \text{ for } W_n \ll L_p. \quad \text{(b)}$$ (5.37)

The excess electron density in the P region, for the condition that the width of the N region is much greater than L_n, is analogously given by

$$n' = n - n_{0p} = n_{0p}\left[\exp\left(\frac{qV_a}{kT}\right)-1\right]\exp\left(\frac{x+x_p}{L_n}\right)$$ (5.38)

where n' is the excess electron density in the P region, n is the electron density in the P region, and n_{0p} is the equilibrium value of the electron density in the P region.

Sketches of the carrier distributions are shown in Fig. 5.15, where $p_n(0)$ is the value of p at $x = x_n(x' = 0)$ and $n_p(0)$ is the value of n at $x = -x_p$ $(x'' = 0)$. The relations in Eqs. (5.37) for the minority carrier densities are valid at equilibrium as well as for forward and reverse bias. For reverse bias, V_a is negative and from Eq. (5.37) the value of p at $x = x_n$ is zero. Also, zero is the value of n at $x = -x_p$.

Currents Crossing Junction

Since we have concluded that minority carriers cause only a diffusion current, the expression for the hole current density in N becomes

$$J_p = -qD_p\frac{dp}{dx} = -qD_p\frac{dp'}{dx}$$ (5.39)

By differentiating Eqs.(5.37) and replacing the derivatives in Eq.(5.39), we obtain

$$J_p(x') = \frac{qD_pp_{0n}}{L_p}\left[\exp\left(\frac{qV_a}{kT}\right)-1\right]\exp\left(\frac{-x'}{L_p}\right) \text{ for } W_n \gg L_p \quad \text{(a)}$$

$$J_p(x') = \frac{qD_pp_{0n}}{W_n}\left[\exp\left(\frac{qV_a}{kT}\right)-1\right] \text{ for } W_n \ll L_p \quad \text{(b)}$$ (5.40)

At $x' = 0$ or $x = x_n$, the edge of the transition layer, Eq.(5.40(a)), becomes

$$J_p(x_n) = \frac{qD_pp_{0n}}{L_p}\left[\exp\left(\frac{qV_a}{kT}\right)-1\right] \text{ for } W_n \gg L_p$$ (5.41)

---

*Details shown in the last section of this chapter.

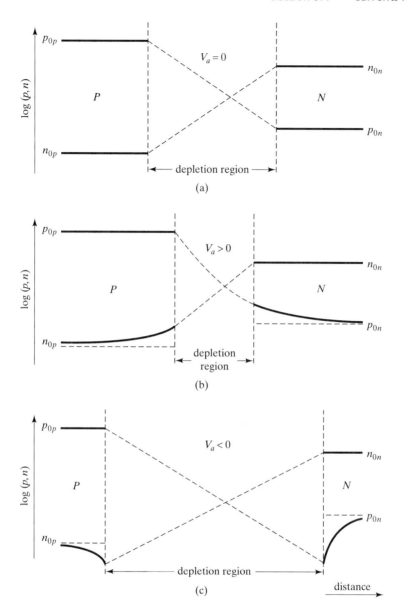

Figure 5.15   Distributions of carrier densities at (a) equilibrium, (b) forward bias for $W_n \gg L_p$, and (c) reverse bias. Note the change in the width of the depletion layer with bias (highly exaggerated).

where D_p, p_{0n} and L_p refer to holes in the N region and W_n is the length of the neutral N region.

An analogous derivation for the electron current density at the edge of the transition layer in the P region gives

$$J_n(-x_p) = \frac{qD_n n_{0p}}{L_n}\left[\exp\left(\frac{qV_a}{kT}\right)-1\right] \text{ for } W_p \gg L_n \tag{5.42}$$

where D_n, n_{0p}, and L_n refer to electrons in the P region, and W_p is the length of the neutral P region. Since we have assumed that the minority carrier currents are continuous across the transition layer, then the total current density across the junction, which is also the total current density throughout the diode, becomes $[J_p(x_n) + J_n(-x_p)]$.

The diode current becomes

$$I = A[J_p(x_n) + J_n(-x_p)]$$

$$I = A\left[\frac{qD_p p_{0n}}{L_p} + \frac{qD_n n_{0p}}{L_n}\right]\left[\exp\frac{qV_a}{kT}-1\right] \quad \begin{array}{l} \text{for } W_n \gg L_p \\ \text{and } W_p \gg L_n \end{array} \tag{5.43}$$

For a reverse-biased diode, with V_a negative, the expression for the current is $I = -I_S$, where I_S is the *reverse saturation current* given by

$$I_S = qA\left[\frac{D_p p_{0n}}{L_p} + \frac{D_n n_{0p}}{L_n}\right] \quad \begin{array}{ll} \text{for} & W_n \gg L_p \\ \text{and} & W_p \gg L_n \end{array} \tag{5.44}$$

When $W_n \ll L_p$, the hole current, given by Eq.(5.40(b)), is constant throughout the N region. If it is also assumed that the width of the P region is much smaller than L_n, then the diode current expression given by Eq.(5.43) applies—except that L_p and L_n are replaced by the widths, W_n and W_p, of the N and P regions respectively.

For a P$^+$N diode, $N_A \gg N_D$ so that $n_{0p} \ll p_{0n}$, the second term within the first set of brackets in Eq. (5.43) and Eq. (5.44), becomes negligibly small compared to the first term.

Sketches of the distributions of current components are shown in Fig. 5.16. We have assumed that there is no recombination in the depletion layer, hence the minority carrier current densities at $x = x_n$ for J_p and at $x = -x_p$ for J_n have the same values as those at $x = -x_p$ and $x = x_n$ respectively. In the region where the current is carried by majority carriers, we obtain the majority current density as the difference between the total current density and the minority carrier current density. The total current density throughout the diode, J, is determined by the summation of J_n and J_p in the depletion region.

In the following example, we illustrate the application of Eq.(5.43).

EXAMPLE 5.2

Given a silicon diode that has the following properties:
N_A in P $= 2 \times 10^{17} \text{cm}^{-3}$
N_D in N $= 2 \times 10^{16} \text{cm}^{-3}$
Junction area $= 2 \times 10^{-3} \text{cm}^2$

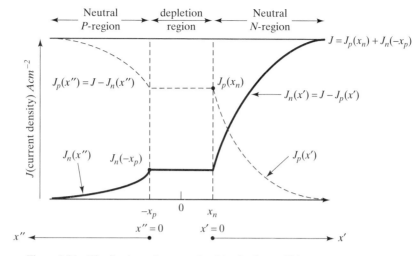

Figure 5.16 Distributions of current densities for forward bias.

Lifetime of holes in N region = 10μs
Lifetime of electrons in P region = 50μs
D_n of electrons in P region = 34 cm²/s.
D_p of holes in N region = 13 cm²/s.
    For V_a = 0.66, volt calculate: $J_n(-x_p), J_p(x_n)$ and the total diode current. Operation is at 300K.

Solution
The intrinsic carrier density for silicon at 300K is 1×10^{10}cm⁻³. Because N_A and N_D are much greater than $n_i, n_{0n} \approx N_D = 2 \times 10^{16}$cm⁻³ and $p_{0p} \approx N_A = 2 \times 10^{17}$cm⁻³. Therefore,

$$p_{0n} = \frac{n_i^2}{n_{0n}} = \frac{1 \times 10^{20}}{2 \times 10^{16}} = 5 \times 10^3 \text{ cm}^{-3}$$

$$n_{0p} = \frac{n_i^2}{p_{0p}} = 5 \times 10^2 \text{cm}^{-3}$$

Using Eq.(5.34), we calculate

$$L_p = (D_p \tau_p)^{1/2} = (10 \times 10^{-6} \times 13)^{1/2} = 11.4 \times 10^{-3}\text{cm}$$

Similarly, $L_n = (D_n \tau_n)^{1/2} = 4.1 \times 10^{-2}$cm.
    By using Eqs.(5.41) and (5.42), we calculate J_p and J_n, keeping in mind that x_n is at $x' = 0$ and $-x_p$ is at $x'' = 0$.

$$J_p(x_n) = \frac{1.6 \times 10^{-19} \times 13 \times 5 \times 10^3}{11.4 \times 10^{-3}} \left[\exp\left(\frac{0.66}{.0259}\right) - 1 \right]$$

$$J_p(x_n) = 0.1064 \text{ A/cm}^2$$

and

$$J_n(-x_p) = 0.00774 \text{ A/cm}^2$$

The total diode current becomes

$$I = A\,[J_p(x_n) + J_n(-x_p)] = 0.228\text{mA}$$

It is worthwhile noting that the much larger hole component of the current is a result of the much higher doping of the P region as compared to that of the N region.

The Current Loop

In the analysis of the currents in the diode, we have concentrated on the diffusion of minority carriers when a forward bias is applied to the diode. Naturally, the inquisitive student will ask: What happens to the carriers once outside the depletion region, in the neutral regions, at the metallic contacts, and in the outside circuit?

Consider first a forward-biased diode with the positive terminal of the battery connected to the metallic contact of the P region and the negative terminal connected to the metallic contact of the N region. The metallic contact is an abundant source of electrons and thus the metal-semiconductor contact is a mechanism for exchanging electrons and holes. A sketch of the diode showing carrier motion is shown in Fig. 5.17.

At the P side, electrons are drawn by the positive terminal of the battery to be circulated in the wires external to the diode. The electrons come from the metal contact and these are replenished by other electrons from the P semiconductor. Those electrons that were carried away from the semiconductor contact created holes behind them. These holes travel to the depletion layer edge where they diffuse to the N region. In their quick journey in the P region, some of these holes recombine with the minority carrier electrons that have diffused from the N region across the depletion layer.

The holes that diffuse from P to N recombine with electrons supplied from the metal-N semiconductor contact that had been extracted from the metallic contact at the P region. Since electrons are majority carriers in the N region, only a few of them will recombine with holes. The remaining will proceed to the depletion layer to diffuse to the P region.

In Fig. 5.16, we identified J_n in P and J_p in N as minority carrier diffusion currents. We obtained the variation of J_n in N and J_p in P by an indirect method. The question is: What is the mechanism of conduction by which these currents occur?

We refer to the earlier section on the ideal diode equation, in which we assumed that the drift current of minority carriers in the neutral regions was negligible because of the very low electric field intensity in those regions when low injection is assumed. In spite of the very small magnitude of the electric field intensity, it does cause large drift currents, J_n in N and J_p in P, of the majority carriers. The majority carrier densities are much larger than the minority carrier densities so that the product of the majority carrier density and the electric field intensity results in large drift currents.

For a reverse bias, the carriers that cross the depletion layer, and as was shown in Fig. 5.13, are electrons from P to N and holes from N to P. This is in contrast to the forward-biased diode in which electrons diffuse from N to P and holes from P to N.

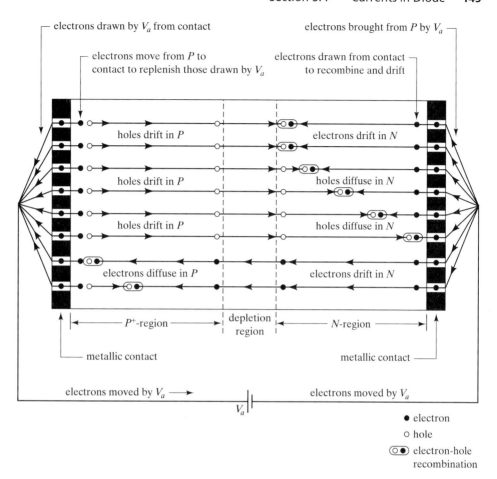

Figure 5.17   Current loop in forward-biased P$^+$N junction diode.

The minority carriers, in reverse bias, drift across the depletion layer (they all fall down the very steep barrier) because of the large electric field that is established.

Of the electron-hole pairs generated on the P side of the depletion layer, electrons fall into the N side and holes migrate towards the metal P semiconductor contact to meet the electrons that were transported there by the applied bias. They recombine there and the resulting hole density is the equilibrium value, whereas at the edge of the depletion layer, the hole density is zero. The electrons generated on the P side drift to the N region, where they migrate to the metallic contact and are attracted to the outside circuit by the positive terminal of the battery, which is now connected to the N region.

Similarly, for the electron-hole pairs generated in the N side of the depletion layer, holes drift down the potential hill across the depletion layer to the contact in P and electrons migrate towards the N semiconductor-contact to enter the external circuit and move towards the metal-P semiconductor contact.

Saturation Current

The current-voltage characteristics of diodes in the forward direction, subject to the assumption of low injection, and for a fixed forward bias, are determined by the magnitude of the reverse saturation current given by Eq.(5.44). For two diodes having identical doping densities and identical minority carrier lifetimes, the current depends on the minority carrier densities and on the mobilities of minority carriers through the diffusion constants. For two such diodes, the current depends on the semiconductor band gap, the mobility of minority carriers, and on the temperature.

A comparison of the characteristics of silicon, germanium, and gallium arsenide diodes having the same areas, the same lifetimes, and the same dopings is shown in Fig. 5.18.

The smaller band gap of germanium results in a higher intrinsic carrier density, thus higher minority carrier densities, p_{0n} and n_{0p}. This causes an increase in the saturation current and hence an increase in the forward and reverse currents. Another reason for the higher current of the germanium diode is the higher mobility of carriers in germanium when compared to that in silicon, which is evidenced in the higher diffusion constant, D, which appears in Eqs. (5.44).

In contrast to germanium and silicon, gallium arsenide has a smaller value of intrinsic carrier density, thus lower minority carrier densities while having a much higher mobility of electrons. It is the difference in the band gaps of the three materials that mainly accounts for the differences in the shapes of the current-voltage characteristics. At room temperature, the band gap of germanium is 0.67eV and of silicon is 1.12eV, whereas it is 1.42eV for gallium arsenide.

Another important factor that influences the diode current is temperature. An increase of the operating temperature exerts the following effects:

- an increase of the intrinsic carrier densities.
- a decrease of the band gaps in accordance with Table 2.2.

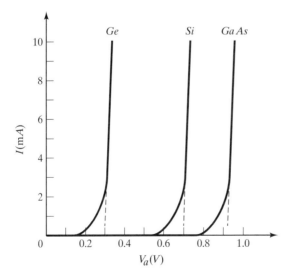

Figure 5.18 Current-voltage characteristics of silicon, germanium, and gallium arsenide diodes.

- a decrease of the mobilities.

The first two effects cause an increase in the current, whereas the decrease of mobility decreases the current. The increase of temperature decreases the value of the exponential in Eq. (5.43), resulting in a decrease of the current. The net result of all the factors is such that an increase of temperature tends to increase the current. This increase of current could result in instability since the power dissipated in the diode acts, in turn, to increase the temperature further.

Boundary Condition at Junction

The expression for the built-in voltage (barrier voltage), V_{bi}, given by Eq.(5.15) relates the equilibrium hole densities on both sides of the space-charge region. This dependence is also obtained in terms of the relevant electron densities. Equation (5.15) can be rewritten as

$$\frac{p_{0p}}{p_{0n}} = \frac{n_{0n}}{n_{0p}} = \exp\frac{qV_{bi}}{kT} \tag{5.45}$$

This expression was derived for equilibrium conditions where each of the hole and electron currents across the junction are zero, making each carrier's diffusion current equal and opposite to that carrier's drift current across the junction.

It is of interest to calculate using the data of Example 5.2, the magnitudes of the diffusion currents at equilibrium. The magnitudes of the hole and electron diffusion current densities across the depletion region, assuming linear distributions of holes and electrons, are

$$J_p(x' = 0) = -qD_p\frac{dp}{dx'} = qD_p\frac{p_{0p} - p_{0n}}{W} \quad \text{(a)}$$

$$J_n(x'' = 0) = qD_n\frac{dn}{dx''} = qD_n\frac{n_{0n} - n_{0p}}{W} \quad \text{(b)} \tag{5.46}$$

where W is the width of the space-charge region and the values of the carrier densities are those at the edges of the region.

REVIEW QUESTIONS

Q5-7 Explain the movement of the energy bands when a diode is forward-biased.

Q5-8 Some holes diffuse throughout the N region. Explain what happens to them when they arrive at the metal contact of that region.

Q5-9 What is meant by low-level injection?

Q5-10 The diode equation (5.43) is sometimes labeled the diffusion equation. Explain.

Q5-11 Define the term "diffusion length" of a minority carrier.

Q5-12 If the diffusion length of holes in N, L_p, is much greater than W_n, what is the shape of the excess hole distribution in N as determined from Eq. (5.37)?

Q5-13 Briefly explain the difference between the I-V relationships of a silicon and a gallium arsenide diode.

HIGHLIGHTS

- In a PN diode at thermal equilibrium, a built-in voltage is generated at the junction that makes N positive with respect to P. Consequently, the drift components for each of the electron and hole currents are balanced by equal and opposite diffusion currents so that each of the electron and hole currents are zero.

- When a voltage is applied that makes P positive with respect to N, the charge in the depletion layer is reduced, the electric field is reduced, and so is the voltage across it. The diode is said to be forward-biased.

- At forward bias, the hole density in N at the edge of the depletion layer is increased and a hole gradient is established throughout the N region. Analogously, electrons in P have a similar distribution. As a result of the gradients of holes in N and electrons in P, holes diffuse in N and electrons diffuse in P.

- To determine the analytical relationship for the current in a diode, expressions for the hole and electron diffusion currents are required at the edges of the N and P regions respectively. Expressions for the diffusion currents are obtained from a knowledge of the distributions of minority carriers in both regions and these distributions are determined from a solution of the continuity equations for these carriers.

- The sum of the hole diffusion current at the edge of the depletion layer in N and the electron diffusion current at the edge of the depletion layer in P results in the exponential I-V equation.

- When the voltage that is applied to the diode is connected to make N positive with respect to P (reverse bias), the voltage across the depletion layer increases, the electric field increases, and the energy barrier increases, making it more difficult for holes to go from P to N and electrons to cross from N to P. It becomes easy for electrons in P and holes in N to cross to the other region. Thus, a very small current results, opposite in direction, to the current caused by the forward bias.

EXERCISES

E5-3   Assume the distribution of excess hole diffusion density in the N regions is given $p' = p'(0)\,(1 - x_n/W_n)$, where $p'(0)$ is the excess hole density at $x_n = 0$ and W_n is the width of the N region.

Derive an expression for the hole diffusion current in N.

E5-4   An N^+P silicon diode has $N_D = 10^{18}\text{cm}^{-3}$ and $N_A = 10^{16}\text{cm}^{-3}$ with $W_p \gg L_n$. Given $\tau_p = \tau_n = 1\mu s$ and $A = 10^{-3}\text{cm}^2$.

Determine the diode currents at (a) $V_a = 0.6V$ and (b) $V_a = 0.66V$

Ans: a) $I = 0.107\text{mA}$

EXAMPLE 5.3

Calculate, for example 5.2, the hole and electron diffusion current densities at thermal equilibrium.

Solution   The expressions for the width of the space-charge region and the barrier voltage are given by Eqs. (5.28) and (5.17) respectively as

$$W = \left[\frac{2\varepsilon(N_A + N_D)}{q\,N_A N_D}V_{\text{bi}}\right]^{1/2} \quad \text{and} \quad V_{\text{bi}} = \frac{kT}{q}\,\ell n\frac{N_A N_D}{n_i^2}$$

The value of V_{bi} is calculated to be 0.811V and W is 2.41×10^{-3}cm so that, at equilibrium, the hole and electron current densities at the edges of the depletion region are calculated using Eq. (5.46) to be

$$J_p(x_{n0}) = \frac{1.6 \times 10^{-19} \times 13 \times (2 \times 10^{17} - 5 \times 10^2)}{2.41 \times 10^{-5}} = 17.26 \times 10^3 \text{A/cm}^2$$

$$J_n(x_{p0}) = 1.6 \times 10^{-19} \times 34 \times \frac{(2 \times 10^{16} - 5 \times 10^3)}{2.41 \times 10^{-5}} = 4.51 \times 10^3 \text{A/cm}^2$$

These values are much greater than those calculated for forward bias in Example 5.2.

It is quite evident that the diffusion currents at equilibrium are many orders of magnitude greater than those at forward bias for low injection. It follows that the drift currents at equilibrium have values that are much larger than the currents at forward bias. *Thus, the values of the diffusion currents calculated at forward bias represent the difference between large diffusion and drift components.*

Using the results and conclusions of this section, we will verify the boundary condition that we used in Eq.(5.35), namely that

$$p'(0) = p_{0n} \left[\exp\left(\frac{qV_a}{kT}\right) - 1 \right]$$

where $p'(0) = (p(0) - p_{0n})$ is the excess hole density in N at the edge of the space-charge region.

The assumption of low-level injection has resulted in a hole current across the junction that is the difference between two large hole current components at equilibrium, namely J_p (drift) and J_p (diffusion).

Equation (5.15) was derived assuming, in Eq. (5.12), that at equilibrium the drift and diffusion currents are equal. Since that assumption is also valid at forward and reverse bias, we replaced V_{bi} (at equilibrium) in Eq. (5.15) by $(V_{bi} - V_a)$ so that Eq. (5.15) may be rewritten as

$$V_j = V_{bi} - V_a = \frac{kT}{q} \ell n \frac{p(-x_p)}{p(x_n)} \tag{5.47}$$

so that

$$p(-x_p) = p(x_n)\left(\exp\frac{qV_{bi}}{kT}\right)\exp\left(\frac{-qV_a}{kT}\right) \tag{5.48}$$

Since from Eq.(5.15), $\exp\dfrac{qV_{bi}}{kT} = \dfrac{p_{0p}}{p_{0n}}$, we have

$$p(-x_p) = p(x_n)\frac{p_{0p}}{p_{0n}}\exp\left(\frac{-qV_a}{kT}\right) \tag{5.49}$$

The value of the hole density in P at the edge of the depletion layer is, for all practical purposes, p_{0p} at low injection. So, $p(-x_p) = p_{0p}$ and from Eq. (5.49)

$$p(x_n) = p_n(0) = p_{0n} \exp\frac{qV_a}{kT} \tag{5.50}$$

The excess hole density at $x = x_n, p'(0)$ is $p(0) - p_{0n}$, so we can write

$$p'(0) = p_{0n}\left[\exp\frac{qV_a}{kT} - 1\right]$$ (5.51)

A similar relation can be obtained for electrons in the P region.

In the next chapter, we will compare the actual diode characteristics with those described by the relations derived here. We will then discuss the causes for the deviations between the two.

General Equation for Hole Distribution in the N Region of the PN Junction Diode

The continuity equation, (Eq. (5.31), can be written as

$$\frac{d^2p'}{dx^2} = \frac{p'}{\tau_p D_p}$$ (5.52)

where p' is the excess hole distribution in N and $L_p = (D_p \tau_p)^{0.5}$
The solution to this equation is

$$p' = B_1\left[\exp\left(-\frac{x'}{L_p}\right)\right] + B_2\left[\exp\left(\frac{x'}{L_p}\right)\right]$$ (5.53)

where
$$p'(0) = p_{0n}\left[\exp\left(\frac{qV}{kT}\right) - 1\right]$$ (5.54)

Using the boundary conditions in Eq. (5.53) we have

$$p'(0) = B_1 + B_2$$ (5.55)

$$p'(W) = 0$$ (5.56)

We substitute the relation of Eq. (5.56) in Eq. (5.53) to get

$$B_1 = -B_2\left[\exp\left(\frac{2W_n}{L_p}\right)\right]$$ (5.57)

Substituting Eq. (5.56) into Eq. (5.57) and solving for B_1 and B_2, we have

$$B_2 = \frac{p'(0)}{1 - \left[\exp\left(\frac{2W_n}{L_p}\right)\right]}, \quad B_1 = \frac{-p'(0)\left[\exp\left(\frac{2W_n}{L_p}\right)\right]}{1 - \left[\exp\left(\frac{2W_n}{L_p}\right)\right]}.$$ (5.58)

Substituting these values for B_1 and B_2 into Eq. (5.53), we have

$$p' = \frac{p'(0)\left[\exp\left(\frac{x'}{L_p}\right) - \exp\left(\frac{2W_n - x'}{L_p}\right)\right]}{1 - \left[\exp\left(\frac{2W_n}{L_p}\right)\right]}.$$ (5.59)

This equation can be simplified for two cases.

Case I: For $W \gg L_p$. Eq. (5.59) becomes:

$$p' = \frac{p'(0)\left[\exp\left(\dfrac{x'}{L_p}\right)\right]}{-\exp\left(\dfrac{2W_n}{L_p}\right)} + p'(0)\left[\exp\left(-\dfrac{x'}{L_p}\right)\right] \qquad (5.60)$$

and this can in turn be rewritten as

$$p' = p'(0)\left[\exp\left(-\dfrac{x'}{L_p}\right)\right] + \frac{p'(0)\left[\exp\left(-\dfrac{x'}{L_p}\right)\right]}{-\exp\left(\dfrac{2(W_n - x')}{L_p}\right)} \qquad (5.61)$$

For values of $(W_n - x') > 3L_p$, Eq. (5.61) reduces to

$$p' = p'(0)\left[\exp\left(-\dfrac{x'}{L_p}\right)\right] \qquad (5.62)$$

Case II: $W_n \ll L_p$. For this case, we shall expand the exponential in Eq. (5.59) into a series, so that we have

$$p' = \frac{p'(0)\left[1 + \dfrac{x'}{L_p} - \left(1 + \dfrac{2W_n - x'}{L_p}\right)\right]}{1 - \left(1 + \dfrac{2W_n}{L_p}\right)} \qquad (5.63)$$

We have neglected higher order terms than the first, since $x' \ll L_p$. We rewrite Eq. (5.63) as

$$p' = \frac{p'(0)\left[1 + \dfrac{x'}{L_p} - 1 - \dfrac{2W_n}{L_p} + \dfrac{x'}{L_p}\right]}{-\dfrac{2W_n}{L_p}} = \frac{2p'(0)\left(\dfrac{x' - W_n}{L_p}\right)}{-\dfrac{2W_n}{L_p}}$$

or

$$p' = p'(0)\left(1 - \frac{x'}{W_n}\right) \qquad (5.64)$$

where $p'(0)$ is given by Eq. (5.54).

PROBLEMS

For all diodes assume: $W_n \gg L_p$ and $W_p \gg L_n$, unless otherwise indicated. Also in P and N regions $p_{0p} = N_A$ and $n_{0n} = N_D$ respectively. $T = 300K$

5.1 An abrupt junction PN germanium diode has $N_A = 10^{17}\text{cm}^{-3}$ and $N_D = 10^{15}\text{cm}^{-3}$. Assuming all impurities are ionized and at $T = 300\text{K}$ determine, at equilibrium:

    a)   The built-in voltage V_{bi}.

    b)   The widths of the depletion region, x_{n0} and x_{p0}.

    c)   The electric field at $x = 0$.

5.2 Assume abrupt junction PN diodes having $N_A = 10^{17}\text{cm}^{-3}$ and $N_D = 10^{14}\text{cm}^{-3}$. Calculate the built-in voltage at $T = 300\text{K}$ for:

    a)   A silicon diode.

    b)   A germanium diode.

5.3 At $T = 300\text{K}$, the P region of an abrupt junction diode has a resistivity of 0.1 ohm-cm and the N region has a resistivity of 1 ohm-cm. Assume $n_i = 2.5 \times 10^{13}\text{cm}^{-3}$, $\mu_n = 3600\text{cm}^2/\text{V-s}$ and $\mu_p = 1700\text{cm}^2/\text{V-s}$. Determine the built-in voltage V_{bi}.

5.4 The forward current across the depletion layer of a PN junction diode consists of electrons from N to P and holes injected from P to N. The ratio of the hole current crossing the junction to the total current is known as the injection efficiency. Determine an expression for the injection efficiency as a function of:

    a) N_A/N_D.

    b)   The ratio of the conductivity of P to that of N.

    Assume $\tau_p = 0.1\tau_n$, and $\mu_n = 2.5\ \mu_p$.

5.5 For a forward-biased P^+N abrupt junction diode, $W_n \gg L_p$, $L_p = 1\mu\text{m}$, and at $x = x_n$, $(x' = 0)$ the ratio of the hole current to the electron current, I_p/I_n, is 100 in the steady state. Determine I_p/I_n at $x' = 1\mu\text{m}$.

5.6 A germanium P^+N diode operating at $T = 300\text{K}$ has the following properties: $\tau_p = \tau_n = 10\mu\text{s}$, $N_A = 10^{19}\text{cm}^{-3}$, $N_D = 10^{16}\text{cm}^{-3}$, $D_p = 39\text{cm}^2/\text{s}$, $D_n = 26\text{cm}^2/\text{s}$ and the area $A = 1.25 \times 10^{-4}\text{cm}^2$. Determine:

    a)   The diode's current when 0.2V is applied in the forward direction.

    b)   Repeat part (a) but in the reverse direction.

    c)   Repeat part (a) for a silicon diode having the same properties as the Ge diode.

    d)   Repeat part (b) for the silicon diode.

5.7 A PN^+ silicon diode has $N_D = 10^{18}\text{cm}^{-3}$ and $N_A = 10^{16}\text{cm}^{-3}$, $\tau_p = \tau_n = 1\mu\text{s}$ and $A = 1.2 \times 10^{-4}\text{cm}^2$. Determine at $T = 300\text{K}$:

    a)   The reverse saturation current.

    b)   The current when the applied voltage in the forward direction is 0.7V.

    c)   The current when the applied voltage in the reverse direction is 0.7V.

5.8 An N^+P diode has the following properties: $N_D = 10^{16}\text{cm}^{-3}$, $N_A = 10^{14}\text{cm}^{-3}$, $A = 10^{-3}\text{cm}^2$, $W_n = 2\mu\text{m}$, $W_p = 200\mu\text{m}$, $\tau_n = \tau_p = 0.2\mu\text{s}$, $D_n = 20\text{cm}^2/\text{s}$, and $D_p = 10\text{cm}^2/\text{sec}$. Plot at $T = 300\text{K}$ the individual electron and hole currents in P, as a function of distance, for $V_a = 0.5\text{V}$, given $n_i^2 = 10^{20}\text{cm}^{-6}$.

5.9 A P^+N junction diode has the following properties: $N_A = 10^{18}\text{cm}^{-3}$, $N_D = 10^{16}\text{cm}^{-3}$, $A = 10^{-2}\text{cm}^2$, $\mu_p = 2000\text{cm}^2/\text{V-s}$, $\mu_n = 4000\text{cm}^2/\text{V-s}$, $L_p = 2 \times 10^{-2}\text{cm}$ and $L_n = 3 \times 10^{-2}\text{cm}$. Given $n_i = 10^{13}\text{cm}^{-3}$ and the relative permittivity $\varepsilon = 16$, determine:

    a)   The conductivities of the P and N regions.

    b)   The built-in voltage.

    c)   The reverse saturation current of the diode.

    d)   The diode current for a forward bias of 0.25V.

e)   The width, W, of the depletion layer when a reverse bias of 10V is applied.

5.10 A PN junction diode has a reverse saturation current of $1\mu A$ at $T = 300K$. Determine the applied voltage for currents of

a)   1mA.

b)   10mA.

5.11 We will repeat Prob. 5.10 except that we will account for the IR drops. A PN junction diode has a reverse saturation current of $1\mu A$. The resistivity of the P region 0.05 ohm-cm and that of the N region is 0.2 ohm-cm. Each region is 1mm long and has an area $A = 0.5mm^2$. Include the IR drops in the P and N regions and find the forward voltage, V_a, for

a) $I = 1mA$.

b) $I = 10mA$.

5.12 A PN junction silicon diode has resistivities for the N and P regions, $\rho_n = 0.2$ ohm-cm and $\rho_p = 1\Omega$-cm. Given that the lifetime of minority carriers in P is 10^{-6} sec. the lifetime of holes in N is 10^{-8}sec and $A = 10^{-3}cm^2$:

a)   Calculate the density of minority carriers at the edge of the depletion region in N when the applied forward voltage is 0.6V.

b)   Plot the values of the majority and minority carrier currents as functions of the distance from the junction, on both sides of the depletion layer, for $V_a = 0.6V$.

c)   Locate the plane in N at which the majority-carrier current equals the minority-carrier current for $V_a = 0.6V$.

5.13 An abrupt junction silicon diode has $N_A = 10^{17}cm^{-3}$ and $N_D = 10^{15}cm^{-3}$. For electrons in P, $\mu_n = 800cm^2/V\text{-}s$ and $\tau_n = 0.1\mu s$, and for holes in N, $\mu_p = 480cm^2/V\text{-}s$ and $\tau_p = 1\mu s$. Use $A = 10^{-3}cm^2$:

a)   Determine at $T = 300K$ the diode current for $V_a = -10V$, $V_a = -0.1V$ and $V_a = 0.4V$.

b)   Assume that the mobility and lifetime do not change with temperature. Repeat part (a) for $T = 400K$.

5.14 For a P^+N diode, use the results obtained in the text to determine:

a)   An expression for the excess minority charge Q_p stored in the N region assuming $W_n \gg L_p$.

b)   An expression for the hole current at the edge of the depletion layer in N, at $x = x_{n0}$, in terms of Q_p.

5.15 The limit of low-level injection is normally assumed to be when the minority carrier density at the edge of the depletion layer in the lower doped region becomes equal to one tenth the majority carrier density in that region. For a silicon diode having $N_A = 10^{17}cm^{-3}$ and $N_D = 10^{15}cm^{-3}$, determine the value of the applied voltage at which the limit of low-level injection is reached.

5.16 A P^+N germanium diode has $N_A = 10^{18}cm^{-3}$ and $N_D = 10^{15}cm^{-3}$. Determine:

a)   The excess hole concentration at the edge of the depletion layer in N, at $x = x_{n0}$, for $V_a = 80mV$.

b)   At what value of V_a is the limit of low injection reached.

5.17 Certain PN junctions have a doping profile that is known as linearly graded, as shown in the figure, such that $(N_D - N_A) = ax$ in the depletion region. Assume symmetrical doping so that $x_n = W/2$ and $x_p = -W/2$.

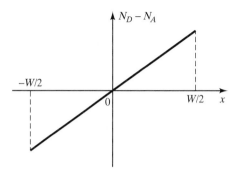

Determine:

    a)   An expression for the electric field distribution in the depletion layer.

    b)   An expression for V_{bi}.

    c)   An expression for the depletion layer width, W.

5.18 A P$^+$N abrupt junction silicon diode has $N_A = 10^{17}\text{cm}^{-3}$ and $N_D = 5 \times 10^{15}\text{cm}^{-3}$. Given τ_p in N $= 0.1\mu\text{s}$, τ_n in P $= 0.02\mu\text{s}$, and $A = 10^{-4}\text{cm}^2$, determine:

    a)   The reverse saturation current.

    b)   For $V_a = V_{bi}/2$, the excess minority carrier concentration at $x' = 0$ ($x = x_n$) and at $x' = 1\mu\text{m}$ into N.

    c)   For $V_a = 0.75V_{bi}$, the injected minority carrier currents on both sides of the depletion layer.

5.19 For a PN silicon junction diode having $N_A = 10^{16}\text{cm}^{-3}$ and $N_D = 10^{13}\text{cm}^{-3}$, determine V_{bi} for:

    a) $T = 300\text{K}$.

    b) $T = 450\text{K}$.

5.20 One of the shortcomings of semiconductor devices is their sensitivity to temperature. Semiconductors with narrow band gaps have a higher intrinsic carry density. In this problem and in Prob. 5.21, you are to establish the effect of temperature on the reverse saturation current.

    (a)   Assuming that the effects on I_S of changes in band gap energy and mobility are negligible, show that the reverse saturation current of a diode is given by

$$I_S = C\,T^3 \exp\left(-E_g/kT\right)$$

where C is a constant that depends on diode properties.

    (b)   Calculate the factor by which I_S increases as the temperature increases from 27°C to 100°C for

    i)   A Ge diode

    ii)   An Si diode

5.21   **a)**   Show that the fractional change in the reverse saturation current of a diode per unit change in temperature is given by

$$\frac{1}{I_S}\frac{dI_S}{dT} = 3/T + \frac{E_g}{kT^2}$$

b)   Determine the fractional change in the reverse saturation current at T = 300K for a germanium diode and a silicon diode.

5.22   **a)**   Show that the fractional change in the forward current of a diode, operated at a given voltage V_a, is given approximately by

$$\frac{1}{I}\frac{dI}{dT} = \frac{E_g - qV_a}{kT^2}$$

b)   Calculate the fractional change in the forward current of a diode operated at a bias of 0.2V at T = 300K for a germanium diode and a silicon diode.

5.23   **a)**   Show that the expression for the electric field in the P region of a silicon PN junction diode at equilibrium is given by

$$\mathcal{E}_x = \frac{kT}{q}\frac{1}{p}\frac{dp}{dx}$$

b)   If the impurity distribution in P is given by $p = 10^{18}\exp^{-x'}/0.4$, where x' is in microns, calculate the magnitude of $\mathcal{E}_x$ at $x' = 0$ in the P region. The point $x' = 0$ occurs at the edge of the depletion layer.

c)   Compare the electric field in the P region to that in the depletion layer by calculating the maximum value of the electric field in the depletion region of an abrupt junction diode at equilibrium if $N_A = 10^{18}\text{cm}^{-3}$ and $N_D = 10^{15}\text{cm}^{-3}$.

chapter 6

FABRICATION TECHNOLOGY

6.0 INTRODUCTION

The use of semiconductors has had a profound impact on the electronics industry and the consequent introduction of integrated circuits has had a major impact on everyday life. The microminiaturization of electronics circuits and systems and their concomitant application to computers and communications represent major innovations of the twentieth century. These have led to the introduction of new applications that were not possible with discrete devices.

The simultaneous formation of many integrated circuits on a single silicon wafer followed by the increase of the size of the wafer to accommodate many more such circuits served to significantly reduce the costs while increasing the reliability of these circuits.

Whereas the electronics engineer was previously concerned with the design of circuits using discrete elements, the engineer is now involved with the ubiquitous interaction between the circuit and the fabrication process, which itself influences the circuit design, thus forming an integral design feedback loop. Design engineers are now required to design the systems, the logic, the circuits, and the layout of the integrated circuits on a wafer. The ingenuity in the design of economically competitive circuits that meet the requirements of both speed and power dissipation is partly a result of the engineer's expertise in the allocation of space on a silicon chip. Real estate on the wafer and on the chip has become a prime commodity.

Before the student proceeds to the study of the operation and characteristics of other semiconductor devices, he or she should be aware of, first, the materials and the processes used in the fabrication of integrated circuits and devices, second, of the layout of the devices on a silicon chip, and, third, of the dimensions involved, in

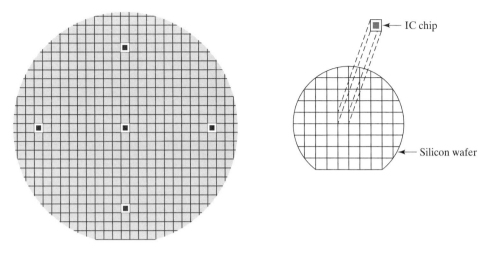

Figure 6.1   Illustration of wafer and chip.

particular, in integrated circuits. The main reason for introducing the subject of fabrication early on is that while we use simple sketches to describe the operation of the device in later chapters, the student will be cognizant of, and can visualize, the actual construction of the device so that she or he may better understand the properties and the limitations of these devices. A photograph of a wafer containing hundreds of dice (chips) and a drawing of a chip are shown in Fig. 6.1. The identical chips, each of which may vary in area from 10 to over 100 mm^2, may contain up to several million devices.

In this chapter, we will first describe the process by which boules or ingots of silicon are obtained. This process results in the formation of solid cylindrical-shaped ingots. It is from these that wafers are sawed off.

Following this, we describe the various processes that are used in the fabrication of devices. Since we have already studied the PN junction diode in the previous chapter, we end this chapter by applying the processes to illustrate the fabrication of a diode, a capacitor, and a resistor for an integrated circuit.

In Chapters 8, 11, and 12, wherein the other major devices are studied, we will apply these processes to the fabrication of a bipolar transistor, a metal-semiconductor field-effect transistor, and a metal-oxide-semiconductor field-effect transistor.

6.1 WHY SILICON?

Semiconductor devices are made in one of two forms: either as single *discrete* units, such as a diode or a transistor, or in conjunction with other circuit elements making up an *integrated circuit*. Integrated circuits may be *monolithic*, whereby transistors, diodes, resistors, and capacitors are fabricated and interconnected on the same silicon chip or they may be *hybrid*. In hybrid circuits, some of the circuit elements are in discrete form and others are interconnected on a chip with the discrete elements connected externally to those formed on the chip.

The fundamental processes in the fabrication of discrete devices and integrated circuits are the same. In this chapter, we will explain and illustrate the various processes that are required to form the devices and the integrated circuits.

The fabrication of semiconductor devices has been based on the use of silicon as the premier semiconductor. Two other semiconductors, germanium and gallium arsenide, present special problems while silicon has certain specific advantages not available with the others.

At 300K, silicon has a band gap of 1.12eV, while germanium's band gap is 0.66eV. Because of this small band gap, the intrinsic carrier density of germanium at $T = 300K$ is about $2.5 \times 10^{13} cm^{-3}$. At temperatures of about 400K, this density becomes $10^{15} cm^{-3}$, which is comparable to the lower range of doping densities used. This property limits its use to low temperature applications at less than 350K.

The other semiconductor of major interest is gallium arsenide. In spite of its attractive electrical properties, gallium arsenide crystals have a high density of crystal defects, which limit the performance of devices made from it. Furthermore, compound semiconductors, such as GaAs (in contrast to elemental semiconductors such as Si and Ge) are much more difficult to grow in single crystal form. Both silicon and germanium do not suffer, in the processing steps, from possible decomposition that may occur in compound semiconductors such as GaAs.

On the plus side, GaAs has a low-field electron velocity that is larger than silicon so that electron devices using GaAs are faster than those using Si. Also, GaAs has a lower saturation electric field than Si so that GaAs devices have a smaller power-delay product. Devices made from substrates of GaAs tend to have smaller parasitic capacitances, which contribute to their speed advantage over silicon devices. Another advantage of GaAs results from its direct band gap, which makes it possible to provide certain functions not possible in Si, such as coherent and incoherent light emission.

A major advantage of silicon, in addition to its abundant availability in the form of sand, is that it is possible to form a superior stable oxide, SiO_2, which has superb insulating properties and, as we shall see later, provides an essential and excellent ingredient in the fabrication and protection of devices.

Lastly, at the present time, silicon remains the major semiconductor in the industry.

6.2 THE PURITY OF SILICON

The starting form of silicon, which manufacturers of devices and integrated circuits use, is a circular slice known as a *wafer*. The wafers are cut from single cylindrical ingots of silicon, with wafer diameters, varying from 10 to 20cm and expected to reach 30cm in the not too distant future. The wafer thickness is of the order of several hundred microns. Large diameter wafers are very cost-effective because of the larger number of integrated circuits they accommodate.

Silicon is found in abundance in nature as an oxide in sand and quartz. A number of processes are required to convert the sand into silicon wafers. To be useful in fabrication, silicon must be in crystalline form—very pure, free of defects, and uncontaminated. There is no other industry that places demands as severe as those required by the semiconductor industry on the purity of silicon.

In the following discussion, we will determine an estimate of the expected purity of silicon. We start with the expected results.

Normally, and for the greatest number of applications, the dopant densities in devices vary from 10^{15} to $10^{19} cm^{-3}$. To be effective, these densities must be very precisely controlled. Impurities that are inherently present in the silicon or that may be accidentally introduced, whether in the bulk (volume) form or on the surface of the material, interfere to a serious extent with the operation of the fabricated devices. In order for the impurities to be at a sufficiently low and acceptable level, their density should be no greater than two or three orders of magnitude lower than $10^{15} cm^{-3}$. At this level, the density of unwanted impurities should not exceed $10^{13} cm^{-3}$. We determined in Chapter 1 that there are approximately 10^{22} atoms per cubic centimeter in silicon. We are requiring, therefore, that there be no more than one unwanted impurity in 10^9 silicon atoms. This is a purity of one in a billion.

Silicon From Sand

Since we need silicon in crystal form for integrated circuit fabrication, the question is: How do we find it or obtain it?

Silicon, as the element, is not found in nature. It is, however, found abundantly in nature in the form of silicon dioxide, which constitutes about 20 percent of the earth's crust. Silicon is commonly found as quartz or sand. Therefore, we have to first convert silicon dioxide into silicon.

One might consider reducing SiO_2 by the addition of hydrogen, but this is not possible because SiO_2 is a very stable compound. There are several methods available for obtaining silicon and the most common is to first refine silicon dioxide chemically with carbon in an arc furnace at very high temperatures resulting in Si and CO_2. The CO_2 evaporates as a gas leaving impure silicon, which is 90–99 percent pure and is known as metallurgical grade silicon. The next step is to purify the silicon.

To purify silicon, a common method is to first produce silicon tetrachloride by burning the silicon:

$$Si + 2C\ell_2 \rightarrow SiC\ell_4$$

Silicon tetrachloride is a liquid that can be distilled and be made of very high purity. The $SiC\ell_4$ is subjected to hydrogen reduction to produce silicon. More commonly, the metallurgical grade silicon is purified by combining it with hydrochloric acid to form trichlorosilane, $SiHC\ell_3$. The trichlorosilane is reduced by hydrogen to form very pure silicon, known as *semiconductor grade silicon,* which is vapor deposited on high purity silicon rods.

At this stage, the solid is polycrystalline composed of many small crystals (of submicron dimensions) having random orientation and containing many defects. The silicon contains one unwanted impurity atom in about 10^9 atoms of silicon. The purity is excellent. For silicon to be used in the fabrication of devices, however, it must be nearly perfect and crystalline in nature. We, therefore, now need to produce single crystals of silicon. This is done by a method known as crystal growth. Crystal growth consists of converting a random oriented or polycrystalline material into one that is orderly and crystalline.

The easiest approach is to melt silicon and let it freeze. One method to carry this out is in a growth process known as the *Czochralski process*. In this process, crystalline silicon, to which a dopant is added, is grown.

6.3 THE CZOCHRALSKI GROWING PROCESS

The Melt and the Dopant

The equipment setup for this process is shown in Fig. 6.2. To grow crystals, one starts with very pure semiconductor grade silicon, which is melted in a quartz-lined graphite crucible. The melt is held at a temperature of 1690K, which is slightly greater than the melting point (1685K) of silicon. The surrounding heaters and heat shield establish a carefully controlled temperature with the center of the melt being the coolest.

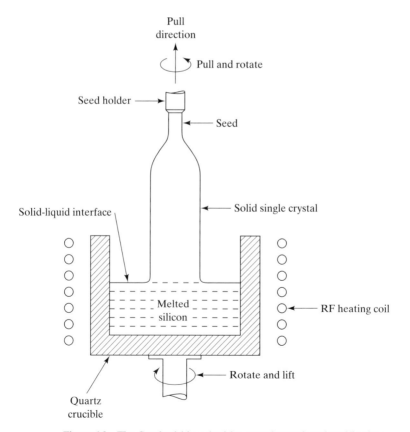

Figure 6.2   The Czochralski method for crystal growth and purification.

A precisely controlled quantity of the dopant is added to the melt; added boron makes P silicon; added phosphorous makes N silicon. One assumes that the density of impurities to be added is determined accurately by the desired resulting conductivity. However, the problem is not as simple as it may seem. When a material freezes, the concentration of impurities incorporated in the solid is usually smaller than the concentration in the liquid. The ratio of the concentration of impurities in the solid, C_O, to that in the liquid, C_ℓ, is known as the equilibrium segregation coefficient k_O,

$$k_O = C_O / C_\ell$$

There is also a limit to the amount of impurity that can be added as the concentration in the melt must not exceed about 2 percent, otherwise single crystal growth is hampered. This together with the segregation coefficient limit the amount of doping in the crystal.

The following example illustrates the mass of a dopant required for a certain doping density.

EXAMPLE 6.1

A silicon ingot that should contain 10^{16} phosphorus atoms/cm^3 is to be grown by the Czochralski method.

a)   Determine the concentration of phosphorus atoms in the melt to give the required concentration in the ingot.

b)   The crucible initially contains 50Kg of molten silicon. Determine how many grams of phosphorus should be added:
Given: for phosphorus $k_O = 0.35$, density of silicon = 2.53g/cm^3 and atomic weight of phosphorus = 30.975g/mole, Avogadro's number = 6.023×10^{23} atoms/mole.

Solution

a) $k_O = C_O/C_\ell, C_O = 10^{16}cm^{-3}, C_\ell = 2.85 \times 10^{16}$ atoms/cm^3

b)   volume of silicon = $50 \times 10^3/2.53 = 1.976 \times 10^4$cm^3
Number of phosphorus atoms = $2.85 \times 10^{16} \times 1.976 \times 10^4 = 5.63 \times 10^{20}$
Amount of phosphorus $= \frac{5.63 \times 10^{20} \text{ atoms} \times 30.975\text{g/mole}}{6.023 \times 10^{23} \text{ atoms/mole}} = 28.95$mg.

We observe that a very small amount of phosphorus, ≈ 0.03g, is needed to dope 50Kg of silicon.

It is important to note that for a given impurity there is a maximum density of the impurity, at a given temperature, that is allowed in or can be absorbed by a crystal. This maximum concentration is known as *solid solubility*. Because the solubility decreases with temperature, if an impurity is introduced at its maximum concentration and the temperature is reduced, the crystal precipitates the excess impurity to achieve equilibrium.

Seed Crystal

After having set up the melt, a seed crystal (a small highly perfect crystal), attached to a holder and possessing the desired crystal orientation, is dipped into the melt and a small portion is allowed to melt. Very slowly, the seed is rotated and pulled up while, at the same time, the crucible is rotated in the opposite direction. The molten semiconductor attaches itself to the seed and it becomes identical to the seed in structure and orientation.

As the seed is pulled up, the melted material that is attached to the seed solidifies (freezes). Its crystal structure becomes the same as that of the seed and a larger crystal is formed. By this method, cylindrical single crystal bars of silicon are produced. As the molten silicon solidifies on the seed, the purity of the silicon is improved as most of the impurities tended to remain in the liquid and melt as the melted silicon gradually solidifies.

The desired silicon bar diameter is obtained by controlling both the temperature and the pulling speed. In the final process, when the bulk of the melt has been grown, the crystal diameter is decreased until there is a point contact with the melt. The resulting ingot is cooled and removed to be made into wafers. The ingots have diameters as large as 200mm, with latest ones approaching 300mm. The ingot length is of the order of 100cm.

Ingot Slicing and Wafer Preparation

The ingot surface is ground throughout to an exact diameter and the top and bottom portions are cut off. Following this, circular wafers are sliced off the ingot with a high speed diamond saw. The wafer thicknesses vary from 0.4 to 1.0mm.

Slicing the wafers to be used in the fabrication of integrated circuits is a procedure that requires precision equipment. The object is to produce slices that are perfectly flat and as smooth as possible, with no damage to the crystal structure. The wafers need to be subjected to a number of steps known as lapping, polishing, and chemical etching. The wafers are first lapped with a suitable abrasive, such as diamond, to remove the irregularities introduced by the sawing. They are also chemically etched to produce flat and parallel surfaces and finally polished to a mirror-like finish.

The wafers are cleaned, rinsed, and dried for use in the fabrication of discrete devices and integrated circuits. It is interesting to note that the final wafer thickness is about one third less than that after the sawing.

The growth of GaAs crystals is much more complex than that of silicon. The largest commercially available wafers are about 10cm in diameter. One reason for this is that the wafers are brittle and may crack. Furthermore, GaAs crystals contain a high concentration of crystal defects that can degrade the device yield significantly.

In the next several pages, we will explain and illustrate with sketches, where necessary and relevant, the major operations required for the fabrication of circuits and devices on a wafer of silicon. Having done that, we will apply the knowledge gained to the fabrication of a PN junction diode. As we study the major devices in the chapters following, we will refer to the operations involved in their fabrication and illustrate the fabrication of a typical device.

6.4 FABRICATION PROCESSES

The category of processes that are used in the fabrication of devices and integrated circuits are the following:

- Oxidation
- Diffusion
- Ion Implantation
- Photolithography
- Epitaxy
- Metallizations and interconnections.

We will now consider each process separately and apply some of these to the formation of a diode, in this chapter, and to the fabrication of transistors, in later chapters.

The basic fabrication process is known as the *planar process*, in which the introduction of impurities and metallic interconnections is carried out from the top of the wafer. A major advantage of the planar process is that each fabrication step is applied to all identical circuits and devices on each of the many wafers at the same time.

It is important to initially emphasize that the fabrication requires an extremely clean environment in addition to the precise control of temperature and humidity.

Thermal Oxidation

The process of oxidation consists of growing a thin film of silicon dioxide on the surface of the silicon wafer. In the planar process, all operations are carried out from the top surface. It becomes necessary to shield certain regions of the surface so that dopant atoms, by diffusion or ion implantation, may be driven into other selected regions. The formation of a silicon dioxide layer is shown in Fig. 6.3 and its shielding effect is illustrated in Fig. 6.6. Silicon dioxide, as we shall see later, plays an important role in making this possible. Furthermore, an SiO_2 layer serves as a passivating or protective layer on the silicon surface to protect the devices during subsequent processing.

The commonly used silicon dopants, such as boron, phosphorous, arsenic, and antimony, have very low diffusion coefficients (diffuse with great difficulty) in SiO_2. Because of this, SiO_2 is used as a shield against infiltration of these dopants. On the other hand, these dopants diffuse very easily if the surface is silicon.

Oxidation is accomplished by placing the silicon wafers vertically into a quartz boat in a quartz tube, which is slowly passed through a resistance-heated furnace, in the presence of oxygen, operating at a temperature of about 1000°C. The oxidizing agent may be dry using dry oxygen or wet using a mixture of water vapor and oxygen. The oxide growth rate in the dry process is much slower but it produces an oxidized layer that has excellent electrical properties. The whole operation is

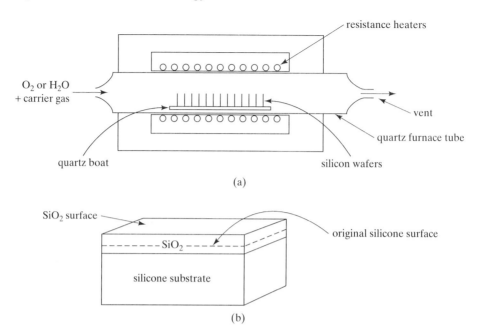

Figure 6.3 (a) Thermal oxidation system and (b) growth of SiO_2.

controlled by microprocessors, which monitor both the gas flow sequence and the furnace temperature. A thermal oxidation system, together with an example of SiO_2 growth, are shown in Fig. 6.3.

A silicon surface oxidizes very rapidly so that even at room temperature a layer of silicon dioxide about 2nm thick is produced on the wafer. In the thermal oxidation process, the thickness of the oxide varies from 0.1 for VLSI gate oxide to one micron. From a consideration of densities and relative molecular weights of Si and SiO_2, an oxide film of thickness x_0 consumes $0.44x_0$ of the silicon. This relationship is determined in the example that follows.

EXAMPLE 6.2

Determine the thickness of silicon that is consumed when a silicon dioxide layer of thickness x_0 is grown on the surface by thermal oxidation.

Solution The volume of one mole of Si or SiO_2 is the ratio of its molecular weight to its density, determined as,

$$\text{the volume of 1 mole of } SiO_2 = \frac{60.08 \text{g/mole}}{2.21 \text{g/cm}^3} = 27.23 \text{ cm}^3$$

$$\text{the volume of 1 mole of Si} = \frac{28.09 \text{g/mole}}{2.33 \text{g/cm}^3} = 12.055 \text{ cm}^3$$

Since 1 mole of SiO_2 uses up 1 mole of silicon over the same area, we have

$$\frac{\text{volume of 1 mole of silicon}}{\text{volume of 1 mole of SiO}_2} = \frac{\text{thickness of Si} \times \text{area}}{\text{thickness of SiO}_2 \times \text{area}} =$$

$$\frac{\text{thickness of Si}}{x_0} = \frac{12.055}{27.23} = 0.44$$

Thus, to grow 200 Angstroms of SiO_2, a layer of 88 Angstroms of silicon is used up.

Once the layer of SiO_2 has been formed on the surface of the wafer, it is selectively removed (etched) from those surfaces where impurities are to be introduced and kept as a shield, for the underlying silicon surface, where no dopants are to be allowed.

Oxide layers are relatively free from defects and provide stable and reliable electrical properties.

Etching Techniques

Etching is the process of selective removal of regions of a semiconductor, metal, or silicon dioxide.

There are two types of etchings: wet and dry. In wet etching, the wafers are immersed in a chemical solution at a predetermined temperature. In this process, the material to be etched is removed equally in all directions so that some material is etched from regions where it is to be left. This becomes a serious problem when dealing with small dimensions.

In dry (or plasma) etching, the wafers are immersed in a gaseous plasma created by a radio-frequency electric field applied to a gas such as argon. The gas breaks down and becomes ionized. Electrons are initially released by field emission from an electrode. These electrodes gain kinetic energy from the field, collide with, and transfer energy to the gas molecules, which results in generating ions and electrons. The newly generated electrons collide with other gas molecules and the avalanche process continues throughout the gas, forming a plasma. The wafer to be etched is placed on an electrode and is subjected to the bombardment of its surface by gas ions. As a result, atoms at or near the surface to be etched are removed by the transfer of momentum from the ions to the atoms.

Diffusion

This process consists of the introduction of a few tenths to several micrometers of impurities by the solid-state diffusion of dopants into selected regions of a wafer to form junctions. Most of these diffusion processes occur in two steps: the *predeposition* and the *drive-in* diffusion. In the predeposition step, a high concentration of dopant atoms are introduced at the silicon surface by a vapor that contains the dopant at a temperature of about 1000°C. More recently, a more accurate method of predeposition, to be explained later, and known as *ion implantation*, is used.

At the temperature of 1000°C, silicon atoms move out of their lattice sites creating a high density of vacancies and breaking the bond with the neighboring atoms. The impurity atoms, which are incident on the surface, move into the silicon because

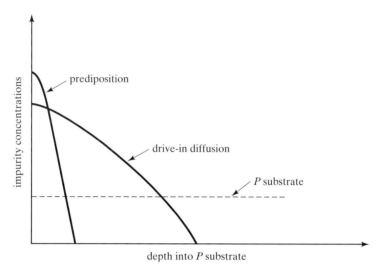

Figure 6.4 N-Impurity distributions into P-substrate, for predeposition and drive-in diffusion.

of their concentration gradient and into the locations that the silicon atoms vacated. Predeposition tends to produce, near the silicon surface, a shallow but heavily doped layer. The second step of diffusion, drive-in, is used to drive the impurity atoms deeper into the surface, without adding any more impurities, thus reducing the surface concentration of the dopant. A sketch of the resulting impurity distributions is shown in Fig. 6.4.

Common dopants are boron for P-type layers and phosphorus, antimony, and arsenic for N-type layers. Diffusion, however, is rarely performed using the pure elements themselves. Rather, compounds of the elements are used and impurities may be introduced from either solid, liquid, or gaseous sources.

Boron and phosphorus have two desirable properties: (1) they have a high diffusion rate (diffuse easily and quickly) into silicon and low diffusion rate in SiO_2 and (2) they are both highly soluble in silicon. We refer to SiO_2 since, prior to carrying out the process of diffusion, windows are opened, in a previously deposited layer of SiO_2, for the impurities to diffuse in. The open windows are in silicon but the adjacent regions are shielded with silicon dioxide. The process of window opening is covered later.

A typical arrangement of the process of diffusion is shown in Fig. 6.5. The wafers are placed in a quartz furnace tube that is heated by resistance heaters surrounding it. So that the wafers may be inserted and removed easily from the furnace, they are placed in a slotted quartz carrier known as a boat. The wafers are mounted on their side, as illustrated in the figure.

To introduce a phosphorus dopant, as an example, phosphorus oxychloride ($POCl_3$) is placed in a container either inside the quartz tube, in a region of relatively low temperature, or in a container outside the furnace at a temperature that helps maintain its liquid form. For a P dopant, boron is used. The proper vapor pressure is maintained by a control of the temperature.

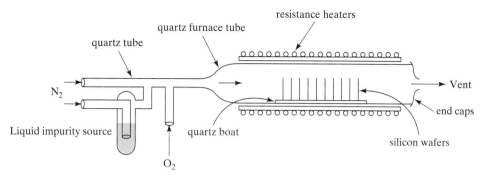

Figure 6.5   Physical layout of equipment used in diffusion.

Nitrogen and oxygen gas are made to pass over the container. These gases carry the dopant vapor into the furnace, where the gases are deposited on the surface of the wafers. These gases react with the silicon, forming a layer on the surface of the wafer that contains silicon, oxygen, and phosphorus. At the high temperature of the furnace, phosphorus diffuses easily into the silicon.

So that the dopant may be diffused deeper into the silicon, the drive-in step follows. This is done at a higher temperature of about 1100°C inside a furnace, similar to that used for predeposition, except that no dopant is introduced into the furnace. The higher temperature causes the dopant atoms to move into the silicon more quickly. Diffusion depth is controlled by the time and temperature of the drive-in process. By precise control of the time and temperature (to within 0.25°C), accurate junction depths of fraction of a micron can be obtained. Diffusion of dopant into silicon is illustrated in Fig. 6.6.

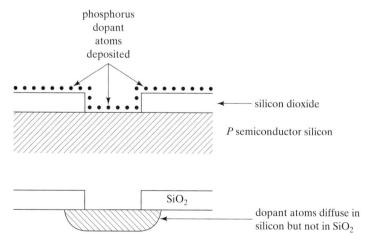

Figure 6.6   Diffusion of dopant atoms in silicon.

Expressions for the Diffusion of Dopant Concentration

The process of diffusion of dopants is similar to that of the diffusion of holes and electrons that was discussed in Chapter 4. We will relate expressions for the rate of dopant diffusions. First, we define relevant terms:

C_s = dopant concentration in cm^{-3} at surface ($x = 0$).

x = diffusion depth into the substrate (cm).

t = duration of diffusion (s).

D = diffusion coefficient (cm^2/s) $\propto \exp^{-k/T}$, k is a constant.

L = diffusion length = $\sqrt{Dt}$.

$C(x,t)$ = dopant concentration at depth x and time t.

Assuming that the diffusion coefficient is independent of doping concentration, the diffusion equation is given by

$$\partial C/\partial t = D\frac{\partial^2 C}{\partial x^2} \tag{6.1}$$

By using the appropriate boundary conditions, it can be shown that the doping concentration C_1, after the *predeposition step* at a depth x and after time t_1, is given by

$$C_1(x,t_1) = C_s \, \text{erfc} \left(x/\sqrt{4D_1 t_1}\right) \tag{6.2}$$

where the error function and the error function complement are defined as

$$\text{erf}(x) = 2/\sqrt{\pi} \int_0^x e^{-y^2} dy \tag{6.3}$$

and

$$\text{erfc}(x) = 1 - \text{erf}(x) \tag{6.4}$$

These functions are tabulated in mathematical handbooks. We list below a few of their properties:

$$\text{erf}(0) = 0$$

$$\text{erf}(\infty) = 0$$

$$\frac{d}{dx}\text{erf}(x) = (2/\sqrt{\pi})e^{-x^2}$$

$$\int_0^\infty \text{erfc}(x)dx = 1/\sqrt{\pi}$$

The expression for C after the drive-in diffusion becomes

$$C_2(x,t_2) = C_s(2/\pi \sqrt{D_1 t_1/D_2 t_2}) \, e^{-x^2/4D_2 t_2} \tag{6.5}$$

Sample calculations and plots are shown in Fig. 6.7 for phosphorus predeposition at 1000°C for 8 minutes followed by a drive-in diffusion at 1250°C for 32 minutes. The

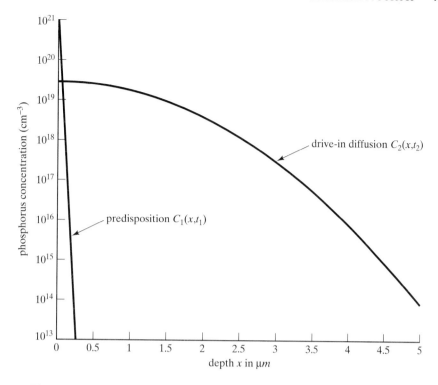

Figure 6.7   Impurity profiles for predeposition and drive-in diffusion.

relevant diffusion coefficients at the given temperatures are $D_1 = 2.5 \times 10^{-14} \text{cm}^2/\text{s}$ and $D_2 = 2.5 \times 10^{-12} \text{cm}^2/\text{s}$ and the concentration at $x = 0$ is $C_s = 10^{21} \text{cm}^{-3}$.

Ion Implantation

This is a process of introducing dopants into selected areas of the surface of the wafer by bombarding the surface with high-energy ions of the particular dopant.

To generate ions, such as those of phosphorus, an arc discharge is made to occur in a gas, such as phosphine (PH_3), that contains the dopant. The ions are then accelerated in an electric field so that they acquire an energy of about 20keV and are passed through a strong magnetic field. Because during the arc discharge unwanted impurities may have been generated, the magnetic field acts to separate these impurities from the dopant ions based on the fact that the amount of deflection of a particle in a magnetic field depends on its mass.

Following the action of the magnetic field, the ions are further accelerated so that their energy reaches several hundred keV, whereupon they are focused on and strike the surface of the silicon wafer. At this time, the ion current is of the order of 1mA.

As is the case with diffusion, the ion beam is made to penetrate only into selected regions of the wafer by a process of masking, which will be discussed later.

On entering the wafer, the ions collide with electrons and with nuclei of silicon atoms, and lose their energy. The depth of penetration is about 0.1 to 1 micron. The higher the energy of the ions and the smaller their mass, the greater is the depth of penetration.

Ion implantation has the following advantages:

- Doping levels can be precisely controlled since the incident ion beam can be accurately measured as an electric current.
- The depth of the dopant can be easily regulated by control of the incident ion velocity. It is capable of very shallow penetrations.
- Extreme purity of the dopant is guaranteed.
- The doping uniformity across the surface can be accurately controlled.
- Because the ions enter the solid as a directed beam, there is very little spread of the beam, thus the doping area can be clearly defined.
- Since this is a low-temperature process, the movement of impurities is minimized.

This process has one major shortcoming in that it may create considerable crystal damage because of the collisions of the high energy ions with the silicon. There are two types of collisions: First, electronic collisions take place between the incident electrons and the target atoms; second, nuclear collisions occur involving the nuclei of the incident and target atoms. Such damage results in inferior performance of devices made by this process. The damage may be so extensive that it transforms the silicon from a crystalline to an amorphous structure (Sec. 1.1).

If the damage is not extensive, the lattice structure is restored by the process of *annealing*. In one process, known as furnace annealing, the wafer is heated in an inert atmosphere at a temperature of 600 to 1000°C for about 30 minutes. This restores the crystalline arrangement of the silicon.

In another process, known as rapid thermal annealing (RTA), optical radiant energy is generated and delivered to the surface of the wafer. The energy is generated by a tungsten-halogen lamp at a wavelength of 0.3 to 4 micrometers in a quartz enclosure. Because of the nature of the process, the quartz walls do not acquire the light energy that is directed at the wafer and thus the wafer is not in thermal equilibrium with the walls of the system. As a result, this process serves to considerably reduce the annealing time compared to that required in the furnace.

Another shortcoming of ion implantation is the investment in the equipment mainly because of its complexity.

Photomask Generation

The whole process of integrated circuit fabrication consists of identifying selected regions of each circuit (or dies) of the wafer surface into which identical dopant or metallic interconnections are made, while protecting the other regions of the wafer surface. To carry out one of the many processes of oxidation, diffusion, ion implantation, or epitaxy, a separate mask or mini mask is required for each operation whose function is to expose the selected regions and protect the others. There may be hundreds of identical dice (or ICs) on a wafer with each circuit containing hundreds of thousands, or millions, of devices. Identical steps are carried out simultaneously for each process, such as the diffusion of the N regions of a diode on one or several circuits, to be repeated over the whole wafer. For each process, a separate mask is needed.

The mask production starts with a drawing using a computer-assisted graphics system with all the information about the drawing stored in digital form. Commands from the computer are prepared that drive a pattern generator, which uses an electron beam to write (photoengrave) the particular pattern, for one or several dice, on a glass plate covered with a thin chromium film. When the glass plate is prepared for one or several dice on the wafer, it is known as a *reticle*. A mask usually refers to a glass plate that contains a pattern for the whole wafer. The reticle pattern is projected onto the wafer and a *wafer stepper* is used that reduces the reticle circuit onto the photoresist-covered (see next section) wafer that steps across over the surface until the entire array of circuits is built up.

The use of a single mask for all the circuits on a wafer is not feasible for printing very small ($<1\mu$m) features because of alignment problems resulting from the heat that the wafer is exposed to, which causes slight distortion of the surface. Such systems are still in use, however, for fabricating simple logic circuits and analog devices such as LEDs.

Photolithography

In this process, the image on the reticle is transferred to the surface of the wafer. This is done to open identical windows so that the diffusion process, for example, may take place in all identical regions of the same IC and for all ICs on the wafer. As an illustration, we assume that the first reticle is used over an oxidized surface.

To transfer the pattern, the wafer is coated with a light-sensitive photoemulsion, known as *photoresist*. By applying about 1cm^3 of the liquid to the wafer surface and spinning the wafer very rapidly, a uniform film, about 1 micrometer thick, of photoresist is formed over the oxidized surface of the wafer. After this, the following steps, also shown in Fig. 6.8, are taken to open a window on the wafer:

1. The wafer is baked at 100°C to solidify the resist on the wafer.
2. The reticle is placed on the wafer and aligned by computer control.
3. The reticle is exposed to ultraviolet light with the transparent parts of the reticle passing the light onto the wafer. The photoresist under the opaque regions of the reticle is unaffected.
4. The exposed photoresist is chemically removed by dissolving it in an organic solvent and exposing the silicon dioxide underneath. This is a process very similar to that used in developing photographic film.
5. The exposed silicon dioxide is then etched away using hydrofluoric acid, which dissolves silicon dioxide and not silicon. The regions under the opaque part of the reticle are still covered by the silicon dioxide and the photoresist.

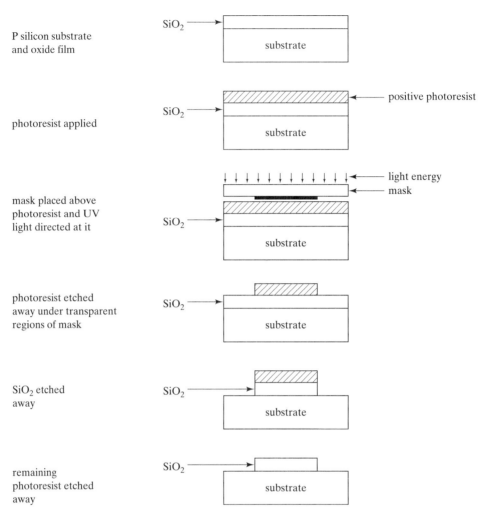

Figure 6.8 Steps in oxidation and window opening.

6. The photoresist under the opaque regions of the reticle is stripped using a proper solvent and the silicon dioxide is exposed.

All surfaces are protected except those covered by silicon only in which diffusion or ion implantation is to take place. The surfaces covered by silicon dioxide do not permit any entry of dopants.

The photoresist used in this discussion is labeled a positive resist, whereby windows are opened wherever the ultraviolet light passes through the transparent parts of the mask. Negative resist is also used and remains on the surface, which is exposed to UV light, and windows are opened under the opaque parts of the mask.

There is a practical limit, at linewidths of about 1–2 micrometers, to using ultraviolet light. The printing of smaller features can be accomplished by using very short wavelength radiation such as electron-beam or x-ray lithography.

We indicated earlier that the reticle or mask is placed in direct contact with the wafer. It may be possible that the wafer has few irregular particles at the surface of the crystal or it may have some dust particles. These particles stick to the mask and cause defects in the surface of each successive operation. This problem is cured by a process known as *proximity printing*, wherein the mask is separated from the wafer by a distance of about 10–20 micrometers.

Epitaxial Growth

Epitaxy is the process of the *controlled growth* of a crystalline doped layer of silicon on a single crystal substrate. The processes of diffusion and ion implantation, which were earlier described, produce a layer at the surface that is of higher doping density than that which existed before the dopant was added. It is not possible by these methods to produce, at the surface, a layer of lower concentration than exists there. This can, however, be accomplished by the method of epitaxy. In the processes of diffusion and ion implantation, a dopant is driven into a substrate of doped silicon. In epitaxy, a layer of doped silicon is deposited on top of the surface of the substrate. Normally, this single crystal layer has different type doping from that of the substrate.

Epitaxy is used to deposit N on N^+ silicon, which is impossible to accomplish by diffusion. It is also used in isolation between bipolar transistors wherein N^- is deposited on P. It may also be used to improve the surface quality of an N substrate by depositing N material over it. The system for growing an epitaxial layer is shown in Fig. 6.9.

In this system, silicon wafers are placed in a long boat-shaped crucible made of graphite. The boat is placed in a long cylindrical quartz tube, which has inlets and

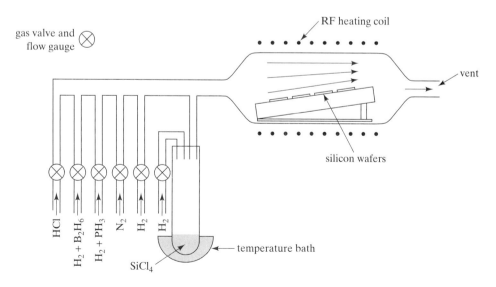

Figure 6.9 System for growing an epitaxial layer.

outlets for the gases. The tube is heated by induction using the heating coils wound around the tube.

All the chemicals that are introduced and that take part in the reactions are in the form of gases, hence the process is known as Chemical Vapor Deposition (CVD). The epitaxial layer is grown from the vapor phase onto the silicon, which is in the solid state. The thickness of the layer varies from 3 to 30 microns and the thickness of the layer and its doping content are controlled to an accuracy of less than 2 percent. The reactions are carried out at a temperature of approximately 1200°C. The high temperature is necessary so that the dopant atoms can acquire a sufficient amount of energy to allow them to move into the crystal to form covalent bonds and become an extension of the single crystal lattice. Because the layer is grown on the substrate, epitaxy is a growth technique where the crystal is formed without reaching the melting point of silicon.

We list below, and with reference to Fig. 6.7, the sequence of operations involved in the process:

1. Heat wafer to 1200°C.
2. Turn on H_2 to reduce the SiO_2 on the wafer surface.
3. Turn on anhydrous $HC\ell$ to vapor-etch the surface of the wafer. This removes a small amount of silicon and other contaminants.
4. Turn off $HC\ell$.
5. Drop temperature to 1100°C.
6. Turn on silicon tetrachloride ($SiC\ell_4$).
7. Introduce dopant.

A number of different chemical reactions can be used to deposit the epitaxial layer. Silane (SiH_4) or $SiC\ell_4$ can be used with the following reactions:

$$SiH_4 \rightarrow Si + 2H_2$$
$$SiC\ell_4 + H_2 \rightarrow Si + 4HC\ell$$

A silicon layer can be produced from silane by the addition of heat, while silicon tetrachloride requires a reduction by hydrogen.

To grow a layer of N-type silicon, very small amounts of impurities, such as PH_3, AsH_3, or SbH_3, are introduced simultaneously with the gases. Diborane (B_2H_6) is used to form a P-layer of silicon. During the epitaxial layer deposition, the dopant atoms decompose and they become part of the layer.

Thus, epitaxy provides a means for accurately controlling the doping profile in order to optimize the performance of devices and circuits.

It is extremely important to repeat one major consideration in the preparation of processes and devices, namely that the wafers must be very smooth and clean. Dirt particles and irregularities can have damaging effects on the properties of the devices. As indicated earlier, and in all these operations, the wafers are thoroughly cleaned before they are placed in a boat. Once inside the quartz tube, the cleaning process is carried out by using nitrogen to flush the air out and hydrochloric acid is made to pass over the wafers in order to etch away a very thin layer of the surface.

Metallization and Interconnections

After all semiconductor fabrication steps of a device or of an integrated circuit are completed, it becomes necessary to provide metallic interconnections for the integrated circuit and for external connections to both the device and to the IC. The requirement that must be met by the interconnections is that they have low resistance to minimize both the voltage drops on the lines as well as the capacitances between the lines so as to reduce delay times. The connections must also make *ohmic contacts* to semiconductors in the devices such as the P and N regions of a PN junction diode. An ohmic contact is one that exhibits a very low resistance, allowing currents to pass easily in both directions through the contact.

The high conductivity of aluminum makes it the metal of obvious choice, particularly in silicon-based devices. It also has the following advantages:

1. easy to evaporate
2. can be easily etched
3. not expensive
4. adheres well to silicon dioxide

There are a variety of methods for depositing aluminum on silicon substrates, and we shall briefly present three common methods, which are: resistance heating, electron beam heating, and sputtering. In resistance heating, the source of the heated element and the silicon substrate are located in an evacuated chamber. The source is a small piece of aluminum attached to a coil of tungsten, which serves as the heater. The heated filament with a high melting point remains solid while the aluminum (with a small addition of silver or copper) is vaporized. The aluminum molecules travel to the substrate where they condense, depositing an aluminum layer on the surface of the silicon. A photolithographic masking and etching method, using a phosphoric acid (H_3PO_4) solution or a dry etching technique, is used to remove the metal from regions where it is not wanted. A typical interconnection between two diffused layers is shown in Fig. 6.10.

Another method of generating vaporized aluminum is to place the aluminum in a crucible into a vacuum chamber together with the substrate. The aluminum is

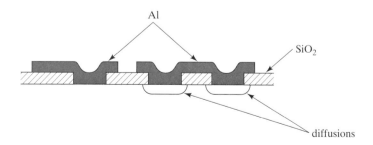

Figure 6.10   An aluminum interconnection between two diffused silicon regions.

subjected to a high intensity electron beam formed by an electron gun, which vaporizes the aluminum that travels to the wafer. By the use of a mask and photolithography, the aluminum is deposited on the identified regions on the wafer surface.

In the sputtering process, the material to be deposited is placed in a container maintained at low pressure in the vicinity of the substrate. The material to be deposited is labeled the cathode or target, while the anode is the substrate. A DC or a radio-frequency high voltage is applied between anode and cathode. This high voltage ionizes the inert gas in the chamber. The ions are accelerated to the cathode (in this operation the anode is usually grounded) where, by impact with the aluminum target, atoms of aluminum are vaporized. A gas of aluminum atoms is generated and deposited on the surface of the wafer.

Following the deposition of aluminum, the silicon wafers are placed in a furnace to solidify the connections so that low resistance metallic contacts are made.

The interconnections between elements of an integrated circuit are made by aluminum lines having a thickness of about 0.5 μm. These are laid on top of the silicon dioxide layer, which covers the surface of the wafer. By using photolithography, openings are made in the silicon dioxide so that the aluminum layer is connected to the silicon or to the ohmic contact on the silicon. In very complicated integrated circuits, it is necessary to have two or three vertically stacked layers of interconnections separated by silicon dioxide layers. The interconnecting lines terminate at aluminum pads from which connections to the outside are made.

The connections from the chips to the outside world are produced by the metallization pattern around the periphery of the chip. These are known as bonding pads and they are connected by wires to the package, as shown in Fig. 6.11. The bonding pads are about 100 μm square and the bonding wires are made of gold with a diameter of about 25 μm. Following these connections, the individual dice are encapsulated and hermetically sealed in a variety of packages.

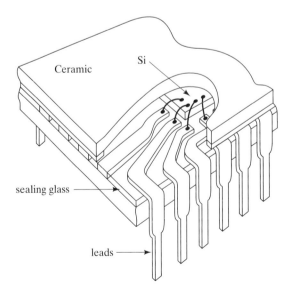

Figure 6.11 Typical die package. *Source: J. Mayer and S. Lau, Electronic Materials Science: For Integrated Circuits in Si and GaAs*, Macmillan (1990).

It is important to maintain the resistance of the interconnections to small values. A large value for this resistance combined with parasitic capacitances can result in a time constant RC, which may end up, in a superfast device, to be the limiting factor in the overall speed of a circuit.

Ohmic Contacts

It cannot be assumed that depositing aluminum on a doped semiconductor forms a very good ohmic contact. An ideal ohmic contact exhibits a perfectly linear relationship between current and voltage with the plotted current-voltage characteristic passing through the origin. In certain cases, and as we shall discuss in later chapters, a contact between aluminum and a semiconductor may result in a rectifying contact having I-V characteristics of a diode, whereby current conduction is only in one direction.

On P-type silicon having a concentration of $10^{16} cm^{-3}$, or more, aluminum forms a good ohmic contact. However, aluminum on lightly doped N-type silicon forms a rectifying contact. In order to prevent the formation of such a contact, an N^+ diffusion is placed over the N-type silicon.

6.5 PLANAR PN JUNCTION DIODE FABRICATION

To illustrate the various steps in the fabrication of a discrete PN junction diode, we show in Fig. 6.12 a series of drawings that include most of the processes discussed earlier. We list below the various steps in the fabrication of a PN junction diode. *It is important to indicate here that in integrated circuits, where all interconnections and device terminals are made at the surface, a diode is formed from a bipolar transistor by placing a short-circuit between two of the three terminals of the transistor (collector to base).*

Figure	Process Description
(a)	An N^+ substrate grown by the Czochralski process is the starting metal $\approx 150 \mu m$ thick.
(b)	A layer of N-type silicon $(1–5 \mu m)$ is grown on the substrate by epitaxy.
(c)	Silicon dioxide layer deposited by oxidation.
(d)	Surface is coated with photoresist (positive).
(e)	Mask is placed on surface of silicon, aligned, and exposed to UV light.
(f)	Mask is removed, resist is removed, and SiO_2 under the exposed resist is etched.
(g)	Boron is diffused to form P region. Boron diffuses easily in silicon but not in SiO_2.
(h)	Thin aluminum film is deposited over surface.
(i)	Metallized area is covered with resist and another mask is used to identify areas where metal is to be preserved. Wafer is etched to remove unwanted metal. Resist is then dissolved.

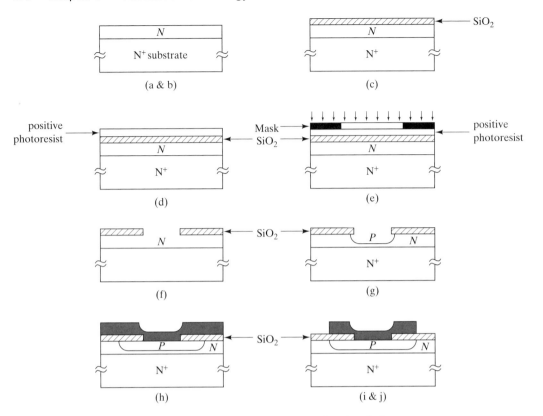

Figure 6.12   Typical steps in the fabrication of a planar PN junction diode.

(j)   Contact metal is deposited on the back surface and ohmic contacts are made by heat treatment.

In the following chapters, we will be discussing other devices. Wherever appropriate and necessary, reference to the particular fabrication techniques will be illustrated.

6.6 FABRICATION OF RESISTORS AND CAPACITORS IN ICs

Resistors

In integrated circuits, resistors are usually made of impurities that are diffused into a semiconductor, which is of opposite polarity. They are made by and from the processes that are used to form devices. Figure 6.13(a) shows a resistor made of a P region diffused into an N-epitaxial layer and to the ends of which metallic contacts are made. The section of the resistor, as dictated by the diffusion, is very nearly rectangular in shape, as shown in Fig. 6.13(b).

The resistance of the layer is given by

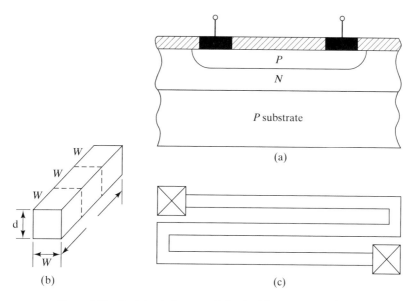

Figure 6.13   Resistors in integrated circuits (a) cross-section, (b) rectangular dimensions, and (c) meander pattern.

$$R = \rho L / Wd$$

where ρ is the average resistivity of the layer in ohm-cm, d is its thickness, L is its length and W is its width.

Resistances in monolithic circuits are defined by a term known as the *sheet resistance*. The sheet resistance has units of ohms per square, it being the resistance of a square having $W = L$.

Assuming 100 to 200 ohms per square, practical resistors may have values ranging from 100 to several kilohms. Higher resistances are obtained by using a meander pattern, as shown in Fig. 6.11(c). The major problem with resistors of high values is that they tend to occupy a large area on the chip. A resistor of 50 Kohms uses up an area of the semiconductor that may be occupied by hundreds of transistors. Ion implantation can be used, however, to generate lightly doped regions suitable for precision high-value resistors.

Capacitors

One type of capacitor in monolithic circuits is made by using the capacitance formed between the P and N regions of a reverse-biased diode. Such a capacitance is shown in Fig. 6.14.

These capacitors, just like resistors, are formed in the same diffusion processes that are used to form devices. As we will see in a later chapter, bipolar transistors are made of three regions in which either of the two PN junctions may be used as capacitors, whereby the breakdown voltage of the capacitor may vary considerably from one to the other.

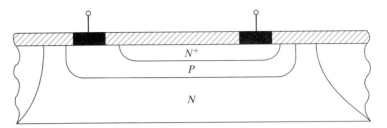

Figure 6.14   A capacitor in a monolithic circuit.

The disadvantage of junction capacitors is the dependence of the capacitance on the voltage applied to the junction. Capacitors that are voltage-independent can be formed from metal insulator N^+ semiconductor layers, as used in MOS structures. Metal oxide semiconductor junctions and devices are studied in Chapters 12 and 13.

PROBLEMS AND QUESTIONS

6.1 A silicon crystal is to be grown by the Czochralski process and is to have in the melt an arsenic concentration of 5×10^{15} atoms/cm³. The segregation coefficient of arsenic in silicon is 0.3. Determine the initial arsenic concentration in the crystal.

6.2 A silicon crystal is to be grown by the Czochralski process and is to contain 5×10^{15} boron atoms/cm³. Given k_O for boron is 0.8:

a) Determine the initial concentration of boron atoms in the melt to produce the required density.

b) If the initial amount of silicon in the crucible is 50kg, how many grams of boron should be added?

6.3 A crystal of silicon is to be grown using the Czochralski process. The melt contains 10kg of silicon to which is added 1mg of phosphorus. Given k_O (phosphorus) = 0.35, atomic weight of silicon = 28.09, atomic weight of phosphorus = 30.97, and density of phosphorus = 0.35g/cm³, determine the initial dopant concentration in the solid at the beginning of the growth if the atomic density of silicon is 5×10^{22} cm⁻³.

6.4 In the diffusion process, Q is defined as the total number of atoms per unit area of the semiconductor.

$$Q(t) = \int_0^\infty C(x,t)\,dx$$

where $C(x,t) = C_s \operatorname{erfc}\left[x/(2\sqrt{Dt})\right]$

Use the equation for the predeposition concentrations to determine an expression for Q in terms of $D, t,$ and C_s.

6.5 A certain process requires a boron predeposition diffusion step into an N-type wafer. The wafer has been uniformly doped, prior to this step, by 1×10^{15} phosphorus atoms/cm³. The predeposition step is carried out at 1000°C for 30 minutes. The diffusion coefficient, D, of boron in silicon at 1000°C is 1.78×10^{-14}cm²/s and the solid solubility of both boron and phosphorus in silicon is assumed to be 3×10^{20}/cm³ at the relevant temperature. Determine:

a) The diffusion length for this step.

b)   The surface concentration C_s after this step.

c)   The number of boron atoms/cm², Q, after this step.

Note that the dopant density cannot exceed the solid solubility in silicon.

6.6 The predeposition step of Problem 6.5 is followed by a drive-in diffusion step at 1050°C for 4 hours. Given the diffusion coefficient of boron in silicon at 1050°C is 17.3 × 10⁻¹⁴cm³/s, determine:

a)   The surface concentration after the drive-in step.

b) Q after the drive-in step.

c)   The junction depth after the drive-in step, recalling that the phosphorus concentration is 1×10^{15}/cm³.

Note that since no new dopant is introduced during the drive-in step, the Q at the initiation of this step is the same as that at the completion of the predeposition step.

6.7 List the principal advantages and disadvantages of ion implantation.

6.8 Why are two diffusion steps normally used when one would be less costly?

6.9 What is the main reason for the use of epitaxy in the fabrication of microcircuits?

chapter 7

LIMITATIONS TO IDEAL DIODE THEORY

7.0 INTRODUCTION

In the previous chapter, we explained the operation of the PN junction diode and determined the ideal diode expression relating the diode current to the voltage that is applied.

Upon comparing experimentally determined characteristics with those predicted by the current-voltage exponential relationship, we observe deviations between the two. These deviations occur mainly at the low-current and the high-current ends for forward bias, and in the reverse-bias regions of the characteristics. We also observe a sudden increase of current in the reverse bias region of the experimental characteristic, labeled a breakdown phenomenon, that is not predicted by our simple theory.

Our emphasis in this chapter is on the explanation of the physical processes that cause the actual characteristics to deviate from the exponential theoretical characteristics. The major differences arise because of the assumptions that we made in deriving the relations between the current in the diode and the voltage applied.

We will also study the switching properties of the diode and the relations between these properties and the capacitances inherent within.

7.1 DEVIATIONS IN THE FORWARD REGION OF THE CHARACTERISTIC

Review of Assumptions

In deriving the ideal current-voltage characteristic, and in the earlier subsection, "Assumptions for Ideal Diode Equation," we made the following major assumptions:

1. No recombination of carriers takes place in the depletion region when a forward bias is applied to the diode.

2. No generation of carriers occurs in the depletion region during reverse bias.

3. Low injection. This refers to a range of diode currents that causes the highest value of minority carrier density, which occurs at the edge of the depletion region at forward bias, to be much smaller than the equilibrium majority carrier density.

4. Coupled with assumption 3 is the assumption that the voltage applied to a diode appears totally across the depletion region. Because there is no voltage change in the regions outside the depletion region, no electric field exists there and hence no charges. These regions are therefore labeled the neutral regions. This justifies the process of calculating the currents in these regions by using the diffusion of minority carriers (refer to the earlier section, *Electric Field in N and P Regions*).

In the absence of any recombination in the depletion region, we conclude that there is no change in the magnitude of the minority carrier currents as they traverse the depletion region. We determined the ideal diode equation by finding the current due to holes at the edge of the depletion region in N and the current due to electrons at the edge of the depletion region in P. Since these currents were assumed to be constant throughout the depletion region, their sum determined the total diode current, which of course is constant throughout the diode.

Recombination Currents

By relaxing the first assumption, requiring that no recombination takes place in the depletion region of the diode when a forward bias is applied, we will show that the actual I-V characteristic of the diode at low voltages deviates from the ideal relations derived in Chapter 5. Recombination takes place in the depletion region because the carrier densities are greater than their equilibrium values.

In fact, we will show that expressions for the ideal and the actual total diode current may be considered and derived as being made up of electron or hole recombination currents, both in the neutral regions and in the depletion region.

We identify the three regions of the forward-biased diode, illustrated in Fig. 7.1, as the neutral regions 1 and 2, and the depletion region 3.

Figure 7.1 is similar to Fig. 5.17, except that we are including carrier recombinations in all three regions. We will trace the path of the total flux of electrons, I/q, consisting of a number F of electrons per unit time drifting from the P metal contact in the external circuit through the voltage supply to the N metal contact. To complete the loop, for a constant current I at a certain voltage through the diode, these F electrons will disappear completely by the time they reach the end of the metal contact that is in touch with the P region. A new batch of F electrons starts out again.

Obviously, these electrons must disappear by recombining with holes. Some of these electrons recombine with holes in N, that diffuse into the N region, within one hole diffusion length from the end of the depletion layer $x = x_n$. Some of the remaining electrons recombine with holes, which are sent from the P region, inside

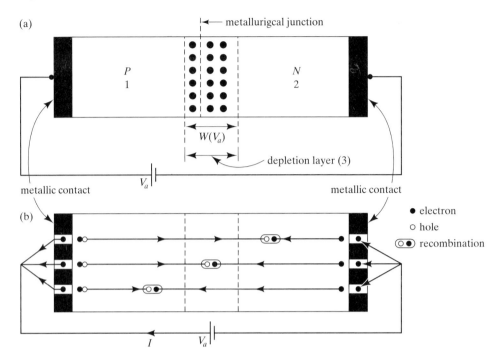

Figure 7.1   (a) A forward-biased diode and (b) diode currents made up of recombination currents.

the depletion region. The rest of the F electrons recombine with holes in the P region within one electron diffusion length from $x = -x_p$. While we have used electrons in the above explanation, one can arrive at the same result by considering the motion and recombination of holes within the diode.

Based on the above discussion, it is just as accurate to derive the ideal I-V relations for the forward-biased diode by considering only the two components of the electron or hole recombination currents that occur in the neutral regions. We will not carry that out but we will use the results in Eq. (5.43) for the ideal I-V diode relations in which recombination in the depletion region was excluded. The total current was obtained as the sum of the electron diffusion current at $x = -x_p$, which in the absence of recombination is the same at $x = x_n$ and the hole diffusion current at $x = x_n$. The ideal diode current, I_I, is given by Eq. (5.43) as

$$I_I = I_S \left[\exp\left(\frac{qV_a}{kT}\right) - 1 \right] \tag{7.1}$$

where I_S is determined from Eq. (5.44) by replacing p_{0n} by n_i^2/N_D and n_{0p} by n_i^2/N_A as

$$A \, q \, n_i^2 \left[\frac{D_p}{(L_p N_D)} + \frac{D_n}{(L_n N_A)} \right].$$

An expression for the recombination current in the depletion layer, I_R, has been obtained* as

$$I_R = I_{RO} \left[\exp \left(\frac{qV_a}{2kT} \right) - 1 \right] \tag{7.2}$$

The term I_{RO} is given by $(Aqn_iW/2\tau_0)$ where τ_0 is assumed to represent the lifetime of holes and electrons in the depletion layer and W is the width of the depletion layer, which is a function of the applied voltage.

Interestingly enough, Eq. (7.2) has been shown to represent also the generation current, I_{gen}, in the depletion layer when the diode is reverse-biased. For that, $I_{gen} = -I_{RO}$.

The expression for the total diode current becomes

$$I = I_I + I_R = I_S \left[\exp \left(\frac{qV_a}{kT} \right) - 1 \right] + I_{RO} \left[\exp \left(\frac{qV_a}{2kT} \right) - 1 \right] \tag{7.3}$$

In a reverse-biased diode, the total current becomes $I = -(I_S + I_{RO})$.

Comparison of Real and Ideal Diode Characteristics

A comparison of the I-V characteristics, predicted by the ideal diode equation, Eq. (7.1), and the actual diode equation, Eq. (7.3) for a forward-biased silicon diode, is shown on semilog scales in Fig. 7.2.

Low-Injection

We observe in the figure that at low values of forward bias, the current predicted by the ideal diode equation is smaller than the actual current, whereas at high values of voltages, the current given by the ideal equation is larger than the actual current. Let us first consider the low-voltage region.

The difference between the two currents at low voltages is the recombination current in the depletion region given by Eq. (7.2). As a result of this recombination, more electrons enter the depletion region from N than leave and similarly for holes from P. Thus, the total hole or electron current entering the depletion region is larger than the hole or electron current leaving it (the difference in either case being the recombination current).

The complete expression for the diode current includes, in addition to the ideal diode equation, (which represents only the sum of the recombination currents in the two neutral regions) a term that accounts for the recombination current inside the depletion region. We label the two currents that make up the ideal equation as diffusion currents, whereas I_R in Eq. (7.2) is labeled the recombination current.

We observe, in Fig. 7.2, that for a silicon diode at room temperature, the recombination current dominates at small forward voltages and, at voltages larger

---

*From Tyagi, *Introduction to Semiconductor Materials and Devices*, p. 198, copyright © Wiley, (1991). Reprinted by permission of John Wiley & Sons, Inc.

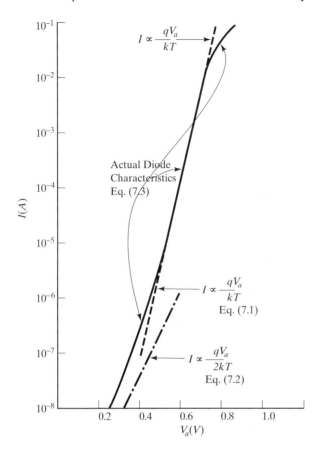

Figure 7.2 Real and ideal diode forward characteristics for a silicon diode at $T = 300$K.

than about 0.4V, the diffusion current is dominant. This is indicated by the change in the slope from 1/2 to 1, of the curves, as the voltage increases. The reason for the dominance of the recombination current at low values of voltage is that I_{RO} in Eq. (7.3) is greater than I_S by a factor of approximately three orders of magnitude for a silicon diode at $T = 300$K. The exact voltage at which the transition between the two slopes occurs depends on the temperature as exhibited through the dependence of the two currents on the intrinsic carrier density n_i. It is also important to point out that the width W of the depletion layer, which appears in I_{RO}, is itself a function of the applied voltage—it being larger at lower values of voltage, thus contributing to making the recombination current larger.

An empirical relation of the forward I-V characteristic for a forward-biased diode based on Eq. (7.3) and using the *ideality factor* η has been determined* as

$$I \propto \exp \frac{qV_a}{\eta kT} \tag{7.4}$$

---

*From S. M. Sze, *Physics of Semiconductor Devices*, p. 92, copyright © Wiley (1981). Reprinted by permission of John Wiley & Sons, Inc.

where $\eta = 1$ for diffusion current only and $\eta = 2$ for recombination current only. When both currents contribute to the total current, η varies between 1 and 2.

High Injection

At values of forward bias greater than about 0.7V in silicon, the diode I-V characteristic, shown in Fig. 7.2, departs from being a straight line with slope of q/kT to a curve indicating a decrease in the current for an increase of voltage when compared to that given by the ideal diode equation. The characteristic in this region is influenced by two phenomena: high-level injection and bulk resistance.

High-level injection is assumed to occur when the minority carrier density at the edge of the depletion layer is comparable to the equilibrium majority carrier density.

The relative carrier densities at low-level injection and high-level injection for a P^+N diode are shown in Fig. 7.3.

In low injection regime, the minority carrier density at the edge of the depletion layer is about six orders of magnitude smaller than the equilibrium majority carrier density. At high injection, the difference becomes negligible, as we illustrate for holes in N, where $N_D = 10^{16} \text{cm}^{-3}$:

At equilibrium: $n_{0n} \cong 10^{16} \text{cm}^{-3}, p_{0n} = 1 \times 10^4 \text{cm}^{-3}$

For 0.4V bias: $p_n(0) \approx 10^{10} \text{cm}^{-3}$, at the edge of the depletion layer in N.

For 0.7V bias: $p_n(0) \cong 1.1 \times 10^{16} \text{cm}^{-3}$, and $\Delta p_n(0) \cong 1.1 \times 10^{16} \text{cm}^{-3}$

                       $n_{0n} \cong 10^{16} \text{cm}^{-3}$, and $\Delta n_n(0) \cong \Delta p_n(0)$

so that, $n_n(0) = 2.1 \times 10^{16} \text{cm}^{-3}$, at the edge of the depletion layer in N.

The increase of the minority carrier concentration, at high injection levels, results from the increased forward bias and, to maintain neutrality, causes an increase in the majority carrier concentration. The increase of both majority and minority carrier concentration leads to higher currents and thus increased voltage drops (IR) in the neutral regions so that the total applied voltage is dropped across both the depletion and the neutral regions. Furthermore, both the mobility and lifetime are functions of the majority carrier concentrations in the N region. We observe, at voltages exceeding 0.7V in silicon, a slower increase in the current with increased voltage.

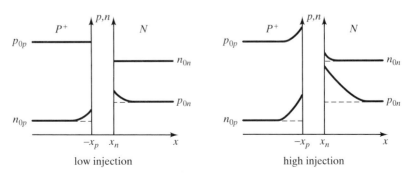

Figure 7.3   Carrier densities at low and high injection.

An approximate expression for the current at high injection level has been obtained as*

$$I = \left(\frac{2AqD_p n_i}{L_p}\right) \exp\left(\frac{qV_d}{2kT}\right) \tag{7.5}$$

where $V_d = V_a - IR$ and R is the total series resistance of the neutral regions. One can conclude that the slope of the I-V curve becomes $1/2$ instead of 1 assuming that the plot is made as a function of V_d. Equation (7.5) is valid only for a P^+N step junction.

It is worthwhile noting that the increase in the densities of both majority and minority carriers throughout the P and N regions, resulting from the increased bias, causes a small increase in the conductivities of these regions or a decrease in the resistivities. However, the decrease in the resistances of the bulk regions is much smaller than the increase in the current, hence the result is a relatively larger voltage drop.

Electric Field in N and P Regions

In our derivation of the ideal diode equation, we excluded any calculation of majority carrier currents. The reason is, we have no analytical method of determining an expression for the electric field in the so-called neutral regions, although the existence of an electric field is basic to the operation of the diode. This existence is verified by the following explanation: Starting at the P side of the depletion layer, electrons diffuse towards the metallic contact of the P region. In diffusing, they recombine with holes until, at the contact, the density of electrons has decreased to the equilibrium value. Because of recombination, the electron current in P has its largest value at the edge of the depletion layer and decreases to zero at the P metallic contact. The total diode current in P has, however, another component, the hole current within the P region. The hole current in the P region is caused mainly by the drift of holes away from the metal contact. The motion of holes away from the contact is a result of the acceleration acquired from the electric field.

A similar explanation holds for the motion of majority electrons in the N region.

The minority-carrier drift current is assumed to be negligible because the drift current of holes in the N region, given by $(q\,\mu_p\,pA\,\mathscr{E})$, is much smaller than the drift current of electrons in N, $(q\,\mu_n\,nA\,\mathscr{E})$, due to the difference in carrier concentrations.

7.2 REAL DIODE CHARACTERISTICS IN THE REVERSE DIRECTION

A comparison of the experimentally observed reverse characteristic with that predicted by the ideal diode equation is shown in Fig. 7.4.

The actual reverse characteristic exhibits two major deviations from that predicted by the ideal diode equation. First, the equation predicts that for a reverse bias

---

*From M.S. Tyagi, *Introduction to Semiconductor Materials and Devices*, p. 205, copyright © Wiley, (1991). Reprinted by permission of John Wiley & Sons, Inc.

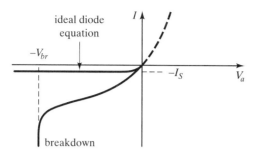

Figure 7.4 Actual reverse characteristics compared to that predicted by the ideal diode equation.

of the order of $4kT/q$, as is evident from Eq. (7.1), the current saturates at a value of $-I_S$, whereas the actual characteristic indicates an increase of reverse current with an increase in the reverse bias. Second, and beyond that, at certain voltage labeled $(-V_{br})$, the current increases rapidly unless limited by an external resistance. This voltage is known as the *breakdown voltage*.

An increase of the reverse bias applied to a diode has the following effects: First, the minority carrier densities at the edges of the depletion layer become zero for all practical purposes (p_{0n} exp $q\,V_a/kT$, $V_a < 0$) since the barrier at the junction has become so steep that minority carriers are immediately swept across the junction to the other region. Away from the vicinity of the depletion layer, the minority carrier density approaches its equilibrium value. Second, the depletion layer becomes wider as the reverse bias increases. Third, the potential barrier separating the two regions becomes steeper ($V_{bi} - V_a$, $V_a < 0$) and the maximum value of the electric field intensity, which occurs at the metallurgical junction, increases.

The high energy barrier at the junction makes it impossible for majority carriers to climb over, while minority carriers in or near the vicinity of the junction easily fall down this barrier. This causes the minority carrier density in the depletion region to be smaller than the equilibrium value. The result is an increase of the generation rate of carriers, a process that is opposite to the increased recombination rate in the forward-bias case. This process causes an increased reverse current and is enhanced by a wider depletion layer as the reverse voltage increases.

The ideal diode equation predicts an almost constant current, $-I_S$, for a reverse bias applied to the diode. The actual characteristic indicates a non-zero slope of the reverse characteristic for voltages less than breakdown. To analytically determine the reverse current, we assume a long-base diode and an applied reverse voltage of about $4kT/q$. The total reverse current is obtained from Eq. (7.3) as

$$I = -(I_S + I_{RO}) \tag{7.6}$$

where I_S and I_{RO} are defined following Eqs. (7.1) and (7.2). We note that while I_S is independent of voltage, I_{RO} increases with an increase of the reverse bias. The increased reverse bias increased the width of the depletion region W, resulting in an increase of the reverse current. The term I_{RO} accounts for the increased recombination current at low values of forward voltage and also accounts for an increase in the generation current with increasing reverse bias.

Junction Breakdown

As the reverse bias applied to a diode increases, both the width of the depletion region and the electric field in that region increase. In general, the upper limit on the reverse voltage is placed by the phenomenon of breakdown that occurs at a certain critical value of the electric field. At breakdown the reverse current increases very rapidly with miniscule increases of the reverse voltage, as shown in Fig. 7.4.

There are two types of breakdown: *Zener* breakdown and *avalanche* breakdown. Zener breakdown results from the tunnelling of a large number of electrons through the energy barrier, causing a large current. Avalanche breakdown results from impact ionization of atoms by electrons that have acquired a high value of kinetic energy from the high electric field in the depletion region. We will discuss avalanche breakdown first.

Avalanche Breakdown

At a critical value of the electric field in the depletion region, a sudden increase of the reverse current caused by impact ionization occurs. The avalanche or snowballing effect that occurs in the depletion region and resulting from impact ionization is illustrated in Fig. 7.5, where a large reverse bias has been applied to the diode. While both electrons and holes take part in impact ionization, to simplify the explanation, we consider the motion of electrons in the depletion region. We recall that the kinetic energy of an electron is measured by its energy separation above the bottom of the conduction band E_C. An electron at E_C has zero kinetic energy.

When an electron in the depletion layer is accelerated by the high electric field caused by a large reverse bias, it gains kinetic energy. If the electron gains kinetic energy equal to or greater than the band gap energy, E_g, and it collides with the lattice, a covalent bond is broken. The breaking of a covalent bond, which is equal to the elevation of an electron from the valence band to the conduction band, results in the generation of an electron-hole pair, as shown in Fig. 7.5 for electron Number 1.

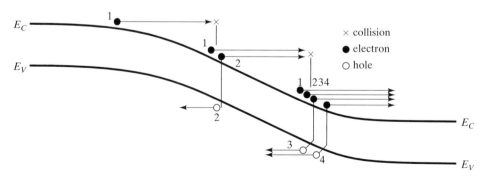

Figure 7.5 Illustration of impact ionization for electrons only. Each electron generates a hole and an electron. Generation of electron-hole pairs by hole ionization is not shown.

The two electrons, electron Number 1 and the one resulting from the collision, are in turn accelerated by the field, gain kinetic energy $KE > E_g$, collide with the lattice, and generate two additional electron-hole pairs. While we have shown the consequences of electrons colliding only with the lattice, it is important to recognize that the holes behave in an analogous manner in causing impact ionization. The avalanche process of carrier generation by collision results in a very large number of carriers and hence a large increase in the current.

Because of the avalanche process, the current entering the depletion layer is multiplied by a factor M, known as the *multiplication factor*, as it crosses the layer. If without any avalanche process the reverse current is I_{RO}, then the actual reverse current, I_R, is determined from

$$M = I_{out}/I_{in} = I_R/I_{RO} \tag{7.7}$$

where I_{out} and I_{in} refer to exit from and entry into the depletion region. An expression for M, as a function of an empirically determined exponent n, is given as

$$M = \frac{1}{1 - \left(\left|\frac{V_R}{V_{br}}\right|\right)^n} \text{ for } 2 < n < 6 \tag{7.8}$$

where V_R is the applied reverse voltage and $-V_{br}$ is the breakdown voltage of the junction.

The use of the word breakdown is unfortunate as it might convey the impression that a destruction of the diode has taken place. The diode is destroyed only when its rated power dissipation is exceeded. Therefore, this will depend on the current at breakdown. The current is normally restricted to be less than the maximum allowable value by the addition of a series resistor. The product of the maximum allowable value of the current and the voltage at breakdown must be less than the rated power dissipation of the diode.

Expression for the Breakdown Voltage

It is evident from Fig. 7.4 that an increase in the reverse bias increases the current and causes breakdown to take place. The applied voltage causes a high electric field, which causes the avalanche breakdown as it accelerates the carriers, which in turn produce new carriers.

It appears then that there is a critical value of the electric field, $\mathscr{E}_{cr}$, which causes the breakdown. The critical field is usually assumed to be a constant for a certain semiconductor, although it varies slightly with the doping.

We will now relate the magnitude of the breakdown voltage, V_{br}, to the diode constants, assuming that the critical value of the field is a constant.

We note from Fig. (5.10) that the electric field maximum occurs at $x = 0$ and has value, at equilibrium, determined from Eq. (5.25) as $-2V_{bi}/W$ and W is given by Eq. (5.28). For biasing conditions, we replace V_{bi} by V_j ($V_{bi} - V_a$), W by Eq. (5.28), and solve (5.25) for $\mathscr{E}_{max}$ as

$$\mathscr{E}_{max} = -[2 V_j q (N_A N_D)/\varepsilon(N_A + N_D)]^{1/2} \tag{7.9}$$

For $N_A \gg N_D$,

$$\mathcal{E}_{max} = -\left[\frac{2\,q\,V_j\,N_D}{\varepsilon}\right]^{1/2} \tag{7.10}$$

where $\mathcal{E}_{max}$ is determined by the region with the lower doping density.

We assume that breakdown occurs at $-V_{br}$, where $\mathcal{E}_{cr} = \mathcal{E}_{max}$ and also that $V_{bi} \ll |V_a|$, so that $V_a = V_{br}$. Solving for V_{br}, we have

$$V_{br} = \frac{\varepsilon\,\mathcal{E}_{cr}^2}{2qN_D} \tag{7.11}$$

For a constant critical field, we note that the magnitude of the breakdown voltage is inversely proportional to the doping of the weaker region of the junction diode.

We will illustrate the calculation of the breakdown voltage by example 7.1, while example 7.2 is used to illustrate the dependence of the multiplication factor on the reverse voltage.

EXAMPLE 7.1

The doping densities of an abrupt-junction silicon PN diode are $N_A = 10^{17}$ atoms/cm^3 and $N_D = 8 \times 10^{15}$ atoms/cm^{-3}. Determine the breakdown voltage if the critical field is $3 \times 10^5 V/$cm. Avalanche breakdown takes place when the maximum electric field intensity in the depletion region is equal to the critical value.

Solution   The expression for the maximum value of the electric field intensity is given by Eq. (7.9) as

$$|\mathcal{E}_{max}| = \left[\frac{2q\,N_A\,N_D\,V_j}{\varepsilon\,(N_A + N_D)}\right]^{0.5}$$

where $V_j = V_{bi} - V_a$, $\varepsilon = \varepsilon_r\varepsilon_o$, ε_r for silicon is 11.8 and $\varepsilon_o = 8.85 \times 10^{-14}$F/cm.

We substitute the values so that,

$$3 \times 10^5 = \left[\frac{2 \times 1.6 \times 10^{-19} \times 10^{17} \times 8 \times 10^{15} \times V_j}{1.04 \times 10^{-12} \times (10^{17} + 8 \times 10^{15})}\right]^{0.5}$$

Solving for V_j, we obtain

$$V_j = |V_{bi} - V_a| = 39.48V$$

The voltage that must be applied is calculated after V_{bi} is found from Eq. (5.17):

$$V_{bi} = \frac{kT}{q}\ell n\,\frac{N_A N_D}{n_i^2} = .0259\,\ell n\,\frac{8 \times 10^{32}}{10^{20}} = 0.77V$$

Thus, a reverse voltage having magnitude $(39.48 - 0.77 = 38.7)$ will cause breakdown in the diode.

EXAMPLE 7.2

In the absence of any multiplication of carriers when a reverse bias is applied to a diode, the magnitude of the reverse current is the saturation current I_S. To obtain an estimate of the effect of multiplication, calculate for the diode of Example 7.1, whose breakdown voltage, V_{br}, is 38.7V, the values of the multiplication factors for reverse voltages of 10, 20, 30, 38.3 and 38.6. Let $n = 3$ in the expression for M.

Solution   From Eq. (7.8), M is given as

$$M = \frac{1}{1 - \left[\left|\dfrac{V_a}{V_{br}}\right|\right]^n}$$

where For $V_a = -10\text{V}$, $M = 1.0175$
$\quad\quad\quad\quad V_a = -20\text{V}$, $M = 1.1601$
$\quad\quad\quad\quad V_a = -30\text{V}$, $M = 1.872$
$\quad\quad\quad\quad V_a = -38.3\text{V}$, $M = 32.58$
$\quad\quad\quad\quad V_a = -38.6\text{V}$, $M = 129.33$

It is important to note the dramatic sudden increase of the multiplication factor, which is the ratio of the reverse current leaving the depletion layer to that entering, as the breakdown voltage is approached. Increasing the applied voltage by 0.3V, from 38.3V to 38.6V, increased the current ratio from 32.58 to 129.33. A smaller value of n causes sharper increases in M as $-V_{br}$ is approached. For $n = 2$ and $V_a = -38.6\text{V}$, M has a value of 193.

Zener Breakdown

The physical mechanisms that result in Zener breakdown are completely different from those causing avalanche breakdown. The only similarity between the two is the general shape of the reverse characteristic of the diode. They have two major differences. First, Zener breakdown in a silicon diode takes place at reverse voltages of the order of 5 volts or less. Second, Zener breakdown characteristic exhibits a more abrupt rise in current at breakdown. The difference in the shape of the characteristic is shown in Fig. 7.6.

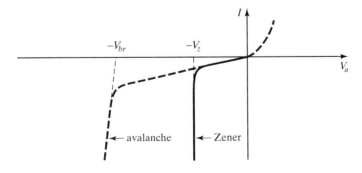

Figure 7.6   The characteristic of avalanche and Zener breakdown compared.

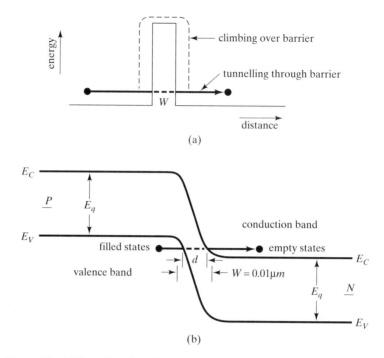

Figure 7.7 (a) Tunnelling through a barrier and (b) tunnelling across a junction.

The breakdown in Zener diodes is a result of the phenomenon of *tunnelling*. The illustration of tunnelling through a barrier and tunnelling in a diode are shown in Fig. 7.7. (Refer to Example 1.4.)

Tunnelling may take place when electrons in the valence band of the P region of a diode are opposite empty conduction band energy levels in N. The probability of tunnelling increases as the width of the depletion layer is reduced. We observe from Eq. (5.28) that the width of the depletion layer decreases substantially as the doping of the two regions is made very large of the order of 10^{18}cm^{-3}. With reverse bias, the decreased width of the depletion layer results in a high electric field intensity.

It is possible to have the relevant energy levels aligned for any PN junction diode by the application of the proper reverse bias, but tunnelling and Zener breakdown will not occur unless the depletion layer is reduced by high doping so as to cause an electric field intensity in silicon of about 10^6V/cm. It is worthwhile noting that the actual separation, d, between the relevant energy levels in P and N is smaller than the depletion layer width W. This effect is shown in Fig. 7.7(b).

Commercially available diodes have breakdown voltages that range from one volt to hundreds of volts. These diodes are fabricated to possess certain breakdown voltages and are very useful in voltage regulator applications. While they are all known collectively as *Zener diodes*, silicon diodes with breakdown voltages of 5 volts or less are normally of the Zener type, the others are of the avalanche type.

REVIEW QUESTIONS

Q7-1 Briefly explain the mechanism by which the recombination current, in a certain voltage range, exceeds the current predicted by the ideal diode equation.

Q7-2 Briefly explain the effect of high-level injection on the diode current.

Q7-3 How does an increase of diode reverse bias increase the diode reverse current?

Q7-4 Explain the role of the avalanche process in breakdown.

Q7-5 Clearly explain the main requirements for Zener breakdown to occur.

HIGHLIGHTS

- Actual diode I-V characteristics deviate from those predicted by the ideal diode equation in both the forward and reverse directions.
- In the forward direction, the actual characteristics establish a higher current at low forward voltages ($\leqslant 0.3$V) because of recombination in the depletion layer. At the high end of the forward region ($\geqslant 0.7$V), the actual current is lower because of the effects of high-level injection and voltage drops in the neutral regions.
- The ideal relations show a very small constant current in the reverse directions, whereas the actual characteristics indicate both an initial increase of current with increase of voltage and a rather sudden jump to an extremely large current. This large current can only be limited by a series resistance in the circuit. This sudden large current leads to breakdown.
- There are two types of breakdown. In silicon diodes the breakdown that takes place at reverse voltages of 5V or less is known as Zener breakdown. At higher voltages, breakdown is of the avalanche type. Zener breakdown results from the tunnelling of carriers across the junction.
- The probability of the transition of an electron through a barrier (tunnelling) is a strong function of the barrier thickness. Tunnelling is only significant in highly doped semiconductors where the depletion region is very narrow and the electric field is very high.

EXERCISES

E7-1   a) Determine the value of the recombination current in a PN silicon diode at $T = 300$K operating at a voltage of 0.3V given $N_A = 10^{17}$cm^{-3}, $N_D = 10^{15}$cm^{-3}, $A = 10^{-5}$cm^2, $W = 0.8$ µm, and $\tau_p = \tau_s = 1$µs.

b) Calculate the diode current using the ideal equation

Ans: a) $I = 2 \times 10^{-10}$ A, b) $I = 6.52 \times 10^{-10}$ A

E7-2   A diode has a breakdown voltage of 20V. Use $n = 3$ in Eq. (7.8) to determine the multiplication factor at a reverse voltage of 19V.

Ans: $M = 7$

E7-3   The critical field for breakdown in silicon is 3×10^5V/cm. Calculate the doping of the N side of a P$^+$N diode if breakdown is to occur at 20V.

Ans: $N_D = 1.47 \times 10^{16}$cm^{-3}.

7.3 CAPACITANCES OF THE DIODE

A PN junction diode possesses two intrinsic capacitances. One of these, the *transition capacitance*, or junction capacitance, has physical characteristics similar to those of a parallel-plate capacitor. The expression that defines the transition capacitance in terms of physical parameters is identical to that of the parallel-plate capacitor.

The second capacitance, known as the *storage capacitance*, or diffusion capacitance, also possesses the charge-storing property of a capacitor but has no other physical characteristics similar to the parallel-plate capacitor.

A simple parallel-plate capacitor has capacitance that depends only on the physical dimensions and on the dielectric constant of the insulator separating the plates. It is known as a linear capacitance because a plot of the charge Q on each plate versus voltage is a straight line passing through the origin. The diode capacitances, on the other hand, are nonlinear, indicating that the capacitance is a function of the applied voltage and it is given by

$$C = \frac{dQ}{dV} \tag{7.12}$$

The transition capacitance exists when a diode is reverse-biased or forward-biased, whereas the storage capacitance occurs only when the diode is forward-biased.

The Transition (Junction) Capacitance

The depletion layer of a diode consists of positively ionized donors and negatively ionized acceptor atoms. We refer to Figs. 5.9 and 5.10, which show the distribution of the ions in the depletion layer. The depletion layer extends deeper into the lightly doped region, so that for $N_A \gg N_D$, $x_p \ll x_n$. However, the charges contained within both parts of the depletion layer are equal as indicated by Eq. (5.19).

The change in the voltage, ΔV_a, applied to the diode causes the width of the depletion layer to change, as shown in Fig. 7.8, resulting in a change of charge ΔQ. Because the relation between the charge and the applied voltage, as we shall see, is nonlinear, we define a transition capacitance or junction capacitance C_j as

$$C_j = \frac{dQ_s}{dV_j} \tag{7.13}$$

where $V_j = V_{bi} - V_a$ and $dV_j = -dV_a$ since V_{bi} is a constant. The quantity of charge stored in each side of the depletion layer is given by Q_s.

The magnitude of the charge in the depletion layer is obtained from Eq. (5.19) modified for a biased condition as

$$Q_s = Aq N_D x_n = Aq N_A x_p \tag{7.14}$$

Using x_p from Eq. (5.27) in Eq. (7.14) and replacing V_{bi} by V_j, which becomes $(V_{bi} - V_a)$, the expression for Q_s is written as

$$Q_s = A\left(\frac{2q\varepsilon V_j N_A N_D}{N_A + N_D}\right)^{1/2} \tag{7.15}$$

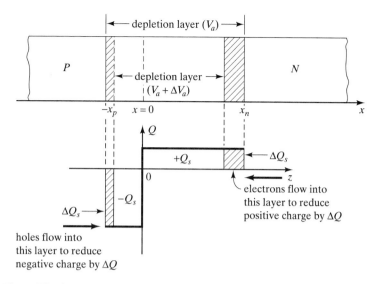

Figure 7.8 Storage charge change and a decrease in the depletion layer width result from positive ΔV_a bias. The distances x_p and x_n represent the width of each of the depletion layers in P and N respectively.

The expression for the transition capacitance becomes

$$C_j = \frac{dQ_s}{dV_j} = A\left(\frac{q\,\varepsilon\,N_A\,N_D}{2(N_A + N_D)}\right)^{0.5} 1/V_j^{0.5} \tag{7.16}$$

For $N_A \gg N_D$, C_j is simplified to

$$C_j \approx A\left(\frac{q\varepsilon\,N_D}{2}\,\frac{}{V_j}\right)^{0.5} = A\left[\frac{q\varepsilon\,N_D}{2(V_{bi} - V_a)}\right]^{1/2} \tag{7.17}$$

It is to be noted that with a larger V_j at reverse bias, C_j is smaller than the value at forward bias.

When V_{bi} in Eq. (5.28) is replaced by V_j, for $N_A \gg N_D$ and $x_p \ll x_n$ the depletion layer width, x_n, is

$$x_n \approx \left(\frac{2\,\varepsilon\,V_j}{qN_D}\right)^{1/2} \tag{7.18}$$

For biased conditions, x_n and x_p replaced x_{n0} and x_{p0} respectively. Voltage V_j becomes

$$V_j \approx \frac{x_n^2 qN_D}{2\varepsilon} \tag{7.19}$$

Using Eq. (7.19) in Eq. (7.17), we have

$$C_j = \frac{\varepsilon A}{x_n} \tag{7.20}$$

Hence, the transition capacitance is a voltage-dependent capacitance (through its dependence on x_n). Such a capacitor finds application in the electronic tuning circuit of a television receiver.

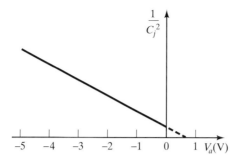

Figure 7.9   Plot of $1/C_j^2$ vs. V_a.

It is interesting to note that, in contrast to a parallel-plate capacitor in which the charges reside on the plates, the charges of transition capacitances are located in the depletion layer.

A plot of $1/C_j^2$ versus V_a using Eq. (7.17) and shown in Fig. 7.9 yields a straight as in test line having an intercept of V_{bi} on the V_a axis. The slope can be used to determine the dopant density N_D.

Example 7.3 illustrates the dependence of the transition capacitance on the applied voltage. We note the larger capacitance at forward bias.

EXAMPLE 7.3

The doping densities of an abrupt-junction silicon PN diode are $N_A = 10^{17}$atoms/cm^3 and $N_D = 8 \times 10^{15}$atoms/cm^3. The area of the junction is 2×10^{-5}cm^2.

Calculate the junction capacitances at: a) zero bias, b) reverse bias of 6V, and c) forward bias of 0.7V.

Solution   The expression for the junction capacitance of an abrupt junction is given by Eq. (7.16) as

$$C_j = A\left(\frac{q\,\varepsilon\,N_A N_D}{2(N_A + N_D)}\right)^{0.5} \times 1/V_j^{0.5}$$

where $V_j = V_{bi} - V_a$. For zero bias, $V_j = V_{bi}$, and $C_j = C_{j0}$.

a) $C_{j0} = 2 \times 10^{-5}\left(\dfrac{1.6 \times 10^{-19} \times 1.04 \times 10^{-12} \times 10^{17} \times 8 \times 10^{15}}{2(1.08 \times 10^{17})}\right) \times \dfrac{1}{V_{bi}^{0.5}}$

$$V_{bi} = .0259\,\ell n\,\frac{8 \times 10^{32}}{10^{20}} = 0.77\text{V}$$

$$C_{j0} = 0.56\text{pF}$$

b) At $V_a = -6$V, $V_j = 0.77 + 6 = 6.77$V

$$C_j = C_{j0}\left(\frac{V_{bi}}{V_{bi} - V_a}\right)^{0.5} = 0.56 \times \left(\frac{0.77}{6.77}\right)^{0.5} \text{pF}$$

$$C_j = 0.188\text{pF}$$

c) At $V_a = 0.7$V, $V_j = 0.77 - 0.7 = 0.07$V

$$C_j = 0.56\left(\frac{0.77}{0.07}\right)^{0.5} \text{pF} = 1.85\text{pF}.$$

We conclude that for the same diode, the transition capacitance in forward bias is much larger than that in reverse bias. The reason is that the width of the depletion region is much smaller when a diode is forward-biased.

Storage Capacitance

The application of a forward bias, V_a, to a diode has the following consequences: a reduction in the barrier height, a reduction in the width of the space-charge layer, and an injection of majority carriers across the depletion layer into the opposite region where they are stored as excess minority carriers. The density of the excess stored minority carriers increases with an increase of the forward bias.

In Eq. (5.37(a)) we determined an expression for the excess hole density distribution in the N region. By relocating the origin at $x' = 0, (x = x_n,)$ the expression of Eq. (5.37(a)) for the excess hole density in N, for $W_n \gg L_p$, becomes

$$p'_n = p_{0n}\left(\exp\frac{qV_a}{kT} - 1\right)e^{-x'/L_p} \tag{7.21}$$

The excess hole charge Q_B stored in the N region determined for a large $W_n \gg L_p$, is given by

$$Q_B = qA\int_0^\infty p'\,dx'$$

$$= qA\,L_p\,p_{0n}\left(\exp\frac{qV_a}{kT} - 1\right) \tag{7.22}$$

The storage capacitance C_s, also labeled the diffusion capacitance becomes

$$C_s = \frac{dQ_B}{dV_a} = \frac{A\,q^2}{kT}\,L_p\,p_{0n}\exp qV_a/kT \tag{7.23}$$

We observe that the storage capacitance increases exponentially with the forward bias. In addition to the storage capacitance, and as we found earlier, a forward-biased diode has a transition capacitance as well. The transition capacitance, in accordance with Eq. (7.20), of a forward-biased diode has a larger value than that of a reverse-biased diode since the depletion layer is narrower. However, for a forward-biased diode the storage capacitance is significantly larger than the transition capacitance.

It is essential to point out that the charge storage we have referred to is that of excess minority carriers. The P and N regions are neutral at low injection because of the presence of excess majority carriers drawn from the metal contact. There is no actual net space charge in the neutral regions as exists in the depletion region.

The storage capacitance is a measure of the change of the area under the minority carrier distribution as the voltage changes. This requires a time delay as measured by the capacitance.

EXAMPLE 7.4

The doping densities of an abrupt-junction silicon diode are $N_A = 10^{17}$ atoms/cm^3 and $N_D = 10^{15}$ atoms/cm^{-3}. The cross sectional area of the diode is 2×10^{-5} cm^2 and the lifetime of holes in the N region is 0.1μs. Given the diffusion constant for holes in N is 16cm^2/s, calculate at room temperature the storage capacitance at a) $V_a = 0.6$V and b) $V_a = 0.65$V.

Solution   Since $N_A \gg N_D$, the storage capacitance is given by the expression for C_s in the N region as

$$C_s = \left(\frac{Aq^2 L_p p_{0n}}{kT} \exp \frac{qV_a}{kT} \right)$$

The equilibrium density of holes in the N region, p_{0n}, is given by

$$p_{0n} \cong \frac{n_i^2}{N_D} = \frac{10^{20}}{10^{15}} = 10^5 / \text{cm}^3$$

The diffusion length for holes is

$$L_p = (D_p \tau_p)^{0.5} = (16 \times 0.1 \times 10^{-6})^{0.5} = 1.26 \times 10^{-3} \text{cm}$$

a) At $V_a = 0.6$V, C_s becomes

$$C_s = 179\text{pF}$$

b) At $V_a = 0.65$V, C_s becomes

$$C_s = 1234.3\text{pF}$$

The results of the above example and those of example 7.3 indicate that, in the forward direction, the storage capacitance is much larger than the transition capacitance.

7.4 SMALL-SIGNAL EQUIVALENT CIRCUIT

We define the incremental resistance of the diode as

$$r_d = \frac{dV_a}{dI} \tag{7.24}$$

By using the exponential relationship of Eq. (5.43) for the current of the diode, and assuming $N_A \gg N_D$ so that $n_{op} \ll p_{on}$ and $V_a \gg kT/q$, the expression becomes

$$I = (qAD_p p_{0n}/L_p) \exp qV_a/kT \tag{7.25}$$

The incremental resistance is found as

$$r_d = \left(\frac{dI}{dV_a} \right)^{-1} = \frac{L_p}{(q/kT)qA\, D_p\, p_{0n} \exp\, (qV_a/kT)} = (kT/q)I \tag{7.26}$$

By taking the product of the expressions for the incremental resistance and the storage capacitance, using Eqs (7.23) and (7.26) we have

$$r_d C_s = \frac{L_p^2}{D_p} = \tau_p \tag{7.27}$$

We note that this time constant is equal to the lifetime of holes in the N region. The following example illustrates the dramatic decrease of the incremental resistance as the forward bias is increased.

EXAMPLE 7.5

For the diode of Example 7.4, calculate the incremental resistance at a) 0.6V and b) 0.65V.

Solution   The simple expression that can be used to calculate the incremental resistance of a diode, where $N_A \gg N_D$, is given by Eq. (7.27) as

$$r_d C_s = \tau_p.$$

Using the values for C_s determined in Example 7.4 and substituting the value of $0.1\mu s$ for the lifetime, we have

$$\text{for } V_a = 0.6V,$$

$$r_d = 558.6 \text{ohms}$$

$$\text{and for } V_a = 0.65V,$$

$$r_d = 81 \text{ ohms.}$$

Equivalent Circuit of the Diode

A circuit containing the capacitances and incremental resistance can be used to replace the diode for incremental variations of the voltage or current. When, for example, an incremental change in the applied voltage is made, by circuit analysis, one can then determine the corresponding change in the current but can also determine the time response of the diode, such as how long it takes the diode current to reach a certain fraction of its final value. The response to a sinusoidal voltage or current can be determined using circuit analysis. As we have just seen, the values of the circuit elements depend upon the operating DC voltage which is applied to the diode. Such a circuit is known as a *small-signal equivalent circuit.*

Small signal equivalent circuits for the diode in the reverse and forward directions are shown in Fig. 7.10. The circuit for the reverse-biased diode includes the

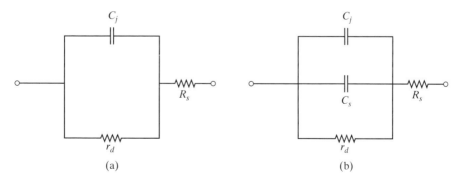

Figure 7.10   Equivalent circuit of diode (a) reverse-biased and (b) forward-biased.

junction capacitance C_j and an incremental resistance whose expression is determined from Eq (7.26) as

$$r_d = \frac{kT}{q\left(I_S \exp \dfrac{qV_a}{kT}\right)} \tag{7.28}$$

Because the diode operates at relatively high voltages in the reverse direction, the incremental resistance is very high and may often be assumed to be infinite.

In addition to the parallel combination of the capacitance and the incremental resistance, a resistance is added in series with the combination. This is the bulk resistance, R_s of the neutral regions. It is calculated as the product of the resistivity and the length of each region divided by the cross sectional area. It is evident that in both the forward and reverse directions the resistivity changes as the current changes. Thus, complicated relations are required to calculate this resistance.

The equivalent circuit in the forward direction includes both the storage and transition capacitances and the incremental resistance. This incremental resistance is a very small quantity. On the other hand, the effect of the series resistance, R_s, becomes quite important in the forward direction and mainly at high values of current in the high injection mode.

7.5 PROPERTIES OF THE SHORT-BASE DIODE

We will label the base of a diode as the region that has the weaker doping. It is the N region of a P^+N junction diode. A short-base PN diode is one in which the base width, W_n, is much smaller than the diffusion length, L_p, of minority carriers, $W_n \ll L_p$.

Referring to Eq. (5.37(b)), which was obtained by a series expansion of the exponential and shifting the origin to $x' = 0$ ($x = x_n$), the expression for the excess hole density in N becomes

$$p'(x) = p_{0n}\left[\exp \frac{qV_a}{kT} - 1\right]\left(1 - \frac{x'}{W_n}\right) \tag{7.29}$$

A sketch of this equation is shown in Fig. 7.11, indicating a straight line distribution.

The expression for the current of a short-base diode is given by the diffusion current anywhere in the N region, since the slope of the excess carriers is constant throughout the base. A constant current implies zero recombination, which is expected since we have assumed that W_n is much smaller than a diffusion length. The forward diffusion current, for $N_A \gg N_D$, becomes

$$I \approx -qD_p A\frac{dp'}{dx'} = \frac{qAD_p}{W_n}p_{0n}\left[\exp\left(\frac{qV_a}{kT}\right) - 1\right] \tag{7.30}$$

Upon comparing Eq. (7.30) for the short-base diode with that of the corresponding long-base diode (Eq. 7.25), we note that the current in the short-base diode exceeds that in the long-base diode by the factor L_p/W_n.

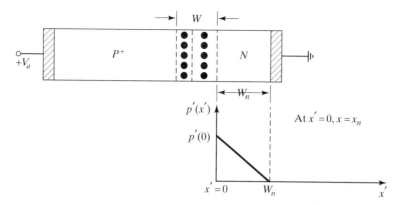

Figure 7.11 Excess hole density distribution $p'(x')$ in the base of a short-base diode ($W_n \ll L_p$).

We will now determine expressions for the storage capacitance and the incremental resistance of a short-base diode. The expression for the transition capacitance is the same as that given for the long-base diode.

The excess minority carrier charge stored in the N region is given by the product of q and the area of the triangle, shown in Fig. 7.11, as

$$Q_B = q\,A\left(\frac{p'(0)}{2}\right) \times W_n \cong q\,A\,p_{0n}\frac{W_n}{2}\left(\exp\frac{qV_a}{kT} - 1\right) \qquad (7.31)$$

The storage capacitance becomes

$$C_s = \frac{dQ_b}{dV_a} = \frac{q^2}{kT}\frac{A\,W_n\,p_{0n}}{2}\exp\frac{qV_a}{kT} \qquad (7.32)$$

The incremental resistance of the short-base diode is given by

$$r_d = \left(\frac{dI}{dV_a}\right)^{-1} = \frac{W_n}{\left(\dfrac{q}{kT}\right)q\,A\,D_p\,p_{0n}\exp\dfrac{qV_a}{kT}} \qquad (7.33)$$

The product $r_d C_s$ becomes

$$r_d C_s = \frac{W_n}{2D_p} \qquad (7.34)$$

On comparing the equations in the forward direction for the currents and for the storage capacitances of the short and long-base diodes having the same p_{0n}, A, V_a, and $N_A \gg N_D$, we have

$$I(\text{short}) = \frac{L_p}{W_n}I(\text{long}) \qquad \text{(a)}$$

$$C_s(\text{short}) = (W_n/2L_p)\,C_s(\text{long}) \quad \text{(b)} \qquad (7.35)$$

Therefore, the short-base diode has a higher current and a smaller storage capacitance than the long-base diode.

Upon comparing Eq. (7.23) for the storage capacitance of the long-base diode, Eq. (7.32) for the capacitance of the short-base diode, and the conditions of Eqs. (7.35), we observe that the capacitance of the short-base diode is much smaller than that of the long-base diode. This is a result of the fact that much less excess charge is stored in the base of the short-base diode, as shown in Eqs. (7.22) and (7.31). This excess charge of minority carriers accumulates when the diode is conducting in the forward direction.

For the same forward voltage, V_a, and hence for the same forward current, less charge is stored in the base of the short-base diode because the width of the base is much smaller than a diffusion length. We recall that the long-base diode is defined as one where the width of the base is equal many diffusion lengths.

In the next section, we will consider the time it takes a diode to switch from a reverse state to forward and back to reverse. We will conclude that this time is made up mainly of the time taken to turn the diode OFF from an ON position. This turn-off time depends directly on the amount of excess charge that is stored in the base while the diode is in the forward state, and hence on the storage capacitance.

7.6 DIODE SWITCHING CHARACTERISTICS

Junction diodes are used as switching elements in a variety of applications. When used as a switch, the diode is said to be ON when it is operating in the forward direction and OFF when it is reverse-biased in the region of the characteristic that precedes breakdown. The performance of a diode as a low-current switch is measured mainly by the time increment it takes to switch from one state to another. Furthermore, an ideal switch is required to have zero resistance in the forward ON state and infinite resistance in the OFF state. This resistance is the ratio of the steady-state voltage to the steady-state current.

We will determine, in this section, the physical parameters that determine the switching times.

In our analysis, we assume a P^+N junction diode so that the major determinant of excess carrier storage is the excess hole density storage in the N region. We therefore neglect the operations in the P region.

Turn-ON Time

To determine the ON time, we consider a long-base diode, $W_n \gg L_p$, in the circuit of Fig. 7.12(a) that is in the reverse-biased state having a $V_D = -V_a$ and a negligibly small reverse current $-I_s$. The procedure we will follow consists in determining the relation between the excess hole charge stored in N and the forward current, I_F, that will be applied at $t = 0$. To turn the diode ON, we can apply a positive voltage pulse or a positive current pulse. Because of its simplicity, we will use a current pulse formed in the circuit of Fig. 7.12(a) by moving the switch to position A, where a forward voltage $V_F(\approx 20V)$ is in series with a large resistance R_F. The positive current $I_F \approx V_F/R_F$ for all $t > 0$, $|V_a| \ll V_F$, and V_a is V_D at large t.

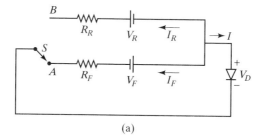

(a)

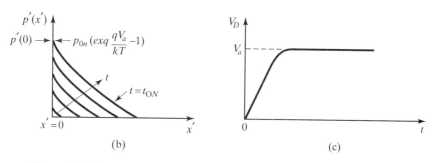

(b)   (c)

Figure 7.12   (a) Switching circuit of the diode, (b) excess hole buildup in the N region, and (c) diode voltage buildup. The voltage V_a is the steady-state value and V_D is the instantaneous value of the diode voltage.

Since the diode current is changed instantaneously from $-I_S$ to I_F, the forward voltage V_D becomes zero in a time of approximately 10^{-10}s since it takes very little time for electrons in the N region and holes in the P region to move to neutralize the donor and acceptor ions. We will therefore begin counting time, $t = 0$, from $V_D = 0$.

Because the forward current is constant $I_F = (qAD_p)(dp/dx')$, the slope of the hole density distribution at the N side of the depletion layer is constant while the hole density builds up, with time, inside the N region as holes diffuse from P to N and as shown in Fig. 7.12(b). The current supplies holes for both the buildup of excess holes and to feed the recombination with electrons. The buildup is terminated when the hole density at the N side of the depletion layer reaches the value that is determined by the voltage across the diode, which corresponds to the current I_F as determined from the static I-V characteristic of the diode. At that time, the hole density has reached the value shown at $t = t_{0n}$ in Fig. 7.12(b).

Since the forward current delivers a certain number of hole-coulombs per unit time, and knowing the final accumulated number of hole-coulombs, neglecting recombination, then the turn-ON time is determined by the relation between the current and the total of excess hole charge accumulated.

The equation, for the hole current density, Eq. (5.41), at the edge of the depletion layer, labeled $J_p = J_F$, and the equation for the excess stored hole charge, Eq. (7.22), are repeated here as

$$J_F = I_F/A = q\,D_p\,p_{0n}/L_p\,[\exp{(q\,V_a/kT)} - 1] \quad \text{(a)}$$

$$Q_B = q A L_p p_{0n} [\exp(q V_a/kT) - 1] \qquad \text{(b)} \qquad (7.36)$$

By solving for Q_B in terms of I_F, we have

$$Q_B = (L_p^2/D_p)I_F = \tau_p I_F \qquad (7.37)$$

The result in Eq. (7.37) demonstrates that the smaller the amount of final stored charge, the faster the device turns ON since I_F is fixed by the circuit. Based on our earlier definition and in accordance with Eq. (7.37), the turn-ON time is therefore equal to the lifetime of holes τ_p in a long-base diode.

The relation between total stored charge and the current, in a short-base diode is determined from Eqs. (7.30) and (7.31) to be

$$Q_B = I_F W_n^2/2D_p \qquad (7.38)$$

The factor $W_n^2/2D_p$ can be shown to represent the transit time of a hole in the N region. It is the time it takes a hole to travel from the edge of the depletion layer in N to the N metal contact.

Therefore, the turn-ON time for a long-base P^+N diode is the lifetime of holes in the N region and the turn-On time for a short-base P^+N diode is the transit time of a hole in N.

Turn-OFF Time

To turn a diode OFF, it is necessary to remove the excess charges stored by the forward current. The diode will be OFF when the current through it is $-I_s$.

At $t = 0$, the switch in Fig. 7.12(a) is almost instantaneously moved to position B so that a reverse bias is applied to the diode. The instant before moving the switch, at $t = 0^-$, the diode is in a forward state, with current I_F, and $V_D = V_a$ as seen in Fig. 7.12c, which is much smaller than V_R. The current is instantly changed from its forward value I_F to a constant reverse value, as seen in Fig 7.13b, and given by

$$I_R = \frac{V_R + V_a}{R_R} \cong \frac{V_R}{R_R} \qquad (7.39)$$

As long as there is a large number of holes near the edge of the depletion layer in N, the reverse current will be large and holes will diffuse in the direction from N to P, with the hole density in N at the edge of the depletion layer having a positive slope (since $I_D = -I_R = -qD_pA \, dp/dx$). This process continues until the excess hole density at the junction is reduced to zero, so that the hole density at that point becomes equal to the thermal equilibrium value, causing V_D to be zero. The time it takes the excess hole density to become zero ($p = p_{0n}$) is known as the storage time, t_s, as shown in Fig. 17.13. This is the time at the end of the constant current phase of the reverse current transient. As more holes are transported back to the N region by diffusion, the hole density at $x = x_n$ drops below the equilibrium value, V_D becomes negative and opposes V_R in order to reduce the magnitude of the current to be less than I_R, as illustrated in the figures.

As more holes are moved to N, the hole density at the junction becomes much smaller than the equilibrium value, V_D becomes large and negative, and the current

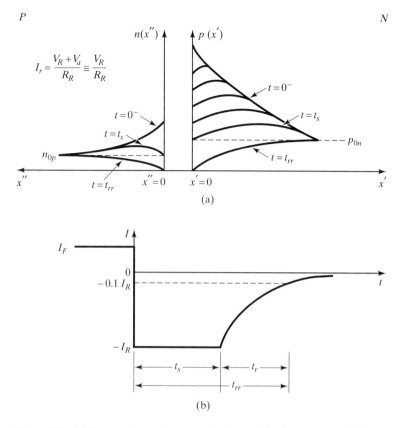

Figure 7.13 (a) Decay of minority carriers in base and (b) the current variation corresponding to the hole decay.

approaches $-I_S$. The time for the current to fall from $-I_R$ to $-0.1I_R$ is known as the *recovery time*, t_r. The current $0.1I_R$ is used as a reference in place of $-I_S$. The total turn-off time, $t_{rr} = t_r + t_s$, is called the *diode reverse recovery time*.

An approximate expression for the storage time, t_s, has been obtained* for a long-base diode as

$$t_s = \tau_p \, \ell n \left[1 + \frac{I_F}{I_R} \right] \tag{7.40}$$

The recovery time, t_r, is, in effect, the time taken to charge the junction capacitance to the reverse voltage, V_R. An estimate of t_r is given as the time constant $2.3R_RC_j$, where C_j is the mean junction capacitance between zero reverse bias and reverse bias at $-V_R$.

The total turn-off time becomes

$$t_{rr} = t_r + t_s \tag{7.41}$$

---

*From M. S. Tyagi, *Introduction to Semiconductor Materials and Devices*, p. 257, copyright © Wiley (1991). Reprinted by permission of John Wiley & Sons, Inc.

We conclude that the storage time in a long-base diode is reduced by a large reverse current and by a small carrier lifetime. For a short-base diode, the storage time is reduced by a large reverse current and by a reduced N region width. A large reverse current sweeps the holes out faster and a short lifetime causes a fast reduction of the carriers by rapid recombination. A small forward current causes a small stored charge, thus causing a small-storage charge removal time. A small width, W, reduces the storage time in a short-base diode by reducing the total hole charge storage.

The lifetime of minority carriers can be reduced by doping the semiconductor with an element that introduces additional recombination centers. This is accomplished by doping silicon with gold. A disadvantage of the reduction of minority carrier lifetime in a long-base diode or the reduction of N region width in a short-base diode is the increased reverse saturation current of the diode.

We can now relate the switching times of a diode to the capacitances. To turn a diode OFF, the excess minority carriers stored must be carried away, which in effect causes a discharge of the storage capacitance and a charging of the transition capacitance. The time taken to turn a diode ON is essentially the time required to discharge the junction capacitance while the voltage across the diode reverts from a large negative value to a small forward bias. This is accompanied by the charging of the storage capacitance.

REVIEW QUESTIONS

Q7-6   Explain why a small capacitance is preferred when a diode is used as a switch.

Q7-7   Which diode capacitance is significant in the forward directions and which one is significant in the reverse direction?

Q7-8   Why and how can the diode transition capacitance be used in the tuning of a television receiver?

Q7-9   Of what significance is the equivalent circuit of the diode?

Q7-10   In accordance with Eq (7.40), a large I_F increases the storage time and a large I_R decreases it. Explain the effect of each current.

HIGHLIGHTS

- The capacitance of a semiconductor device is nonlinear in that the capacitor charge is not directly proportional to the voltage, as is the case in a linear capacitor. A nonlinear capacitance is defined as the ratio of the change in the charge to the change in the voltage that produced it (dQ/dV).

- Positive and negative charges are found in the space-charge layer of a diode. They constitute a transition (junction) capacitance, as the change in the width of the layer does not vary directly with the applied voltage.

- The storage capacitance, on the other hand, refers to the rate of change of charge storage of minority carriers as the voltage is changed.

- A small-signal equivalent circuit is made up of circuit elements that represent the diode when incremental changes are made in the applied voltage. The diode circuit is made up of the transition and storage capacitances and the dynamic resistance dV/dI of the diode.

- When the diode is used as a switch, a low turn-ON time is favored by a small lifetime for a long-base diode and by a small transit time for a short-base diode.

EXERCISES

E7-4   A silicon abrupt-junction diode has a junction capacitance of 10pF at a reverse bias of 8.8V and 20pF at a reverse bias od 1.6V. Determine the built-in voltage.

$$\text{Ans: } V_{bi} = 0.8V$$

E7-5   A P$^+$N diode has the following properties:

$\tau_p = 2.5\mu s$, $W_n = 0.5\mu m$, and $D_p = 10 \text{ cm}^2/\text{sec}$. Determine:
a)   the transit time of the holes in the N region.
b)   the turn-ON time for this diode and the approximate turn-ON time if $W_n = 10\mu m$.

$$\text{Ans: } \quad \text{a) } 1.25 \times 10^{-10}\text{s}$$

$$\text{b) } 1.25 \times 10^{-10}\text{s}, 2.5 \times 10^{-6}\text{s}$$

PROBLEMS

7.1 The effect of recombination in the depletion layer of a diode is studied by comparing the magnitudes of the recombination current and the diffusion current. A P$^+$N silicon diode, at $T = 300$K, has $N_A = 10^{16}\text{cm}^{-3}$, $N_D = 5 \times 10^{15}\text{cm}^{-3}$, and $A = 10^{-3}\text{cm}^2$. The lifetimes in both P and N are assumed to be 0.1μs. Determine the ideal and total diode current for,

a) $V_a = 0.1V$
b) $V_a = 0.5V$

7.2 The low-voltage characteristics of the PN junction diode can be modeled by two diodes in parallel, as shown in the figure.

Plot on semilog paper I vs. V_a, at T = 300K, as V_a increases from 0.01V to 0.75V. Assume $I_S = 10^{-15}$A, $I_{R0} = 10^{-13}$A, and that I_{R0} does not vary with V_a.

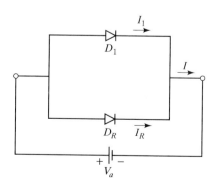

7.3 Use the diode parameters given in Prob. 7.1 to determine; at $T = 300K$,

    (a)   the excess hole density in the base at the emitter junction for an applied voltage of 0.6V.

    (b)   the voltage at which high injection is reached.

7.4 A PN silicon diode has $N_A = 10^{16}\text{cm}^{-3}$, $N_D = 5 \times 10^{15}\text{cm}^{-3}$, $\tau_n = \tau_p = 0.1\mu s$, and $A = 10^{-3}\text{cm}^2$. Determine, at $T = 300K$:

    (a)   the junction capacitance at zero applied voltage.

    (b)   the storage capacitance at $V_a = 0.5V$.

    (c)   the junction capacitance at $V_a = -10V$.

7.5 A silicon P$^+$N junction diode has $A = 10^{-2}\text{cm}^2$. The relation between the junction capacitance and the voltage is given by

$$1/C_j^2 = 7.48 \times 10^{18}(13 - 20V_a)$$

Determine, at $T = 300K$,

    (a)   the built-in voltage V_{bi}.

    (b)   the doping density in N.

    (c)   the doping density in P.

7.6 Measurements on a silicon P$^+$N junction diode operating at $T = 300K$ yield the following results:

At $V_a = -4.3V$ $C_j = 20pF$
At $V_a = -0.55$ $C_j = 40pF$

Given that $N_A = 10^{18}\text{cm}^{-3}$ and $A = 2 \times 10^{-3}\text{cm}^2$, determine:

    (a)   the built-in voltage.

    (b) N_D.

7.7 A P$^+$N step junction diode has a breakdown voltage of 500V. The critical electric field is $3 \times 10^5 \text{V/cm}$. Determine, at $T = 300K$;

    (a) N_D.

    (b)   the depletion layer thickness at breakdown.

7.8 An abrupt junction silicon PN diode, operating at $T = 300K$, has $N_A = 5 \times 10^{15}\text{cm}^{-3}$, $N_D = 10^{15}\text{cm}^{-3}$, and $A = 10^{-4}\text{cm}^2$. Determine for a critical field of $3 \times 10^5 \text{V/cm}$

    (a)   the applied voltage at which the low level injection assumption is violated.

    (b)   the breakdown voltage.

    (c)   the junction capacitance at breakdown.

7.9 Repeat problem 7.8 for $N_A = 10^{18}\text{cm}^{-3}$ and $N_D = 10^{16}\text{cm}^{-3}$.

7.10 For a P$^+$N junction diode, determine the effect of an increase of N_D on the following:

    (a) $\mathscr{E}(x = 0)$

    (b) C_j

    (c) C_s

    (d) τ_p in N

    (e) τ_n in P

    (f)   depletion layer thickness

Briefly explain your reasoning.

7.11 **(a)** If the series resistance, R_S, of the neutral regions and metal contacts of a diode is 300 ohms, determine the current at which the applied voltage deviates from that predicted by the ideal diode equation by 10 percent (given $I_S = 10^{-14}$A).

(b) Repeat part (a) if $R_S = 5$ ohms.

7.12 An ideal silicon N$^+$P long-base diode has $N_D = 10^{18}$cm^{-3}, $N_A = 10^{16}$cm^{-3}, $\tau_p = \tau_n = 10^{-8}$s, and $A = 10^{-4}$cm^2. Determine, at $T = 300$K:

(a) the small-signal incremental resistance at a forward bias of $0.1, 0.5$, and 0.7V.

(b) repeat part (a) for a reverse bias of $0.1, 5$, and 20V.

7.13 Use the parameters of the diode in problem 7.12 to calculate each of the junction capacitances, C_j, and the diffusion capacitance C_s for:

(a) forward bias of $0.1, 0.5$, and 0.7V.

(b) reverse bias of 0.5 and 20V.

7.14 In a silicon PN junction, the doping density is linearly graded and symmetrical from $-W/2$ to $+W/2$, where $N_D - N_A = Bx$. Derive the expression for the junction capacitance showing that

$$C_j = A \left[\frac{q\, B^2}{12(V_{bi} - V_a)} \right]^{1/3}$$

chapter 8

BIPOLAR TRANSISTORS I: CHARACTERISTICS AND FIRST-ORDER MODEL

8.0 INTRODUCTION

A transistor is a device made from semiconductor material and has three terminals. Current is made to flow through the semiconductor from one terminal to another terminal, while the third terminal controls the flow of this current.

There are two types of transistors: the bipolar junction transistor and the field-effect transistor. The major difference between the two lies in the mechanism by which control of the current is achieved.

The two major applications of transistors are amplification and switching. Current, voltage and power can be amplified using transistors in the proper circuit. For digital applications, a transistor can be made to switch between the ON and the OFF states in a very short duration of time.

We will study the structure, operation, and characteristics of bipolar junction transistors in this chapter.

8.1 STRUCTURE AND BASIC OPERATION

The essential physical structure of a PNP transistor is shown in Fig. 8.1(a), with the three regions and their corresponding depletion layers at the two junctions. While this model will be used to analyze the operation and characteristics of the bipolar junction transistor (BJT), actual cross sections of devices are shown in Figs. 8.1(b) and 8.1(c).

The BJT is composed of three doped regions of a semiconductor in contrast to the two regions of a PN junction diode. The middle region, which is sandwiched

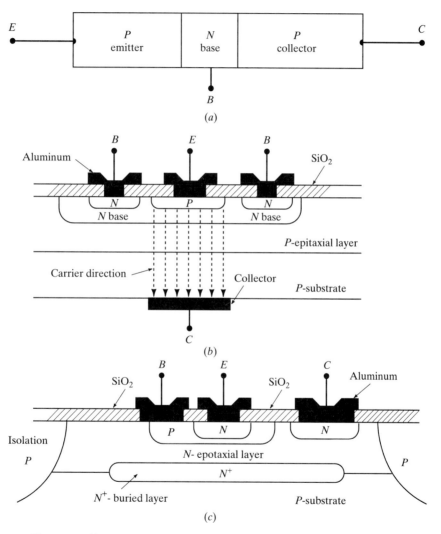

Figure 8.1   (a) Two-dimensional representation of the structure of a PNP BJT; (b) a cross section of a discrete PNP BJT with approximate hole paths and current direction indicated by dotted lines; and (c) a cross section of an IC NPN BJT.

between two regions having the same type of doping, has the opposite type of doping so that a transistor may be of PNP or NPN type.

The BJT is labeled as bipolar, in reference to the two types of carriers, electrons and holes, which take part in the conduction process. Conduction in the field-effect transistor is carried out by means of only one type of carrier; hence, it is known as a unipolar device.

The three regions of the BJT are the emitter, the base, and the collector. These three regions form two interacting PN junction diodes. In the most basic operation of the device, known as the normal or forward active operation, the emitter-base junction is forward-biased and the collector-base junction is reverse-biased. In this state, majority carriers from the emitter diffuse into the base, where a very small percentage of them recombine (because the diffusion length of these carriers is much greater than the width of the base) and the rest are swept into the collector by the reverse-biased junction. If the width of the base is much greater than the diffusion length, then all the carriers would recombine in the base and the transistor effectively becomes a combination of two diodes back to back and the collector current effectively becomes the reverse saturation current of one of the diodes.

The word *transistor* is an acronym for transfer-resistor, where *transfer* refers to the relation of the output to the input. A low resistance is seen at the forward-biased diode where the ratio of voltage to current is relatively low, while the second diode exhibits high resistance as obtained from the ratio of voltage to current.

Figure 8.1(b) represents a cross section of a PNP discrete transistor. We use the term *discrete* for a transistor that is separately packaged, whose three terminals are accessible, in contrast to an integrated circuit transistor. The drawing in Fig. 8.1(a) represents the section shown by the dotted lines in Fig. 8.1(b), with the orientation shifted by 90 degrees.

In the majority of applications, BJTs are incorporated into an *integrated circuit (IC)*, also known as a *microchip*. The IC incorporates combinations of transistors, resistors, and capacitors, all fabricated simultaneously on a single chip of crystal silicon. Hundreds of thousands, and more recently several millions, of devices are fabricated on a single silicon chip. We show in Fig. 8.1(c) a cross section of an integrated circuit NPN BJT. All connections on the chip are made at the surface of the IC and consequently all currents are confined to a very thin region at the surface.

The normal direction of current is shown by the dotted arrows in Fig. 8.1(b). In the discrete transistor, the emitter and base contacts are located at the top, the collector contact is shown at the bottom, and the carriers cross part of the base region, which is normal to the direction of flow. The purpose of the two base contacts is to reduce the ohmic base resistance.

While we will use the single model of Fig. 8.1(a) in our discussion, the actual construction of a BJT in an IC is the one to keep in mind. We observe that all three contacts to the transistor in an integrated circuit are made from the top. One obvious reason for this is that circuit connections to other elements in the IC chip need to be made in one plane.

8.2 FABRICATION OF THE BIPOLAR INTEGRATED CIRCUIT TRANSISTOR

We now consider the fabrication of an NPN silicon BJT on an integrated circuit using the processes that we discussed in Chapter 6. While our discussion will be focused on the BJT, it is understood that the surface of a whole wafer is being processed.

The starting material is a boron-doped wafer on a very small area where we will form a BJT. The base on which the transistor is made is known as the substrate and its function is to act as the mechanical support for the device. The reason for the use of a P substrate for the NPN will be clarified when the term *isolation* is discussed. This substrate has a resistivity of 3–10 ohm-cm with a thickness between 250 and 400μm for wafers having diameters over 100mm.

The fabrication steps up to, and including, the metal contacts to the three regions are illustrated in Fig. 8.2.

First, a layer of SiO$_2$, about 5000Å thick, is deposited on the surface of the substrate by thermal oxidation. Using the *first mask* and the photo lithographic process, windows are opened in the oxide for the buried layer. The N$^+$ buried layer is diffused to a depth of about 3μm, followed by oxide removal. This layer serves to collect the carriers that have crossed the base on their way to the collector terminal, as shown in Fig. 8.1(c). It serves as a sub-collector and is used to reduce the collector ohmic resistance, as we will explain later.

After the buried layer is diffused, the wafer is stripped of all oxide to permit the next deposition. It is to be noted that during the subsequent high-temperature processes, the buried layer tends to diffuse out.

The second operation is the deposition of a phosphorus-doped N-epitaxial layer on the whole wafer and in which all devices are made. This layer has a resistivity of 0.1 to 1 ohm-cm, with a thickness of 0.5 to 5μm for high-speed digital circuit applications and 10–20μm for linear analog circuits. A layer of silicon dioxide about 5000 to 10,000Å thick is grown thermally on the surface of the epi-layer.

Since the collector of the NPN is N type, and so are the collectors of adjacent transistors, there is an obvious need to isolate the collectors from each other. This is accomplished by what is known as P isolation. To start with, each transistor is placed in an N island, or tub, as shown in Fig. 8.3, with a P-isolation region enveloping each tub and extending from the surface of the wafer all the way to the P substrate. To guarantee isolation of NPN translators, the substrate is connected to the most negative point on the circuit, so that reverse-biased junctions are formed between collectors of adjacent transistors, since the collectors are normally at a positive potential.

The *second mask* is used to etch windows for the *isolation regions*, which are formed by the subsequent diffusion of boron extending from the surface down to the substrate. An N-epitaxial layer separates the isolation regions, thus serving as the tub in which each transistor is formed. The diffusion of the isolation region is followed by oxidation of the wafer surface.

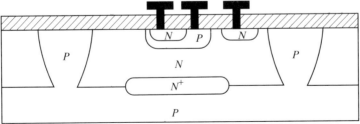

Figure 8.2   Steps in the fabrication of an NPN BJT.

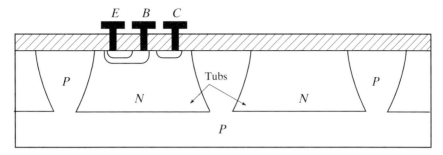

Figure 8.3   Transistor tubs.

Mask number three is used to open a window for the P-type *base* of the transistor. P-type diffusion or ion implantation is driven to form the base to a depth of about 2–3μm. This is followed by an oxide layer.

The *fourth mask* is used to open windows in the oxide for the N^+ *emitter* and the *collector contacts*. The phosphorus or arsenic diffusion is driven to a depth of about 2μm. The need for an N^+ collector contact is to form a good ohmic contact. As we will see in the discussion of metal-semiconductor contacts, these contacts may be rectifying or ohmic. The ohmic contact permits easy current flow in both directions. To form a good ohmic contact to an N material, an N^+ region is needed between the metal on top and the N region. Following the N^+ diffusion, an oxidation layer is formed over the entire wafer surface.

Mask number five is used to open *windows* for the formation of *metallic contacts* to the transistor terminals. Then, an aluminum thin film 0.5 to 1μm thick is deposited, by evaporation or sputtering, on the top surface of the IC wafer.

Assuming that complete circuits are to be formed on the surface of the wafer, the *sixth mask* is used to define the *interconnection pattern* in the circuits. These interconnections are etched into the metal that has been deposited on the surface.

To protect the surface of the wafer from moisture and chemical contamination, a passivation layer on the surface is required. The material is a phosphorus-doped oxide, which is deposited on the surface.

Contacts to the integrated circuits are made on pads that are located on the periphery of the IC chip. Since the IC chip will be bonded to an IC package, connections are to be made from the package leads to the bonding pads on the IC chip. *Mask number seven* is used to define the *bonding holes* over the aluminum pads for external connections.

Following the seven masks, the circuits are tested by a computer-controlled system and all faulty chips are identified and marked. The wafer is then sawed into chips, which are bonded on IC packages. Gold wires about 25μm in diameter are used to connect the package leads to the bonding pads on the chip.

Additional detail on the buried layer is illustrated in Fig. 8.4. In Fig. 8.4(a), we show the path taken by the carriers on their way from the emitter to the base and to the collector. This path is considerably longer than the path in a discrete BJT, shown in Fig. 8.16, and because of this the collector series resistance, labeled the parasitic resistance, is quite large and is of the order of hundreds of ohms. To reduce this

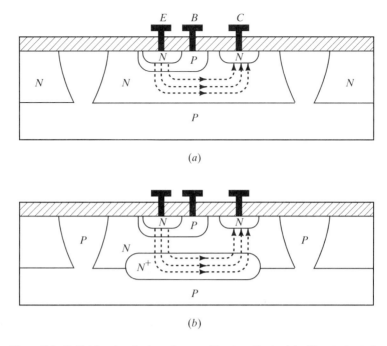

Figure 8.4   Path taken by electrons from emitter to collector (a) without a buried layer through N-epilayer and (b) with a low resistivity buried layer.

resistance, we have placed the low resistivity buried layer in Fig. 8.5(b) into the path of the carriers, which acts as a sub-collector. The use of the buried layer reduces the collector resistance by as much as a factor of 20. The result of this is to improve the quality of the transistor, defined by a term known as the figure of merit, or gain-bandwidth product, by the same factor.

8.3 TERMINOLOGY, SYMBOLS, AND REGIONS OF OPERATION

Terminology and Symbols

The symbols for a PNP and an NPN transistor are shown in Fig. 8.5. The arrow on the emitter lead identifies the actual direction of the current. In a PNP transistor, holes are normally made to move from the emitter to the collector, hence the arrow on the emitter terminal points into the lead labeled E. The arrow on the NPN emitter lead points outward, indicating that the actual direction of the current is out of the emitter as a result of the flow of electrons from emitter to collector.

In this book, we will use the PNP structure in the analysis of the operation and in the development of the analytical relations for the BJT. The main reason for this is that the PNP follows directly from the PN junction diode discussions and the equations pertaining to it developed in the previous two chapters. To apply the information of this chapter to the NPN transistor, the following is needed: First, replace the references to carriers such that the terms, holes and electrons, and their

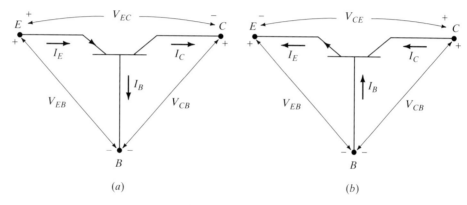

Figure 8.5   (a) PNP symbol and (b) NPN symbol.

relevant constants are replaced by electrons and holes and their constants respectively. Second, reverse polarities of all voltages and the directions of all currents.

Modes of Operation

The bipolar junction transistor has two PN junctions: the emitter-base and collector-base junctions. There are, therefore, four possible combinations of biasing of these junctions, as shown in Fig. 8.6: both forward-biased junctions, both reverse-biased junctions, and one forward-biased with another reverse-biased junction.

Different sets of currents of the transistor will result for each set of biasing. These combinations of currents and voltages, corresponding to each quadrant of Fig. 8.6, will be labeled as *modes of operation*. These modes are named: *active, saturation, cutoff*, and *inverse active*.

The *active mode* is most commonly used in analog circuits. This corresponds to the fourth quadrant of Fig. 8.6, where the emitter-base junction is forward-biased

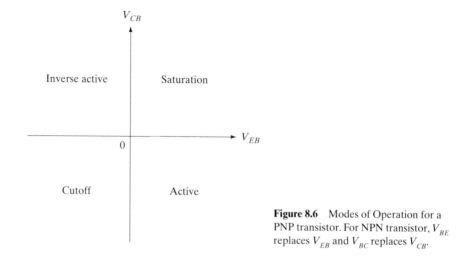

Figure 8.6   Modes of Operation for a PNP transistor. For NPN transistor, V_{BE} replaces V_{EB} and V_{BC} replaces V_{CB}.

and the collector-base junction is reverse-biased. For a PNP transistor in the active region, V_{EB} is positive and V_{CB} is negative, while for a NPN transistor V_{BE} is positive and V_{BC} is negative. Bipolar transistors used in amplifier circuits are made to operate in this region, where the largest voltage and current amplification are obtained.

In the *saturation mode*, which corresponds to the first quadrant in Fig. 8.3, both junctions are forward-biased. When the transistor is to be used as a switching device, this mode corresponds to the ON condition of the switch because the ratio of the voltage to the current is small, hence simulating a low resistance.

When both junctions are reverse-biased, the transistor currents are so small that this condition is referred to as *cutoff* and corresponds to the third quadrant of Fig. 8.3. Since the voltages causing these very small currents are non-zero, the resulting resistances are very large, indicating an open-circuit. A transistor operating as a switch in the OFF position is in cutoff.

The fourth mode, known as the *inverse-active mode*, is the one least used and corresponds to the second quadrant of Fig. 8.3. In this mode, the collector takes the role of the emitter and the emitter becomes the collector. Certain digital circuits use transistors that are made to operate in this mode.

The doping of the emitter is much greater than the doping of the base, which in turn may be greater than the doping of the collector. These differences in doping are dictated by the requirements of the amplification properties of the transistor, which take place when the device is operating in the active mode. In a later section, we will discuss the effect of the doping on the currents. Because of the differences in both the cross sectional areas and doping, the collector and emitter are not interchangeable. This interchange happens only when the device is operating in the inverse-active mode, where the device practically loses all its amplifying properties.

8.4 CIRCUIT ARRANGEMENTS

In circuit applications, the transistor is connected so that it presents to the rest of the circuit, two input terminals and two output terminals. This is done by having one of the three transistor terminals common to the input and output circuits. As a result, there are three possible circuit connections for the bipolar transistor, as shown in Fig. 8.7. They are the *common-base, common-emitter*, and *common-collector* connections. The most commonly used connection is the common-emitter type, with the common-collector being the one least commonly used. In Fig. 8.7, actual directions of currents are shown.

8.5 TRANSISTOR CURRENTS IN THE ACTIVE REGION

We will use the two-dimensional representation of the Fig. 8.1(b), repeated in Fig. 8.8, to analyze the sources of the currents in the PNP bipolar junction transistor. There are three terminal currents and three voltages so that we may write Kirchhoff's current and voltage laws as

$$I_E = I_B + I_C \qquad \text{(a)}$$

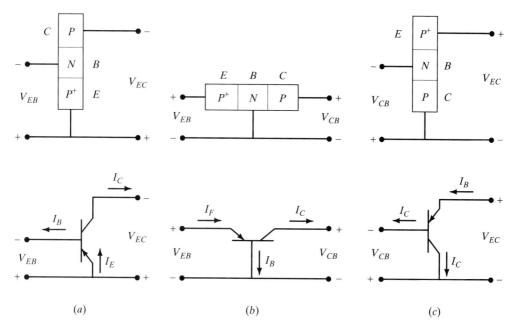

(a)   (b)   (c)

Figure 8.7 (a) Common-emitter, (b) common-base, and (c) common-collector transistor connection for a PNP transistor. For an NPN transistor, all currents are in opposite directions to those shown above and the subscripts to all voltage are interchanged.

$$V_{EB} + V_{BC} + V_{CE} = 0 \quad \text{(b)} \tag{8.1}$$

For a BJT operating in the active mode, and, as shown in Fig. 8.3, V_{EB} is positive and V_{CB} is negative, resulting in a forward-biased emitter-base junction and a reverse-biased collector-base junction and $V_{EB} = -V_{BE}, V_{CB} = -V_{BC}, V_{CE} = -V_{EC}$.

Emitter Current

For a forward-biased emitter-base junction, and just as we determined in the PN junction diode, holes injected from the emitter diffuse into the base and electrons injected from the base diffuse into the emitter region. The sum of these two currents forms the emitter current I_E, as

$$I_E = I_{Ep} + I_{En} \tag{8.2}$$

where I_{Ep} represents the hole current component and I_{En} the electron current component of the emitter current.

The electrons that cross from base to emitter and that form I_{En} recombine with holes in the emitter. For typical devices, the electron density at the metal emitter contact is the thermal equilibrium value, as shown in Fig. 8.8(b). At the emitter metallic contact, electrons are forced out of the contact by the DC voltage, V_{BB}, into the external circuit, thus freeing excess holes. Some of these holes, which are majority carriers in the emitter, recombine with the electrons arriving from the base.

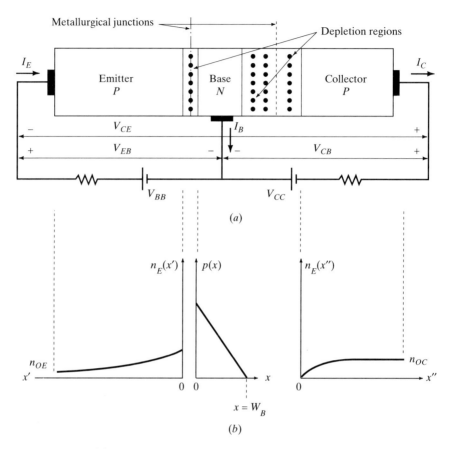

Figure 8.8 (a) Two-dimensional model of the PNP transistor in the active region and (b) profiles of minority carrier distributions.

The rest are injected into the base as a result of the forward-biased emitter-base junction.

Collector Current

The holes that are injected into the base from the emitter face either of two possibilities: one, if the width of the base is much greater than the diffusion length of holes (which depends on the lifetime of holes as minority carriers in the base), then all the holes will recombine in the base. If, on the other hand, the width of the base is much smaller than the diffusion length of holes, then the great majority of the holes reach the reverse-biased collector junction. The former condition makes the transistor two PN junction diodes back to back, since no interaction between the emitter and collector takes place. The latter conditions make it possible for the holes that originated in the emitter to cross into the collector and contribute to the collector current. Why will the holes cross into the collector?

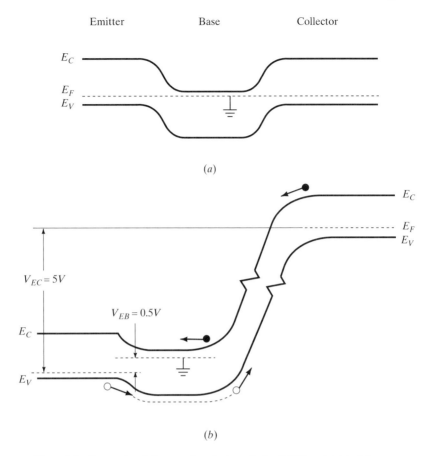

Figure 8.9   Energy-level diagrams for electrons in the PNP transistor in (a) thermal equilibrium (zero bias) and (b) active region. Because the diagrams are drawn for electrons, electrons find it easy to roll down the steep hill. Simultaneously, holes find it just as easy to bubble up the same steep hill. It is assumed that in diagram (b) $V_{EB}(-V_{BE})$ is of the order of 0.5V and $V_{EC}(-V_{CE})$ is of the order of 5V.

The answer is found in the energy level diagrams shown in Fig. 8.9, where under reverse-bias the electric field in the collector-base junction provides an easy path for minority carriers (for holes to cross from the base into the collector and electrons to cross from collector to base).

The holes originated in the emitter and diffused in the base to be injected into the collector. The reverse bias across the collector-base depletion layer has decreased the density of electrons and holes to values that are lower than their thermal equilibrium. Hence, thermal generation exceeds recombination and electrons and holes are generated in the layer. The holes join those arriving from the base and cross into the collector.

The electrons generated in the collector-base depletion layer roll down the steep hill from collector to base, as shown in Fig. 8.9(b). This current of electrons, which is of the order of picoamperes when the collector current is in mA, is also known as the *leakage current*.

We now have a collector current made of two components where I_{Cp} is the current of holes bubbling up the hill from base to collector and I_{Cn} is the current of the electrons that roll down the steep hill from collector to base

$$I_C = I_{Cp} + I_{Cn} \tag{8.3}$$

Base Current

The base current, I_B, consists of three components—all shown in Fig. 8.10.

The first component, I_{B1}, is the I_{En} of the emitter current, which consists of electrons that diffuse from base to emitter, and which is directed out of the base terminal. The second component, $I_{B2} = I_{rec}$, is produced by the motion of electrons from the base lead that cross into the base in order to recombine with some of the holes that are diffusing from the emitter to the collector. The direction of this current is also out of the base terminal. The current of electrons from collector to base, I_{Cn}, represents the third component of the base current, so that the total base current becomes

$$I_B = I_{B1} + I_{B2} - I_{B3} \tag{8.4}$$

where $I_{B3} = I_{Cn}$ and it is directed into the base.

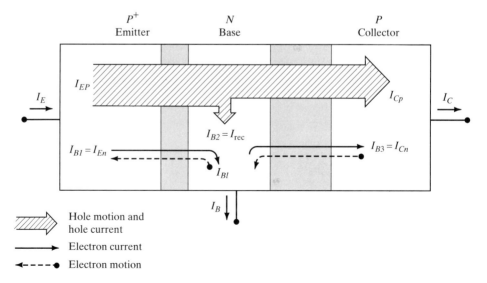

Figure 8.10   Current components in the active region.

In reviewing this section, we have not highlighted the reason for our interest in this transistor. This is the subject of our next section.

8.6 THE BJT AS A CURRENT AMPLIFIER

Approximations to Base Current

By making assumptions that are quite valid in modern-day silicon BJTs, we will demonstrate that a BJT in the common-emitter connection is in fact a linear current amplifier.

We have concluded earlier that carrier motion in a BJT, which is operating in the active mode, results in five distinct currents. For the PNP transistor there is, first, a current due to holes that diffuse from the emitter to the base, which makes up the overwhelming part of the emitter current. Second, there is a current of holes that diffuse in the base and this current makes up practically all of the collector current. Lastly, there are three components of the base current: the current due to electrons injected from the base to the emitter labeled I_{B1}; the current I_{B2}, which results from electrons supplied by the base contact to recombine with some of the holes that are traversing the base to the collector; and the current I_{B3}, which consists of a generation current originating at the reverse-biased collector-base junction (which sends electrons into the base). Simultaneous to the generation electrons going into the base, generation holes are driven by the electric field into the collector. However, this current due to holes is negligible compared to the current of holes that have diffused into the base and are collected by the collector. The current of electrons, I_{B3}, may be lumped partly with the electrons that recombine in the base and partly with the electrons that are injected into the emitter.

We have therefore reduced the base current to two currents: one of electrons injected from the base that diffuse in the emitter and the second a current of electrons that recombine with holes in the base.

In the BJTs of the 1960s, the recombinations component of the base current was considered to make up the larger portion of the base current because of the combination of large basewidths, of the order of 10 microns, and the low lifetime of minority carriers in the base. In today's transistors with basewidth of less than one micron and with long lifetimes, the recombination current can be safely neglected for a very broad range of transistors.

Therefore, we have limited the effective base current to the current of electrons that are injected into the emitter. While this is valid for the purpose of our discussion here, in a later section we will derive complete expressions for I_C, I_B, and I_E, in terms of the physical properties of the transistor.

Base Current as the Control Current

As we shall see later, the base current defined in this section as consisting of I_{B1} only is of the order of 1 percent of the collector current. In spite of its small size, its importance lies in the fact that it is the control current supplied to the input of the common-emitter BJT.

Let us establish why we label I_{B1} as the control current. We can safely assume that in the great majority of transistors, the width of the base is much smaller than the diffusion length of holes in the base of the PNP device, so that the P$^+$N emitter-base junction has properties that are similar to the short-base diode discussed in the last chapter. Therefore, the profile of the hole density distribution in the base is a straight line having slope $-p(0)/W_B$, so that the hole current at the collector junction in a CE device is given by

$$I_C \cong I_{Ep} = (-qA_E D_p)dp/dx = qA_{Ep}p(0)/W_B \qquad (8.5)$$

$$= (qA_E D_p p_0/W_B) \exp(qV_{EB}/kT) \qquad (8.6)$$

where A_E is the emitter junction area, p_0 is the equilibrium hole density in the base, $p(0)$ is the hole density in the base at the emitter-junction, W_B is the effective width of the base, and D_p is the diffusion constant of holes in the base.

The base current, I_{B1}, calculated at the emitter side of the junction, is a diffusion current in an emitter whose width is assumed to be many times the diffusion length of electrons, L_{nE}. This current is given by

$$I_B = I_{B1} = (qA_E D_{nE})dn/dx' = qAD_{nE}n_E(0)/L_{nE} =$$

$$(qA_E D_{nE}n_{0E}/L_{nE}) \exp(qV_{EB}/kT) \qquad (8.7)$$

where $n_E(0)$ is the electron density at the emitter side of the junction, n_{0E} is the equilibrium electron density in the emitter, L_{nE} is the diffusion length of electrons in the emitter, D_{nE} is the diffusion constant of electrons in the emitter, and A_E is the emitter area.

From Eqs. (8.6) and (8.7), we conclude that by fixing the base (input) current to a CE BJT circuit, V_{EB} is fixed. If V_{EB} is fixed, the collector current, I_C, is also fixed, so that the ratio of I_C to I_B is given by

$$I_C/I_B = \frac{D_p p_0 L_{nE}}{D_{nE}n_{0E}W_B} \qquad (8.8)$$

Because I_C is directly related to I_B, independent of the voltage V_{EB}, a linear relationship exists between I_C and I_B, and this makes the BJT in the CE connection a *linear current amplifier*.

Fixing I_B or V_{BE}?

From our discussion, it is apparent that either I_B or V_{EB} could be fixed to obtain the current amplification. By fixing V_{EB}, problems may arise, as illustrated in Fig. 8.11(a). Because of the exponential dependence of I_B on V_{EB}, attempting to fix I_B through the fixing of V_{EB} by a voltage source represents a fragile situation, since a small change in V_{EB} causes a large change in I_B. The voltage supply, V_{EB}, is not the problem. It is the shifting of the characteristic due to a change in temperature that is of concern.

A superior arrangement, used to fix I_B, is to place a battery in a series with a resistor across the input of the CE arrangement. By adjusting either the battery voltage, V_{BB}, or the resistor, R_B, a stable current, I_B, can be fixed, as shown in Fig.

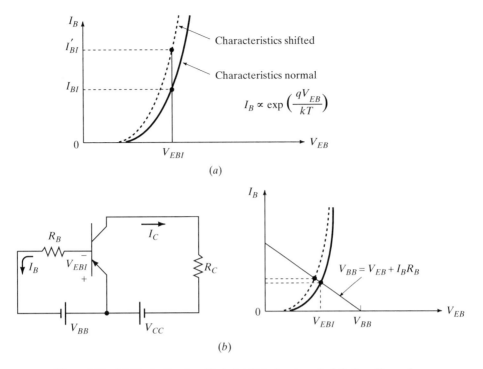

Figure 8.11 (a) Illustrating the effect of shift in the characteristic $I_B - V_{BE}$ on I_B and (b) a battery in a series with the resistor used to fix I_B. Note the effect of the shift is minimized.

8.11(b). The corresponding value of V_{EB} is determined at the intersection of the circuit equation, $I_B = (V_{BB} - V_{EB})/R_B$, and the characteristic curve of the input of the device.

8.7 TRANSISTOR PARAMETERS

The terminal currents of the PNP transistor operating in the active mode are summarized below as

$$I_E = I_{En} + I_{Ep} \qquad \text{(a)}$$
$$I_C = I_{Cp} + I_{Cn} = I_{Cp} + I_{CBO} \quad \text{(b)} \qquad (8.9)$$
$$I_B = I_{En} + I_{rec} - I_{CBO} \qquad \text{(c)}$$

We will identify the term I_{CBO} by its subscripts as follows: The first two subscripts refer to the two terminals between which the current is measured and the third subscript refers to the state of the third terminal (O for open). The current, I_{CBO}, is therefore the collector-to-base current with the emitter open. It includes the electrons and holes that are generated in the depletion region and swept into the base and the collector respectively.

We recall that I_{En} and I_{rec} are directed out of the base terminals and I_{CBO} is directed into the base and out of the collector terminal.

We define the *emitter injection efficiency*, γ, as the ratio of the hole current injected from the emitter to the base to the total emitter current defined by Eq. (8.9(a)).

$$\gamma \equiv \frac{I_{Ep}}{I_E} \tag{8.10}$$

We also define the *base transport factor*, δ, to be the ratio of the hole current entering the collector to the hole current entering the base from the emitter as

$$\delta = \frac{I_{Cp}}{I_{Ep}} \tag{8.11}$$

The product $\gamma\delta$ equals I_{Cp}/I_E and this product is known as the *DC common-base current gain*, α, and given as

$$\alpha = \gamma\delta = I_{Cp}/I_E \tag{8.12}$$

It is obvious that in a good BJT, the value of α is very close to unity and varies usually from 0.99 to 1.

By incorporating Eq. (8.12) in Eq. (8.9(b)), we have

$$I_C = \alpha I_E + I_{CBO} \tag{8.13}$$

To introduce I_B in Eq. (8.13), we replace I_E by $(I_B + I_C)$ and we solve for I_C, so that

$$I_C = \frac{\alpha}{1 - \alpha} I_B + \frac{I_{CBO}}{1 - \alpha} \tag{8.14}$$

We define two new symbols, β and I_{CEO}, so that Eq. (8.14) becomes

$$I_C = \beta I_B + I_{CEO} \tag{8.15}$$

where β and I_{CEO} are given by

$$\beta \equiv \alpha/(1 - \alpha) \quad \text{(a)}$$
$$I_{CEO} = I_{CBO}/(1 - \alpha) \quad \text{(b)} \tag{8.16}$$

The symbol I_{CEO} refers to the collector-to-emitter current with the base open-circuited. For $I_B = 0, I_C = I_{CEO}$. This current is larger than I_{CBO} because the electrons generated at the reverse-biased C-B junction are swept into the base whereupon they diffuse into the emitter causing a larger diffusion of holes from the emitter into the base to proceed to the collector. It is as if these holes represent a magnified current compared to the electron current. The E-B junction is slightly forward-biased by V_{EC}, causing the diffusion of electrons and holes. Since α is almost unity, β is a large number known as the *DC common-emitter current gain*. This number, for a good BJT, is 100 or more. The relations in Eqs. (8.1) through (8.16) apply as well to the NPN transistor provided that actual directions for currents are used.

In the following example, we will carry out calculations for transistor currents and transistor parameters.

EXAMPLE 8.1

Given a PNP transistor that has the following current components: $I_{Ep} = 2\text{mA}$, $I_{En} = 0.01\text{mA}$, $I_{Cp} = 1.98\text{mA}$, and $I_{Cn} = 0.001\text{mA}$, determine: a) the base transport factor, b) the injection efficiency, c) α and β, d) I_B, I_{CBO}, and I_{CEO}, and e) repeat part (c) for $I_{Cp} = 1.99\text{mA}$.

Solution

a) The base transport factor δ is

$$\delta = I_{Cp}/I_{Ep} = 1.98/2 = 0.99$$

b) The injection efficiency γ is calculated as

$$\gamma = I_{EP}/I_E = I_{EP}/(I_{EP} + I_{EN}) = 2/(2 + 0.01) = 0.995$$

c) $\alpha = \delta\,\gamma = 0.985$

$$\beta = \alpha/(1 - \alpha) = 65.67$$

d) $I_B = I_E - I_C, I_E = I_{En} + I_{Ep} = 2.01\text{mA}, I_C = 1.981\text{mA}, I_B = 0.029\text{mA}$

$$I_{CBO} = I_C - \alpha I_E = 1.15\mu\text{A}$$

$$I_{CEO} = I_{CBO}/(1 - \alpha) = 76.67\mu\text{A}$$

e) The base transport factor becomes $= 1.99/2 = 0.995$

$$\alpha = \delta\,\gamma = 0.995 \times 0.99502 = 0.990$$

$$\beta = \alpha/(1 - \alpha) = 99.49$$

8.8 GRAPHICAL CHARACTERISTICS AND MODES OF OPERATION

In Fig. 8.12, we display the output graphical characteristics for the PNP BJT together with the relevant circuits of the connections. These are the characteristics of interest when the BJT is used in amplification or switching. Complete analytical dependencies for both the input and output variables in the PNP common-emitter and common-base connections are represented by the following relations:

$$\text{CE} - \text{input}, V_{EB} = f(I_B, V_{EC}) \quad \text{(a)}$$
$$\text{output}, I_C = f(I_B, V_{EC}) \quad \text{(b)}$$

(8.17)

$$\text{CB} - \text{input}, V_{EB} = f(I_E, V_{BC}) \quad \text{(a)}$$
$$\text{output}, I_C = f(I_E, V_{BC}) \quad \text{(b)}$$

(8.18)

If either the CE or CB relations were available as complete expressions, then by using Kirchhoff's current and voltage laws, one can determine expressions for each of the three transistor currents as a function of any two of the voltages so that either of the above relations completely define the I-V characteristics of the transistor. It is important to point out that, no matter which connection of the BJT is used,

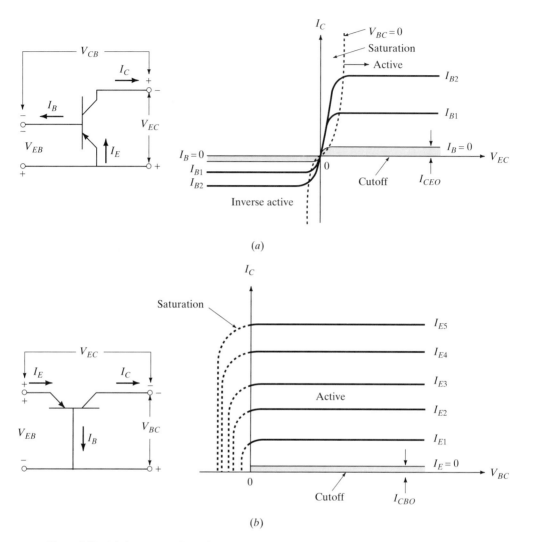

Figure 8.12   (a) Common-emitter circuit and output characteristics of the PNP transistor
(b) Common-base circuit and output characteristics of the PNP transistor.

and which variables are the dependent and independent quantities, the I-V characteristics of the transistor are uniquely determined. Analogously, either connection of the BJT represents the I-V characteristics completely.

Modes of Operation

Both the CE and CB characteristics exhibit the three major modes of operation, namely: active, saturation, and cutoff. In certain switching circuits, the BJT in the CE connection is in the inverse active mode shown in the third quadrant of those characteristics of Fig. 8.12(a).

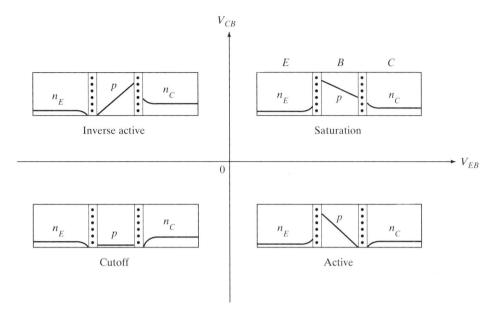

Figure 8.13   Distributions of minority carriers in the four operation modes of the PNP BJT.

In analyzing the diode currents in Chapter 6, we used the distribution of minority carriers in the P and N regions. In Fig. 8.13, we have displayed the distributions of minority carriers in each of the four operation modes of the BJT. These will be used in discussing the characteristics of the transistor. In sketching the minority carriers distributions, we have assumed that the width of the base is much smaller than the diffusion length of holes in the base and that the widths of the emitter and collector regions are much greater than the diffusion length of electrons. Just as with the diode, the values of the minority carriers densities at the junctions are given by

$$\text{In the base at } x = 0, p(0) = p_0 \exp{(qV_{EB}/kT)} \quad \text{(a)}$$
$$\text{In the base at } x = W_B, p(W_B) = p_0 \exp{(qV_{CB}/kT)} \quad \text{(b)} \tag{8.19}$$

$$\text{In the emitter at } x' = 0, n_E(0) = n_{OE} \exp{(qV_{EB}/kT)} \quad \text{(a)}$$
$$\text{In the collector at } x'' = 0, n_C(0) = n_{OC} \exp{(qV_{CB}/kT)} \quad \text{(b)} \tag{8.20}$$

Based on the above, we can determine expressions for the various currents by assuming that the BJT currents may be obtained from the diffusion of the minority carriers.

CE Active Mode

With I_B fixed, V_{EB} fixed, $p(0)$ fixed, and assuming that the distribution of hole density in the base is linear, as shown in Fig. 8.13, the hole current in the base is equal to the hole current in the collector (since there is no recombination in the base) and this is equal to the collector current, I_C. As long as the E-B junction is forward-biased and the C-B junction is reverse biased ($V_{CB} < 0$ or, $V_{BC} > 0$), I_C and I_B are given approximately by Eqs. (8.6) and (8.7). For a constant I_B, I_C is fixed so long as V_{EC} has a value that makes $V_{BC} > 0$, as determined by

$$V_{EC} = V_{EB} + V_{BC} = V_{EB} - V_{CB} \qquad (8.21)$$

The active mode is terminated when $V_{EC} = V_{EB}$ and hence $V_{BC} = 0$. With $V_{BC} < 0$, the C-B junction is forward-biased and the device is in the saturation mode. Thus, $V_{BC} = 0$ is the dividing line between active and saturation modes.

CE Saturation Mode

In the saturation mode, the C-B junction is forward-biased causing an increase in the hole density at $x = W_B$, which for a fixed I_B and V_{EB} reduces the slope of the hole density distribution, thus decreasing the collector current, as shown in Fig. 8.12(a). As V_{EC} is decreased further, for a fixed I_B, V_{CB} becomes more positive (more forward bias on C-B junction) and the slope of the hole density in the base decreases, thus I_C decreases.

At larger values of I_B, hence larger values of V_{EB}, the dividing line between the active and saturation modes, which occurs at $V_{BC} = 0$, moves to lower values of V_{EC}. This mode represents the ON position when the BJT is used as a switch.

CE Cutoff Mode

When both junctions are reverse-biased, the base current becomes negative. We consider first the condition at the boundary between positive and negative I_B, at $I_B = 0$. With the applied V_{EC}, the emitter is positive with respect to the collector. In this condition, V_{BC} is positive, the C-B junction is reverse-biased, and electrons are injected from collector to base. Since the electrons cannot exit through the base, they neutralize some of the donor ions at the E-B junction, attracting holes from the emitter in order to neutralize acceptor ions. The emitter-base junction is, therefore, slightly forward-biased, causing a gradient of holes in the base that diffuse to the collector. Because of the high doping of the emitter, the relatively smaller number of electrons that left the collector cause a larger number of holes to be injected into the base and diffuse to the collector. This collector current, which is an amplified form of the electron current, is labeled I_{CEO}. The collector to emitter current with the base open is the collector current at $I_B = 0$ and represents the edge of cutoff.

For $I_C < I_{CEO}$, both junctions are reverse biased and the collector current is very small. This mode represents the OFF position when the transistor is used as a switch.

CE Inverse Active Mode

The inverse active mode occurs when the emitter-base junction is reverse-biased and the collector-base junction is forward-biased. The roles of the emitter and collector are thus interchanged, with the former collector emitting the holes (in the PNP BJT) and the former emitter (new collector) collecting them. The general shape of the characteristic in this mode, shown in the third quadrant of Fig. 8.12(a), is identical to that in the first quadrant, except that the collector current for the same base current (as in the active mode) is much smaller than that of the first quadrant. The main reason for the smaller collector current is that the doping of the new emitter is considerably smaller than the former emitter. This is evident from Eq. (8.8), repeated here by replacing the parameters of the emitter by those of the collector, as

$$I_C/I_B \text{ (inverse active)} = \frac{A_E D_p p_0 L_{nC}}{A_C D_{nC} n_{OC} W_B} \tag{8.22}$$

Upon comparing Eq. (8.22) to Eq. (8.8), we note that $n_{OC} \gg n_{OE}$ because of the doping of the emitter $N_{AE} \gg N_{AC}$. Furthermore, $A_C \gg A_E$. This makes I_C/I_B quite small and slightly greater than unity.

CB Active Mode

In the CB active mode, I_E is the relevant input current and is much larger than the input current to the CE mode. The base current consists of electrons injected from the base into the emitter and the collector current consists of holes diffusing from the emitter into the base. The portion of the emitter current due to holes diffusing in the base is almost equal to the collector current, as shown in the characteristics of Fig. 8.12(b). An increase in I_E brought about by an increase in the slope of the hole density in the base results, for a fixed V_{BC}, from an increase of I_B. Both of these correspond to a linear increase in I_C that is almost equal that of I_E.

At higher values of I_E, for a fixed V_{BC}, a higher I_B is required, which results in a larger I_C with $I_C \cong I_E$.

CB Saturation Mode

To the left of $V_{BC} = 0$, the collector junction is forward-biased and the transistor is in saturation ($V_{BC} < 0$). Holes are injected from the collector to the base and these are, for ($V_{EB} > 0, I_B > 0$) in a direction that is opposite to the direction of holes that originate in the emitter and end up in the collector. As a result, the collector current is the difference between the two hole currents crossing the collector junction and is smaller as V_{BC} becomes more negative. At this time, the emitter current is constant. The larger the forward bias on the C-B junction, the larger the injection of holes from collector to base. For the same I_E, I_C is smaller. At larger values of I_E, which indicates more holes injected from base to collector, and to obtain the same I_C, more holes have to be injected from collector to base, requiring a larger forward bias on the C-B junction (larger V_{CB}).

CB Cutoff

In the CB cutoff mode, $I_E = 0$ represents the edge of cutoff. The collector current is labeled I_{CBO}. With the emitter open-circuited and $V_{BC} > 0$, I_C becomes the reverse saturation current of the C-B junction. This current is smaller than the reverse current of a PN diode because for $I_E = 0$, the hole gradient in the base at the emitter junction is zero. This reduces the hole gradient at $x = W_B$, which becomes smaller than when the E-B junction is short-circuited with $V_{EB} = 0$.

REVIEW QUESTIONS

Q8-1 Identify the junction voltages and their signs in each of the four modes of operation for an NPN transistor.

Q8-2 Identify the components of each of the three terminal currents for an NPN transistor.

Q8-3 What is meant by emitter efficiency?

Q8-4 What is meant by base transport factor?

Q8-5 State, in equation form, the relationship between the three voltages for a PNP transistor.

Q8-6 Why is the collector current in a PNP transistor in the inverse active region much smaller than in the active region?

Q8-7 Why does the collector current decrease when the transistor is operated in the saturation region?

Q8-8 What is the function of the buried layer?

Q8-9 Why are the devices placed in tubs?

HIGHLIGHTS

- The transistor is labeled a transresistance because in the normal or active mode of operation, the emitter-base junction is forward-biased at a low E-B voltage, resulting in a small input resistance. The collector-base junction is reverse-biased at a high C-B voltage, resulting in a large resistance.
- Transistor output characteristics are displayed in the common-emitter connection with I_C vs. $|V_{CE}|$ at various values of I_B, or in common-base connection with I_C vs. $|V_{CB}|$ at various values of I_E. The same information is contained in both.
- When used in an amplifier circuit, the BJT is operated in the active mode. When used in a switching circuit, the transistor is switched between cutoff and saturation.
- An important property of a transistor is that in normal operation, small changes in the base-emitter voltage cause large changes in the collector current.
- The alpha of a transistor in the forward active region is the ratio of the collector to the emitter current, while beta is the ratio of the collector to the base current.

EXERCISES

E8-1 The following components of the currents have been determined as: $I_{En} = 2.712 \times 10^{-6}$A, $I_{Ep} = 0.678$mA, $I_{Cn} = 9.4 \times 10^{-15}$A, and $I_{Cp} = 0.6779$mA.
Determine: a) the injection efficiency, b) the transport factor, c) alpha, d) beta, and e) I_{CEO}

Ans: a) 0.996, b) 0.99985, c) 0.99585, d) 240

E8-2 Sketch the energy band diagram for the electrons in an NPN transistor for (a) thermal equilibrium and (b) saturation.

8.9 ANALYTICAL RELATIONS FOR THE CURRENTS

Assumptions and Procedure

In the preceding sections, we have described the operation of the bipolar junction transistor and determined certain relations basic to its operation in the active mode. In this section, we will derive complete expressions for the currents subject to the following assumptions:

- Low-level injection.
- The electric field intensity in the bulk regions outside the depletion regions is so small that the drift current of minority carriers is neglected.
- No recombination and generation in the depletion regions.
- The widths of the emitter and collector regions are much greater than the diffusion length of minority carriers so that the minority carrier densities have their equilibrium values at the contacts.
- The collector area is much larger than the emitter area so as to collect all holes crossing the collector junction.
- Each of the three bulk regions is assumed to be uniformly doped and both junctions are considered to be step junctions so that the change in impurity density, from one region to another, is abrupt.

For the PNP cross section shown in Fig. 8.14, we will use the symbols shown below the figure. The procedure we will follow in determining expressions for the currents in terms of V_{EB} and V_{CB} is identical to the one we used for the diode in Chapter 5. Again, we determine expressions for the minority carrier distributions in the emitter, base, and collector. Using these distributions, we will then derive expressions for the emitter and collector currents, as determined from the minority carrier currents at both ends of each depletion layer.

We draw the attention of the reader to the use of primed symbols to refer to the excess density, over that of equilibrium, of minority carriers. We also note the zero axes for distances are $x = 0$ in the base, $x' = 0$ for the emitter, and $x'' = 0$ for the collector.

Emitter Current

Just as in the case of the diode, we will determine the expression for the emitter current from the distributions of the minority carrier currents at the edges of the emitter-base depletion layer.

We start with the continuity equation for holes in the steady-state in the base of a PNP transistor, given by Eq. (4.36), in one dimension, as

$$-\frac{p - p_0}{\tau_p} - \frac{1}{q}\frac{\partial J_p}{\partial x} = 0 \qquad (8.23)$$

where p is the hole density anywhere in the base, p_0 is the equilibrium hole density in the base, τ_p is the lifetime of holes in the base, and J_p is the hole current density

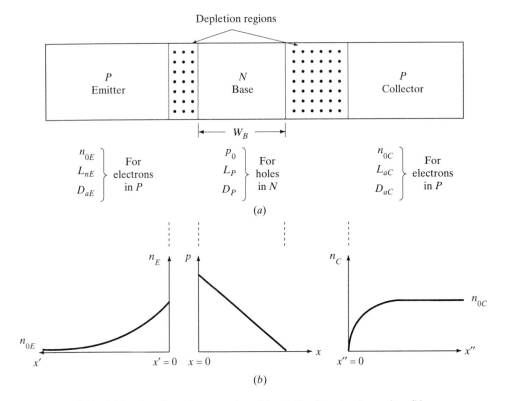

Figure 8.14   (a) Section of transistor together with notation for minority carriers. (b) Minority carrier distributions for active mode operation. The vertical scale represents the relevant carrier and the subscripts E and C refer to the emitter and collector respectively where the hole density in the base is represented without subscript. The equilibrium values in the emitter and collector are n_{OE} and n_{OC}, and p_O in the base.

anywhere in the base. Subject to the assumptions we made, the hole current in the base is a diffusion current so that we can rewrite Eq. (8.23) as

$$-\frac{p - p_0}{\tau_p} + D_p \frac{d^2p}{dx^2} = 0. \tag{8.24}$$

We define the excess hole density p' as $(p - p_0)$ and use $L_p = \sqrt{D_p \tau_p}$, so that Eq. (8.24) becomes

$$\frac{d^2p'}{dx^2} = \frac{p'}{L_p^2} \tag{8.25}$$

The general solution to Eq. (8.25) can be written as

$$p' = B_1\, e^{x/L_p} + B_2 e^{-x/L_p} \tag{8.26}$$

The low-level boundary conditions are

$$p'(0) = p_0 \left[\exp\left(qV_{EB}/kT\right) - 1\right] \qquad (a)$$

$$p'(W_B) = p_0\left[\exp\left(qV_{CB}/kT\right) - 1\right] \quad \text{(b)} \tag{8.27}$$

Applying the boundary conditions to Eq. (8.26), we determine the expression for the hole density $p = p' + p_0$

$$p = p_0 + p_0\left[\exp\left(\frac{qV_{EB}}{kT}\right) - 1\right]\left[\frac{\sinh\left(W_B - x\right)/L_p}{\sinh\left(W_B/L_p\right)}\right]$$

$$+ p_0\left[\exp\left(qV_{CB}/kT\right) - 1\right]\left[\frac{\sinh(x/L_p)}{\sinh(W_B/L_p)}\right] \text{ for } 0 \leqslant x \leqslant W_B \tag{8.28}$$

where W_B is the base width.

The portion of the emitter current due to holes, I_{Ep}, is found from

$$I_{Ep} = (-qD_pA)(dp/dx) \text{ at } x = 0$$

so that

$$I_{Ep} = \frac{q\,A\,D_p p_0}{L_p}\left\{\left[\exp\left(qV_{EB}/kT\right) - 1\right]\left[\coth\left(W_B/L_p\right)\right]\right.$$

$$\left. - \left[\exp\left(qV_{CB}/kT\right) - 1\right]/\left[\sinh\left(W_B/L_p\right)\right]\right\} \tag{8.29}$$

Now we will determine an expression for the electron current component of the emitter current. The distribution of excess electron density in the emitter region n'_E is determined from an equation similar to Eq. (8.25), written as

$$\frac{d^2n'_E}{dx'^2} = \frac{n'_E}{L_{nE}} \tag{8.30}$$

where $n'_E = n_E - n_{OE}$ and n_{OE}, as shown in Fig. 8.11, is the equilibrium value of the electron density in the emitter. The emitter width is assumed to be much greater than L_{nE} so that the boundary conditions are:

$$n'_E = n'_E(0) = n_{OE}\left[\exp\left(\frac{qV_{EB}}{kT}\right) - 1\right], \text{ at } x' = 0 \quad \text{(a)}$$

$$n'_E = 0 \text{ at } x' = \infty \quad \text{(b)} \tag{8.31}$$

The solution to Eq. (8.30) is of the form

$$n'_E = n'_E(0)\exp\left(-x'/L_{nE}\right) \tag{8.32}$$

The electron current component of the emitter current is

$$I_{En} = -q\,A\,D_{nE}\frac{dn'_E}{dx'}\text{ (at } x' = 0) = \frac{q\,A\,D_{nE}}{L_{nE}}n_{0E}\left[\exp\left(qV_{EB}/kT\right) - 1\right] \tag{8.33}$$

The total emitter current is

$$I_E = I_{Ep} + I_{En}$$

$$= \left[(q\,A\,D_p p_0/L_p)(\coth\left(W_B/L_p\right)) + (q\,A\,D_{nE}n_{0E}/L_{nE})\right]\left[\exp\left(qV_{EB}/kT\right) - 1\right]$$

$$- \left[q\,A\,D_p p_0/(L_p\sinh\left(W_B/L_p\right))\right]\left[\exp(qV_{CB}/kT) - 1\right] \tag{8.34}$$

Collector Current

The collector current is made of two components: the hole current crossing the collector junction from base to collector I_{CP} and the current of the electrons that cross this junction from collector to base I_{CN}. Therefore, $I_C = I_{CP} + I_{CN}$, where I_{CP} is determined by using the hole density distribution in the base at the edge of the C-B junction from Eq. (8.28).

$$I_{Cp} = -q \, A \, D_p \frac{dp}{dx} \text{ at } x = W_B$$

$$I_{Cp} = \frac{q \, A \, D_p p_0}{L_p} \left\{ [\exp(q \, V_{EB}/kT) - 1] \frac{1}{\sinh(W_B/L_p)} \right.$$

$$\left. - [\exp(q \, V_{CB}/kT) - 1] \coth(W_B/L_p) \right\} \tag{8.35}$$

To determine I_{CN}, we will use the distribution of excess electron density in the collector region, n'_C, in an equation similar to Eq. (8.30), as

$$\frac{d^2 n'_C}{dx''^2} = \frac{n'_C}{L_{nC}} \tag{8.36}$$

where $n'_C = n_C - n_{OC}$ and n_{OC} is the equilibrium electron density in the collector. The boundary conditions for the electron distribution in the collector are

$$n'_C = n'_C(0'') = n_{OC} [\exp(q \, V_{CB}/kT) - 1] \text{ at } x'' = 0 \qquad \text{(a)}$$
$$n'_C = 0 \text{ at } x'' = \infty \qquad \text{(b)} \tag{8.37}$$

The solution to Eq. (8.36) is

$$n'_C = n'_C(0)\exp(-x''/L_{nC}) \tag{8.38}$$

The electron component of the collector current is

$$I_{Cn} = q \, A \, D_{nC} \frac{dn'_C}{dx''} \text{ (at } x'' = 0) = \frac{-q \, A \, D_{nC}}{L_{nC}} n_{OC} [\exp(q \, V_{CB}/kT) - 1] \tag{8.39}$$

The sum of the current in Eqs. (8.35) and (8.39) forms the collector current

$$I_C = \left[\frac{q \, A \, D_p p_0}{L_p \sinh(W_B/L_p)} \right] [\exp(q \, V_{EB}/kT) - 1]$$

$$- \left[\frac{q \, A \, D_p p_0}{L_p} \coth(W_B/L_p) + \frac{q \, A \, D_{nC} n_{0C}}{L_{nC}} \right] [\exp(q \, V_{CB}/kT) - 1] \tag{8.40}$$

The base current, I_B, is

$$I_B = I_E - I_C \tag{8.41}$$

By substituting Eqs. (8.34) and (8.40) in Eq. (8.41), we obtain the expression for the base current. This expression includes three terms: the electron current across the emitter junction, the electron current across the collector junction, and a

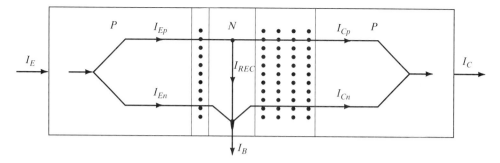

Figure 8.15   Transistor currents and their components. Arrows indicate actual directions.

third current equal to the difference between the hole current crossing the emitter junction and the hole current crossing the collector junction. The third term is the recombination current in the base and results from the electrons that the base lead must supply to recombine with holes. The transistor currents and their components are shown in Fig. 8.15.

It is important to emphasize that, subject to the assumptions that were made at the beginning of the section ***Assumptions and Procedure***, the equations that we have derived for the transistor currents are applicable in all four regions of operation. Later in this chapter, we will simplify these relations and illustrate their applications through examples.

We show in Fig. 8.16 the distributions of minority carriers in the emitter, base, and collector for the active, saturation, and cutoff mode.

In a subsequent section, we will further investigate the recombination current in the base.

So far, the relations derived for the currents in the PNP transistor can be made applicable to the NPN transistor when the following changes are made:

REPLACE

D_p by D_n, p_0 by n_0, and L_p by L_n

D_{nE} by D_{pE}, n_{0E} by p_{0E}, and L_{nE} by L_{pE}

D_{nC} by D_{pC}, n_{0C} by p_{0C}, and L_{nC} by L_{pC}

V_{EB} by V_{BE} and V_{CB} by V_{BC}

We will now use Eqs. (8.34) and (8.40) to calculate the values of currents and their corresponding components. We have two objectives. First, we present the reader with an order of magnitude of the values of the terms in the equations that we have derived. Second, we want to highlight the effect of a decrease of the doping of the emitter, and later, of an increase in the width of the base, on the current gain of the transistor, and on the magnitudes of the currents.

EXAMPLE 8.2

To illustrate the application of the relations derived so far, we consider a PNP silicon transistor having $N_{AE} = 10^{17} \text{cm}^{-3}$, $N_{DB} = 10^{16} \text{cm}^{-3}$, and $N_{AC} = 10^{15} \text{cm}^{-3}$. Assume an effective cross section

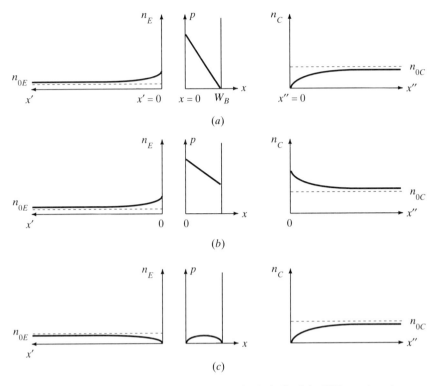

Figure 8.16   Distribution of minority carriers in the bulk of the PNP transistor for (a) active, (b) saturation, and (c) cutoff operation. The width of the base is assumed here to remain constant at W_B in all modes of operation.

area A = 10^{-4}cm², a base width $W_B = 1\mu$m, and minority carrier lifetimes in each of three regions at $\tau = 10^{-6}$s.

For active mode operation, and at $V_{EB} = 0.63$V, calculate:

a) I_{Ep}, I_{En}, and I_E
b) I_{Cp}, I_{Cn}, and I_C
c) The three components of I_B
d) The beta of the transistor

Solution   We determine the minority carrier densities in the three regions by using $n_i^2 = 10^{20}$cm⁻⁶ for silicon:

$$n_{OE} = 10^3 \text{cm}^{-3}, \qquad p_O = 10^4 \text{cm}^{-3}, \quad \text{and} \quad n_{OC} = 10^5 \text{cm}^{-3}$$

The mobilities are calculated from the relations in Eqs. (4.15) and (4.16), the diffusion constants are then determined using the Einstein relation and the diffusion lengths are calculated using $L = (D\tau)^{1/2}$

$\mu_{nE} = 826\text{cm}^2/\text{V-sec}$,	$\mu_p = 447\text{cm}^2/\text{V-sec}$,	$\mu_{nC} = 1330\text{cm}^2/\text{V-sec}$
$D_{nE} = 21.4\text{cm}^2/\text{sec}$,	$D_p = 11.6\text{cm}^2/\text{sec}$,	$D_{nC} = 34.5\text{cm}^2/\text{sec}$
$L_{nE} = 4.62 \times 10^{-3}\text{cm}$,	$L_p = 3.4 \times 10^{-3}\text{cm}$,	$L_{nC} = 5.87 \times 10^{-3}\text{cm}$

The factors multiplying the exponentials in Eqs. (8.34) and (8.40) are calculated to be

$$\frac{qAD_{nE}n_{OE}}{L_{nE}} = 7.41 \times 10^{-17}\text{A}, \frac{qAD_pp_0}{L_p} = 5.45 \times 10^{-16}\text{A}, \frac{qAD_{nC}n_{OC}}{L_{nC}} = 9.4 \times 10^{-15}\text{A}$$

For $W_B/L_p = \dfrac{10^{-4}}{3.4 \times 10^{-3}} = 0.0294117$:

$$\sinh(W_B/L_p) = 0.0294159, \coth(W_B/L_p) = 34.0099$$

For $V_{EB} = 0.63\text{V}, \exp(qV_{EB}/kT) = 3.66 \times 10^{10}$. Active mode operation requires V_{CB} to be negative and its magnitude to be much greater than kT/q.

a) Using Eqs. (8.29) and (8.33) for I_{Ep} and I_{En} respectively, we have

$$I_{Ep} = .678\text{mA}$$
$$I_{En} = 2.712 \times 10^{-6}\text{A}$$
$$I_E = I_{Ep} + I_{En} = 0.681\text{mA}$$

b) The components of the collector current, I_C, are calculated by using Eqs. (8.35) and (8.39) as

$$I_{Cp} = 0.678\text{mA}$$
$$I_{Cn} = 9.4 \times 10^{-15}\text{A}$$
$$I_C = 0.678\text{mA}$$

c) The three components of the base current are:

$$I_{B1} = I_{En} = 2.712 \times 10^{-6}\text{A}$$
$$I_{B2} = I_{Ep} - I_{Cp} = 0.293 \times 10^{-6}\text{A}$$
$$I_{B3} = I_{Cn} = 9.4 \times 10^{-15}$$
$$I_B = I_{B1} + I_{B2} - I_{B3} = 3 \times 10^{-6}\text{A}$$

d) $\beta = \dfrac{I_C}{I_B} = 225.7$

It is important to compare the magnitudes of the three components of the base current. It is obvious that the collector-base leakage current, I_{B3}, is negligibly small. Furthermore, the recombination current, for the ratio of W_B to L_p used is about one tenth of the base current. This is quite true in modern silicon transistors where the dominant component of the base current is the one resulting from majority carrier electrons crossing from base to emitter.

EXAMPLE 8.3

Use the data of Example 8.2 except replace the doping of the emitter by $N_{AE} = 5 \times 10^{16}\text{cm}^{-3}$ to calculate the value of beta.

Solution   Assuming that the lifetime of minority carriers is unchanged, the relevant terms for minority carriers in the emitter become

$$n_{OE} = 2 \times 10^3\text{cm}^{-3}, \mu_{nE} = 1016\text{cm}^2/\text{V-s}, D_{nE} = 26.3\text{cm}^2/\text{s}$$

$$L_{nE} = 5.13 \times 10^{-3}\text{cm and } \frac{qAD_{nE}n_{OE}}{L_{nE}} = 1.6405 \times 10^{-16}\text{A}$$

By repeating the calculations of Example 8.2 we have

$$I_E = 0.684\text{mA}$$

I_C is unchanged at 0.678mA

$$\beta = \frac{I_C}{I_E - I_C} = 107.5$$

We observe the dramatic decrease of beta as the emitter doping was reduced. The reduction of the doping increased the equilibrium density of electrons in the P emitter and consequently increased I_{En}. The increase of I_{En} served to increase I_E and I_B but had no effect on the value of I_C.

EXAMPLE 8.4

Repeat parts (a) and (b) of Example 8.2 except for a base width $W_B = 2\mu\text{m}$.

Solution   The change in W_B causes changes in the hyperbolic functions of W_B/L_p. The value of W_B/L_p is now 0.0588235.

We evaluate the hyperbolic functions as

$$\sinh(W_B/L_p) = 0.0588574, \coth(W_B/L_p) = 17.01962$$

a) By using the above relations and the terms calculated in Example 8.2, we have

$$I_E = 0.342\text{mA}, \qquad I_C = 0.3389\text{mA}$$

b) $\beta = \dfrac{I_C}{I_E - I_C} = 102.7$

An increase in the width of the base caused a decrease in the slope of the hole density at the base side of the depletion layer. This resulted in a decrease of both I_{Ep} and I_{Cp} and hence decreases in the emitter and collector currents. Both the injection efficiency and the transport factor decreased with a larger relative decrease in the injection efficiency. This caused a sizeable decrease in the value of beta.

Relations for the NPN Transistor

By making the substitutions listed earlier in this section relevant, emitter and collector current equations for the NPN transistor in the active region are determined below and in accordance with the reference, as shown in Fig. 8.17. Applications of these equations are illustrated in Example 8.4.

Relations for the NPN transistor corresponding to the PNP device equations, given by Eqs. (8.29), (8.33), (8.35), and (8.39), are shown below

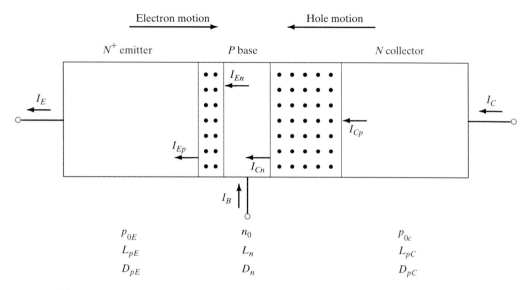

Figure 8.17 Reference currents in NPN BJT and symbols identifying minority carriers in the three regions.

$$\frac{I_{En}}{\text{in B at E} - \text{B junction}} = \frac{qAD_n n_0}{L_n}\left\{\left[\exp\frac{q\,V_{BE}}{kT} - 1\right]\coth\frac{W_B}{L_n}\right.$$

$$\left. - \left[\exp\frac{q\,V_{BC}}{kT} - 1\right]\frac{1}{\sinh\dfrac{W_B}{L_n}}\right\} \quad (8.42)$$

$$\frac{I_{Ep}}{\text{in E at E} - \text{B junction}} = \frac{q\,AD_{pE}p_{OE}}{L_{pE}}\left[\exp\frac{q\,V_{BE}}{kT} - 1\right] \quad (8.43)$$

$$\frac{I_{Cn}}{\text{in B at C} - \text{B junction}} = \frac{qAD_n n_0}{L_n}\left\{\left[\exp\frac{q\,V_{BE}}{kT} - 1\right]\frac{1}{\sinh\dfrac{W_B}{L_n}}\right.$$

$$\left. - \left[\exp\frac{q\,V_{BC}}{kT} - 1\right]\coth\frac{W_B}{L_n}\right\} \quad (8.44)$$

$$\frac{I_{Cp}}{\text{in C at C} - \text{B junction}} = -\frac{qAD_{pC}p_{0C}}{L_{pC}}\left[\exp\frac{q\,V_{BC}}{kT} - 1\right] \quad (8.45)$$

where $I_E = I_{En} + I_{Ep}$ and $I_C = I_{Cn} + I_{Cp}$.

EXAMPLE 8.5

An NPN silicon transistor has the following properties: $N_{DE} = 10^{17} \text{cm}^{-3}$, $N_{AB} = 10^{16} \text{cm}^{-3}$, $N_{DC} = 10^{15} \text{cm}^{-3}$, $A = 10^{-4} \text{cm}^2$, and $W_B = 1 \mu\text{m}$. Assume that the minority carriers' lifetime in each of the three regions is $\tau = 10^{-6}$. Use $n_i^2 = 10^{20} \text{cm}^{-6}$.

For active mode operation use $V_{BE} = 0.63\text{V}$ and $|V_{BC}| \gg V_{BE}$, calculate:

a) $I_{En}, I_{Ep},$ and I_E

b) $I_{Cn}, I_{Cp},$ and I_C

c) I_B

d) β

Solution   For minority carriers: $p_{OE} = 10^{20}/10^{17}$, $n_0 = 10^{20}/10^{16}$ and $p_{OC} = 10^{20}/10^{15}$, mobilities are calculated from Eqs. (4.15) and (4.16) $D = \mu kT/q$ and $L = (D\tau)^{0.5}$.

	Emitter	Base	Collector
cm^{-3}	$p_{OE} = 10^3$	$n_0 = 10^4$	$p_{OC} = 10^5$
cm^2/V-s	$\mu_{pE} = 350.5$	$\mu_n = 1258.3$	$\mu_{pC} = 460$
cm^2/s	$D_{pE} = 9.078$	$D_n = 32.59$	$D_{pC} = 11.91$
cm	$L_{pE} = 3.013 \times 10^{-3}$	$L_n = 5.7 \times 10^{-3}$	$L_{pC} = 3.45 \times 10^{-3}$

$$W_B/L_n = 1/57, \ \sinh(W_B/L_n) = 0.017544759, \ \coth\left(\frac{W_B}{L_n}\right) = 57$$

a) $I_{En} = 1908.47 \mu\text{A}, I_{Ep} = 1.76438 \times 10^{-4}\text{A}, I_E = 1910.2343 \mu\text{A}$

b) $I_{Cn} = 1908.358 \mu\text{A}, I_{Cp} = 5.523 \times 10^{-15}\text{A}, I_C = 1908.35 \mu\text{A}$

c) $I_B = 1.8639 \times 10^{-6}\text{A}$

d) $\beta = \dfrac{I_C}{I_B} \cong 1024$

The high current gain obtained in this example illustrates the superiority of the NPN BJT. All data in this example are the same as those used in the PNP BJT of Example 8.2, yet β_F is more than four times greater. This is mainly a result of the differences in the mobilities of the minority carriers in the base.

Recombination Current in the Base

We have indicated earlier that both the common-base current gain, α, and the common-emitter current gain, β, are made large by reducing the ratio of the width of the base to the diffusion length of minority carriers in the base. Reduction of this ratio decreases the recombination current in the base. Let us digress briefly to investigate this by developing an expression for the hole density in the base of a PNP transistor based on the condition for which alpha and beta were defined, namely operation in the active mode.

Operation in the active region results when V_{EB} is positive and V_{CB} is negative. We assume that both have magnitudes that are much greater than kT/q. Using these conditions, Eq. (8.28) can be approximated by

$$p/p(0) \cong \frac{\sinh[(W_B - x)/L_p]}{\sinh W_B/L_p} \tag{8.46}$$

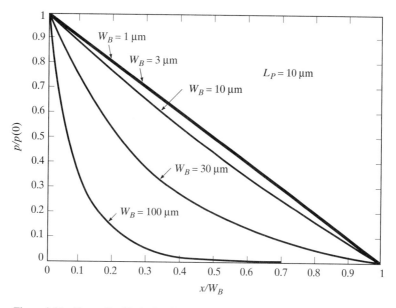

Figure 8.18 Normalized hole density distribution in the base of a PNP transistor as a function of normalized distance in the base for a diffusion length $L_p = 10\mu m$.

where $p(0)$ is the hole density at $x = 0$ in the base and is given by $p_0 \exp (q V_{EB}/kT)$.

Calculations based on the above relation are plotted in Fig. 8.18. The ordinate is the normalized hole density and the abscissa is normalized distance. Curves are plotted for a diffusion length of 10 microns and various values of base widths.

It is quite evident from the plots that for $(W_B/L_p) \gg 1$ the distribution is exponential, while for $W_B \ll L_p$, the distribution approaches a straight line. This can be concluded by setting $W_B \ll L_p$ and hence $(W_B - x) \ll L_p$ in Eq. (8.46). The first two terms of the Taylor series expansion of the hyperbolic sine are given as

$$\sinh u = u + u^3/6$$

For a small u, $\sinh u$ can be replaced by u so that the expression for the hole density becomes

$$p = p(0) \frac{(W_B - x)/L_p}{W_B/L_p} = p(0)(1 - (x/W_B)) \tag{8.47}$$

A perfectly linear distribution of hole density in the base makes $I_{Ep} = I_{Cp}$, hence zero recombination current. The constant slope makes the second derivative of the excess hole density distribution zero, which from Eq. (8.25) implies that either p' is zero ($p = p_0$), which is obviously wrong, or that L_p is infinite, which would result from an infinite lifetime, which is also impossible.

In fact, by making $W_B \ll L_p$, the hole density distribution approaches, but never becomes, a straight line, for there will always be some recombination.

Expressions for Alpha and Beta

To obtain expressions for α and β in terms of the physical constants of the transistor, we will simplify Eqs. (8.29), (8.33), and (8.35) for operation in the active mode ($V_{EB} > 0$ and $V_{CB} < 0$) as

$$I_{Ep} = \frac{q\,A\,D_p\,p_0}{L_p\,\sinh(W_B/L_p)}\{[\exp(q\,V_{EB}/kT)\,(\cosh(W_B/L_p)) + 1]\} \qquad (8.48)$$

$$I_{En} = \frac{q\,A\,D_{nE}\,n_{0E}}{L_{nE}}\exp(q\,V_{EB}/kT) \qquad (8.49)$$

$$I_{Cp} = \frac{q\,A\,D_p\,p_0}{L_p\,\sinh(W_B/L_p)}[\exp(q\,V_{EB}/kT) + \cosh(W_B/L_p)] \qquad (8.50)$$

We further simplify the above expressions by assuming that, in a good transistor, $(W_B/L_p) \ll 1$ and hence $\cosh(W_B/L_p)$ is close to unity. Since $V_{EB} \gg kT/q$, the unity in Eq. (8.48) and the cosh term in Eq. (8.50) become negligible and Eqs. (8.48) and (8.50) may be written with $p(0)$ replacing $p_0 \exp(q\,V_{EB}/kT)$, as

$$I_{Ep} = (q\,A\,D_p\,p(0)/L_p)\coth(W_B/L_p) \qquad (8.51)$$

$$I_{Cp} = q\,A\,D_p\,p(0)/(L_p\,\sinh W_B/L_p) \qquad (8.52)$$

The transport factor, δ, becomes

$$\delta \equiv I_{Cp}/I_{Ep} = \text{sech}(W_B/L_p)$$

A further simplification is made by expanding $\text{sech}(W_B/L_p)$ and assuming $W_B/L_p \ll 1$ so that the transport factor becomes

$$\delta = \frac{1}{1 + \dfrac{(W_B/L_p)^2}{2}} \qquad (8.53)$$

We conclude again that a transport factor of almost unity is achieved by making $W_B \ll L_p$. This is consistent with our earlier conclusion.

The injection efficiency has been defined as

$$\gamma \equiv \frac{I_{Ep}}{I_{Ep} + I_{En}}$$

By substituting for I_{Ep} and I_{En} from Eqs. (8.51) and (8.49), and rearranging terms, the injection efficiency becomes

$$\gamma = \frac{1}{1 + \dfrac{D_{nE}\,L_p\,n_{0E}\,\tanh(W_B/L_p)}{D_p\,L_{nE}\,p_0}} \qquad (8.54)$$

For $(W_B/L_p) \ll 1$, $\tanh(W_B/L_p) \cong W_B/L_p$. Using this simplification and replacing the ratio of the equilibrium carrier densities, n_{0E}/p_0, by the ratio of the doping of the base to the emitter, the injection efficiency becomes

$$\gamma = \frac{1}{1 + \dfrac{D_{nE}\, W_B\, N_{DB}}{D_p\, L_{nE}\, N_{AE}}} \tag{8.55}$$

The injection efficiency is increased mainly by a high emitter doping relative to the doping of the base.

The common-base current gain, α, is the product of the transport factor and the injection efficiency, and the common-emitter current gain, β, is $\alpha/(1 - \alpha)$. For active mode operation, we label alpha as α_F and beta as β_F with the subscript referring to forward. This is in contrast to the special region where the roles of the emitter and collector are interchanged and labeled inverse-active mode. For the inverse active mode, alpha and beta will be labeled α_R and β_R.

We will now use the expressions in Eqs. (8.53) and (8.55) to determine an expression for β_F, defined as $\alpha_F/1 - \alpha_F$, where $\alpha_F = \delta\gamma$.

The expression for β_F becomes

$$\frac{1}{\beta_F} = \frac{1 - \alpha_F}{\alpha_F} = \frac{1}{2}\left(\frac{W_B}{L_p}\right)^2 + \frac{D_{nE}W_B N_{DB}}{D_p L_{nE} N_{AE}} + \frac{1}{2}\left(\frac{W_B}{L_p}\right)^2 \frac{D_{nE}W_B N_{DB}}{D_p L_{nE} N_{AE}} \tag{8.56}$$

EXAMPLE 8.6

Use the data of Example 8.2 and Eq. (8.56) to calculate a value for β_F.

Solution   For $W_B = 1\mu\text{m}$ and $L_p = 34\mu\text{m}$, $\dfrac{1}{2}\left(\dfrac{W_B}{L_p}\right)^2 = \dfrac{1}{2312}$

The second term in the expression is calculated to be $1/250$ and the third term is $1/(250 \times 2312)$ so that

$$\frac{1}{\beta_F} = \frac{1}{2312} + \frac{1}{250} + \frac{1}{250 \times 2312}$$

$$\beta_F = 225$$

The value of β_F, calculated in Example 8.5, is a little higher than the average value of transistors used in current and voltage amplifiers operating in the common-emitter connection. The question is then, what are typical values of β_F and what major parameters influence those values?

The discrete silicon transistors used in current and voltage amplifiers have a normal β_F range from a low of approximately 75 to a high of 300. These are nominal values and the actual values may vary over a range of plus or minus 20 percent of these. Their ability to dissipate power is quite small and hence they are of small physical size. Their power dissipation is of the order of 100 milliwatts.

For applications in power amplifiers, where the AC output power is in watts or tens of watts, transistors may have a beta of about 10 since their function is to transfer power rather than amplifying voltage. These transistors have a large size and large surface area since, in acting as agents of power transfer from the DC source to the load, they are required to dissipate a sizable fraction of the power delivered by the DC power.

For certain applications, such as in the input stage of an integrated-circuit, operational amplifier, transistors may have a beta of approximately 10,000. These are known as superbeta transistors.

For silicon transistors and as was shown by the results of Example 8.5, control over the design of the beta is achieved through control of the ratio of the base width to the diffusion length of minority carriers and control over the ratio of N_{AE} to N_{DB}.

It is quite evident from the above calculation that the central control on the value of β_F is achieved by increasing the value of the injection efficiency given by Eq. (8.54). The advantage of a small value of (W_B/L_p) has been exhausted in modern day transistors by the ability to manufacture devices whose base widths are a fraction of the diffusion length. Furthermore, these small base widths have probably reached the limits of manufacturing technology.

8.10 EBERS-MOLL MODEL

The basic equations for the transistor currents, Eq. (8.34) for I_E and Eq. (8.40) for I_C, are general enough so as to apply for all four combinations of V_{EB} and V_{CB} and thus applicable in all four regions of the characteristics. We will now simplify Eqs. (8.34) and (8.40) by assuming that $W_B \ll L_p$.

The hyperbolic functions in these equations are expanded into series as

$$\cosh u = 1 + u^2/2 + \ldots$$

$$\sinh u = u + u^3/6 + \ldots$$

For small values of u, they become

$$\cosh u \cong 1, \sinh u \cong u, \coth u = 1/u$$

Using the simplifications for the hyperbolic functions, Eqs. (8.34) and (8.40) become

$$I_E = q\,A\left[\frac{D_p\,p_0}{W_B} + \frac{D_{nE}\,n_{0E}}{L_{nE}}\right][\exp\,(q\,V_{EB}/kT) - 1] \quad \text{(a)}$$

$$-\frac{q\,A\,D_p\,p_0}{W_B}[\exp\,(q\,V_{CB}/kT) - 1] \quad (8.57)$$

$$I_C = \frac{q\,A\,D_p\,p_0}{W_B}[\exp\,(q\,V_{EB}/kT) - 1] \quad \text{(b)}$$

$$-\,q\,A\left[\frac{D_p\,p_0}{W_B} + \frac{D_{nC}\,n_{0C}}{L_{nC}}\right][\exp\,(q\,V_{CB}/kT) - 1]$$

We now define the following:

$$I_{EF} \equiv q\,A\left[\frac{D_p\,p_0}{W_B} + \frac{D_{nE}\,n_{0E}}{L_{nE}}\right][\exp\,(q\,V_{EB}/kT) - 1] \text{ (a)}$$

$$I_{CR} \equiv q\,A\left[\frac{D_p\,p_0}{W_B} + \frac{D_{nC}\,n_{0C}}{L_{nC}}\right][\exp\,(q\,V_{CB}/kT) - 1] \text{ (b)} \quad (8.58)$$

$$\alpha_F \equiv \left(\frac{D_p\,p_0}{W_B}\right)\Big/\left[\frac{D_p\,p_0}{W_B} + \frac{D_{nE}\,n_{0E}}{L_{nE}}\right] \quad \text{(c)}$$

$$\alpha_R \equiv \left(\frac{D_p\,p_0}{W_B}\right)\Big/\left[\frac{D_p\,p_0}{W_B} + \frac{D_{nC}\,n_{0C}}{L_{nC}}\right] \quad \text{(d)}$$

We note that the expression for α_F in Eq. (8.58(c)) is identical to the expression found by taking the product of Eqs. (8.53) and (8.54) and letting $W_B/L_p \ll 1$ so that $\tanh W_B/L_p \cong W_B/L_p$.

The general expressions for I_E, I_C, and I_B for a bipolar junction transistor using Eqs. (8.57) and (8.58) become

$$I_E = I_{EF} - \alpha_R I_{CR} \qquad \text{(a)}$$

$$I_C = \alpha_F I_{EF} - I_{CR} \qquad \text{(b)} \qquad (8.59)$$

$$I_B = I_E - I_C = I_{EF}(1 - \alpha_F) + I_{CR}(1 - \alpha_R) \qquad \text{(c)}$$

These equations are known as the *Ebers-Moll equations* for the PNP transistor. Equations (8.59) also apply to the NPN transistor except that appropriate modifications are made to the terms defined in Eqs. (8.58). Recall that these equations are valid for the currents' reference directions adopted (which are actual directions) for a device operating in the active mode.

If we denote the prefactor of the exponential in Eq. (8.58(a)) by I_{ES} and the similar term of Eq. (8.58(b)) by I_{CS}, we obtain the following equality

$$I_{ES}\,\alpha_F = I_{CS}\,\alpha_R \equiv I_{SM} \qquad (8.60)$$

where $I_{SM} = D_p p_0/W_B$.

The Ebers-Moll equations for I_E and I_C can then be written from Eqs. (8.58) and (8.59) as

$$I_E = I_{ES}[\exp(q\,V_{EB}/kT) - 1] - \alpha_R I_{CS}[\exp(q\,V_{CB}/kT) - 1] \qquad \text{(a)}$$

$$I_C = \alpha_F I_{ES}[\exp(q\,V_{EB}/kT) - 1] - I_{CS}[\exp(q\,V_{CB}/kT) - 1] \qquad \text{(b)} \qquad (8.61)$$

We also define

$$\beta_F \equiv \alpha_F/(1 - \alpha_F) \quad \text{(a)}$$

$$\beta_R = \alpha_R/(1 - \alpha_R) \quad \text{(b)} \qquad (8.62)$$

By setting $V_{CB} = 0$ in Eqs. (8.61), the ratio of I_C to I_E becomes α_F and likewise the ratio of I_C to I_B becomes β_F. We can, therefore, define β_F as the common-emitter short-circuit (B to C) current gain and α_F as the common-base short-circuit current gain. It is to be noted that only three parameters of the transistor are necessary and sufficient to solve for the currents given the voltages. These three parameters are β_F, β_R, and I_S. The parameters β_F and β_R are used to calculate α_F and α_R, and by using I_{SM} in Eq. (8.60), I_{ES} and I_{CS} are available.

A circuit model for the transistor using Equations (8.59) and (8.61) is shown in Fig. 8.19.

REVIEW QUESTIONS

Q8-10 Identify the currents and voltages in the common-base and common-emitter connections.

Q8-11 Why is the base width an important dimension in the quality of a transistor?

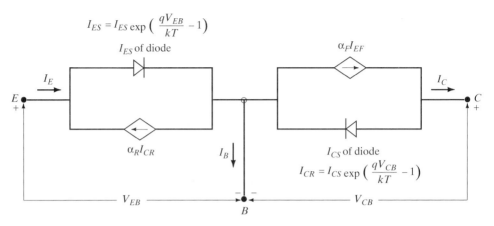

Figure 8.19   The Ebers-Moll model for the PNP transistor. Current directions are actual.

Q8-12 Write the Ebers-Moll equations for an NPN transistor.

Q8-13 Identify three physical parameters that contribute to a high beta transistor.

Q8-14 Use the Ebers-Moll equations to define α_F and β_F.

Q8-15 Use the Ebers-Moll equations to define α_R and β_R.

HIGHLIGHTS

- Subject to the assumption that there is no generation or recombination in the depletion regions, analytical relations for the components of the three currents of a BJT are derived using diffusion currents expressions.

- The above is accomplished by solving the continuity equation for the minority carriers in each of the emitter, base, and collector subject to the boundary conditions in each region.

- The boundary conditions in the emitter and collector are functions of the emitter-base and collector-base voltages respectively and assuming that the excess minority carrier densities at the metal contacts are zero. In the base, the boundary conditions are determined by the emitter-base and collector-base voltages.

- The emitter current and the collector current components of the PNP, I_{En}, I_{Ep}, I_{Cn}, and I_{CP} are determined as diffusion currents at each of the emitter-base and collector-base junctions. The recombination current in the base, I_{REC}, is the difference between I_{Ep} and I_{Cp}.

- The emitter current is given by the sum of I_{En} and I_{Ep}, the collector current is the sum of I_{Cn} and I_{Cp}, and the base current is the algebraic sum of I_{En}, I_{Cn}, and I_{REC}.

- The Ebers-Moll relations are expressions for the emitter and collector currents as functions of the emitter-base and collector-base voltages and in terms of $\alpha_F, \alpha_R, I_{ES}$, and I_{CS}.

- The Ebers-Moll relations are applicable in all modes of operation of the device.

EXERCISES

E8-3 Consider a PNP bipolar transistor at 300K with the following parameters:

$$N_{AE} = 10^{18} \text{cm}^{-3} \quad N_{DB} = 10^{16} \text{cm}^{-3} \quad N_{AC} = 10^{15} \text{cm}^{-3}$$
$$D_{nE} = 52 \text{cm}^2/\text{s} \quad D_p = 40 \text{cm}^2/\text{s} \quad D_{nC} = 116.5 \text{cm}^2/\text{s}$$
$$\tau_{nE} = 10^{-8} \text{s} \quad \tau_p = 10^{-7} \text{s} \quad \tau_{nC} = 10^{-6} \text{s}$$
$$W_B = 4\mu\text{m} \quad A = 0.1 \text{mm}^2$$

Determine: a) α_F, b) β_F, c) I_{ES}, d) α_R, e) β_R, f) I_{CS}

Ans: a) 0.99283, c) 1.61×10^{-13}A, d) 0.48, f) 3.3×10^{-13}A

To emphasize their generality, we will use the Ebers-Moll equations to derive Eqs. (8.13) and (8.15). In the active region of operation, V_{CB} is negative and $|V_{CB}| \gg kT/q$. Using this, solving for the first term of Eq. (8.61)(a) and replacing for it in Eq. (8.61)(b), we have

$$I_C = \alpha_F I_E + I_{CS}(1 - \alpha_F \alpha_R) = \alpha_F I_E + I_{CBO} \tag{8.63}$$

where I_{CBO} has been defined as the collector-to-base current with the emitter open-circuited ($I_E = 0$). Equation (8.63) represents the common-base characteristics of the PNP BJT in the active region, as shown in Fig. 8.12(b). The common-emitter characteristics in the active region are obtained from Eq. (8.63) by replacing I_E by $(I_B + I_C)$ and solving for I_C to obtain

$$I_C = (\alpha_F/1 - \alpha_F)I_B + I_{CBO}/1 - \alpha_F = \beta_F I_B + I_{CEO} \tag{8.64}$$

where I_{CEO} has been defined as the collector-to-emitter current with the base open-circuited.

Equation (8.64) represents the common-emitter characteristics of the PNP BJT in the active region, as shown in Fig. 8.12(a).

We will now determine numerical values for the four constants in Eqs. (8.57) using the data of Example 8.2.

EXAMPLE 8.7

Use the information given in Example 8.2 to calculate I_{ES}, I_{CS}, α_F, and α_R.

Solution   The constants I_{ES} and I_{CS} are defined by the prefactors in Eqs. (8.58)(a) and (b):

$$I_{ES} = qA\left[D_p p_0/W_B + D_{nE} n_{0E}/L_{nE}\right]$$

$$I_{CS} = qA\left[D_p p_0/W_B + D_{nC} n_{0C}/L_{nC}\right]$$

Using the values of Example 8.2, we obtain

$$I_{ES} = 18.6341 \times 10^{-15} \text{A and } I_{CS} = 27.9637 \times 10^{-15} \text{A}.$$

The values for α_F and α_R are calculated to be

$$\alpha_F = 0.996 \text{ and } \alpha_R = 0.663.$$

The corresponding values of β_F and β_R are

$$\beta_F = 249 \text{ and } \beta_R = 1.97.$$

The value of β_F is different from the one calculated in Example 8.2 for two reasons. First, the hyperbolic functions used in Example 8.2 have been simplified and second, a small inaccuracy in calculating α_F leads to large errors in β_F.

We observe very small values for α_R and β_R. These are to be expected since the roles of the emitter and collector have been interchanged.

We caution the reader that in all of these calculations, we have used the same value for the areas of the base, emitter, and collector. The area of the collector is generally much larger than that of the emitter and this difference has a significant tendency to cause a further reduction in the value of β_R.

The Ebers-Moll relations in Eqs. (8.61) are applicable to the NPN transistor, when the relations for α_F, α_R, I_{ES}, and I_{CS} are modified in accordance with the corresponding changes made in Eqs. (8.42) to (8.45), provided also that V_{BE} replaces V_{EB} and V_{BC} replaces V_{CB}.

Subject to the assumptions that have been made in arriving at these equations, it is important to realize that these relations are valid in all four regions of operation, namely: active, saturation, cutoff, and inverse active.

In the following example, we will illustrate the numerical values that result from using the Ebers-Moll relations in the four regions.

EXAMPLE 8.8

Use the results of Example 8.6 to calculate the values of the currents in the four regions of operation.

For the active region use $V_{EB} = 0.63\text{V}, |V_{CB}| \gg \dfrac{kT}{q}$, and $V_{CB} < 0$

For the saturation region use $V_{EB} = 0.63\text{V}$ and $V_{CB} = 0.53\text{V}$ assuming $V_{EC} = 0.1\text{V}$

For the cutoff region use $|V_{EB}| = |V_{CB}| \gg \dfrac{kT}{q}, V_{EB} < 0$, and $V_{CB} < 0$

For the inverse active region use $|V_{EB}| \gg \dfrac{kT}{q}, V_{CB} = 0.63\text{V}, V_{EB} < 0$

Solution   We determined the values for the constants in the Ebers-Moll equation as

$I_{ES} = 18.6341 \times 10^{-15}\text{A}$ $I_{CS} = 27.9637 \times 10^{-15}\text{A}$

$\alpha_F = 0.996$ $\alpha_R = 0.663$

For the active region: For the cutoff region:

$I_E = 0.6827\text{mA}$ $I_E = -0.0941 \times 10^{-15}\text{A}$

$I_C = 0.680\text{mA}$ $I_C = 9.404 \times 10^{-15}\text{A}$

$I_B = 0.0027\text{mA}$ $I_B = -9.31 \times 10^{-15}\text{A}$

$\beta_F = 249$

For the saturation region: For the inverse active region:

$I_E = 0.6684\text{mA}$ $I_E = 0.6792\text{mA}$,

$I_C = 0.658437\text{mA}$ $I_C = -1.0245\text{mA}$

$I_B = 0.009996\text{mA}$ $I_B = 0.344\text{mA}$

$I_C/I_B = 66$

The new I'_C (negative of old I_E) $= -0.6792\text{mA}$
The new I'_E (negative of old I_C) $= 1.0245\text{mA}$
$I'_C/I'_B = 1.97 = \beta$

It is worthwhile noting the very small ratio of I'_C to I_B. Furthermore, it can be shown that the magnitude of V_{EC} at saturation in the inverse active region is much smaller than the corresponding magnitude in the active region.

We note that it is in the third quadrant of the $I_C - V_{EC}$ characteristics of Fig. 8.9(b) that inverse active operation is displayed. In this region, I_B is positive whereas the actual directions of the emitter and collector currents are the reverse of our actual references.

The bipolar junction transistor is operated in the inverse active mode in only certain applications in digital switching circuits. As we indicated earlier, the value of β_R is very small because of the relative dopings of the various regions and also because of the much smaller area of the emitter (which is the collector in the inverse active mode) when compared to that of the collector.

PROBLEMS

8.1 For an N^+PN silicon transistor,

    a) draw the energy level diagram at equilibrium, clearly identifying E_C, E_V, and E_F.

    b) repeat part (a) when the device is biased in the active region with $\left|V_{BC}\right| = 10V_{BE}$.

8.2 A silicon PNP transistor is operating in the active region with $V_{EB} = 0.668V$ and $V_{CB} = -2V$. Given that $N_{AE} = 4 \times 10^{17}cm^{-3}$, $N_{DB} = 2 \times 10^{16}cm^{-3}$, $N_{AC} = 2 \times 10^{14}cm^{-3}$, $W_B = 5\mu m$, $A = 1.8 \times 10^{-4}cm^2$, and $\tau_p = \tau_n = 1\mu s$, determine at $T = 300K$,

    a) I_{Ep}

    b) I_{Cp}

    c) I_{En}

    d) I_{Cn}

    e) the recombination component of the base current

8.3 Determine β_F for Problem 8.2.

8.4 A silicon NPN BJT has $N_{DE} = 10^{18}cm^{-3}$, $N_{AB} = 10^{16}cm^{-3}$, and $N_{DC} = 10^{15}cm^{-3}$. Assume: i) the base width is much smaller than the diffusion length of minority carriers in the base, ii) the widths of the emitter and collector are much greater than the diffusion lengths of minority carriers in the respective regions.

    a) Sketch the minority carrier distributions in the base for saturation and cutoff and identify, by cross-hatching, the minority carrier storage charge area when the device is switched from cutoff to saturation. At saturation $V_{BE} = 0.8V$ and $V_{BC} = 0.6V$.

    b) Give the values of the minority carrier densities in the base at both junctions.

8.5 A silicon PNP BJT has $N_{AE} = 2.5 \times 10^{18}cm^{-3}$, $N_{DB} = 2 \times 10^{17}cm^{-3}$, $N_{AC} = 10^{16}cm^{-3}$, $A = 10^{-3}cm^2$, $W_B = 1\mu m$, $\tau_{nE} = 1\mu s$, $\tau_{nC} = 1\mu s$, and $L_{pB} = 10\mu m$. Determine, at $T = 300K$, the collector current in the active region for $V_{BC} = 2V$ and,

    a) $V_{EB} = 0.62V$

    b) $I_B = 2.5\mu A$.

8.6 A PNP silicon BJT has $N_{AE} = 10^{18}cm^{-3}$, $N_{DB} = 10^{16}cm^{-3}$, $N_{AC} = 10^{15}cm^{-3}$, the metallurgical base width $= 1.0\mu m$, and $A = 3mm^2$. For $V_{EB} = 0.5V$ and $V_{CB} = -5V$, determine at $T = 300K$, the effective width of the neutral base.

8.7 For the device of Problem 8.6, given $\tau_{nE} = 10^{-8}$s, $\tau_{pB} = 10^{-7}$s, and $\tau_{nC} = 10^{-6}$s. Determine:

   a) I_{Cp}

   b) I_{En}

   c) I_{Cn}

8.8 For the device of Problem 8.7, determine:

   a) I_E

   b) the emitter efficiency

   c) the transport factor

   d) I_C

   e) I_B

8.9 Assume that the minority carrier distribution in the base of a PNP BJT is linear when the BJT is operating in the active region with $|V_{BC}| \gg |V_{BE}|$ and $|V_{BE}| \gg kT/q$. Derive expressions for:

   a) the component of I_E due to carriers crossing from emitter to base.

   b) the charge density stored in the base as a function of the collector current density.

   Assume injection efficiency = 1 and $I_{Cn} = 0$.

8.10 The Ebers-Moll model of the BJT includes three device properties only, namely: I_{SM}, α_F, and α_R. Briefly explain how you would determine these for an NPN device from measurements carried out on the BJT.

8.11 A silicon PNP transistor is operating at 300K and has: $N_{AE} = 1 \times 10^{18}$cm^{-3}, $N_{DB} = 1 \times 10^{17}$cm^{-3}, $N_{AC} = 1 \times 10^{15}$cm^{-3}, $L_{nE} = L_{pB} = L_{nC} = 10\mu$m, $W_B = 1\mu$m, and $A = 1$mm^2. Determine the collector current, I_C, when the BJT is operating in the active mode for

   a) $V_{EB} = 0.6$V

   b) $I_E = 2.55$mA

8.12 Briefly explain why I_C/I_B, for an NPN BJT, operating in saturation, decreases as V_{CE} is reduced. Does I_C decrease? Does I_B increase?

8.13 A silicon NPN BJT has $N_{DE} = 5 \times 10^{17}$cm^{-3}, $N_{AB} = 10^{16}$cm^{-3}, $W_B = 5\mu$m, and $\tau_{pE} = \tau_{pC} = 1\mu$s. Determine β_F at $T = 300$K for

   a) $\tau_{nB} = 1\mu$s

   b) $\tau_{nB} = 10\mu$s

8.14 A silicon NPN BJT has $N_{AB} = 10^{16}$cm^{-3} and $\tau_{pB} = 1\mu$s. Determine W_B so that the transport factor is 0.995 at $T = 300$K. Assume $(W_B/L_p) \ll 1$.

8.15 The expression for V_{CESAT} of a BJT in saturation is given by

$$|V_{CESAT}| = V_T \ell n \frac{1/\alpha_R + \beta_S/\beta_R}{1 - \beta_S/\beta_F}$$

where $\beta_S = I_C/I_B$. Use the Ebers-Moll model equations and the equality of $\alpha_F I_{ES}$ and $\alpha_R I_{CS}$ to derive the above expression.

8.16 For an NPN BJT having $\alpha_F = 0.99$ and $\alpha_R = 0.5$, plot V_{CESAT} versus I_C/I_B as I_C/I_B varies from 20 to 95. $T = 300$K.

8.17 An NPN BJT at $T = 300$K has $\alpha_F = 0.995$ and $\alpha_R = 0.15$ at $I_C = 1$mA. Determine I_B for V_{CESAT} equal to:

a)   0.2V

b)   0.1V

8.18 A symmetrical silicon PNP BJT has $N_{AE} = 10^{18}\text{cm}^{-3}$, $N_{DB} = 5 \times 10^{16}\text{cm}^{-3}$, $W_B = 2\mu\text{m}$, $\tau_{pB} = 10ns$, $\tau_{nE} = 1\mu\text{s}$, $A = 0.01\text{cm}^2$, and the device is operating in the active region. For $V_{EB} = 0.65$V and $V_{CB} = -2$V, determine at $T = 300$K:

   a) I_C

   b) β_F

8.19 An NPN BJT has $N_{DB}\ 3 \times 10^{18}\text{cm}^{-3}$, $N_{AB} = 5 \times 10^{16}\text{cm}^{-3}$, and $N_{DC} = 5 \times 10^{15}\text{cm}^{-3}$. Also $W_B = 1\mu\text{m}$, $A = 10^{-3}\text{mm}^2$ and the lifetime of minority carriers in all 3 regions is 0.1μs. For operation in the active region, and at $T = 300$K, determine I_C for $V_{BE} = 0.5$V.

8.20 **a)** An N$^+$PN BJT is operating in the inverse active region. Sketch the minority carrier distribution in the base and in the new emitter.

      b) Using the distribution of minority carriers, briefly explain why $\alpha_R \ll \alpha_F$.

8.21 A diode is formed in an integrated circuit by placing a short-circuit from collector to base of a transistor. For such a diode:

   a) Sketch the minority carrier density in the base.

   b) Based on storage time considerations, briefly explain why such a connection is preferable to other diode connections made from transistors.

8.22 A diode on an IC can also be formed between the base and emitter of a transistor while the collector is open-circuited.

   a) Sketch the minority carrier density in the base.

   b) Using the carrier distribution and carriers' motions, explain how the collector current becomes zero.

8.23 Use the Ebers-Moll equations to derive expressions for V_{EB} and V_{CB} of a PNP transistor in terms of I_C, I_E, I_{SM}, β_F, and β_R.

chapter 9

BIPOLAR TRANSISTORS II: LIMITATIONS, SWITCHING, AND MODELS

9.0 INTRODUCTION

In the previous chapter, we discussed the principle of operation of the BJT and derived current-voltage relations subject to certain assumptions. Although these relations are remarkably accurate over a broad range of currents and voltages, secondary effects, which are present in actual transistor characteristics, were excluded.

The properties of real transistors diverge from the characteristics we derived because of the following effects:

1. Changes in the effective width of the base as a result of the changes in the reverse bias applied to the collector-base junction.
2. Multiplication of carriers in the collector-base depletion layer at a high reverse bias. Breakdown at the junction.

The two effects stated above can be discerned by a comparison of the ideal and the actual graphical characteristics. Other important effects, which do not appear in the graphical or analytical characteristics discussed so far, are:

1. Decrease of the current gain, β_F, at very low and very high injection.
2. Response times of the BJT related to the switching of the transistor between the ON and OFF states.
3. The circuit modelling of the transistor when it is subjected to small changes in the applied voltages. The circuit includes the capacitances of the BJT that become extremely important in the operation of the transistor at high frequencies.

The contents of this chapter include three major topics. We will first study the actual static characteristics of the BJT and explain the factors, properties, and processes that account for the differences between the ideal characteristics that were considered in the last chapter and the actual characteristics. The second and third subjects are the switching properties of the BJT and its small-signal equivalent circuit. These two subjects represent the two major applications of the transistor, its use as a switch, and as an amplifier.

In the second section, we examine the switching characteristics of the BJT and determine the turn-ON and turn-OFF times as functions of its properties. We then develop a small-signal equivalent circuit of the BJT, which replaces the transistor in the analysis of amplifier circuits. The equivalent circuit components will be determined, using the analytical static characteristics, by allowing incremental variations in the voltages applied to the transistor. The circuit will include capacitances that are intrinsic to the operation of the transistor. These capacitances account for the time and phase delays in the output quantities of the BJT and are responsible for the variation of the gains of an amplifier as the frequency of the input signal is changed.

In a short section, we will subsequently evaluate the performance of a BJT as a current amplifier based on a factor known as the gain-bandwidth product.

Finally, we briefly explain the basis for the model used in simulating the BJT in the computer program SPICE.

9.1 EFFECTS OF LIMITATIONS ON STATIC CHARACTERISTICS

Experimentally observed characteristics for a PNP transistor in the common-base and common-emitter output characteristics are shown in Fig. 9.1.

A comparison of the above characteristics with those shown in Fig. 8.12 reveals the following differences:

- The collector current for a fixed I_B in the active region of the common-emitter connection increases gradually with an increase of the emitter-to-collector voltage. Since I_B is dependent mainly on V_{EB}, an increase in V_{EC} results from an increase in the magnitude of the reverse bias (V_{BC}), since V_{EC} is equal to $V_{EB} + V_{BC}$. All of the above currents and voltages are positive for a PNP transistor.

- At voltages preceding breakdown, the collector current increases rapidly with the increase of the reverse bias, indicating a multiplicative effect.

- At a certain collector-to-base voltage, in the common-base characteristics, and, at a corresponding emitter-collector voltage in the common-emitter characteristics, breakdown occurs. It is worth noting, from Fig. 9.1, the large difference between the breakdown voltage in the C-B connection, BV_{CBO}, and that in the C-E connection, BV_{CEO}.

We will now explain the differences between the ideal and actual characteristics by relating these differences to the phenomena listed in the introductory section.

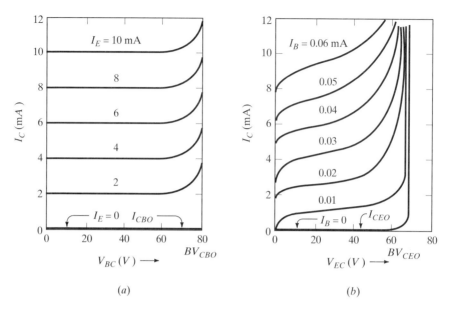

Figure 9.1 Output characteristics of the PNP transistor in the (a) common-base and (b) common-emitter connection.

Increase of Collector Current with V_{EC} in Forward Active Region

In our study of the diode, we observed a change of the width of the depletion layer as the applied voltage changes. An increase of forward voltage causes the layer to become narrower and an increase of the magnitude of the reverse bias causes an increase of the width of the layer.

An increase of the base-collector voltage of a PNP transistor increases the width of the depletion layer and this reduces the effective width of the base. For a fixed value of I_B, as for the CE connection generated by a fixed V_{BE}, an increase of V_{BC} causes an equal increase of V_{EC}.

The value of the minority carrier density at the base side of the emitter-base depletion layer is fixed by V_{EB}. The minority carrier density at the base side of the collector-base depletion layer is almost zero. Thus, an increase of V_{EC} caused by an increase of V_{BC} increases the width of the depletion layer and thus reduces the effective base width while maintaining practically the same values of minority carrier densities at the extremities of the new base. This results in an increase of the gradient of the linear hole density distribution in the base, as shown in Fig. 9.2, and an increase in the collector current. This accounts for the positive slope of the constant I_B curves in the active region of the common-emitter characteristics, shown in Fig. 9.1(b) and Fig. 9.3. It is interesting to note that the extrapolations of the constant I_B curves to the negative V_{EC} axis all meet at one point in the axis. This point is at $V_{EC} = -V_A$.

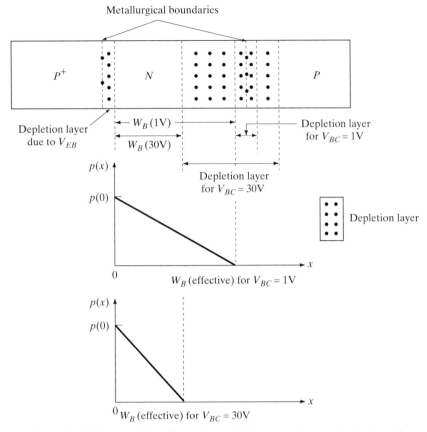

Figure 9.2   Reduction of the effective base width causes an increase in the slope of the hole density distribution in the base. V_{EB} is assumed to be constant.

In conjunction with the intersection of the constant I_B curves at $-V_A$, as shown in Fig. 9.3, we also observe an increase in the slope of the constant I_B curves with an increase of I_B. To explain this change, we will derive an expression for the conductance $\partial I_C / \partial V_{EC}$ at a constant V_{EB}. Since the base current, in most modern

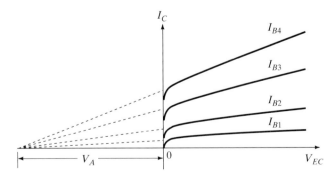

Figure 9.3   The projections of the constant I_B curves meet at $-V_A$.

day discrete BJTs and in integrated circuits BJTs, consists mainly of electrons injected from the base to the emitter in a PNP BJT, a constant V_{EB} therefore corresponds to a constant I_B.

In the active region of the PNP BJT, the hole density distribution in the base is assumed to be linear, as shown in Fig. 9.2. The collector current can therefore be written as

$$I_C = -qAD_p dp/dx = qAD_p p(0)/W_B = (qAD_p/W_B)p_0 \exp qV_{EB}/kT \quad (9.1)$$

The conductance, $\partial I_C/\partial V_{EC}$, can be written as

$$\partial I_C/\partial V_{EC} = (dI_C/dW_B)(dW_B/dV_{EC}) \quad (9.2)$$

By using Eq. (9.1), we find

$$dI_C/dW_B = -qAD_p p(0)/W_B^2 = -I_C/W_B \quad (9.3)$$

Equation (9.2) can then be written as

$$\partial I_C / \partial V_{EC} = (-I_C / W_B)dW_B/dV_{EC} \quad (9.4)$$

Since V_{EB} is constant, one can assume that $dW_B/dV_{EC} = dW_B/dV_{BC}$ because it is V_{BC} that causes the change in the width of the base. An increase in V_{EC} decreases the base width, thus making $dW_B/dV_{EC} < 0$ and the $\partial I_C / \partial V_{EC}$ positive.

The student may wonder as to what would be the effect of increasing V_{EB} on the base width and the collector current. An increase of V_{EB} decreases the depletion layer width at the E-B junction, thus increasing the width of the base while it increases the hole density in the base at the junction. The increase of the base width due to V_{EB}, for practical values of V_{EB}, is negligible, whereas the increase of hole density at the base side of the junction and the electron density at the emitter side of the junction lead to an increase of the collector and base current respectively.

Since V_{BE} is constant and I_B is constant, Eq. (9.4) predicts an increase of I_C with V_{EC} in the active region, for a constant I_B. It also predicts that at higher values of V_{EB}, thus higher values of I_B and I_C, the slope of the $I_C - V_{CE}$ curves is larger.

This phenomenon of an increase in the collector current with an increase of V_{EC}, as shown in Fig. 9.3, was first described by Early and it is known as the *Early effect*. This effect is also known as *base-width modulation* and V_A is the base width modulation factor or the Early voltage. Extrapolation of the characteristics of Fig. 9.3 back intersects the V_{EC} axis at V_A, where $V_A \equiv I_C/(\partial I_C/\partial V_{EC})$. Substitution of Eq. (9.4) in the expression for V_A gives

$$V_A = -W_B \, dV_{EC}/dW_B \quad (9.5)$$

Referring to the Ebers-Moll equation for I_C, assuming operation in the active region so that the second term is neglected, and replacing $\alpha_F I_{ES}$ by I_S, Eq. (8.61(b)) is modified to include the Early effect. Thus, for a PNP transistor

$$I_C = I_S [\exp (q V_{EB}/kT) - 1][1 + V_{EC}/V_A] \quad (9.6)$$

For an NPN transistor, V_{EB} and V_{EC} are replaced by V_{BE} and V_{CE} respectively.

Carrier Multiplication and Breakdown

At the collector-base junction, an upper limit is set on the magnitude of V_{BC} by avalanche breakdown. This breakdown voltage is labeled BV_{CBO} on the common-base characteristics of Fig. 9.1. The first two subscripts refer to the collector-to-base with the third subscript indicating that the third terminal, the emitter, is open-circuited. On the common-emitter characteristics, breakdown occurs at a smaller voltage, labeled BV_{CEO}, referring to the collector-to-emitter breakdown voltage with the open base.

At collector-to-base voltages below the breakdown value, the energy level diagram at that junction is very steep. At breakdown, and as shown in Chapter 7 for the diode, the holes that are crossing the junction from base to collector and the electrons crossing in the opposite direction acquire sufficient energy from the high electric field in the depletion layer, causing ionizing collisions with the lattice. These collisions generate electron-hole pairs and each new carrier causes further ionizing collisions. This multiplicative avalanche process increases the number of carriers crossing the junction and results in an increase of the collector current near breakdown as shown in Fig. 9.4.

We determined in Chapter 7 that the current in a reverse-biased PN junction is multiplied by the factor M near breakdown because of the avalanche process. For the common-base BJT connection, the current I_C, arriving at the reverse biased base-collector junction, when operating near breakdown, can be written by using Eq. (8.13) as

$$I_C = M(\alpha_F I_E + I_{CBO}) \qquad (9.7)$$

The multiplication factor M has been determined empirically and referred to in Chapter 7. It is modified for a PNPBJT to be

$$M = 1/[1 - (V_{BC}/BV_{CBO})^n] \qquad (9.8)$$

where BV_{CBO} is the magnitude of the avalanche breakdown voltage for the junction. The subscripts, CBO, refer to the collector-base voltage with the emitter open. Consequently, with $I_E = 0$, Eq. (9.7) becomes

$$I_C = M I_{CBO} \qquad (9.9)$$

The current approaches a very high value as $V_{BC} \rightarrow BV_{CBO}$ and $M \rightarrow \infty$. The transistor current can be limited, even at breakdown, by inserting a resistor in series in the external collector-base circuit so that the power dissipated in the BJT is below the rated power. Breakdown in the common-base connection is shown in Fig. 9.4(a).

The collector current with $I_E = 0$ (also labeled I_{C0}) is considerably smaller than the reverse current of a PN junction diode having properties that are identical to those of the collector-base region. The reason for this is that when I_E is zero, the hole density gradient at the edge of the depletion layer in the base is zero, thus reducing the slope of the hole density slope at $x = W_B$. On the other hand, in integrated circuits, diodes are made by shorting the base to the emitter of a BJT ($V_{EB} =$

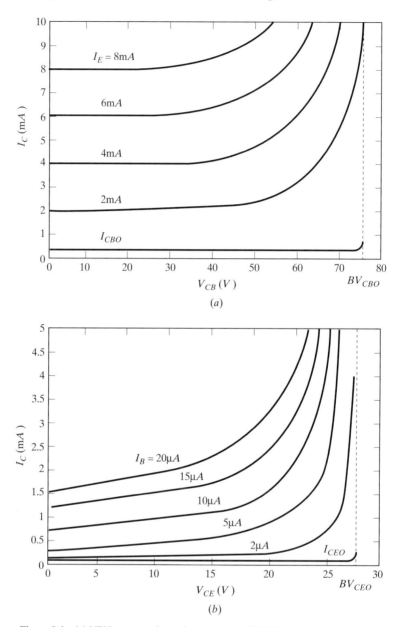

Figure 9.4 (a) NPN common-base characteristics. (b) NPN common-emitter characteristics.

0), which makes the hole density at $x = 0$ in the base equal to the equilibrium value, p_0, while the hole density at $x = W_B$ is almost zero.

For a transistor in the common-emitter connection, the breakdown voltage is labeled BV_{CEO}, as the voltage between collector and emitter with the base open.

Common emitter characteristics are shown in Fig. 9.4(b). By replacing I_E in Eq. (9.7) by $(I_C + I_B)$ and solving for I_C, we obtain

$$I_C = M(\alpha_F I_B + I_{CBO})/(1 - M\alpha_F) \qquad (9.10)$$

Since breakdown is measured at $I_B = 0$, the collector current becomes

$$I_C = M I_{CBO}/(1 - M\alpha_F) \qquad (9.11)$$

Breakdown for this connection takes place, for $I_B = 0$ as well as for all other values of I_B, when $M\alpha_F$ is unity, which implies that M is slightly larger than unity. By assuming that $V_{EC} \cong V_{BC}$, labeling the voltage V_{BC} in Eq. (9.8) at breakdown as BV_{CEO}, and setting $M\alpha_F$ equal to unity, we have

$$\alpha_F / \left[1 - \left(\frac{BV_{CEO}}{BV_{CBO}} \right)^n \right] = 1 \qquad (9.12)$$

This results in

$$BV_{CEO}/BV_{CBO} = (1 - \alpha_F)^{1/n}$$

so that

$$BV_{CEO} \cong BV_{CBO}/\sqrt[n]{\beta_F} \qquad (9.13)$$

Near breakdown, the collector current for nonzero values of I_B, calculated at the same V_{EC}, is larger as I_B is increased. At values of V_{EC} away from breakdown, where $M = 1$, Eq. (9.11) indicates that the collector to emitter current with the base open is I_{CEO}, which is considerably larger than I_{CBO}, and is given by $I_{CBO}/(1 - \alpha_F)$. Relative magnitudes of BV_{CBO} and I_{CBO} for the common-base connection and the corresponding quantities for the common-emitter connection are shown in Fig. 9.5.

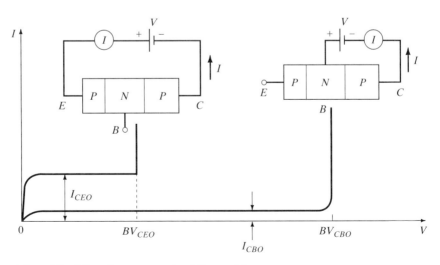

Figure 9.5 BV_{CBO}, I_{CBO} compared to BV_{CEO} and I_{CEO}.

From Eq. (9.13) and Fig. 9.5, we note that the breakdown voltage for a BJT in the common-emitter connection is much smaller than that in the common-base connection. We have plotted in Fig. 9.4 the common-base and common-emitter characteristics for an NPN BJT operating in the active region. We assumed that the BJT has $\beta_F = 79$, $V_A = 50V$, $n = 4$, and $BV_{CBO} = 80V$. The breakdown voltage in the common-emitter characteristics, BV_{CEO}, is calculated to occur at $V_{CB} = 26.75V$. The mechanism responsible for this difference results from the current-amplifying property of the common-emitter connection. As V_{BC} becomes large and multiplication of carriers takes place at the collector junction, a hole by the avalanche process produces an electron-hole pair with the primary and secondary holes moving into the collector. The electron is swept into the base by the electric field at the junction. To preserve charge neutrality, the excess electrons swept into the base invite excess hole injection from the emitter. The new holes diffuse to the collector junction, resulting in an increase of the collector current, thus causing a regenerative process.

It is important to realize that the change in the slope of the common-emitter characteristics of Fig. 9.4(b) is a consequence of two phenomena. At voltages, V_{CE}, of about 10V in Fig. 9.4(b), the increase in the slope is a result of the Early effect. Beyond that, the collector current increases due mainly to the multiplication of carriers, at the reverse-biased collector junction, as breakdown is approached.

The following example illustrates the calculation of breakdown voltages.

EXAMPLE 9.1

A silicon PNP BJT has a collector doping $N_A = 3 \times 10^{15} \text{cm}^{-3}$ and is much less than the doping N_D of the base. Given that $\beta_F = 100$, $n = 4$ and $\mathcal{E}_{cr} = 3 \times 10^5 \text{V/cm}$, determine:

   a) BV_{CBO} **b)** BV_{CEO}

Solution

   a) From Eq. (7.9), we have

$$\mathcal{E}_{max} = [2 V_j q N_A N_D / \varepsilon (N_A + N_D)]^{0.5}$$

where $V_J = V_{bi} - V_a$ and for $|V_a| \gg V_{bi}$ $V_j = |V_a| = V_{br}$

Since $\mathcal{E}_{cr} = \mathcal{E}_{max}$, we solve the above equation for V_{br} and since $N_A \ll N_D$

$$V_{br} = \mathcal{E}_{cr}^2 \varepsilon / 2q N_A$$

$$BV_{CBO} = V_{br} = \frac{9 \times 10^{10} \times 11.8 \times 8.85 \times 10^{-14}}{2 \times 1.6 \times 10^{-19} \times 3 \times 10^{15}} = 98.73V$$

   b) $BV_{CEO} = BV_{CBO}/(100)^{0.25} = 31.22V$

Punchthrough

An interesting destructive phenomenon, distinct from avalanche breakdown, may take place if the depletion layer on the base side of the collector-base junction extends, because of the reverse bias, so far into the base that it reaches the emitter-base junction before avalanche breakdown can occur. This phenomenon is known as *punchthrough*. As soon as the collector-base depletion layer reaches the emitter-base depletion layer, the emitter and collector P regions of a PNP transistor are joined by one depletion layer providing a highly conductive path from emitter to collector, which can lead to currents that damage the BJT. At punchthrough, the base is depleted of mobile carriers and loses its controlling action on the collector with a corresponding reduction in barrier height for hole injection from the emitter, as shown in Fig. 9.6. The current increases rapidly and is limited by the external circuit resistance and whatever small resistances are offered by the emitter and collector bulk materials. In most transistors, avalanche breakdown precedes punchthrough. However, punchthrough may precede avalanche breakdown if the base has very small width or relatively low doping.

9.2 EFFECTS AT VERY LOW AND HIGH INJECTION

The curves illustrating the changes in I_B and I_C, over a broad range of changes in V_{EB}, as shown in Figs. 9.7 and 9.8, can be divided into three regions.

In the moderate collector current level (ideal region) of the curves, both I_B and I_C are diffusion currents and vary with a slope of exp ($q\, V_{EB}/kT$). At lower values of V_{EB}, the base current is dominated by the component that accounts for recombination in the emitter-base depletion layer and which is larger than the diffusion component.

In the third region, which corresponds to values of V_{EB} that are greater than those of the ideal region, two effects cause a reduction in the current gain β_F. One of these effects occurs at the emitter junction and the other is at the collector junction. At the emitter junction and at high values of I_C, conductivity modulation of the base

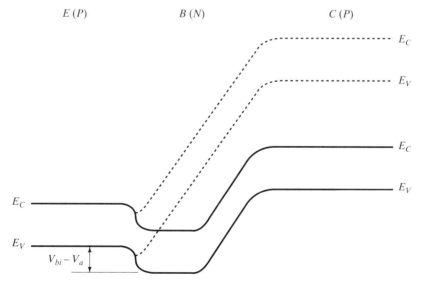

Figure 9.6 PNP energy band diagrams. Solid lines show operation in the active mode. Dashed lines refer to punchthrough.

and current crowding under the emitter are the factors that cause a reduction in the current gain. At the collector-base junction, the widening of the base region when I_C increases causes a reduction in the rate of increase of the collector current.

Very Low Injection and Current Gain

In the depletion layer of the forward-biased emitter-base junction the carrier densities, because of injection of carriers from both sides, are greater than the thermal equilibrium values. As a result, the rate of recombination exceeds the generation rate of the carriers. At low values of V_{EB}, of the order of 0.4V and less, in a PNP silicon transistor operating in the active region, all currents are small and the depletion region is quite wide. At these values of the base current, the recombination component in the depletion layer is no longer negligible.

We recall that we determined, for a PNP transistor, the electron component of the emitter current, I_{En}, from the slope of the electron density on the emitter side of the emitter-base depletion layer. This current forms the most important of the three components of the base current. Because of recombination in the depletion layer, the base current component that we have assumed equal to I_{En}, on the base side of the depletion layer, is in fact greater than I_{En}. This requires a higher base current. The collector current, which consists mainly of I_{Ep}, has not changed. At low values of V_{EB} and hence low values of the base current, the recombination component of this current in the depletion layer becomes significant. Furthermore, and of much less importance, the low values of V_{EB} cause a wider depletion layer that increases the probability of recombination.

The higher base current results in a decrease of β_F at low values of the *collector* current. A plot of log I_C and log I_B, as a function of V_{EB}, is shown in Fig. 9.7. In the plot, β_F is shown to be constant in the active region over a wide range of *collector* currents. Depending on the particular transistor, the region of constant β_F may extend over six decades of collector current.

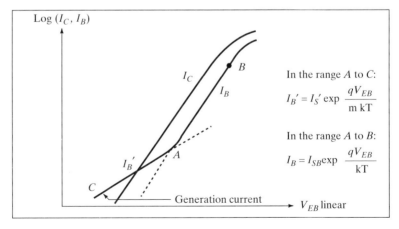

Figure 9.7   Plot of log I_C and log I_B versus V_{EB} showing degradation of I_C/I_B at a low and high emitter-base bias.

At moderate current levels for operation in the active mode ($V_{EB} \gg kT/q$ and $V_{CB} < 0$), the expression for I_C may be approximated by I_{Ep} in Eq. (8.29). The expression for I_B is set equal to I_{En} in Eq. (8.33). Assuming $W_B/L_p \ll 1$ so that coth $(W_B/L_p) \cong L_p/W_B$ and sinh $W_B/L_p \approx W_B/L_p$

$$I_C = (q\,A\,D_p p_0/W_B)\,\exp q\,V_{EB}/kT \quad \text{(a)}$$

$$I_B = (q\,A\,D_{nE} n_{0E}/L_{nE})\,\exp q\,V_{EB}/kT \quad \text{(b)}$$

$$(9.14)$$

At low current levels, the expression for I_C is unchanged, but I_B increases to I_B' due to the additional recombination current. The expression for the recombination current in the depletion layer of a forward-biased diode, given by Eq. (7.2), is modified for the emitter-base junction with $V_{EB} \gg kT/q$ to become

$$I_{\text{rec}} \cong \left(\frac{qAn_i\,W}{2\tau_0}\right)\,\exp q\,V_{EB}/2kT \qquad (9.15)$$

where W is the width of the depletion layer and τ_0 is the lifetime of electrons in the layer.

The new expression for I_B' (shown in Fig 9.7) at low current levels becomes

$$I_B' = (q\,AD_{nE}n_{0E}/L_{nE})\,\exp qV_{EB}/kT + (q\,An_iW/2\tau_0)\,\exp q\,V_{EB}/2kT \quad (9.16)$$

To relate the expression for the total base current to the plot of I_B in Fig. 9.7, we write

$$I_B' = I_s'\,\exp q\,V_{EB}/mkT \qquad (9.17)$$

where m is an empirical factor obtained by curve fitting and its value varies from one to two, and I_s' is another empirical factor related to the two prefactors in Eq. (9.16).

At low values of V_{EB}, the slope of the I_B curve is $(q\,V_{EB}/kT)$, while the slope of the I_B' curve is smaller. While the slope of I_B' is a result of the exponential term in Eq. (9.17), the term I_s' is greater than the prefactor of I_B in Eq. (9.14(b)) because of the terms in Eq. (9.16), causing I_B' to be larger than I_B.

The increase in I_B at these low levels, compared to the values of I_C, results in a smaller β_F, as illustrated in Fig. 9.8.

High-Level Injection and the Kirk Effect at the Base-Collector Junction

The Kirk effect is the phenomenon that results in the widening of the base region and a corresponding slower increase of I_C as V_{EB} is increased beyond the moderate region. The base current, however, continues to increase at the moderate rate.

We consider first an N^+PN^+ BJT in which the collector current is high. At this value of collector current, the mobile carriers' (electrons) space-charge density is no longer negligible when compared to the fixed-dopant densities; in particular, to the negatively charged acceptor ions in the collector-base space-charge layer. The resulting space-charge density in the base side of the C-B space-charge layer is increased. The space-charge layer on the collector side is so heavily doped that it is

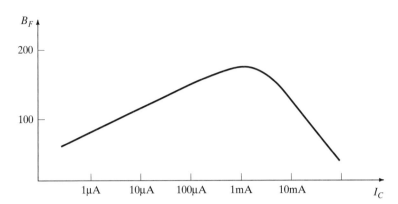

Figure 9.8   Plot of common-emitter current gain, β_F, versus I_C at a fixed $|V_{CE}|$.

not affected. To maintain zero space-charge density through the layer, the layer on the base side must shrink and thus cause an increase of the width of the base leading to a reduced slope in the electron distribution in the base. Hence, a reduction in the collector current.

A few analytical relations for this effect will help the explanation. We start by writing Poisson's equation in the depletion region as

$$d\mathscr{E}/dx = 1/\varepsilon \, [q \, N(x) - I_C/v(x)A_E] \tag{9.18}$$

where $N(x) = (N_D^+ - N_A)$ in the space-charge region (negative acceptor ions in the P base side and positive donor ions in the collector side). The velocity $v(x)$ is that of the electrons, A_E is the emitter area, and $I_C/v(x)A$ represents the mobile electron density of the collector current.

For a constant, V_{CB}, and by neglecting the built-in voltage when compared to V_{CB}, the relation between V_{CB} and the electric field is

$$V_{CB} = -\int_{x_B}^{x_C} \mathscr{E} \, dx \tag{9.19}$$

where $(x_C - x_B)$ is the width of the space-charge layer. In general, and at the values of V_{CB} that are used, the electric field is high enough to cause $v(x)$ to be the saturation drift velocity, v_s, which is of the order of 10^7cm/s.

At low injection levels, the second term inside the brackets of Eq. (9.18), which represents the charge density of the collector current, is negligible. At a particular collector current, labeled the critical current I_o, that term causes the derivative in Eq. (9.18) to become zero. This critical current is given by

$$I_o = q \, N(x) v_s A_E \tag{9.20}$$

The relevance of I_o will be shown later in this section.

Next, we consider a second BJT structure whereby a lightly-doped epitaxial layer separates the base from the highly doped buried layer of the collector, and the configuration is $N^+PN^-N^+$ as shown in Fig. 8.1c. An increase of the space charge density, caused by an increase of collector current, subtracts from the positive epitaxial donor ions. To balance the charges, the depletion layer in the epitaxial side tends to expand with minimal change in the base side. At the same time, and more importantly, an increase of I_C results in an increase of the voltage that is dropped in the undepleted part of the epitaxial region, causing a reduction of the applied reverse bias between the P base and N epitaxial layer. The depletion layer tends to shrink and eventually collapse. At a sufficiently high current, this junction becomes forward-biased, causing a quasi-saturation effect that increases the electron density at this junction in the base. This reduces the slope of the electron distribution in the base and hence reduces the collector current.

In both situations described above, for the $N^+PN^-N^+$ BJT, the effective end result is a widening of the width of the base when I_C exceeds the specific critical value I_o.

An expression for the collector-base space charge region width has been obtained* as

$$x_{CB} = \frac{x_{CO}}{\left[1 + \left(\dfrac{I_C}{I_o} \right) \right]^{0.5}} \tag{9.21}$$

where x_{CO} is the space charge width at $I_C = 0$ and I_o is the critical current defined by Eq. 9.20.

High-Level Injection at the Emitter-Base Junction

Two phenomena take place in the emitter and base regions that tend to cause I_C to increase at a lower rate in the high-injection regime than the moderate regime. First, the increasing minority carrier concentrations in the base causes an equal increase in the majority carrier concentrations. The increase in the electron density in the base of the PNP transistor is reflected by an increase in N_{DB} of Eq. (8.55), thus leading to a reduction in the injection efficiency and hence a reduction in the current gain.

The second effect is the current "crowding" under the emitter, as shown in Fig. 9.9. The lateral motion of the base-current carriers causes a potential drop, due to the base resistance, in the lateral direction. Because of this, different parts of the emitter are biased to different voltages, giving rise to uneven injection. The greater part of the injection occurs from the edge of the emitter that is nearest to the base, causing a localized concentration in the emitter current. This is labeled *emitter crowding*. The current density at that edge exceeds the average value so that conductivity modulation of the base occurs earlier than it would have if the current was spread out over the whole emitter area.

In Sec. 9.2, we have confined our discussions to the differences that exist in the *active mode* of the BJT between the ideal and actual characteristics. Switching of a

*From Muller and Kamins, *Device Electronics for Integrated Circuits*, p. 327, copyright © Wiley (1986). Reprinted by permission of John Wiley & Sons, Inc.

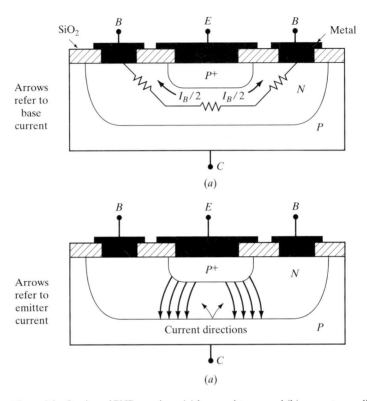

Figure 9.9 Section of PNP transistor (a) base resistance and (b) current crowding.

transistor is related to the other two operating modes, *saturation and cutoff*, which occur in the first quadrant of the BJT common-emitter characteristics. In Sec. 9.3, we investigate the processes that take place when a BJT is switched between cutoff and saturation, and consider the properties that lead to faster switching.

REVIEW QUESTIONS

Q9-1 Briefly define the significance of the multiplication factor M.

Q9-2 Why is punchthrough of concern in modern BJTs?

Q9-3 To increase the punchthrough voltage, the base doping can be increased. What is the negative effect of an increase in the base doping?

Q9-4 Explain the negative effect of reducing the base doping on the base resistance.

Q9-5 What is the effect on the emitter of high base resistance at high current?

Q9-6 Explain why BV_{CBO} is larger than BV_{CEO}.

Q9-7 Briefly explain the shape of the curve β_F versus I_C in Fig. 9.8.

HIGHLIGHTS

- The collector-base voltage has dramatic effects on the common-emitter output characteristics in two regions: at saturation (low values of V_{BC}) and at breakdown (high values of V_{BC}) for a PNP transistor.

- The Early effect is the reduction in the effective width of the base as the collector to emitter voltage is increased. This is accompanied by an increase in the collector current.

- Punchthrough is the process whereby the reverse-bias at the collector-base junction is so large that the depletion layer in the base extends to the emitter-base junction.

- As the collector-base reverse bias is increased in the active region, the collector current increases, due both to the Early effect and the multiplication of carriers in the collector-base junction.

- As $|V_{CB}|$ is increased, the base width is reduced, a higher slope of the minority carriers in the base results, the emitter efficiency increases, and the transport factor is higher.

- Avalanche breakdown occurs when the generation of new carriers is so large that the collector current tends to increase to destructive values.

- Emitter crowding results from the *IR* (where *I* is the base current and *R* is the base resistance) drop of the base current in the direction of base current flow towards the emitter, reducing the effective forward bias in the center of the emitter relative to the edges. Thus, the emitter electron current in an NPN transistor is concentrated at the emitter periphery.

EXERCISES

E9-1   Punchthrough takes place in a certain NPN silicon transistor when $|V_{CB}| = 30$V. Assume the collector doping is much greater than that of the base where $N_A = 10^{15}$cm^{-3}. The relative dielectric constant is 11.8. Determine the approximate value of the base width when $|V_{CB}| = 0$.

Ans: $W_B = 6.25\mu$m

9.3 TRANSISTOR SWITCHING

An ideal switch is a short-circuit in the ON position and an open-circuit in the OFF position. Switching, although not quite, but close to the ideal variety may be achieved in a BJT common-emitter circuit. Operation of the transistor in saturation occurs at a low V_{EC} and high I_C, simulating the ON state (low V_{EC}/I_C), whereas in cutoff, the transistor current is nearly zero, thus resembling an open-circuit. Figure 9.10 shows the circuit and the current drive of a transistor together with its characteristics in order to illustrate switching.

At $i_B = 0$, the BJT is in cutoff and operation is at point A, while saturation corresponds to a high value of I_{B1}, where V_{BC} is virtually zero. The ON state occurs provided I_{B1} is equal to or greater than I_{C1}/β_F or $(V_{CC} - V_{BC})/\beta_F R_L$, which corresponds to point B in Fig. 9.10(b).

Even if i_B is switched in zero time from zero to I_{B1}, the saturation state is not reached instantaneously because the speed of response of the transistor is limited mainly by the storage or diffusion capacitance, which accompanies the storage of minority carriers in the base. The distributions of minority carriers in the base of a PNP transistor are shown in Fig. 9.11 for the cutoff and saturation conditions where it is assumed that the width, W_B, of the base is much smaller than the diffusion length of minority carriers so that the distribution is linear in both the active and

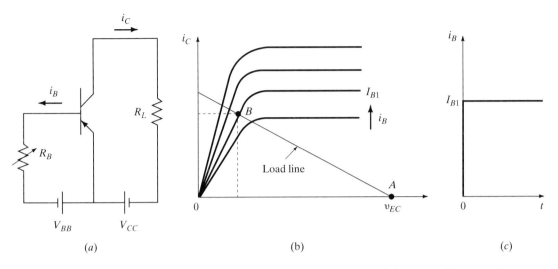

Figure 9.10   Switching of a transistor (a) circuit, (b) BJT characteristics and load line, and (c) base current drive.

saturation modes. It is the storage of minority carriers in the base of a transistor operating in the active and saturation regions that accounts for the diffusion capacitance.

The time delay that occurs in switching may also be related to the transit time of the minority carriers that flow from emitter to collector through the base.

Stored Charge and Transit Time

First, let us relate the magnitude of the collector current to the minority carrier charge stored in the base. We assume a linear distribution of minority carriers, as shown in Fig. 9.12, for operation in the active region.

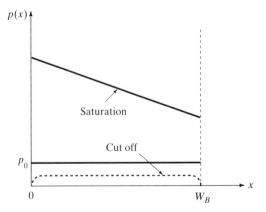

Figure 9.11   Distribution of holes in the base for the BJT in saturation and cutoff.

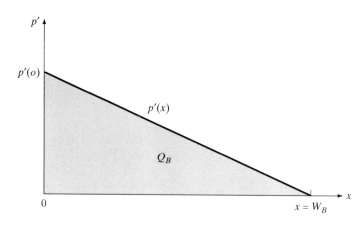

Figure 9.12   Distribution of excess minority carriers in the base of a PNP transistor.

The collector current is related to the excess hole density by

$$I_C = -q\,AD_p\,dp'/dx = q\,AD_p p'(0)/W_B \qquad (9.22)$$

The total excess minority carrier charge stored in the base, Q_B, is the product of qA and the area of the distribution in Fig. 9.12.

$$Q_B = q\,A\,p'(0)\,W_B/2 \qquad (9.23)$$

The ratio of the charge stored to the collector current, which has unit of time, becomes

$$Q_B/I_C = W_B^2/2D_p \equiv \tau_B \qquad (9.24)$$

Therefore, to change the collector current, it is necessary to change the charge stored in the base. If the collector current is I_C, then a charge, Q_B, is accumulated or removed in τ_B time.

Let us now find a relation between τ_B and the approximate time it takes a hole to travel from the emitter junction to the collector junction.

The incremental distance traveled by a hole in time dt is dx, given by

$$dx = v(x)\,dt \qquad (9.25)$$

where $v(x)$ is the velocity of the hole. The time it takes the hole to cross the base becomes

$$t_{tr} = \int_0^{W_B} dt = \int_0^{W_B} dx/v(x) \qquad (9.26)$$

The current of holes is

$$I_p = -q\,A\,D_p\,dp'/dx = q\,A\,D_p p'(0)/W_B \qquad (9.27)$$

This current can also be written in terms of the velocity of the carriers. We also use the linear distribution of Fig. 9.12 to write

$$I_p = q\,A\,p'(x)\,v(x) = q\,A\,v(x)\,p'(0)\,(1 - x/W_B) \tag{9.28}$$

We set the expressions in Eqs. (9.27) and (9.28) equal in order to obtain the expression for $v(x)$ as

$$v(x) = \frac{D_p}{W_B(1 - x/W_B)} = \tau_B \tag{9.29}$$

Equation (9.29) is used in Eq. (9.26) to obtain the transit time

$$t_{tr} = \int_0^{W_B} \frac{W_B(1 - x/W_B)}{D_p}\,dx = \frac{W_B^2}{2D_p} \tag{9.30}$$

We thus conclude that the transit time is equal to, and has the same significance as, τ_B.

Charge Control Relations

The charge control equations represent a powerful tool used in the analysis of the transient behavior of the transistor. They relate the excess charge stored in the base to the base current and the collector current.

The general expression for the excess hole density charge stored in the base of a PNP transistor, Q_B, is given by Eq. (9.31)

$$Q_B = qA \int_0^{W_B} p'(x)dx \tag{9.31}$$

where $p'(x)$ is the excess hole density at x. We will now determine the relation between the base current and the stored charge by referring to the continuity equation formulated in Chapter 4. Equation (4.36) for the continuity of holes in the base is repeated in Eq. (9.32) for one dimension after introducing the factor qA and replacing $(p - p_0)$ by p'

$$\frac{\partial}{\partial t}(p'(x)qA) = -\frac{qAp'(x)}{\tau_p} - \frac{qA\partial J_p}{\partial x} \tag{9.32}$$

We integrate Eq. (9.32) from $x = 0$ to $x = W_B$ as

$$\frac{\partial}{\partial t}\int_0^{W_B} p'(x)qA\,dx = -\int_0^{W_B} \frac{qAp'(x)}{\tau_p}\,dx - \int_0^{W_B} qA\,\partial J_p \tag{9.33}$$

The first integral in Eq. (9.33) represents the rate of change of the stored charge, Q_B, with time; the second integral is the ratio of the stored charge to the lifetime of holes; and the third integral becomes $(i_{Cp} - i_{Cn})$, where $i_{Cp} \cong i_C$ and $i_{Cn} \cong i_E$ so that the difference is $i_B(t)$. Lower case i's refer to instantiations.

The result of Eq. (9.33) is the first charge control equation to be stated as

$$i_B(t) = \frac{Q_B(t)}{\tau_p} + \frac{dQ_B(t)}{dt} \tag{9.34}$$

Equation (9.34) is interpreted as indicating that the rate of change of stored charge in the base $dQ_B(t)/dt$, is determined by two processes: the base current $i_B(t)$, which adds charge to the base, and by the process of recombination that removes charge at the rate Q_B/τ_p.

The stored charge, Q_B, is transferred totally to the collector every τ_B seconds by the collector current where τ_B is the base transit time. If the charge is transferred at this rate, and it is a steady-state charge, then it must be replenished at the same rate. These statements, which are valid in the active region of operation, are expressed in equation form

$$i_C(t) = Q_B(t)/\tau_B \tag{9.35}$$

Equations (9.34) and (9.35) are known as the *charge control equations*.

Turn-ON Time

A transistor is switched from cutoff to saturation by the application of a step of base current, I_{B1}, as shown in Fig. 9.10(c). For $i_B = I_{B1}$, Eq. (9.34) becomes

$$\frac{dQ_B(t)}{dt} = I_{B1} - \frac{Q_B(t)}{\tau_p} \tag{9.36}$$

Equation (9.36) is a first order differential equation requiring one boundary condition.

At $t = 0, Q_B = 0$, so that the solution to Eq. (9.36) is given as

$$Q_B(t) = I_{B1}\tau_p \left[1 - \exp\left(-t/\tau_p\right)\right] \tag{9.37}$$

The collector current for active mode operation is determined by the total charge that must be transferred to the collector every τ_B seconds. Using this definition in Eq. (9.35), the expression for the collector current becomes

$$i_C(t) = \frac{Q_B(t)}{\tau_B} = \frac{I_{B1}\tau_p}{\tau_B}\left[1 - \exp(-t/\tau_p)\right], \text{ for } Q_B \leq Q_{SAT} \tag{9.38}$$

where Q_{SAT} is the value of Q_B at $i_C = I_{SAT}$. The reason for restricting the expression for i_C to the range below Q_{SAT} is that, whereas the charge continues to build up as long as i_B is at I_{B1}, the same collector current saturates at I_{CSAT} that corresponds to Q_{SAT}. Even an increase of i_B beyond the onset of saturation, as shown by Fig. 9.13, causes a slight increase in I_C but a correspondingly large buildup of charge. Sketches of $Q_B(t)$ and $i_C(t)$ are shown in Fig. 9.14.

If the value of i_B is such that it places i_C at the boundary between the active and saturation regions, such as point 1 in Fig. 9.13(a), then Q_B is just equal to Q_{SAT} as determined from the area under plot 1 in Fig. 9.13(b). Equation (9.38) can be written in the steady-state as

$$i_C = i_B \, \tau_p/\tau_B = \beta_F i_B \tag{9.39}$$

If i_B, however, places i_C in deep saturation, then the value of i_C is determined as $(V_{CC} - V_{ECSAT})/R_L$ and V_{EC} for a silicon transistor is approximately 0.2V. We observe that the collector current may be assumed to be approximately constant while the transistor is moving from the edge of saturation to deep saturation. This

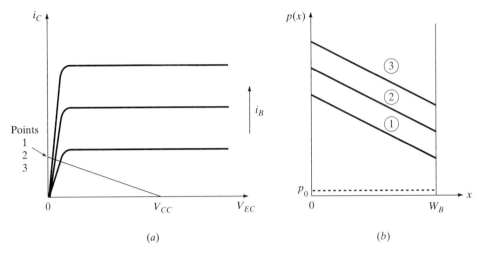

Figure 9.13 Operation in saturation where the (a) change of I_B moves the operating point into deep saturation and (b) a hole density profile.

collector current is I_{CSAT}, which is approximately equal to V_{CC}/R_L since $V_{ECSAT} \ll V_{CC}$. Once the transistor is in saturation, an increase of i_B has negligible effect on the value of i_C. Hence, the ratio of i_C to i_B is β_F only at the edge of saturation, any increase in i_B drives the transistor deeper into saturation and the ratio of i_C to i_B is

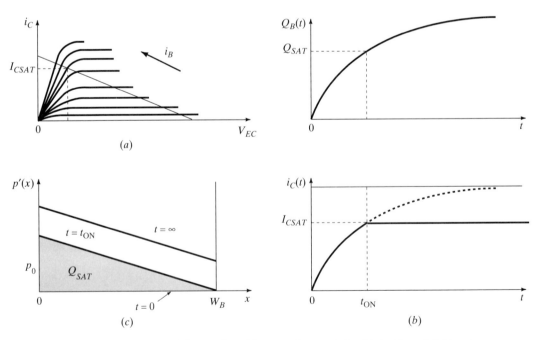

Figure 9.14 Turn-ON of the transistor (a) output characteristics, (b) sketches of $Q_B(t)$ and $i_C(t)$, and (c) excess hole density buildup.

less than β_F. Even though the value of i_C may be I_{CSAT}, Q_B continues to increase beyond Q_{SAT}, as shown in Fig. 9.14.

The transistor is assumed to be turned-ON when i_C is at I_{CSAT}, an expression for which is obtained from Eq. (9.38) at time t_{ON} as

$$I_{CSAT} = \frac{Q_{SAT}}{\tau_B} = \frac{I_{B1}\tau_{BP}}{\tau_B}\left[1 - \exp\left(-t_{ON}/\tau_p\right)\right] \tag{9.40}$$

The turn-ON time is determined from Eq. (9.40) by setting I_{CSAT} equal to V_{CC}/R_L (neglecting V_{CESAT}) so that

$$t_{ON} = \tau_p \, \ell n \frac{1}{1 - (V_{CC}/I_{B1}R_L)(\tau_B/\tau_p)} \tag{9.41}$$

Therefore, a fast turn-ON time is favored by a large i_B, a smaller I_{CSAT} (V_{CC}/R_L), as well as a smaller τ_p.

Turn-OFF Time

To turn a transistor OFF, the excess stored charge in the base must be removed and the collector current must be made almost zero (I_{CE0}). This can be accomplished by setting i_B to zero or better yet by making it negative in order to aid in removing the excess stored charge faster.

By reducing the base current to zero, the excess minority carriers that have been stored in the base decay by recombination and diffusion because there is no i_B to replenish the neutralizing majority carrier charge. For zero i_B, Eq. (9.34) becomes

$$dQ_B/dt = -Q_B/\tau_p \tag{9.42}$$

As shown in the sketches of Fig. 9.15, the hole density decreases and Q_B decreases. The collector current remains at I_{CSAT} until Q_B decreases to Q_{SAT}, at which time i_C decreases exponentially towards zero.

The turn-OFF time, defined as the time required to reduce the collector current to almost zero, is made up of two increments: the time it takes Q_B to reach Q_{SAT}, known as the storage time, t_s, and second, the time t_f, it takes the collector current to reach zero, or more practically to a value of about $0.1\,I_{CSAT}$. The decrease of stored charge and current are shown in Fig. 9.15.

Assume that the base current is made zero at a new $t = 0$. The solution to Eq. (9.42) for $t > 0$ is given by

$$Q_B(t) = Q_B(0)\, e^{-t/\tau_p} \tag{9.43}$$

where $Q_B(0)$ is the total excess charge available in the base at the end of the base current pulse. When $Q_B(t)$ is equal to Q_{SAT} at $t = t_s$, the transistor is at the edge of the active region so that $Q_B(t_s) = Q_{SAT}$. For $t > t_s$, the transistor is in the active region, so that i_C is found from Eqs. (9.35) and (9.43) as

$$i_C(t) = Q_B(t)/\tau_\beta = (Q_{SAT}/\tau_B)\, e^{-t/\tau_p} = I_{CSAT}\, e^{-t/\tau_p} \tag{9.44}$$

The storage time, t_s, is found from

$$Q_B(t_s) = Q_{SAT} = Q_B(0)\, e^{-t_s/\tau_p} \tag{9.45}$$

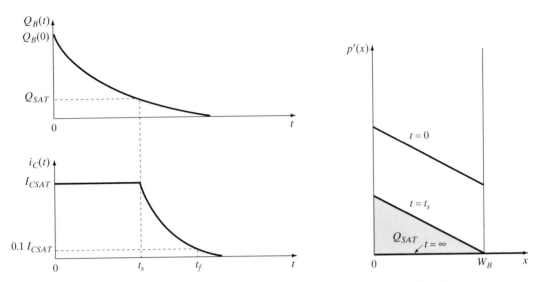

Figure 9.15 Sketches of Q_B, i_C, and $p'(x)$ during the turn-OFF period as i_B is reduced to zero instantaneously.

Solving for t_s, we obtain

$$t_S = \tau_p \, \ell n \, (Q_B(0)/Q_{SAT}) \tag{9.46}$$

Once the transistor returns to the active region at $t = t_s$, the collector current decays exponentially towards zero and is given by Eq. (9.44) as $I_{CSAT} \, e^{-t/\tau_p}$. The time, t_f, it takes the collector current to become $0.1 \, I_{CSAT}$ is found from Eq. (9.44) to be $2.3\tau_p$.

We note from Eq. (9.46) that a small storage time is favored by a small i_B and a smaller τ_p. The smaller i_B causes a smaller charge buildup and a small lifetime causes the stored charge to be removed faster. Adding gold to the base doping reduces the lifetime.

We will now use an example to illustrate the turn-ON and turn-OFF time delays.

EXAMPLE 9.2

The base current pulse into a switching transistor is shown in Fig. 9.E1. Given that for Fig. 9.10, $V_{CC} = 5V$, $R_L = 1K\Omega$, $\tau_p = 0.25\mu s$, $W_B = 2\mu m$, and $D_p = 5cm^2/s$, sketch the waveshapes of the collector current and of the charge Q_B and identify the values at the critical points.

Solution

$$I_{CSAT} \cong V_{CC}/R_L = 5mA$$

For $Q_B = Q_{SAT}$, we use Eq. (9.24) to determine τ_B

$$\tau_B = W_B^2/2D_p = 4ns$$

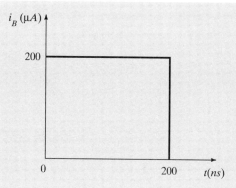

Figure 9.E1   Base current pulse.

From Eq. (9.40) we determine Q_{SAT} as

$$Q_{SAT} = I_{CSAT}\tau_B = 20\text{pC}$$

We now determine the turn-ON time at which $Q_B = Q_{SAT}$, from Eq. (9.41), using

$$\tau_p = 0.25 \times 10^{-6}\text{s}, V_{CC} = 5.0\text{V}, I_{B1} = 200 \times 10^{-6}\text{A}, R_L = 1\text{K}\Omega, \text{ and } \tau_B = 4 \times 10^{-9}\text{s}$$

$$t_{ON} = 127\text{ns}$$

The collector current remains at 5mA and, as long as i_B is greater than zero, the charge Q_B continues to increase until the end of the i_B pulse at $t = 200$ns, at which time Q_B is determined from Eq. (9.37) as

$$Q_B(200\text{ns}) = 200 \times 10^{-6} \times 0.25 \times 10^{-6} [1 - \exp(-200 \times 10^{-9}/2.5 \times 10^{-7})]$$

$$Q_B(200\text{ns}) = 27.5\text{pC}$$

This is the $Q_B(0)$ to be used in Eq. (9.46) to determine the storage time.

After i_B becomes zero, the transistor discharges from 27.5pC towards zero. Once it reaches Q_{SAT}, i_C begins to decay towards zero as shown in the sketches below. How long after i_B becomes zero does it take Q_B to decrease to $Q_{SAT} = 20$pC?

We use Eq. (9.46) to calculate the storage time as

$$t_S = 2.5 \times 10^{-7} \ln(27.5/20) = 79.6\text{ns}$$

Sketches of the time variation of stored charge and collector current are shown in Fig. 9.E2.

It is worth mentioning here, and as we will further study in Chapter 11, that the BJT switching time can be considerably reduced by connecting a metal-semiconductor diode (Schottky diode) from collector to base. The diode clamps the base to the collector and prevents the BJT from going into deep saturation.

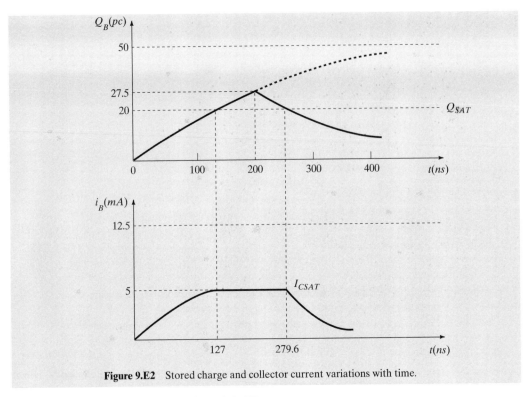

Figure 9.E2   Stored charge and collector current variations with time.

9.4 SMALL-SIGNAL EQUIVALENT CIRCUIT

In the previous section, we investigated the switching properties of a transistor and our interest was restricted to the extremities of the transistor characteristics, namely, cutoff and saturation. In this section, we will be concerned with the active region of operation where the transistor exhibits its amplifying properties. Since amplification is one of the basic functions of the transistor, our objective is to develop a model for the transistor that we can use to study its amplifying properties.

The analytical and graphical static characteristics of the transistor are highly nonlinear. In amplification, we are not only interested in magnifying a voltage or a current, but this must be done while preserving the original shape of the signal. For this to be possible, we must restrict the operation, on the nonlinear characteristics, to linear segments. This can be made possible only by selecting a certain point on the characteristics and allowing the input variations to occur on the slope of the characteristics at that point. Only then can we insure faithful reproduction of the signal we wish to amplify. This is the case for amplification of small signals of voltage and current.

For what we label as power amplifiers, large excursions of the signals over non-linear portions of the characteristics are required to obtain large amounts of output power. Here, the nonlinearities introduced are removed by some filtering means.

In order to study the dynamic AC response of the transistor to a sinusoidal excitation, for example, it becomes necessary to model the transistor by what we label as an AC equivalent circuit, which describes the response of the device to incrementally small variations, small compared to the DC values at the point where

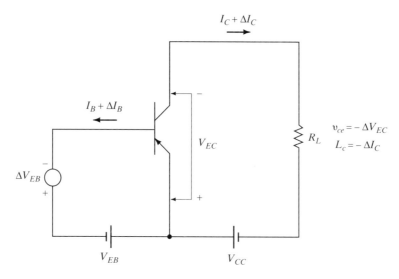

Figure 9.16   Biasing of a PNP transistor in the common-emitter connection.

the slope is used. The AC equivalent circuit is made of lumped elements, such as resistances and capacitances, in addition to a dependent current generator. The current generator models the amplification property of the transistor.

Just as in the derivation of the static characteristics, it is sufficient to study the variations in two of the three currents caused by the variations in two of the three voltages. We will consider the common emitter circuit, which is the circuit of the transistor most commonly used for amplification. This circuit, shown in Fig. 9.16, will be assumed to be biased by the DC voltages, V_{EB} and V_{CC}, to an operating point on the static characteristics. In other words, the values of I_C, I_B, and V_{EC} are fixed.

Since we are causing the small-signal variations to take place on the slope at the operating point, we are in effect linearizing the variations in the excitation and in the response. Because of this the equivalent circuit is labeled a *small-signal linear equivalent circuit*. We are in effect representing the relations between the varying currents and voltages by linear elements in the equivalent circuit. Because of the nonlinear shape of the static characteristics, a change in the location of the operating point is accompanied by a change in the slope at the new point, resulting in changes of the values of the circuit elements.

Our task consists in determining the effects of changes in V_{EB} and V_{CB} on the processes in the transistor and hence on the currents I_B and I_C.

To simplify our analysis, let us make the following assumptions:

a) Operation is in the active mode so that V_{EB} is positive and V_{CB} is negative and the magnitudes of both are much greater than kT/q.

b) The width of the base is much smaller than the diffusion length of minority carriers in the base so that a linear distribution of hole density results. This assumption is quite valid in modern discrete and integrated circuit transistors. The assumption does not exclude recombination in the base.

c) Low-level injection of holes from the emitter into the base.

EFFECTS OF CHANGES IN V_{EB}

Carrier Processes

The equivalent circuit we aim to find consists of lumped passive elements and dependent current generators that will replace the box shown in the common-emitter connection of Fig. 9.17. The lower case symbols, shown below, refer to incremental quantities that replace the changes in the PNP currents and voltages as

$$i_b = -\Delta I_B, i_c = -\Delta I_C, v_{be} = -\Delta V_{EB},$$
$$v_{ce} = -\Delta V_{EC}, v_{cb} = -\Delta V_{BC}$$

In this section, we shall investigate the effects of changes in V_{EB} on the currents in the transistor and we will derive equations for the changes in the currents in terms of the changes in V_{EB}.

Let us review the effects of an increase in V_{EB}.

1. The hole density in the base at the edge of the emitter-base depletion layer increases. To maintain neutrality, the electron density in the base increases. We assume a fixed V_{CB} at the reverse-biased C-B junction so that the excess hole density at $x = W$ is zero. The hole density distribution is shown in Fig. 9.18.
2. The gradient of the hole density increases throughout the base.
3. The hole currents in the base at the emitter and collector junctions increase.
4. The electron current across the emitter-base junction increases.
5. The concentration of holes and electrons stored in the base increases.
6. Because of the increased density of holes in the base, there is a greater rate of recombination.
7. The voltage barrier at the emitter-base junction decreases due to a reduction of the density of ionized atoms in the depletion layer. The base supplies a transient current to cover the ionized acceptors.

How then, do all these changes influence the base and collector currents in Fig. 9.17?

First, the base lead has to supply the additional electrons required by the increased current of electrons, ΔI_{En}, crossing the emitter-base junction. Second, the

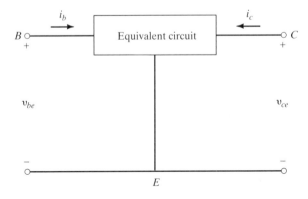

Figure 9.17 Representation of incremental quantities in common-emitter circuit.

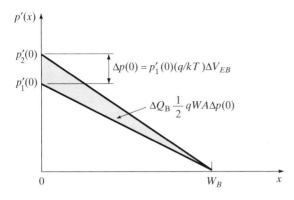

Figure 9.18   Change in the distribution of excess hole density in the base caused by an increase in V_{EB}.

base lead has to supply more electrons to recombine in the base, ΔI_{rec}. Third, the base lead has to supply additional electrons to maintain neutrality in the base. Fourth, the base supplies the electrons that cover the ionized donors in the depletion layer. The first two currents have to be continuously supplied, whereas the last two are transient in nature. Fifth, the collector current increases as a result of the increase in the gradient of the hole density in the base. We will consider first the steady-state current changes. The base current increment becomes

$$i_b = -\Delta I_B = -(\Delta I_{En} + \Delta I_{rec}) \tag{9.47}$$

Small-Signal Currents and Circuit Elements

Because of the linear distribution of hole density in the base and neglecting both the collector-base leakage current and the recombination current of holes when compared to the total collector current, and with reference to Fig. 9.18, we write

$$I_C = I_{Cp} = qAD_p\, p_1'(0)/W_B \quad \text{(a)}$$
$$I_C + \Delta I_C = qAD_p p_2'(0)/W_B \quad \text{(b)} \tag{9.48}$$

By subtracting the terms in Eq. (9.48(a)) from Eq. (9.48(b)), the change in the collector current becomes

$$\Delta I_C = qAD_p\, \Delta p(0)/W_B \tag{9.49}$$

where $\Delta p(0) = \Delta p'(0)$ and

$$\Delta p(0) = p_2'(0) - p_1'(0) = p_0 \left[\exp \frac{q(V_{EB} + \Delta V_{EB})}{kT} - \exp \frac{q\, V_{EB}}{kT} \right]$$

$$= p_0 \exp\left(q\, V_{EB}/kT\right) \left[\exp\left(q\Delta V_{EB}/kT\right) - 1 \right]$$
$$= p_1(0) \left[\exp\left(q\Delta V_{EB}/kT\right) - 1 \right] \tag{9.50}$$

By an expansion of the exponential in Eq. (9.50) and assuming ΔV_{EB} to be much smaller than kT/q, we can now show that

$$\Delta p(0) = p_1(0) \, (q/kT) \, \Delta V_{EB} \tag{9.51}$$

Our last assumption serves as a general guideline to the interpretation of incremental changes. In order to obtain the linear relation of Eq. (9.51), the change in ΔV_{EB} must be made smaller than kT/q or smaller than 26mV at room temperature. By expanding $\exp{(qv_{eb}/kT)}$ where $v_{eb} = \Delta V_{EB}$, we have

$$\exp{(qv_{eb}/kT)} = 1 + qv_{eb}/kT + \frac{1}{2}\left(\frac{qv_{eb}}{kT}\right)^2 \tag{9.52}$$

If $v_{eb} \ll kT/q$, Eq. (9.50) reduces to Eq. (9.51). Thus, the small signal analysis is valid only for $v_{eb} \ll 26\text{mV}$ at room temperature.

The electron current across the emitter junction is given by Eq. (8.33)

$$I_{En} = \frac{qAD_{nE}n_{0E}}{L_{nE}}\left[\exp{(qV_{EB}/kT)} - 1\right] \tag{9.53}$$

We replace the exponential in Eq. (9.53) by its equivalence, $p_1'(0)/p_0$, and we have

$$I_{En} = \frac{qAD_{nE}n_{0E}\,p_1'(0)}{L_{nE}\,p_0} \tag{9.54}$$

By following the procedure of Eqs. (9.48), (9.49), and (9.50), the change in I_{En}, reflected by the change in $p'(0)$, $\Delta p'(0)$, or $\Delta p(0)$, is

$$\Delta I_{En} = \frac{qAD_{nE}n_{0E}\,\Delta p(0)}{L_{nE}p_0} \tag{9.55}$$

We now consider the change in the base recombination current. Although we assumed a linear distribution of hole density, there is still recombination. The recombination current is negligible compared to the collector current but not negligible compared to the base current.

The base recombination current is

$$I_{\text{rec}} = Q_B/\tau_p = qA\int_0^{W_B}\left[(p - p_0)/\tau_p\right]dx \tag{9.56}$$

where Q_B is the excess hole density charge in the base.

By replacing p' by $p_1'(0)(1 - x/W)$ and integrating Eq. (9.56), we have

$$I_{\text{rec}} = \frac{qAp_0W_B}{2\tau_p}\left[\exp{\left(\frac{qV_{EB}}{kT}\right)} - 1\right] \tag{9.57}$$

where $p_1'(0) = p_0'\left[\exp{(q\,V_{EB}/kT)} - 1\right]$.

Using the same procedure employed earlier, Eq. (9.57) becomes

$$\Delta I_{\text{rec}} = (qAW_B/2\tau_p)\,\Delta p(0) \tag{9.58}$$

The change in the base current is the sum of Eqs. (9.55) and (9.58). By using Eq. (9.51), to replace $\Delta p(0)$, we have

$$\Delta I_B = p_1(0)\frac{q}{kT}\Delta V_{EB}\left[\frac{qAD_{nE}n_{0E}}{L_{nE}p_0} + \frac{qAW_B}{2\tau_p}\right] \tag{9.59}$$

We replace ΔI_B and ΔV_{EB} by $-i_b$ and $-v_{be}$ respectively and we also replace $p_1'(0)qA$ by its dependence on I_C from Eq. (9.48(a)). The result is

$$i_b = \frac{q}{kT} I_C \left[\frac{D_{nE} n_{OE} W_B}{D_p p_0 L_{nE}} + \frac{W_B^2}{2\tau_p D_p} \right] v_{be} \tag{9.60}$$

where $\tau_p D_p$ may be replaced by L_p^2 and $p_0 \ll p_1'(0)$.

We also find an expression for i_c by using Eqs. (9.48), (9.49), and (9.51) as

$$i_c = \frac{q}{kT} I_C v_{be} \tag{9.61}$$

The factor multiplying v_{be} has units of conductance. In fact, it is known as the *transconductance* g_m and it is the ratio of the change in the collector current caused by a small-signal change in the base-to-emitter voltage so that

$$g_m = i_c / v_{be} = I_C / V_t \tag{9.62}$$

where $V_t = kT/q = 26\text{mV}$ at room temperature.

The base current increment is written in terms of i_c by using Eqs. (9.60) and (9.62) as

$$i_b = g_m v_{be} / \beta_0 = i_c / \beta_0 \tag{9.63}$$

where β_0 is a dimensionless quantity known as the small-signal common-emitter current gain and determined at the operating point (V_{EB} and V_{CB}) for small variations in v_{be} and i_b from Eqs. (9.60) and (9.62).

$$\beta_0 \equiv \left[\frac{D_{nE} n_{OE} W_B}{D_p p_0 L_{nE}} + \frac{W_B^2}{2L_p^2} \right]^{-1} \tag{9.64}$$

This expression for β_0 agrees very well with Eq. (8.56) for β_F.

Equation (9.63) may be rewritten as

$$i_b = g_m v_{be} / \beta_0 = v_{be} / r_\pi \tag{9.65}$$

where $r_\pi = v_{be} / i_b$ is the ratio of β_0 to g_m and is known as the CE input resistance.

We draw now a small-signal low frequency equivalent circuit for the BJT in the common-emitter connection as shown in Fig. 9.19. This is a low-frequency circuit because capacitive effects are not included.

It is important to note that the equivalent circuit is valid for both the PNP and NPN transistors.

It is important to point out that the values of both g_m and r_x are dependent on the magnitude of the collector current at the quiescent point. Hence, a change in the location of the quiescent point on the static characteristics changes the magnitudes of r_x and g_m and hence β_0. We now need to address the significance of the small-signal current gain, β_0.

We have defined earlier the common-emitter current gain, β_F, to be I_C / I_B. The symbol β_0 is the ratio of the collector small-signal current to the base small-signal current. Although the values are often assumed to be the same, but because of the dependence of β_F on I_C, as shown in Fig. 9.8, it is instructive to determine a more

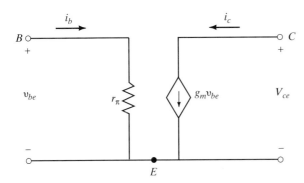

Figure 9.19 Low-frequency common-emitter equivalent circuit of the transistor.

precise definition for β_0. Since β_0 is the ratio of the incremental currents i_c/i_b $(\Delta I_O/\Delta I_B)$ and using $I_C = \beta_F I_B$, we have

$$\beta_0 = \Delta I_C/\Delta I_B = i_c/i_b = \left[\frac{d}{dI_C}(I_C/\beta_F)\right]^{-1}$$

$$= \beta_F/\left[1 - \left(I_C/\beta_F\frac{d\beta_F}{dI_c}\right)\right] \qquad (9.66)$$

If β_F does not vary with I_C, then the last term in Eq. (9.66) is β_F.

A quick method to determine the parameters β_0, g_m, and r_π is to use Eq. (9.62) and the common-emitter graphical characteristics, such as those shown in Fig. 8.12(a). At the operating point, I_C is known, so that g_m is determined by using Eq. (9.62). Since for all practical purposes, β_0 is equal to β_F, the ratio of I_C/I_B at the operating point of the characteristics is β_0. The resistance r_π is then β_0/g_m.

Capacitance Effects

We still have to determine the capacitive effects associated with a change in ΔV_{EB}. There is a *storage capacitance* or diffusion capacitance as defined for the diode in Sec. 7.3. This is a consequence of the associated change in stored charge in the base. We note from Fig. 9.18 that

$$\Delta Q_B = qAW_B\Delta p(0)/2$$

The rate of change of stored charge is accompanied by a transient base current, ΔI_{BS}, which by using Eq. (9.51) and assuming that $p_0 \ll p_1'(0)$ is given by

$$\Delta I_{BS} = \Delta Q_B/\Delta t = qAW_B\Delta p(0)/\Delta t/2 = (q\,A\,W_B p_1'(0))\,(q/kT)\,\Delta V_{EB}/2\Delta t \quad (9.67)$$

We introduce I_C from Eq. (9.48)(a), replace the delta quantities by their incremental equivalents, and replace I_C by its g_m equivalent so that

$$i_{bS} = \Delta I_{BS} = I_C\,(W_B^2/2D_p)\,\frac{q}{kT}\frac{\Delta V_{EB}}{\Delta t} = (W_B^2/2D_p)\,g_m\frac{dv_{be}}{dt} \qquad (9.68)$$

Since the current in a capacitance is $C\,dv/dt$, we can write

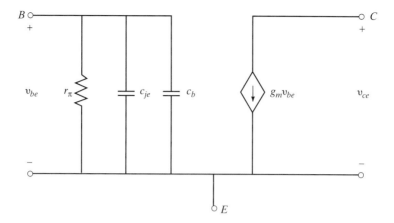

Figure 9.20 Small-signal equivalent circuit based on changes of V_{EB} only.

$$i_{bs} = C_b \frac{dv_{be}}{dt} \tag{9.69}$$

where C_b is the base storage or diffusion capacitance and using Eq. (9.30) is given by

$$C_b \equiv g_m \left(W_B^2/2D_p \right) = g_m \tau_B \tag{9.70}$$

There is an additional component of transient base current, which upon an increase of ΔV_{EB}, is used to neutralize some of the acceptor ions in the depletion layer of the emitter-base junction. This neutralization is required to reduce the depletion layer width that accompanies an increase of V_{EB}. This is a capacitive current defined by

$$i_{bt} = C_{je} \frac{dv_{be}}{dt} \tag{9.71}$$

where C_{je} *is the emitter-base junction capacitance* and has been defined for the diode in Sec. 7.3.

Based on the variations caused by a change in V_{EB}, we can draw the equivalent circuit shown in Fig. 9.20. While the elements shown in the figure may seem to include the major components in the equivalent circuit, there are two additional parameters that are important, one of which is extremely influential in determining the high frequency response of the transistor.

EFFECTS OF CHANGES IN THE MAGNITUDE OF V_{CB}

Carrier Processes

Our emphasis on the active region performance of the transistor may have undermined the influence of the collector-to-base voltage. This voltage has dramatic effects on the saturation region response and on activities at and near breakdown. Even in the active region, as we have seen earlier, an increase of reverse-bias at the collector-base junction has an important, though secondary, effect on the static characteristics of the transistor. The increase of this voltage increases the collector current.

The complete-small signal equivalent circuit of the transistor in the common-emitter connection represents the changes in I_C and I_B brought about by changes in V_{EB} and V_{CB}. The immediate effect of an increase in the magnitude of V_{CB} (higher reverse bias) is to widen the depletion layer at the collector-base junction and thus decrease the effective width of the base, as was shown in Fig. 9.2.

You may wonder why we did not consider base width modulation from the emitter side when V_{EB} changed. The reason is that the change in V_{EB} is very small compared with ΔV_{CB}.

The change in the distribution of hole density brought about by base-width modulation has three consequences that are independent of the effects of ΔV_{EB}. We consider that V_{EB} is fixed and determine the effects of increasing V_{CB} (making it more negative). These effects are as follows.

1. Increase in collector current due to the increase in the gradient of the hole density in the base.
2. Decrease of total stored base charge, which requires a transient component of base current.
3. Decrease of recombination rate, and hence a decrease in the base current, which supplies electrons to recombine in the base.
4. The transient base and collector currents increase the density of ionized atoms in the collector-base depletion layer.

The effects listed in (1) and (3) produce steady-state changes in the currents, whereas those in (2) and (4) are transient in nature.

Collector Current Change

The increase in collector current caused by a change in V_{BC} can be written in terms of the change in the base width using the expression for the collector current $q\,A\,D_p\,p'_1(0)/W_B$

$$\frac{\Delta I_C}{\Delta V_{BC}} = \frac{\partial I_C}{\partial W_B}\frac{\Delta W_B}{\Delta V_{BC}} = \frac{-qAp'_1(0)}{W_B^2}\frac{\Delta W_B}{\Delta V_{BC}}D_p \tag{9.72}$$

A decrease in the base width is accompanied by an increase in the collector current. By introducing I_C, Eq. (9.72) becomes

$$\frac{\Delta I_C}{\Delta V_{BC}} = \frac{-I_C}{W_B}\frac{\Delta W_B}{\Delta V_{BC}} \tag{9.73}$$

From Eq. (9.5) and for $V_{BC} \cong V_{EC}$, the expression for the Early voltage, V_A, is

$$V_A = -W_B\frac{dV_{BC}}{dW_B} \tag{9.74}$$

where V_A is a positive quantity since an increase in V_{BC} for a PNP transistor causes a decrease in W_B. Therefore, replacing ΔV_{BC} and ΔW_B with dV_{bc} and dW_b, we have

$$\Delta I_C = \Delta V_{BC} I_C / V_A \tag{9.75}$$

In terms of small-signal quantities and using Eq. (9.62), we let $i_{c1} = \Delta I_C$.

$$i_{c1} = v_{bc} I_C / V_A = v_{bc} g_m V_t / V_A \tag{9.76}$$

where $g_m = I_C / V_t$ and V_t is kT/q.

An increase in the base-collector voltage decreases W_B and increases i_{c1}. The expression in Eq. (9.76) represents a (dependent) current generator from collector to emitter, which depends on the base- to-collector voltage.

We will include this current generator in the circuit together with the elements resulting from the recombination base current change in the next section.

Recombination Current Change

An increase of V_{BC} causes a reduction in the total minority carrier charge stored in the base, which results in a reduction of the rate of recombination in the base and thus a reduction in the base current. Since the recombination current is defined as Q_B/τ_p, we can write

$$\Delta I_{rec} = \Delta Q_B / \tau_p \tag{9.77}$$

The effect of change in V_{BC} is included in Eq. (9.77) by using Eqs. (9.38) and (9.39) and letting $\beta_0 = \beta_F$,

$$\frac{\Delta I_{rec}}{\Delta V_{BC}} = \frac{\Delta Q_B}{\tau_p \Delta V_{BC}} = \frac{-\Delta (I_C \tau_B)}{\tau_p \Delta V_{BC}} = \frac{-\Delta I_C}{\beta_0 \Delta V_{BC}} \tag{9.78}$$

The negative sign has been introduced to indicate that an increase in V_{BC} causes an increase in I_C whereas ΔI_{rec} decreases.

Using Eq. (9.75) to replace $\Delta T_c / \Delta V_{BC}$ with I_c / V_{-A} in Eq. (9.78) and ΔV_{BC} with v_{bc}, we have

$$i_{rec} = \Delta I_{rec} = -\frac{1}{\beta_0} \frac{\Delta I_C}{\Delta V_{BC}} v_{bc} = \frac{-I_C}{\beta_0 V_A} v_{bc} = -(g_m V_t / \beta_0 V_A) v_{bc} \tag{9.79}$$

We have expressed the change in recombination current in terms of the small signal base collector voltage.

COMPLETE EQUIVALENT CIRCUIT

The change in the stored charge accompanying a change in the collector base voltage causes a storage capacitance. By using Eq. (9.75), we obtain the expression for the capacitance caused by a change in V_{BC}, which is

$$C_{bu} = \frac{-\Delta Q_B}{\Delta V_{BC}} = \frac{\Delta (I_C \tau_B)}{\Delta V_{BC}} = \tau_B \frac{I_C}{V_A} \tag{9.80}$$

As V_{BC} increases, the charge stored in the base of a PNP transistor decreases because of the narrower base width. This corresponds to ΔQ_B being equal to $(-\Delta I_C \tau_B)$. The expression for the transit time, τ_B, from Eq. (9.24) is $W_B^2 / 2D_p$. Using this in Eq. (9.80), we have

$$C_{bu} = \frac{W_B^2 I_C}{2D_p V_A} \tag{9.81}$$

We replace the expression for the transit time by its equivalence from Eq. (9.70) and use the definition of g_m from Eq. (9.62) so that

$$C_{bu} = \frac{C_b}{g_m} \frac{I_C}{V_A} = \frac{kT}{q} \frac{C_b}{V_A} \tag{9.82}$$

The capacitance, C_{bu}, connected from base to collector, is a very small fraction of the base storage capacitance considering that kT/q is 26mV and V_A is in tens of volts.

By using the current generators representing the effects of the collector current change, the base current change, and the storage capacitance, the circuit shown in Fig. 9.20 is modified to include two additional current generators and a capacitance. The new circuit is shown in Fig. 9.21.

We will summarize below the changes that the circuit elements of Fig. 9.21 represent:

$$
\begin{aligned}
r_\pi &= & \text{change in the base current caused by a change in } V_{EB} \\
g_m v_{be} &= & \text{change in the collector current caused by a change in } V_{EB} \\
C_{je} &= & \text{capacitance of the base-emitter depletion layer} \\
C_b &= & \text{base charging capacitance caused by a change in } V_{EB} \\
C_{bu} &= & \text{base charging capacitance caused by a change in } V_{CB} \\
g_m V_t v_{bc}/\beta_0 V_A &= & \text{change in the base recombination current resulting from a} \\
& & \text{change in } V_{CB}
\end{aligned}
$$

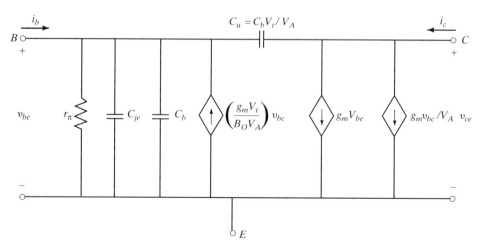

Figure 9.21 High-frequency small-signal equivalent circuit modified to include effects of the changes of V_{CB}.

$g_m V_t v_{bc}/V_A$ = change in the collector current resulting from a change in V_{CB}

We will now reduce the number of current generators by inserting resistances and we will also include parasitic elements to form the complete circuit.

By setting $v_{bc} = v_{be} + v_{ec}$, the current generator i_{c1}, given by Eq. (9.76), is split up into

$$i_{c1} = g_m(V_t/V_A)v_{bc} = g_m(V_t/V_A)v_{be} + g_m \frac{V_t}{V_A} v_{ec} \tag{9.83}$$

The first term is in parallel with, and much smaller than, the current generator $g_m v_{be}$ and may therefore be neglected. The second term may be replaced by a resistance connected from collector to emitter given by

$$r_O = \frac{v_{ec}}{i_{c1}} = \frac{V_A}{g_m V_t} = \frac{V_A}{I_C} \tag{9.84}$$

The term, r_O, is known as the CE output resistance of the transistor.

The expression for the base current i_{b1}, given by Eq. (9.79), may be modeled by the inclusion of a resistance from base to collector, determined as

$$r_u = \frac{v_{bc}}{i_{rec}} = \frac{\beta_0 V_A}{I_C} = \beta_0 r_O \tag{9.85}$$

A collector-base junction or transition-capacitance, C_{jc}, has to be included in the circuit to account for the change in the width of the depletion layer as the collector-base voltage changes. We identify, as well, the ohmic base resistance, r_b, and the ohmic resistances of the emitter and collector as r_e and r_c, with the complete equivalent circuit shown in Fig. 9.22.

Although the circuit in Fig. 9.22 has been obtained for a PNP transistor, it is equally applicable for the NPN transistor.

Typical values for the elements calculated at I_C = 1mA, V_{EC} = 5 volts for a transistor having β_0 = 100, V_A = 50 volts, and τ_β = 0.4ns are shown in the accompanying table.

Small-signal elements of the CE BJT

Symbol	Name	Relevant Equation	Typical Value	Unit
g_m	transconductance	I_C/V_t	38.6×10^{-3}	S
β_0	CE current gain	(9.64)	100	none
r_π	CE input resistance	β_0/g_m	2600	ohms
r_O	CE output resistance	V_A/I_C (9.84)	50,000	ohms
C_b	storage capacitance (V_{EB})	(9.70)	15	pF
C_{je}	E-B junction capacitance	(7.16)	2	pF
C_{bu}	storage capacitance (V_{CB})	(9.80)	.008	pF
C_{jc}	C-B junction capacitance	(7.16)	2	pF
r_b	base ohmic resistance	—	100	ohms
r_e	emitter ohmic resistance	—	2	ohms
r_c	collector ohmic resistance	—	20	ohms
r_u	collector base resistance	(9.875)	5×10^6	ohms

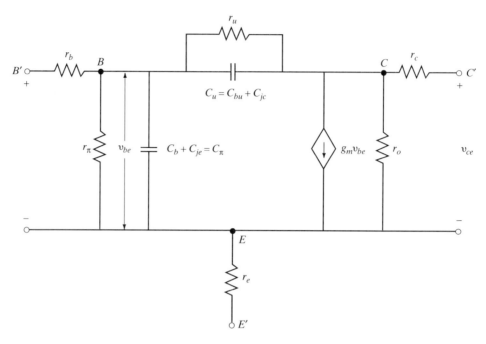

Figure 9.22   Complete high-frequency small-signal equivalent circuit of the BJT.

9.5 FIGURE OF MERIT

A measure of the quality of a high frequency transistor is its figure of merit f_T. As we shall see, f_T is a measure of the ratio of g_m to the total capacitance of the transistor.

By neglecting r_u, r_c, and r_e, we will determine an expression for the short-circuit current gain, $\beta(j\omega)$, of the transistor. It is defined as the ratio of the current I_O in a short-circuit placed at the output, as shown in Fig. 9.23, to an input driving current I_i. We use RMS quantities for I_O, I_p, and V_{be}.

Because of the short-circuit, C_u is in parallel with C_π. By neglecting the current in C_u compared to $g_m V_{be}$, I_O is $g_m V_{be}$ and V_{be} is $I_i(r_\pi)/[1 + j\omega r_\pi(C_\pi + C_u)]$. The current gain is given by

$$I_O/I_i = \beta(j\omega) = \beta_0/[1 + j\omega\, r_\pi(C_\pi + C_u)] \tag{9.86}$$

At high frequencies, the magnitude of the imaginary part of the denominator of Eq. (9.86) is much greater than unity, so that

$$|\beta(j\omega)| = \frac{g_m}{\omega\,(C_\pi + C_u)} \tag{9.87}$$

where $g_m = \beta_0/r_\pi$.

The symbol, f_T, is defined as the frequency at which the magnitude of the short-circuit current gain is unity, so that at $f = f_T$, $|\beta(j\omega)| = 1$ and

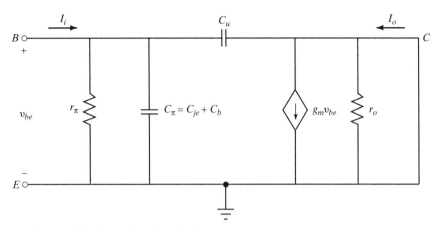

Figure 9.23 Circuit for calculating f_t of a transistor.

$$f_T = \frac{g_m}{2\pi(C_\pi + C_u)} \qquad (9.88)$$

If we define τ_T as $1/\omega_T$, where $\omega_T = 2\pi f_T$, we have

$$\tau_T = \frac{C_\pi + C_u}{g_m} = \frac{C_b}{g_m} + \frac{C_u + C_{je}}{g_m} \qquad (9.89)$$

From Eq. (9.70), C_b/g_m is τ_B, where $\tau_B = W_B^2/2D_p$, so that

$$\tau_T = \tau_B + \frac{C_u + C_{je}}{g_m} \qquad (9.90)$$

The capacitances, C_{je} and C_u, consist mainly of transition capacitances and hence depend, in accordance with Eq. (7.16) on the voltages across the emitter-base and collector-base depletion layers respectively. The storage capacitance, C_b, on the other hand, depends on the quiescent collector current through g_m and on the base transit time. The base transit time, τ_B, is uniquely determined for a given transistor.

It is obvious that low values of the capacitances contribute to an improved high frequency response. A high value of g_m is an indication of a high ratio of small-signal collector current to a small signal input voltage. Hence, the f_T of a transistor is a measure of its current amplification performance at high frequencies. It is also known as the figure of merit of a high-frequency transistor.

REVIEW QUESTIONS

Q9-8 To switch a transistor (PNP) from cutoff to saturation, what should happen to the hole density in the base and how is this accomplished?

Q9-9 Explain clearly the meaning of each of the charge-control equations.

Q9-10 By what mechanism is the turn-ON time reduced?

Q9-11 Why is the BJT small-signal circuit labeled linear?

Q9-12 Briefly define the significance of the frequency f_T.

HIGHLIGHTS

- There are two major fields of application of the BJT: switching circuits and amplifier circuits. While operation is in the active region in amplifiers, switching takes place between cutoff and saturation.

- The switching process is studied by using the two charge control equations. A base current pulse initiates the switching from cutoff to saturation, causing a buildup of minority carrier charge in the base. Rapid turn-ON time is enhanced by a large amplitude of the base current, a smaller saturation collector current, and a small lifetime.

- The turn-OFF time is smaller if the lifetime is smaller, resulting in a fast removal of the minority carrier charge buildup in the base.

- In the use of a BJT in amplifiers and because of the nonlinearity of the BJT characteristics, changes in the input and output variables must be restricted to very small values. To analyze a transistor circuit, an equivalent small-signal circuit is determined, which represents the transistor for incremental changes about an operating point on the output characteristics.

- The quality of a BJT as an amplifier is measured by the gain and high-frequency response. Large gain is enhanced by a large transductance, g_m, and high-frequency operation is improved by smaller storage and junction capacitances.

EXERCISES

E9-2 Calculate the value of r_π for a BJT operating at $I_C = 2\text{mA}$ and whose $\beta_o = 100$.

$$\text{Ans: } r_\pi = 1.3\text{K}\Omega.$$

E9-3 A BJT operating at $I_C = 1\text{mA}$ has $f_T = 500\,\text{MHz}$, $\beta = 100$, and $C_u = C_{je} = 1\text{pF}$, determine C_b.

$$\text{Ans: } C_b = 10.28\text{pF}$$

9.6 NPN TRANSISTORS

In the early part of the previous chapter, we indicated that, in the interest of consistency with the PN junction diode, we would use the PNP transistor in the analysis of the operation, the equations, and the models of the device.

However, it is important to mention that because of circuit yield and economy, integrated circuit fabrication is geared towards the NPN transistor. Although it is quite possible to produce PNP transistors on the same chip as NPN devices, they will not possess the high quality properties of the NPN.

The major advantage of NPN transistors rests in the attractive properties of the minority carrier electrons in the base. Electrons have a higher mobility than holes, hence resulting in faster and higher gain devices. The higher β_F of the NPN in comparison to that of the PNP was confirmed in the examples of Chapter 8.

9.7 THE GUMMEL-POON MODEL

In the early sections of this chapter, we compared actual characteristics of the BJT to the simple relations that we derived in Chapter 8 for the Ebers-Moll model. We also identified the physical processes that are the causes of the discrepancies between actual and derived characteristics.

The Ebers-Moll equations, while representing a simple and elegant model, do not provide very precise results when applied to modern-day minimal geometry transistors. A modified model is needed to represent the second order effects. The results of the modifications are included in the Gummel-Poon model. This model is the basis for the computer simulation program SPICE. SPICE is the acronym for *Simulation Program with Integrated Circuit Emphasis*.

The Gummel-Poon model incorporates three important second order effects. They are:

- Recombination in the emitter depletion layer at low values of emitter-base bias, resulting in a decrease of current gain.
- Decrease of common-emitter current gain at high values of collector current.
- Non-zero slope in the active region of the common-emitter output characteristics, resulting from the Early effect.

The Gummel-Poon relations are derived through a three-step process. First, the Ebers-Moll relations for I_B and I_C are written in terms of I_S, β_F, and β_R. The resulting equations are known as the "transport version" of the Ebers-Moll relations. The second step consists of modeling the increase of base current at low values of emitter-base bias by a superposition of ideal diode and non-ideal diode parameters, as determined from measurements on the transistor similar to those shown in the plot of Fig. 9.5. This constitutes the addition of four new parameters.

The third step is the replacement of I_S by an equivalent majority carrier charge in the base. Both the effects of high injection and base-width modulation influence the majority carrier charge in the base through the emitter and collector injected carriers. These carriers are compensated by an increase in the majority carrier charge needed to neutralize the injected minority carriers. Furthermore, the changes in the emitter-base and collector-base voltages influence the ionized charges through the changes in the depletion layers at the emitter and collector.

We refer to Eq. (8.48) to obtain the expression for I_S, defined as the factor multiplying the V_{EB} exponential as

$$I_S = qAD_p p_0 / W_B = \frac{qAD_p n_i^2}{N_{DB} W_B} = \frac{q^2 A^2 D_p n_i^2}{qAN_{DB} W_B} \tag{9.91}$$

The denominator of Eq. (9.91) may be defined as the majority carrier charge in the base at equilibrium. Since the total charge varies due to the effects of high injection and base-width modulation, we write Eq. (9.91) in terms of a charge Q_G as

$$I_S = \frac{q^2 A^2 D_p n_i^2}{Q_G} \tag{9.92}$$

where

$$Q_G = qA \int_0^{W_B} n(x)dx \qquad (9.93)$$

The symbol Q_{G0} is known as the Gummel number and is defined as

$$Q_{G0} = qA \int_0^{W_B} N_{DB}dx \qquad (9.94)$$

Because of the effects of high injection and base-width modulation, Q_G becomes

$$Q_G = Q_{G0} + Q_E + Q_C + Q_F + Q_R \qquad (9.95)$$

where

Q_{G0} = net majority carrier charge commensurate with space charge neutrality

Q_E and Q_C = immobile ion charges uncovered by the changes in the depletion layer boundaries

Q_F and Q_R = base majority carrier charge needed to neutralize the emitter and collector injected minority carriers.

The high-injection effects are modeled using the charge control relations and the charge-control time constants τ_B and τ_R, while base-width modulation effects are modeled by a forward Early voltage, V_A, and an equivalent, V_B, for reverse operation together with the depletion layers capacitances, C_{je} and C_{jc}.

The process in steps two and three requires the addition of four new parameters. These plus the four required for depletion layer recombination and I_S, β_F, and β_R make up the eleven parameters needed to describe the transistor static characteristics in the Gummel-Poon model. To make the model more complete, the values of r_b, r_c and r_e are added.

Complete numerical analysis using the required parameters is easily carried out by using SPICE.

PROBLEMS

9.1 An NPN silicon BJT has $N_{DB} = 10^{17}\text{cm}^{-3}$, $N_{AC} = 10^{16}\text{cm}^{-3}$, and $W_B = 1.2\mu\text{m}$. It is operating in the active region at $V_{BE} = 0.6\text{V}$ and $T = 300\text{K}$. Determine:

   (a)   the change in the width of the base as V_{CE} changes from 1.5V to 6.5V DC.

   (b)   the corresponding change in the collector current.

9.2 A silicon NPN BJT has $N_{DE} = 10^{18}\text{cm}^{-3}$, $N_{DC} = 10^{16}\text{cm}^{-3}$, $W_B = 0.6\mu\text{m}$, $V_{BE} = 0.7\text{V}$, and $V_{CB} = 5\text{V}$. Given $D_{nB} = 20\text{cm}^2/\text{s}$ and $\tau_{nB} = 5 \times 10^{-3}\text{s}$. When V_{CB} is increased to 10V, the minority carrier diffusion current in the base increases by 20 percent. Determine, at $T = 300\text{K}$,

   (a)   the base doping.

   (b)   the Early voltage V_A.

9.3 For problem 9.1, determine the Early voltage, V_A.

9.4 A silicon NPN BJT has $N_{AB} = 10^{17} \mathrm{cm}^{-3}$, $N_{DC} = 10^{16} \mathrm{cm}^{-3}$, and $W_B = 0.25 \mu \mathrm{m}$. Determine, at $T = 300 \mathrm{K}$,

    (a)   the punchthrough voltage.

    (b)   the average value of the electric field intensity at punchthrough.

    To increase the punchthrough voltage, the base doping can be increased. At what cost would this be?

9.5 An NPN silicon BJT at $T = 300 \mathrm{K}$ has heavy collector doping and $N_A = 10^{16} \mathrm{cm}^{-3}$. Given $W_B = 1.0 \mu \mathrm{m}$. Determine:

    (a)   the breakdown voltage for active operation in the common-base mode. The breakdown field in silicon is $3 \times 10^5 \mathrm{V/cm}$.

    (b)   the punchthrough voltage.

9.6 A silicon PNP BJT at $T = 300 \mathrm{K}$ has $N_{DB} = 2 \times 10^{16} \mathrm{cm}^{-3}$, $N_{AC} = 4 \times 10^{15} \mathrm{cm}^{-3}$, $W_B = 2 \mu \mathrm{m}$, $D_{pB} = 11.2 \mathrm{cm}^2/\mathrm{s}$, and $\tau_{pB} = 0.5 \mu \mathrm{s}$ and $A = 10^3 \mathrm{cm}^2$. Determine the output resistance at $I_C = 1 \mu \mathrm{A}$ and $V_{CB} = -10 \mathrm{V}$.

9.7 For a BJT operating in the common-base active region, derive an expression for the output resistance r_{OC} in terms of I_C, W_B, L_{pB}, and dW_B/dV_{BC}. The output resistance is defined by $1/r_{OC} = dI_C/dV_{BC}|_{I_E}$.

9.8 Show that the transport factor of a BJT is given approximately by

$$\delta \equiv \frac{1}{1 + \tau_B/\tau_P}$$

where τ_B is the transit time of minority carriers in the base and τ_P is the lifetime of minority carriers in the base. Explain the physical significance.

9.9 A base current pulse of $250 \mu \mathrm{A}$ with a duration of $300 \mathrm{ns}$ is used to turn on a silicon PNP BJT in the circuit of Fig. 9.10(a). Given $V_{CC} = 5.2 \mathrm{V}$, $R_L = 1 \mathrm{K}\Omega$, $\tau_{PB} = 1 \mu \mathrm{s}$, $W_B = 5 \mu \mathrm{m}$, and $D_{PB} = 10 \mathrm{cm}^2/\mathrm{s}$, determine, at $T = 300 \mathrm{K}$,

    (a)   the turn-ON time.

    (b)   the storage time.

9.10 **(a)**   Use the Ebers-Moll expression for I_C in the active region of operation of an NPN BJT and assume a small-signal voltage v_i (ΔV_{BE}) is superimposed on V_{BE}. By expanding the exponential show for $v_i \ll V_T$, the small-signal collector current is given by $g_m i_c$.

    (b)   For what maximum approximate value of v_i is the small-signal equivalent circuit valid?

9.11 The transistor in the circuit of Fig. 9.24 is a silicon PNP device operating at $T = 300 \mathrm{K}$, and at $I_C = 1 \mathrm{mA}$, $V_{EB} \cong 0.7 \mathrm{V}$, and $V_{BC} = 5 \mathrm{V}$. The device has $\beta_0 = 200$, $N_{DB} = 10^{17} \mathrm{cm}^{-3}$, $N_{AC} = 10^{16} \mathrm{cm}^{-3}$, $W_B = 0.8 \mu \mathrm{m}$, $V_A = 100 \mathrm{V}$, and $D_{PB} = 10 \mathrm{cm}^2/\mathrm{s}$. Determine:

    (a) g_m

    (b) r_π

    (c) r_O

9.12 For the device of Problem 9.11, determine, for $N_{AE} = 10^{18} \mathrm{cm}^{-3}$, $A = 10^{-3} \mathrm{cm}^2$, and $C_u = 0.2 pF$,

    (a) C_{je}

    (b) C_b

    (c) C_π

    (d) f_T

9.13 A GaAs PNP BJT has $N_{AE} = 10^{20}\text{cm}^{-3}$, $N_{DB} = 10^{16}\text{cm}^{-3}$, $D_{PB} = 30\text{cm}^2/\text{s}$, $\tau_P = 10\mu\text{s}$, $A = 10^{-3}\text{cm}^2$, and $W_B = 12\mu\text{m}$. At $T = 300\text{K}$, and for $I_C = 2\text{mA}$ in the active region, determine:

    (a)   the base storage capacitance C_b.

    (b)   the emitter junction capacitance.

9.14 For Problem 9.13, given $\beta_0 = 100$, and $C_u = 0.25\text{pF}$, determine, at $T = 300\text{K}$,

    (a) r_π

    (b) f_T

chapter 10

JUNCTION FIELD-EFFECT TRANSISTORS

10.0 INTRODUCTION

Field-Effect transistors (FETs) are labeled as such because the primary action is the effect of a transverse electric field on the longitudinal motion of the carriers. Both their construction and operation are considerably different from bipolar transistors. Whereas the currents in a BJT include both holes and electrons, the current in an FET involves one carrier only. In a BJT, the study of the currents is based mainly on the study of the diffusion of the minority carriers. In an FET, the current is a result of carrier drift under the influence of a longitudinal electric field. Finally, FETs are labeled as *unipolar devices* to highlight the fact that the current carriers are either holes or electrons.

In general, the FET has several advantages over the bipolar junction transistor. First, it has a much higher input resistance, thus causing negligible loading of a voltage source connected at the input. Second, the FET is relatively insensitive to temperature and immune to radiation. Third, and in particular for silicon-based devices, it is less noisy. The main disadvantage of the FET is its lower transconductance, g_m, and hence lower gain. The FET is more economical to produce than the BJT as it requires fewer fabrication steps and occupies much less chip area.

Because of its low noise and relative insensitivity to cosmic radiation, the JFET amplifier is primarily used in satellite communications operating in the gigahertz frequency range. It is also used as an amplifier such as in operational amplifiers and in comparators.

There are two basic classes of FETs: the junction FET (JFET) and the metal-oxide-semiconductor FET (MOSFET). The junction in the JFET may be at a semiconductor-semiconductor surface or at a metal-semiconductor surface, and we distinguish the latter by labeling it the MESFET, leaving the JFET designation to the PN junction device.

In this chapter, we will study the PN junction JFET. In later chapters, we will consider the MESFET and the MOSFET. We begin our study with the JFET for two reasons. First, the JFET represents a natural transition from the BJT to the MOS-FET and, second, it is a simpler device to fabricate and to analyze. It must be mentioned here that it is not nearly as widely used as the MOSFET.

10.1 CONSTRUCTION AND OPERATION

Construction and the Basic Functions of the Terminals

We will use the perspective sketch of Fig. 10.1(a) and the cross section in Fig. 10.1(b) to explain the construction of the JFET.

The transistor has three terminals: the source, the drain, and the gate. The *source* is the semiconductor terminal from which carriers are emitted that travel through a semiconductor *channel* to be collected by the *drain*. In the NJFET, the N refers to the type of semiconductor of the channel and the carriers are electrons. A positive voltage applied from drain to source is used to accelerate the electrons through the channel. The third terminal, labeled the *gate*, is used to control the flow of electrons. In the single gate device, the gate is situated above the channel. In the two-gate model, an additional gate is shown below the channel with the two gates operating at the same voltage. The gate, a P^+ semiconductor in the NJFET, is of opposite conductivity to the channel. A negative voltage is applied to the gate with the source grounded. With a positive voltage applied to the drain, with respect to the source, the gate-channel junction is reverse-biased and the resulting electric field is transverse to the direction of motion of the electrons. As the gate voltage is changed, or as the drain-source voltage is changed, the reverse bias across the gate-channel junction changes. The reverse bias changes the width of the depletion layer, thus changing the vertical dimension and the cross sectional area of the channel that is normal to the direction of motion of the electrons travelling from source to drain.

Operation

Effect of gate-to-source voltage   The channel of an NJFET is a bar of N silicon, whose resistance or conductance depends upon its length, L, its width, $2a$, its depth, Z, and its conductivity. For a fixed drain-source voltage, the width of the depletion layer, formed between the gate and the channel, is controlled by a voltage applied between gate and source. Because of the high doping of the P^+ gate and the reverse-bias across the gate-channel junction, the depletion layer extends almost entirely in the channel. The gate becomes the control terminal since by varying the reverse bias, between gate and source, and hence between gate and channel for a fixed drain-source voltage, the width of the channel and therefore the effective cross-sectional area of the channel normal to the carrier flow is changed.

The change in the cross-sectional area of the channel causes a change in its conductance. The process of controlling the conductance of the channel is known as

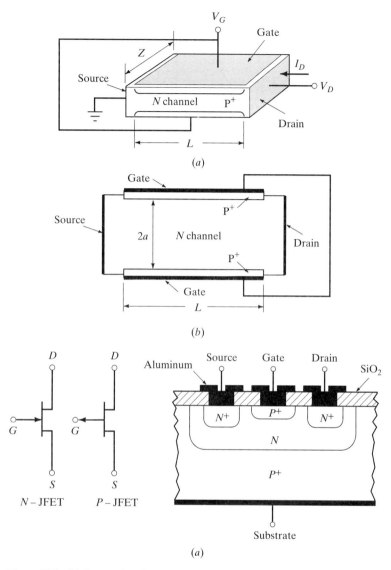

Figure 10.1   (a) Perspective view of an NJFET; (b) simplified physical cross-section; (c) symbol for the N and P JFET; and (d) basic structure of a one-gate model.

conductance modulation. Because of the voltage applied between drain and source, the drift current of electrons moving from source to drain is determined by Ohm's law, it being the ratio of the drain-source voltage to the resistance of the channel between drain and source.

We can conclude from the above that, for a *fixed drain-source voltage*, by making the P⁺ gate more negative with respect to the source, and hence with respect to the channel, the width of the depletion layers increases, the width of the channel as

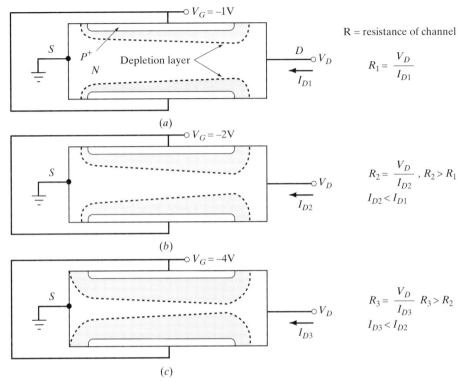

Figure 10.2   Making the gate voltage more negative (a to c) at constant V_D results in decrease of the width of the channel. Note the decrease of width at the source end.

well as its cross sectional area decreases, and its resistance increases, resulting in a decreasing drain current. This effect is illustrated in Fig. 10.2.

Effect of the drain voltage   The previous discussion may have left the impression that the width of the channel is determined only by the gate-source voltage. In fact, the drain-source voltage is an integral part of the process of channel-width modulation.

An increase of the drain voltage, with respect to a grounded source for an NJFET at a fixed negative gate voltage, has obviously no effect on the width of the depletion layer at the source end of the channel. However, as we move from the source towards the drain end of the channel, the reverse bias between the negative gate and the channel, which is positive (with respect to the source because of the positive drain voltage), is increased. This causes an increase in the width of the depletion layer, hence a decrease in the effective width of the channel, and consequently a decrease in the cross-sectional area of the channel. The channel, therefore, has the smallest area at the drain end. This effect is exacerbated as the drain voltage is increased, as shown in Fig. 10.3

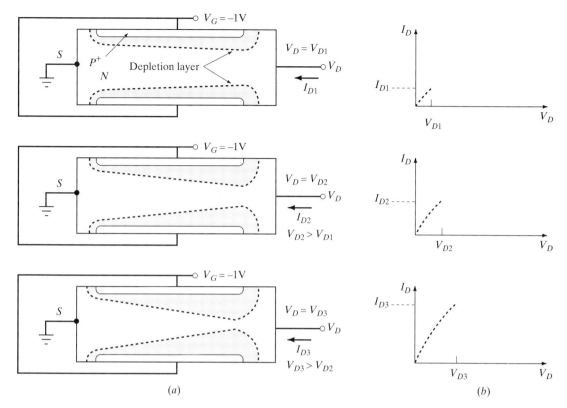

Figure 10.3 Making the drain voltage more positive (top to bottom) results in a (a) decrease of the area of the channel and (b) smaller increase of the drain current with increasing drain voltage, as indicated by the decreasing slope with increasing voltage.

As we indicated earlier, the channel has the smallest area at the drain end. A further increase of the drain voltage causes the top and bottom ends of the depletion layers, in a two gate model, to touch at the drain end. This condition is known as *pinchoff*.

Drain Current Up to Pinchoff   With the gate voltage fixed, an increase of the drain voltage from zero, and also at small drain voltages, causes small changes in the channel resistance so that the drain current increases fairly linearly with the drain voltage.

Two conflicting phenomena, which influence the current in opposite ways, accompany further increase of the drain voltage for a fixed gate voltage. The resistance of the channel increases because of the decrease of the width of the channel and simultaneously the drain voltage, which caused the increase of resistance, has increased. We will later analytically demonstrate that the combined phenomena tend to increase the drain current up to the point where the channel is pinched-off at the drain end.

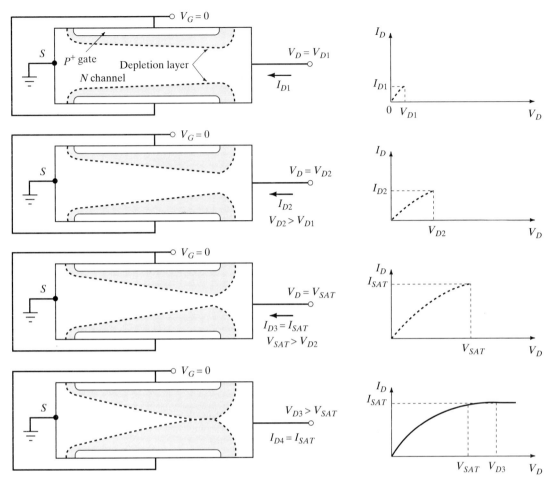

Figure 10.4 (a) Increase of drain voltage at $V_G = 0$ results in pinchoff. (b) Drain current increases with V_D and saturates at I_{SAT} for $V_D \geqslant V_{SAT}$.

Additional increase of the drain voltage, and as shown in Fig. 10.4, causes the drain current to increase at a smaller rate and, eventually, increase of the drain voltage, causes the channel to be pinched at the drain end. The drain voltage at the onset of pinchoff is known as the *saturation drain voltage, V_{SAT}*. We illustrate the onset of pinchoff by the sketches of Fig. 10.4. It is important to point out that pinchoff results from a particular combination of drain and gate voltages.

Beyond Pinchoff The question becomes: Once pinchoff has occurred, what happens to the channel resistance and to the drain current as the drain voltage is increased, so that $V_D > V_{SAT}$?

For voltages greater than the saturation drain voltage, further pinching of the channel takes place and the pinched-off region spreads out in the channel towards the source, effectively isolating the channel from the drain, as shown in Fig. 10.4. The part of the channel that is pinched-off, we label as the isolation region.

The increased drain voltage above V_{SAT}, $V_D - V_{SAT}$, appears across the depleted isolation region so that the voltage drop across the unpinched part of the channel is V_{SAT}.

Above pinchoff, a strong longitudinal (along the channel) electric field appears across the isolation region. This field originates on the positively ionized donors in the depleted isolation region and terminates on electrons in the channel. Such a strong electric field forces electrons to move from the tip of the remaining channel through the isolation region and into the drain. This field attracts all electrons that arrive at the tip. The number of electrons that are available at the point of juncture of the channel with the isolation region is dependent upon the voltage drop $(V_D - V_{SAT})$. Therefore, the current depends on this voltage difference.

If we assume that the length of the pinched-off section of the channel at the drain end is negligibly small compared to the channel length, L, and since the voltage, V_{SAT}, appears across L, then it is reasonable to conclude that the drain current for $V_D > V_{SAT}$ remains constant at the value it has when $V_D = V_{SAT}$. We remind the reader that V_G and V_D are measured with respect to the source, which we have assumed is grounded.

For $V_G = 0$, pinchoff occurs at the drain end at a certain V_D. For negative values of V_G, and since the channel gets pinched-off at a fixed channel-gate voltage, pinchoff occurs at lower values of V_D and consequently the drain current at pinchoff is smaller than that at $V_G = 0$, as shown in the sketch of the characteristic curves of Fig. 10.5. An appropriate question at this time is: What combination of V_G and V_D causes the onset of pinchoff?

At pinchoff, the width of the depletion layer becomes equal to a, half the channel width, shown in Fig. 10.1(b). The channel-to-gate voltage at pinchoff, V_{CG}, is determined from the expression for the width of the depletion layer given by Eq. (5.30). In fact, this voltage is equal to the magnitude of the gate voltage, for $V_D = 0$, at which the channel is completely pinched off.

Since pinchoff occurs at the drain end first at $V_D = V_{SAT}$, the voltage drop from channel to gate at the pinchoff point, determined by Kirchhoff's voltage law, becomes $(V_{SAT} - V_G)$. We therefore define a *pinchoff voltage*, V_p, as

$$V_p = V_{SAT} - V_G \qquad (10.1)$$

We note that the built-in voltage at the junction has not been included. We will consider that in the next section.

For every V_G there corresponds a V_{SAT} at which pinchoff occurs. The difference at pinchoff between V_{SAT} (>0 for an NJFET) and the corresponding V_G (<0 for NJFET) is always V_p. The pinchoff voltage, V_p, is a property of the particular device.

We have also shown in Fig. 10.5, the locus of all the pinchoff points, labeled the pinchoff line, where each point represents the difference between V_{SAT} and V_G. This difference is equal to V_p.

We concluded earlier that, by assuming that the length of the isolation region is much smaller than the channel length L, the drain current remains constant at the value where $V_D = V_{SAT}$. The region on the characteristic curves where $V_D > V_{SAT}$ is

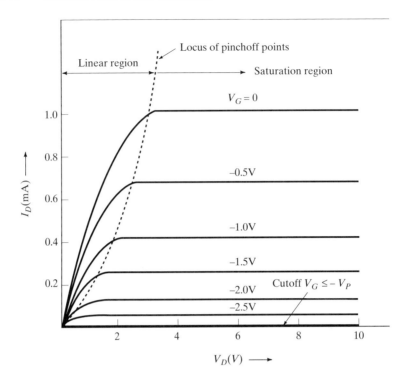

Figure 10.5   Characteristic curves of NJFET for $V_p = 3V$ ($V_{bi} = 0$).

known as the *saturation region*. The region to the left of the pinchoff line in Fig. 10.5 and where $V_D < V_{SAT}$ is labeled the *linear region*, although only for small V_D, is the variation of I_D with V_D being linear. *Cutoff* represents the condition corresponding to $V_{SAT} = 0$ and hence zero I_D.

When the transistor is to be used as an amplifier, the operating point is selected to be in the saturation region. The principal use of the JFET in the linear region is as a variable resistor and, in particular, in the region very near the origin of Fig. 10.5.

Since the gate-channel junction is operating as a reverse-biased diode, minority carrier electrons from the gate cross into the channel. However, because the doping of the P$^+$ gate is so high, the density of electrons in the gate is very small and the number of electrons injected into the channel is negligibly small compared to the number of electrons available in the N channel.

It is to be noted that the gate in the NJFET has been assumed to operate at $V_G \leq 0$. It is possible to operate the gate at a very small positive voltage, equal to or less than the built-in voltage of the gate-channel. However, operating the gate at a higher forward bias defeats the purpose of having the gate act as a control element since a positive gate voltage may form a forward-biased diode between the gate and the channel. This diode then exhibits a low input resistance and causes an unwanted gate current.

10.2 CURRENT-VOLTAGE CHARACTERISTIC EQUATION

Preliminary Conditions

We will now derive the relation between the drain current and the drain and gate voltages. Before we do that, we establish certain basic equations related to the derivation. Equation (5.30) gives, for an abrupt P$^+$N junction, the relation between the depletion layer width and the voltages across the layer. For the dimensions shown in Fig. 10.6 and at pinchoff, the depletion layer width, at the drain end from channel to either of the gates, becomes the distance a and the resulting voltage drop from drain to gate, which is the voltage across the depletion layer and in accordance with Eq. (5.30), becomes

$$V_{DG} = V_{SAT} + V_{bi} - V_G = qN_D a^2/(2\varepsilon) \tag{10.2}$$

where V_{bi} is the built-in voltage drop from channel to gate across the depletion layer, N_D is the channel doping, N_A (gate) $\gg N_D$, and ε is the dielectric constant of the semiconductor.

This drain-to-gate voltage, at which the channel is completely pinched-off, is the pinchoff voltage V_p, so that

$$V_p = qN_D a^2/(2\varepsilon) = V_{SAT} + V_{bi} - V_G \tag{10.3}$$

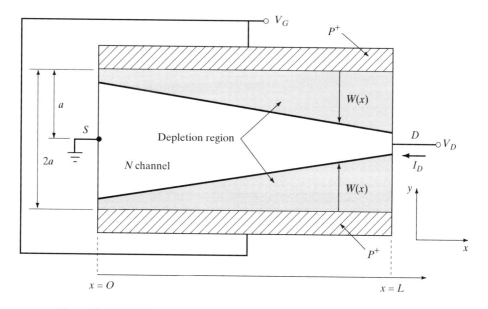

Figure 10.6 NJFET structure with symbols used in derivation.

The expression for the resistance of the channel is given by

$$R(x) = \rho\frac{L}{A} = \frac{L}{\sigma A} = \frac{L}{q\mu_n N_D 2[a - W(x)]Z} \tag{10.4}$$

where μ_n is the mobility of the electrons, $W(x)$ is the width of the depletion region of each of the top and bottom P$^+$N junctions at x, and $2Z[a - W(x)]$ is the cross-sectional area of the channel normal to the direction of motion of electrons from source to drain. We note that the resistance of the channel varies with x because the width of the depletion layer depends on x.

The current-voltage relations we will derive are idealized principally because of the following assumptions:

1. The channel length, L, is assumed to be so large that even in the saturation region of the characteristics, where high values of longitudinal electric fields may exist, the mobility of the electrons is assumed to be constant.

2. The rate of change of the transverse electric field, $d\mathscr{E}_y/dy$, is much greater than $d\mathscr{E}_x/dx$, the rate of change of the electric field in the direction of motion of the electrons, which we label the x-direction. To obtain an exact expression for the depletion layer width, we need to use Poisson's equation in two dimensions since there are variations in the electric field in both the x and y directions. The assumption, $d\mathscr{E}_y/dy \gg d\mathscr{E}_x/dx$, implies that the change in the depletion layer width is a function of the voltage between gate and channel only, and it permits one to use the one-dimensional Poisson equation in place of the two-dimensional equation. This assumption, which simplifies the mathematics considerably, is known as the *gradual-channel approximation*.

Derivation of Current-Voltage Relationship

For the device structure shown in Fig. 10.6, the drift current of electrons that constitutes the drain current I_D shown in Fig. 10.6 is given by

$$I_D = qN_D v(x)A = 2qN_D v(x)Z[a - W(x)] \tag{10.5}$$

where Z is the channel depth, $2[a - W(x)]$ is the channel width at x, and $v(x)$ is the drift velocity of the electrons at a point x in the channel and is given by

$$v(x) = -\mu_n \mathscr{E}_x = \mu_n \frac{dV_x}{dx} \tag{10.6}$$

where dV_x is the voltage drop across an increment dx in the longitudinal direction. The depletion layer width at x is found from Eq. (5.30) by invoking the gradual-channel approximation as

$$W(x) = [2\varepsilon(V_{bi} - V_a)/qN_D]^{1/2} \tag{10.7}$$

where V_{bi} is the built-in voltage across the channel-gate depletion layer and V_a is the voltage drop from gate to channel at any x. The voltage drop V_a is given by

$$V_a = V_G - V_x \tag{10.8}$$

The voltage, V_x, is the voltage drop from a point x in the channel to the source. Using Eq. (10.8), Eq. (10.7) is written as

$$W(x) = [2\varepsilon(V_{bi} + V(x) - V_G)/qN_D]^{1/2} \tag{10.9}$$

By substituting, in Eq. (10.5), for $v(x)$ from Eq. (10.6) and for $W(x)$ from Eq. (10.9), we obtain

$$I_D = 2qN_DZ\mu_n\frac{dV_x}{dx}\left[a - \left(\frac{2\varepsilon}{qN_D}\right)^{0.5}(V_{bi} + V_x - V_G)^{0.5}\right] \tag{10.10}$$

We now separate the variables x and V_x, and integrate along the channel from $x = 0, V_x = 0$ to $x = L$, and $V_x = V_D$ as

$$\int_0^L I_D dx = 2q\mu_n N_D Z \int_0^{V_D}\left[a - \left(\frac{2\varepsilon}{qN_D}\right)^{0.5}(V_{bi} + V_x - V_G)^{0.5}\right]dV_x \tag{10.11}$$

By using the fact that I_D is constant for all x, we perform the integration to obtain

$$I_D L = 2q\mu_n N_D Z\left\{aV_D - \frac{2}{3}\left(\frac{2\varepsilon}{qN_D}\right)^{0.5}[(V_{bi} + V_D - V_G)^{0.5} - (V_{bi} - V_G)^{1.5}]\right\} \tag{10.12}$$

Equation (10.12) shows the dependence of the drain current on the applied voltages, V_D and V_G, on the built-in voltage V_{bi}, and on the physical parameters of the device. Upon rearranging terms, it becomes

$$I_D = 2q\mu_n N_D Za/L\left\{V_D - \frac{2}{3}\left(\frac{2\varepsilon}{qN_D a^2}\right)^{0.5}[(V_{bi} + V_D - V_G)^{1.5} - (V_{bi} - V_G)^{1.5}]\right\}$$

$$\tag{10.13}$$

We introduce V_p from Eq. (10.3), so Eq. (10.13) can be written as

$$I_D = G_0\left[V_D - \frac{2V_p}{3}\left(\frac{V_{bi} + V_D - V_G}{V_p}\right)^{3/2} + \frac{2}{3}V_p\left(\frac{V_{bi} - V_G}{V_p}\right)^{3/2}\right] \tag{10.14}$$

where $G_0 = 2q\mu_n ZaN_D/L$ and $V_p = qN_D a^2/2\varepsilon$. The term G_0 represents the conductance of a channel having width $2a$, depth Z, and length L. Equation (10.14) is valid only in the linear region, up to pinchoff, as either V_G, V_D, or both are allowed to vary. Beyond pinchoff, for $V_D > V_{SAT}$, the drain current is assumed to saturate at the value it has at pinchoff. The drain voltage at pinchoff, V_{SAT}, is determined from Eq. (10.2) to be $(V_p - V_{bi} + V_G)$. This can also be verified from Eq. (10.14) by finding dI_D/dV_D and setting it equal to zero. The characteristic curve for cutoff is found by letting V_{SAT} equal zero, which gives $V_G = V_{bi} - V_p$ at $I_D = 0$.

In the preceding derivation, we have used an idealized two-gate mode. The resulting equation also applies to the one-gate model, whose structure is shown in Fig. 10.1(d). Effect of the gate voltage is illustrated in Fig. 10.7. We now apply the foregoing relations in the following example.

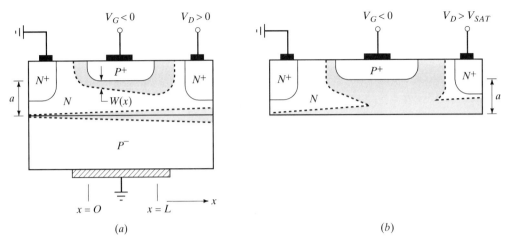

Figure 10.7 Single-gate model showing (a) depletion layer and (b) pinchoff.

EXAMPLE 10.1

An N channel silicon JFET has: $N_A = 10^{18}\text{cm}^{-3}$, $N_D = 10^{16}\text{cm}^{-3}$, a channel length of $25\mu\text{m}$, $Z = 250\mu\text{m}$, and $a = 1\mu\text{m}$. Assume $\mu_n = 1300\text{cm}^2/\text{V-s}$ and determine:

a) The pinchoff voltage V_p.

b) The conductance G_0.

c) The built-in voltage.

d) The drain current at $V_G = 0$ and $V_D = 4\text{V}$.

e) V_{SAT} for $V_G = -2\text{V}$ and for $V_G = 0$.

Solution

a) From Eq. (10.3), $V_p = \dfrac{qN_D a^2}{2\varepsilon}$

The dielectric constant ε for silicon is $11.8 \times 8.85 \times 10^{-14}\text{F/cm}$, so that

$$V_p = \frac{1.6 \times 10^{-19} \times 10^{16} \times (10^{-4})^2}{2 \times 11.8 \times 8.854 \times 10^{-14}} = 7.65\text{V}$$

b) As defined in Eq. (10.14),

$$G_0 = \frac{2q\mu_n Z a N_D}{L} = \frac{2 \times 1.6 \times 10^{-19} \times 1300 \times 250 \times 10^{-4} \times 10^{-4} \times 10^{16}}{25 \times 10^{-4}}$$

$$G_0 = 4.16 \times 10^{-3}\text{S}$$

c) From Eq. (5.17), $V_{bi} = \dfrac{kT}{q} \ln (N_A N_D / n_i^2) = .026 \ln \dfrac{10^{34}}{10^{20}} = 0.838\text{V}$

d) By using Eq. (10.14), we have

$$I_D = 4.16 \times 10^{-3} \left[4 - \frac{2 \times 7.65}{3} \left(\frac{0.838 + 4}{7.65}\right)^{3/2} + \frac{15.3}{3} \left(\frac{0.838}{7.65}\right)^{3/2} \right] = 6.73\text{mA}$$

e) From Eq. (10.2), for $V_G = -2\text{V}$, $V_{\text{SAT}} = V_P - V_{\text{bi}} + V_G = 7.65 - 0.838 - 2 = 4.812\text{V}$, and for $V_G = 0$, $V_{\text{SAT}} = 7.65 - 0.838 = 6.812\text{V}$

Additional Remarks

We remind the reader that for an N channel JFET, V_{bi}, V_D, and V_p are positive quantities, whereas V_G is zero or negative. Equation (10.14) predicts that at $V_D = 0$, the current is zero and all characteristic curves pass through the origin of the $I_D - V_D$ axes. The drain current is also zero for all V_D when $(V_G = V_{\text{bi}} - V_p)$. By assuming that V_{bi} is much smaller than V_p and for $V_G = -V_p$, the current is zero, V_{SAT} is zero, and the locus of the current curve is the V_D axis. Based on Eq. (10.2), the above conclusion indicates that setting $V_G = -V_p$ causes the channel to be completely pinched-off from source to drain at all V_D, its thickness and cross sectional area are both zero, its resistance becomes infinite, and the current is zero.

We define the *threshold voltage or turn-OFF voltage*, V_T, as that value of V_G that makes $V_{\text{SAT}} = 0$ and $I_D = 0$. This value of V_G turns the transistor OFF. The V_D axis represents *cutoff* for $V_G \leq V_T$, as shown in Fig. 10.5, so that the expression for V_T is determined from Eq. (10.3) as

$$V_T = V_{\text{bi}} - V_p \tag{10.15}$$

Since the current in the saturation region has been assumed to be independent of V_D, we determine the expression for that current, I_{SAT}, by substituting the expression for V_{SAT} from Eq. (10.2), $(V_p + V_G - V_{\text{bi}})$, in Eq. (10.14) to yield

$$I_{\text{SAT}} = G_0 \left[V_p - V_{\text{bi}} + V_G - \frac{2V_p}{3} + \frac{2V_p}{3}\left(\frac{V_{\text{bi}} - V_G}{V_p}\right)^{3/2} \right] \tag{10.16}$$

Equation (10.16) shows the dependence of the drain current in the saturation region on the voltages V_G and V_{bi}, and on the physical properties of the device represented by G_0 and V_p.

On the characteristic curves of Fig. 10.5, the cutoff, linear, and saturation regions are clearly identified. The locus of the boundary between the linear and saturation regions is shown by the dotted line, which represents the locus of all the pinchoff points.

REVIEW QUESTIONS

Q10-1 For a fixed V_D, explain how the drain current in an N-channel JFET is reduced as the gate voltage is made more negative.

Q10-2 For a fixed gate voltage, explain how the drain current is increased as the drain voltage is increased.

Q10-3 Briefly define and explain in an equation the significance of the pinchoff voltage.

Q10-4 Explain the reason for using the gradual channel approximation.

Q10-5 What is the difference between the pinchoff voltage and the threshold voltage?

Q10-6 Explain how there is drain current after the channel is pinched off.

HIGHLIGHTS

- The JFET is basically a 3-terminal device, which contains, in its simple form, a terminal from which carriers originate, a channel through which they drift, a second terminal that collects the carriers, and a third terminal that controls the flow of the carriers. The sender is the source, the receiver is the drain, and the control is exerted by the gate.

- The drain, in an N-channel device, is biased positively with respect to the source. The gate, which in the N-channel JFET is a P region, is biased negatively with respect to the source and sits on top of the channel. By controlling the gate-to-drain voltage, hence the gate to the channel voltage, a depletion region is formed in the channel that constricts the depth of the channel throughout its length. This process changes the cross-sectional area of the channel, hence the resistance of the channel, and consequently the current through the channel.

- It is to be recalled that for all combinations of gate and drain voltages, including $V_G = 0$, $V_D = 0$, a depletion region exists and account must be taken of the built-in voltage.

- For very low values of V_D and a variety of values of V_G, the relation between the drain current and drain voltage is linear and the device may be used as a variable resistor. The value of this resistor is controlled by V_G.

- An increase of V_D beyond the linear region for a fixed V_G results in further constriction of the channel and an increase of current accompanies the increase of V_D at a slower rate as V_D increases. This process continues until the channel is pinched-off at the drain end and the current reaches its saturation value I_{SAT}, corresponding to a drain voltage V_{SAT}.

EXERCISES

E10-1 An N-channel double-gate silicon JFET, operating at 300K, has $N_D = 10^{15} \text{cm}^{-3}$, $N_A = 5 \times 10^{18} \text{cm}^{-3}$, $a = 1.5\mu\text{m}$, $L = 10\mu\text{m}$, and $Z/L = 5$. Determine: a) the built-in voltage and b) the pinchoff voltage.

Ans: a) $V_{bi} = 0.816\text{V}$ b) $V_p = 1.72\text{V}$.

E10-2 An N-channel double-gate silicon JFET operating at 300K has $N_D = 10^{16} \text{cm}^{-3}$, $N_A = 10^{19} \text{cm}^{-3}$, $a = 0.5\mu\text{m}$, $L = 25\mu\text{m}$, $Z = 0.05\text{cm}$, and $\mu_n = 1200\text{cm}^2/\text{V} - \text{s}$. Determine: a) the drain current for $V_D = V_p$ with the gate connected to the source, b) the gate voltage at which, for all V_D, the transistor is OFF.

Ans: a) $I_D = 0.21\text{mA}$,   b) $V_G = -1.02\text{V}$.

10.3 CHANNEL CONDUCTANCE AND JFET TRANSCONDUCTANCE

The *channel conductance*, defined as the slope of the $I_D - V_D$ relations at a certain V_G, is

$$g_d \equiv \partial I_D / \partial V_D \big|_{V_G = \text{constant}} \tag{10.17}$$

By differentiating Eq. (10.14) at constant V_G, we obtain the dependence of g_D on V_D and V_G as

$$g_d = G_0 \left\{ 1 - [(V_{bi} + V_D - V_G)/V_p]^{1/2} \right\} \tag{10.18}$$

In the linear region and for values of $V_D \ll (V_{bi} - V_G)$, Eq. (10.18) becomes

$$g_d = G_0 \left\{ 1 - [(V_{bi} - V_G)/V_p]^{1/2} \right\} \tag{10.19}$$

Equation (10.19) represents the slope of the characteristics near the origin, which depends on FET constants and V_G. Thus, the device may be viewed as a variable resistor whose resistance is controlled by V_G. Obviously, in the saturation region, and since the current is constant, the channel conductance is zero.

The transconductance, g_m, relates the change in the drain current to the change of the gate voltage at constant V_D. In the saturation region, we find g_m by differentiating I_{SAT} in Eq. (10.16), with respect to V_G at constant V_D, so that

$$g_m \equiv \left. \frac{\partial I_D}{\partial V_G} \right|_{V_D = \text{constant}} = G_0 \left[1 - \left(\frac{V_{bi} - V_G}{V_p} \right)^{1/2} \right] \tag{10.20}$$

This expression is identical to the equation for the channel conductance given by Eq. (10.19). It is to be observed that the largest value of g_m and the largest value of g_D are obtained when $V_G = 0$. The transconductance, g_m, is an important parameter as it is a measure of the voltage gain obtained when using a BJT or a JFET as an amplifier.

In the following example, we illustrate the application of some of the relations.

EXAMPLE 10.2

For the JFET of Example 10.1, determine:

a) The channel resistance for $V_G = 0$ at $V_D = 0$.

b) The saturation drain current for $V_G = -2V$.

c) The transconductance at $V_G = -2V$.

Solution

a) From Example 10.1, we have

$$V_{bi} = 0.838V, G_0 = 4.16 \times 10^{-3}S, \text{ and } V_p = 7.65V$$

By using Eq. (10.18) at $V_G = 0$ and $V_D = 0$, we find

$$g_d = 4.16 \times 10^{-3} \left[1 - \left(\frac{V_{bi}}{V_p} \right)^{0.5} \right] = 2.78 \times 10^{-3}S$$

The channel resistance $r_d = 1/g_d = 360\Omega$

b) By using Eq. (10.16), we have for the saturation current at $V_G = -2V$

$$I_{SAT} = 4.16 \times 10^{-3} \left[7.65 - 0.838 - 2 - \frac{15.3}{3} + \frac{15.3}{3} \left(\frac{2.838}{7.65} \right)^{1.5} \right]$$

$$I_{SAT} = 3.58mA$$

c) By using Eq. (10.20), we determine the transconductance at $V_G = -2V$

$$g_m = 4.16 \times 10^{-3} \left[1 - \left(\frac{2.838}{7.65} \right)^{0.5} \right] = 1.62 \times 10^{-3}S$$

10.4 SECONDARY EFFECTS

Channel-Length Modulation

In the previous section, we concluded that at the onset of pinchoff, the drain current becomes I_{SAT} and remains at that value as V_D increases beyond V_{SAT}. However, measured characteristics shown in Fig. 10.8 indicate that a gradual and slow increase of drain current accompanies an increase of V_D beyond pinchoff.

The nonzero slope of the $I_D - V_D$ characteristic, which is exhibited in the saturation region, is a result of the decrease of the effective channel length as V_D is increased beyond V_{SAT}. The decrease in effective L, resulting from the extension of the depleted isolation region towards the source, reduces the channel resistance and causes the current to increase. This increase is confirmed by Eq. (10.16), in which G_0 is inversely proportional to L. This effect is in a way analogous to base-width modulation of the BJT and is known as *channel-length modulation*.

Breakdown

Avalanche breakdown occurs in a JFET when the reverse bias on the gate-channel junction, at the drain end of the channel, equals the breakdown voltage of the junction, so that

$$V_{br} = V_D - V_G - V_{bi} \qquad (10.21)$$

where V_{br} is the magnitude of the breakdown voltage determined by the physical properties of the junction and given by Eq. (7.11). Breakdown results in a very sharp

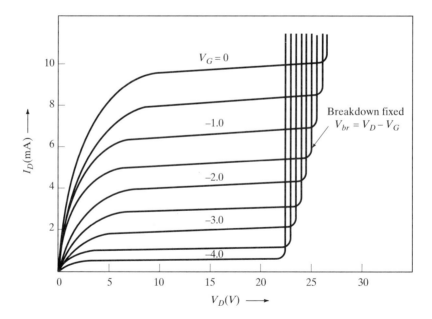

Figure 10.8 Measured characteristics of an NJFET, indicating a nonzero slope in the saturation region and breakdown.

increase of the current and, furthermore, as shown in Fig. 10.8, at more negative values of V_G breakdown occurs at lower values of V_D, as evidenced by Eq. (10.21).

Variation in Mobility

One of the assumptions that we made in deriving the expression for the current-voltage characteristic is that the mobility of the electrons in the channel is constant and therefore not dependent on the voltages applied to the channel.

The assumption is quite valid for long channels, those identified as having $L \gg a$. However, at smaller values of L and for high drain-source voltages in saturation, the electric field intensity in the x-direction of Fig. 10.6 is high and the mobility decreases with increasing field intensity.

This effect was illustrated in Fig. 4.3, which showed the variation of drift velocity with electric field intensity for silicon. We observed that at values of field intensity up to about 10^4V/cm, the drift velocity in silicon increased in a linear manner with an increase in field intensity, so that the mobility was assumed to be constant. At higher values of field intensity, the slope of the curve decreased, indicating a decrease of mobility. At a value of field intensity of approximately 8×10^4V/cm, the velocity saturates at the thermal value and the relationship of $v = \mu\mathcal{E}$ is no longer valid.

Therefore, in a JFET that has a short channel, and for a fixed drain voltage, the higher electric field intensity decreases the mobility. This leads to lower values of drain current through a decrease of G_0 in Eq. (10.14).

Temperature Effects

The major influence of temperature on the current-voltage characteristic is exhibited through the decrease of mobility that accompanies an increase of temperature. Mobility is determined by carrier scattering and the scattering at higher temperatures causes a decrease in the mobility, which results in a small decrease of current at high temperatures. This is in contrast to the effect of temperature on the collector current of a BJT, which increases with temperature and in some cases may lead to thermal runaway of the device.

10.5 SMALL-SIGNAL EQUIVALENT CIRCUIT

The small-signal low-frequency equivalent circuit represents the operation of the transistor as changes in the gate and drain voltages are made about an operating point, on the characteristic, which is determined by I_D, V_G, and V_D. These changes are initiated by a change in V_G, v_g, which causes the changes in I_D and thereby V_D. In general, we write

$$I_D = I_D(V_G, V_D) \tag{10.22}$$

Each of the variables in Eq. (10.22) is the sum of its value at the operating point plus a small incremental change. We illustrate this by the drain current i_D, expressed as

$$i_D = I_D + i_d = I_D(V_D + v_{ds}, V_G + v_{gs}) \tag{10.23}$$

where I_D is the operating point DC current, i_d is the incremental change in the current, v_{ds} and v_{gs} are the incremental changes in the drain and gate voltage respectively, and i_D is the total instantaneous current. By using Eqs. (10.22) and (10.23), the current change becomes

$$i_d = I_D(V_D + v_{ds}, V_G + v_{gs}) - I_D(V_D, V_G) \tag{10.24}$$

By expanding the first term on the right-hand side of Eq. (10.24) and subtracting the second term, Eq. (10.24) becomes

$$L_d = \frac{\partial I_D}{\partial V_G}\bigg|_{V_D}^{v_{gs}} + \frac{\partial I_D}{\partial V_D}\bigg|_{V_G}^{v_{ds}} + \text{higher order terms} \tag{10.25}$$

Equation (10.25) includes non-linear terms represented by the higher order terms. It can be linearized by neglecting the non-linear terms, as they are much smaller than the linear terms.

Upon neglecting the higher order terms in Eq. (10.25), it becomes

$$i_d = g_m v_{gs} + g_d v_{ds} \tag{10.26}$$

where g_m and g_d are the transconductance and the channel conductance respectively, which have been defined in Section 10.3. Using Eq. (10.26), we can draw the small-signal low-frequency equivalent circuit shown in Fig. 10.9. The current and voltages in Eq. (10.26) represent instantaneous values, usually, of sinusoidal waves. The equivalent circuit has been labeled as linear so that a sinusoidal input gate voltage will generate sinusoidal drain current and drain voltage. It is low-frequency since capacitances have been neglected.

10.6 FIGURE OF MERIT OF THE JFET

The Figure of Merit of an active device, such as a transistor, is a measure of both the gain and the high frequency response. The high-frequency response of a device is determined by its internal capacitances. We show in Fig. 10.10 the high-frequency equivalent circuit of the JFET. This circuit is made up of the low-frequency circuit plus the capacitances.

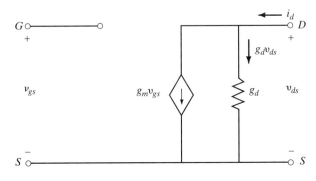

Figure 10.9 Small-signal linear low-frequency equivalent circuit of the JFET.

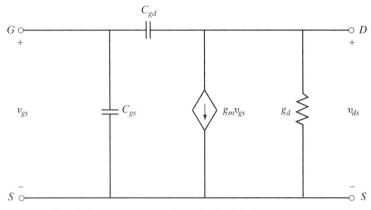

Figure 10.10   High-frequency equivalent circuit of the JFET.

In the high-frequency equivalent circuit of the FET shown in Fig. 10.10, two capacitances are shown: C_{gs} from gate to source and C_{gd} from gate to drain.

Because of the reverse-biased PN junction, a distributed junction capacitance exists all along the channel. It is, however, distributed unevenly due to the shape of the depletion layer. In the equivalent circuit, we have arbitrarily divided this capacitance into two equal parts; the gate-to-source capacitance and the gate-to-drain capacitance. It is actually the gate-to-channel capacitance.

The effect of the capacitances is to limit the high-frequency response. The distributed capacitance of the channel is very difficult to determine and one must therefore obtain an estimate of the frequency limit by approximating the total capacitance and the total channel resistance. In order to change the gate voltage, one must charge this distributed capacitance through the associated channel resistance.

For the *one-gate model*, the capacitance across the PN junction at zero gate voltage and at pinchoff has area equal to ZL and separation of average value of $a/2$, assuming a gradual linear channel depth. Hence, the capacitance C_g is

$$C_g = 2\varepsilon ZL/a \tag{10.27}$$

This capacitance replaces the distributed capacitance, and hence must be located at the center of the channel. In this case, it charges through half the channel resistance. The channel has length L and average area, $Za/2$. Hence, half the channel resistance is given by

$$R = \frac{1}{2}\rho\,\frac{L}{Za/2} = \frac{L}{q\mu_n ZaN_D} \tag{10.28}$$

The charging time constant T_C is therefore

$$T_C = RC_g = 2\varepsilon L^2/q\mu_n a^2 N_D \tag{10.29}$$

The limiting frequency f_t can be approximated by

$$f_t = \frac{1}{2\pi T_C} = \frac{q\mu_n a^2 N_D}{4\pi\varepsilon L^2} \tag{10.30}$$

The expression in Eq. (10.30) can also be obtained by determining the short-circuit current gain and then finding the frequency at which that gain is unity. By placing a short-circuit across the output of Fig. 10.10 and applying a sinusoidal gate current having an RMS value of I_i, we determine the equation for the short-circuit current gain, I_o/I_i, where I_o is the RMS value of the current through a short-circuit placed from drain to source. By neglecting the current through C_{gd}, as compared to $g_m V_{gs}$, where V_g is the RMS value of v_g, the current gain becomes

$$I_o/I_i = g_m/j\omega(C_{gs} + C_{gd}) \tag{10.31}$$

The magnitude of the current gain is unity at the frequency given by

$$f_T = \frac{g_m}{2\pi(C_{gs} + C_{gd})} \tag{10.32}$$

The expression for f_T, given by Eq. (10.32), can be reduced to that given by Eq. (10.30) by making the following substitutions: replace g_m by the maximum value it can have and that occurs at $V_G = 0$, replace $2a$ by a for the one-junction model, neglect the small value of V_{bi}, and replace the sum of the two capacitances by the expression for C_g in Eq. (10.27). The expression for f_T becomes

$$f_T = \frac{q\mu_n a^2 N_D}{4\pi\varepsilon L^2} = f_t \tag{10.33}$$

The above expression is also labeled the *gain-bandwidth product*, f_T, is also known as the *cutoff frequency*.

Compared to the carrier transit time, the RC time constant given by Eq. (10.29) is the feature that sets the high-frequency limit. The transit time of a carrier, however, is dependent upon the important parameters shown in Eq. (10.33). The transit time from source to drain, assuming a constant drift velocity $\mu_n \mathscr{E}_x$ and assuming a uniform field so that $\mathscr{E}_x = V_D/L$, has the very approximate form given by the ratio of the channel length to the velocity as

$$t_{tr} = \frac{L}{v_d} = \frac{L}{\mu_n \mathscr{E}_x} = \frac{L^2}{\mu_n V_D} \tag{10.34}$$

where V_D is the drain-source voltage and v_d is the mean drift velocity. Of course, the derivation of Eq. (10.34) is based on the assumption of constant mobility. In the region where the velocity has saturated at v_s, the transit time becomes L/v_s. The transit time is usually small compared to the RC time constant.

10.7 HIGH-FREQUENCY LIMITATIONS

The high-frequency limit of operation of a JFET is dependent on the dimensions and physical constants of the transistor. To improve the high-frequency response, let us examine Eq. (10.29) and note how the parameters have to vary.

1. Decreasing the channel length, L, decreases the capacitance and increases g_m; hence, there is an improved gain-bandwidth product. From the point of view of transit time, decreasing L has two desirable effects: (a) At a fixed V_D, a lower L increases $\mathscr{E}_x$ and hence increases the drift velocity. (b) The transit time

decreases for a fixed conduction velocity as L is reduced. These two effects account for the presence of L^2 in Eqs. (10.29) and (10.30). The smallest value of L that one can use is limited by two factors:

i) Technological difficulties of fabrication have so far limited the lowest value of L achievable.

ii) At very low values of L, and hence a high value of $\mathscr{E}_x$, the mobility ceases to be a constant and begins to decrease with an increase of V_D.

2. The use of semiconductors with a high mobility of carriers increases f_r. A higher mobility implies a higher velocity and less transit time for the carriers.

3. When we examine Eq. (10.33), we note that if the channel doping is increased, the high-frequency response is enhanced. This is true so long as the channel conductivity is not too high (high mobility is incompatible with high conductivity).

REVIEW QUESTIONS

Q10-7 At values of V_D greater than V_{SAT} and for short-channel JFETs, two phenomena work in opposite directions, one to increase the current and another to decrease it. Explain.

Q10-8 What is the major source of capacitance in the equivalent circuit of the JFET?

Q10-9 Why is the gate-to-source connection open-circuited in the small-signal equivalent circuit?

HIGHLIGHTS

- Increase of V_D beyond pinchoff causes a gradual increase of the current, based on the assumptions listed, because of channel-length modulation.
- Breakdown at the reverse-biased diode between gate and channel takes place beyond saturation when the drain-to-gate voltage causes a sufficiently large electric field that is normal to the direction of current flow.
- The gate voltage controls the drain current just as the emitter-base voltage controls the collector current in a BJT. A BJT has a base current whereas the JFET has a negligible gate current, it being the current of a reverse-biased diode.
- The equivalent circuit of the JFET at low frequencies includes a transconductance g_m and an output conductance g_d. The transconductance is the rate of change of drain current with gate voltage at the DC operating point in the saturation region of the output characteristics. The output conductance is a measure of the change of the drain current as the drain voltage is changed at an operating point. The equivalent circuit is relevant when the device is used as an amplifier where operation is in the saturation region.

EXERCISES

E10-3 A double-gate N-channel silicon JFET operating at 300K has $N_A = 10^{19} \text{cm}^{-3}$, $N_D = 10^{16} \text{cm}^{-3}$, and $a = 1.5\mu\text{m}$. Determine: a) the built-in voltage, b) the effective channel width for $V_G = 0$ and $V_D = 0$, c) the effective channel width for $V_G = -1\text{V}$ and $V_D = 3\text{V}$.

Ans: a) $V_{bi} = 0.894V$ c) $W = 2\mu m$
b) $W_B = 2.32\mu m$

E10-4 a) Determine an expression for the maximum value of g_m of a NJFET as V_G is changed. Neglect the built-in voltage.

b) Compare the ratio of g_m to the current at maximum g_m to that of a BJT. For the NJFET, use the drain current, and for the BJT, use the collector current.

Ans: a) $g_m = G_O$
b) For FET $g_m/I = 3/V_p$ and for BJT $g_m/I = 38.6$

EXAMPLE 10.3

For the silicon NJFET of Example 10.1, determine:

a) The capacitance C_g.

b) The gain-bandwidth product f_T.

Solution

a) From Eq. (10.27) $C_g = \dfrac{2\varepsilon ZL}{a} = \dfrac{2 \times 11.8 \times 8.854 \times 10^{-14} \times 250 \times 10^{-4} \times 25 \times 10^{-14}}{10^{-4}}$

$C_g = 1.305pF$. The corresponding charging resistance is calculated from Eq. (10.28) to be 480.7ohms.

b) From Eq. (10.30)

$$f_T = \frac{q\mu_n a^2 N_D}{4\pi\varepsilon L^2}$$

$$= \frac{1.6 \times 10^{-19} \times 1300 \times 10^{-8} \times 10^{16}}{4\pi \times 11.8 \times 8.854 \times 10^{-14} \times 625 \times 10^{-8}}$$

$$f_T = 253MHz$$

PROBLEMS

Unless otherwise indicated, all devices are double-gate silicon at $T = 300k$.

10.1 An N-channel silicon JFET has a gate doping of $5 \times 10^{18}cm^{-3}$, a channel doping of $10^{17}cm^{-3}$, and width $a = 0.2\mu m$. Determine:

   a) the pinchoff voltage.

   b) the gate bias required to make the width of the undepleted channel equal to $0.15\mu m$ at $V_D = 0$.

10.2 An N-channel silicon JFET has $N_D = 5 \times 10^{15}cm^{-3}$, $N_A = 10^{17}cm^{-3}$, and $a = 1.2\mu m$. Determine:

   a) the built-in voltage.

   b) the pinchoff voltage.

10.3 A P-channel GaAs JFET has $N_D = 5 \times 10^{18}cm^{-3}$, $N_A = 10^{17}cm^{-3}$, and $a = 0.2\mu m$. Determine:

a) the built-in voltage.

b) the pinchoff voltage.

c) the gate bias required to make the width of the undepleted channel equal to 0.15μm at $V_D = 0$.

10.4 An N-channel silicon JFET has $N_D = 5 \times 10^{16} \text{cm}^{-3}$, $N_A = 10^{18} \text{cm}^{-3}$ and $a = 0.3\mu\text{m}$. Determine:

a) the built-in voltage.

b) the value of V_G that will cause pinchoff at the drain for $V_D = 2V$.

c) the width of the undepleted channel for $V_G = -1.5V$, $V_D = 0$.

10.5 Use the data of Prob. 10.4 to determine an expression for the drain resistance of a JFET in terms of G_0 at $V_D = 0$ and

a) $V_G = -V_p/4$

b) $V_G = -V_p/9$

c) $V_G = -2V_p$

10.6 An N-channel silicon JFET has $N_A = 10^{19} \text{cm}^{-3}$, $N_D = 10^{16} \text{cm}^{-3}$, $a = 1\mu\text{m}$, $L = 25\mu\text{m}$, and $Z = 1\text{mm}$. Determine:

a) the built-in voltage.

b) the pinchoff voltage.

c) the drain current at $V_G = -2V$ and $V_D = 3V$.

d) I_{SAT} at $V_G = -2V$.

10.7 Experimentally determined NJFET characteristics in saturation can be roughly approximated by the expression

$$I_{SAT} = I_{DSS} (1 + V_G/V_p)^2 \text{ for } V_G < 0, V_p > 0$$

where I_{DSS} is the saturation drain current at $V_G = 0$. Use the data of Prob. 10.6 to determine I_{DSS} from Eq. (10.14) and repeat part (d) of Prob. 10.6. Comment on the accuracy of this approximation.

10.8 A P-channel silicon JFET has $a = 2.5\mu\text{m}$, $L = 8\mu\text{m}$, $Z = 300\mu\text{m}$, $N_A = 10^{15} \text{cm}^{-3}$, and $N_D = 2 \times 10^{18} \text{cm}^{-3}$. Use $\mu_p = 400 \text{cm}^2/V - s$ in the channel. Determine:

a) The pinchoff voltage.

b) The saturation drain current at $V_G = 1V$.

c) V_{DS} at the current of part (b).

10.9 An N-channel silicon JFET has $Z = 24\mu\text{m}$, $L = 4\mu\text{m}$, $a = 1.2\mu\text{m}$, $N_A = 10^{19} \text{cm}^{-3}$, $N_D = 5 \times 10^{15} \text{cm}^{-3}$. Use $\mu_n = 1200 \text{cm}^2/V - s$.

At $V_G = -3V$, determine:

a) the drain current at saturation.

b) the drain voltage at saturation.

10.10 An N-channel silicon JFET is to be used as a voltage-variable resistor at very low values of V_D. The channel has resistivity of 5 ohm-cm, $\mu_n = 1200 \text{cm}^2/V - s$, $N_A = 10^{17} \text{cm}^{-3}$, $a = 4\mu\text{m}$, $Z = 10\mu m$, and $L = 15\mu\text{m}$. Determine:

a) the minimum resistance achievable.

b) the value of $(-V_G/V_p)$ that is required to double this resistance.

10.11 An N-channel silicon JFET has $N_D = 5 \times 10^{15} \text{cm}^{-3}$, $N_A = 10^{17} \text{cm}^{-3}$, and $a = 1.2 \mu\text{m}$. Determine the ratio Z/L to realize $I_{SAT} = 2\text{mA}$ at $V_G = 0$. Assume the channel resistivity is 3 ohm-cm.

10.12 An N-channel silicon JFET has $V_p = 4\text{V}$, $V_{bi} = 0.8V$ and $I_{DSS} = 1mA (V_G = 0)$.

   a) G_0.

   b)   the saturation current at $V_G = -2V$.

   c)   the transconductance at $V_G = -2V$.

chapter 11

METAL-SEMICONDUCTOR JUNCTIONS AND DEVICES

11.0 INTRODUCTION

Metal-semiconductor junctions have long been used as linearly conductive (ohmic) metallic connections to devices and in integrated circuits. Depending on the metal and on the type of semiconductor, such junctions may pass current easily into and out of a junction (labeled *ohmic contacts*) or they may be rectifying in allowing easy current flow in only one direction. The rectifying property has made possible both the rectifier diode, known as the *Schottky barrier diode*, and the MESFET (metal-semiconductor FET).

We are interested in studying the current voltage characteristics and properties of such a junction. We begin, just as we did with the PN diode, by examining the properties of the contact, such as the potential barrier, the depletion layer, and the mechanisms of carrier transfer. These basic properties are uniquely determined by the energy barriers that are formed when a metal and a semiconductor are brought into intimate contact.

Fabrication of the Schottky diode and the MESFET is discussed in Sec. 11.6.

11.1 ENERGY-BAND DIAGRAMS OF METAL AND N-SEMICONDUCTOR

Before Contact

The band diagrams for a metal, such as gold, and an N-type semiconductor (N silicon) that are not in contact are shown in Fig. 11.1. In order to compare the relevant energy levels within and between the two solids, we seek an energy level that is common to both solids and that is fixed with respect to both. That energy level is the vacuum level E_0. The vacuum level, E_0, is defined as the energy that an electron is

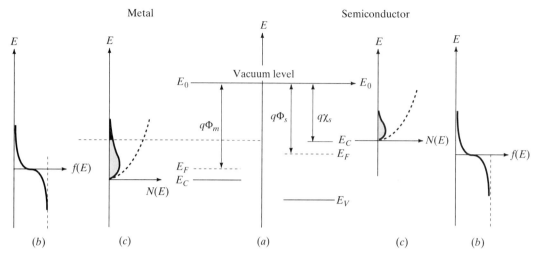

Figure 11.1 (a) Energy levels of metal (gold) and semiconductor (silicon) (not in contact). (b) Fermi function. (c) Carrier density distribution function. Above the E_C level of silicon and relative to E_0, there is a larger density of electrons in silicon than in gold. Crossed areas are filled with electrons..

assumed to have if it were at rest outside and just free of the solid. We remind the reader that E_0 is the zero level of energy and all other energy levels shown in the figures represent negative energies.

The Fermi energy represents the average energy of an electron in the system. For a metal, the Fermi energy is the average energy of the most energetic electrons. The energy difference between the vacuum level and the Fermi level is labeled the *work function* of the solid. The work function is therefore defined as the energy required to move an electron from the Fermi level to E_0, where it is at rest and free of the influence of the solid.

In an extrinsic semiconductor, the location of the Fermi level is determined by the degree of doping of the semiconductor. Therefore, the Fermi level is not located at a fixed level with respect to the conduction and valence bands. In a given metal, however, the Fermi level is located slightly above the bottom of the conduction band and at a fixed energy separation from the vacuum level E_0. In a semiconductor, the bottom of the conduction band, E_C, is located at a fixed energy separation from E_0. This energy difference between E_0 and E_C in the semiconductor is labeled the *electron affinity*, which is denoted by $q\chi_s$, where the symbol χ_s is the Greek letter chi. The affinity is therefore the energy needed to move an electron from the bottom of the conduction band and place it at rest outside the solid.

From Fig. 11.1, we note that the work function for the metal is $(E_0 - q\Phi_m)$ and for the semiconductor is $(E_0 - q\Phi_s)$, where $q\Phi_s$ is the energy location of the Fermi level below E_0. We observe that if Φ_m is less than Φ_s while the solids are separated, then the electrons in the metal have, on the average, a total energy that is higher than that of the electrons in the semiconductor. On the other hand, if Φ_m is greater than Φ_s, the electrons in the semiconductor have an average total energy that is greater than the electrons in the metal. In Fig. 11.1, we have used a metal and a semiconductor such that $\Phi_m > \Phi_s$, so that the work function of N-type silicon is smaller than that of gold ($\Phi_m = 4.75$eV) and the affinity of silicon is 4.15eV.

Also in Fig. 11.1, we have shown the Fermi distribution function variation versus energy, centered at E_F, which shows the probability of occupancy, $f(E)$, of a state, at energy E, by an electron. The density of states distributions functions, $N(E)$ (refer to Sec. 3.1), shown for electrons, have their zero energy level located at the level of the bottom of the conduction bands of the metal and of the semiconductor. The distributions of the densities of electrons are formed from the product $N(E)f(E)$ and are shown within the solid lines of the curves of $N(E)$. In order to compare at the same energy level the densities of electrons in the metal to those of the semiconductor, the darkened areas are shown measured above the bottom of the conduction band of the semiconductor. Since the work function of the semiconductor is smaller than that of the metal and the average energies of electrons in the semiconductor are higher than those of the metal, it stands to reason that above the energy level, E_C, of the semiconductor, the density of electrons in the semiconductor is greater than that of electrons in the metal.

At the instant of contact and since the Fermi level of the silicon is higher than that of the metal, electrons will transfer from the semiconductor into the metal until equilibrium is reached and the Fermi levels are aligned. As a result, and at thermal equilibrium, the semiconductor is charged positively with respect to the metal. The effects of intimate contact on the energy levels and energy bands is shown in Fig. 11.2. After contact and at thermal equilibrium, there will be a continuous flow of electrons at the same rate in both directions with the result that the net current across the junction is zero.

Thermal Equilibrium Conditions of Metal and N-Semiconductor after Contact—Schottky Barrier

As a result of the net transfer of electrons from the semiconductor to the metal, a depletion layer consisting of positively ionized donor atoms is established in the silicon. This is coupled with a layer of excess electrons lining the surface of the metal in contact with the semiconductor. An electric field, directed from semiconductor to metal is thus built and a potential barrier is established. This is accompanied, just as happens in the N region of a PN junction diode, by bending of the energy levels of the conduction band and of the valence band in silicon in the region of the depletion layer. Since the electron affinity is a constant of the solid, the bending of the conduction band is accompanied by an identical bending of the vacuum level in the semiconductor. Both in the metal and in the semiconductor, the locus of constant E_0 is not changed, it is still the energy that an electron has when it is free of the solid.

The bending of the conduction band level of the semiconductor is largest at the surface of contact with the metal where the electric field is largest. Since the

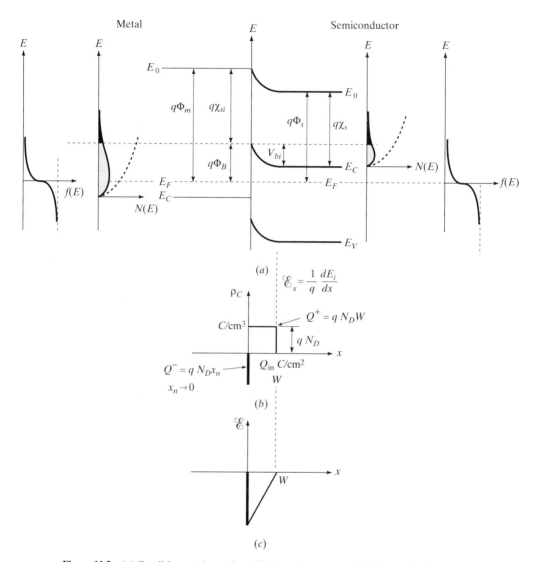

Figure 11.2 (a) Conditions at thermal equilibrium after contact. (b) Charge density distribution. (c) Electric field. Dark areas in the distribution functions represent the electrons that have enough energy to cross the junction.

Fermi energy level is fixed, the increase of energy separation between the conduction band and the Fermi level in the semiconductor near the surface results in reduced electron density at the surface with the metal. This is also the location in the conduction band where the electrons possess the highest energy.

After contact formation, the distribution of the density of electrons above the bottom of the semiconductor conduction band is identical in both the metal and the semiconductor. This is shown by the equal areas above E_C that represent the density of electrons that are able to cross the junction.

After contact and simultaneous to the bending of the bands, an energy barrier qV_{bi} is formed in the semiconductor at the surface of contact with the metal. In contrast to the depletion layer, having width $x_n = W_s$, that is formed in the semiconductor, a surface sheet of electrons is formed in the metal. There can be no depletion layer and no voltage drop in the metal. This is because the metal is assumed to be a perfect conductor with zero resistance, which does not permit the formation of an electric field and the sustaining of a voltage drop.

At thermal equilibrium there will be a flow of electrons in both directions and the net current across the junction is zero. Just as with the PN junction diode, there are potential barriers, which the electrons must overcome, on both sides of the junction. The barrier for electrons in the metal shown in Fig. 11.2 is known as a *Schottky barrier*.

One of the important characteristics of a Schottky barrier is the barrier height $q\Phi_B$, which, as we observe from Fig. 11.2, is the energy difference between the aligned Fermi levels and the semiconductor band edge at the surface with the metal. The Schottky barrier, $q\Phi_B$, represents the energy barrier that electrons in the metal must overcome to move into the semiconductor. This barrier, shown in Fig. 11.2 at the surface of contact, is written for an *ideal contact* as

$$q\Phi_B = q(\Phi_m - \chi_s) \tag{11.1}$$

The barrier for the electrons that are in the bulk of the semiconductor, which prevents them from moving into the metal and which they must overcome, is qV_{bi} and is given by

$$qV_{bi} = q(\Phi_m - \Phi_s) \tag{11.2}$$

The built-in voltage, V_{bi}, can be expressed in terms of the barrier, Φ_B, and the potential difference, Φ_D between the Fermi level and E_C, defined by Eq. (3.18) as (kT/q) $\ell n\ N_c/N_D$, where $n \cong N_D$ so that

$$V_{bi} = \Phi_B - \Phi_D \tag{11.3}$$

This barrier is smaller than the barrier Φ_B, which the electrons in the metal face. However, at thermal equilibrium and because of the larger density of electrons in the metal, as shown by the tail of the distribution curve, there are as many electrons in the metal as there are in the semiconductor that have energies greater than the barrier they face. These electrons are continuously in motion, so that there is an equal current of electrons from the metal to the semiconductor as there is from the semiconductor to the metal.

We italicized the phrase *ideal contact* earlier, prior to Eq. (11.1), to indicate that calculated values for Φ_B, using Eq. (11.1), are in general considerably smaller than actual values. This is because the surface states that are produced as a result of the disruption of the semiconductor crystal lattice when the contact is formed tend to have an important effect on the barrier height.

Measured values for the barrier height, Φ_B, of various metals on silicon and N-GaAs are shown in Table 11.1.

TABLE 11.1 Measured barrier heights for some metals on Si and N GaAs at 300K

	Aℓ	Au	W	PtSi
Si(N)	0.72eV	0.80eV	0.67eV	0.85eV
Si(P)	0.58eV	0.34eV	0.45eV	—
GaAs(N)	0.80eV	0.90eV	0.80eV	—

Source: S.M. Sze, *Physics of Semiconductor Devices*, pp. 291–92, copyright © Wiley (1981). Reprinted by permission of John Wiley & Sons, Inc.

11.2 SCHOTTKY BARRIER DIODE

Rectifying Metal-N Semiconductor Contact

We now consider conditions that ensue when the equilibrium is disturbed by an applied voltage. Since the built-in voltage at equilibrium appears only in the semiconductor and since the metal cannot sustain any voltage drop, any voltage that is applied appears entirely in the semiconductor. Therefore, the barrier height in the metal, which appears in the metal between the Fermi level and the conduction band edge of the semiconductor at the surface, is unchanged. As a consequence, the density of electrons in the metal that have energies greater than this barrier is unchanged from its equilibrium value. This is shown in Figs. 11.3 and 11.4.

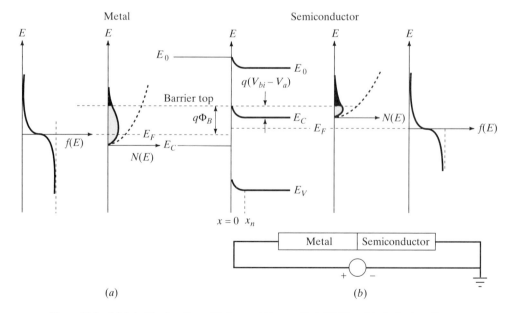

Figure 11.3 (a) Schottky junction with forward bias applied. (b) Circuit to indicate voltage reference. (Dark areas refer to electron density.)

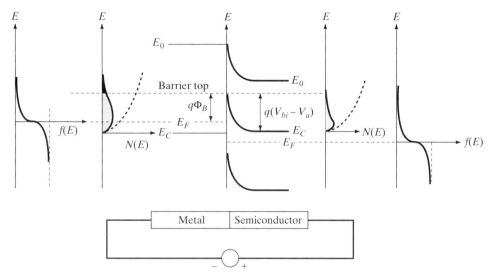

Figure 11.4   (a) Schottky junction with reverse bias applied. (b) Circuit to indicate voltage reference.

An applied voltage causes a change in the bending of the bands in the semiconductor and a corresponding change in the electric field in the semiconductor at the junction with the metal. The barrier in the semiconductor is reduced when a forward bias is applied which makes the metal positive with respect to the semiconductor. This bias reduces the electric field and the degree of bending of the bands, as shown in Fig. 11.3(a). As the barrier is reduced, more electrons, as compared to thermal equilibrium, cross from the semiconductor to the metal. On the other hand, the number of electrons that cross from the metal to the semiconductor is unchanged from the number that crosses at thermal equilibrium because the barrier height, $q\Phi_B$, they have to surmount is unaffected by the applied voltage and unchanged from its thermal equilibrium value of

$$q(\Phi_m - \chi_s).$$

A reverse bias increases the barrier height in the semiconductor, as shown in Fig. 11.4, and reduces the current of electrons compared to that at thermal equilibrium, which cross from the semiconductor to the metal. Because the metal does not sustain a voltage drop, the barrier to electron flow from the metal to semiconductor remains unchanged and hence the rate of the flow of charge from the metal to the semiconductor is the same as it was at thermal equilibrium.

We conclude that with forward bias, the current from the metal to the semiconductor, which results from electrons crossing from the semiconductor (labeled the forward current I_m), increases. With reverse bias applied, the current consists only of the electrons that cross from the metal to the semiconductor and is a reverse current.

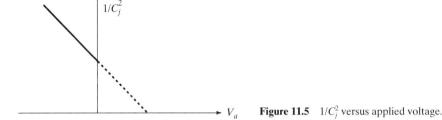

Figure 11.5 $1/C_j^2$ versus applied voltage.

Properties of Depletion Layer

The depletion layer in the semiconductor has properties that are similar to the N region of a P⁺N junction diode. Assuming an abrupt junction, where the charge density in the depletion layer for $0 < x < W$ is qN_D and the charge density and the electric field are zero for $x > W$, the expressions for the depletion layer width and the electric field are obtained from Eqs. (5.30) and (5.20) respectively as

$$W = \sqrt{(2\varepsilon/qN_D)(V_{bi} - V_a)} \text{ for } N_A >> N_D \tag{11.4}$$

$$\mathscr{E}(x) = -(qN_D/\varepsilon)(W - x) \tag{11.5}$$

where the applied voltage, V_a, is positive for forward bias and negative for reverse bias. The symbol W replaces x_n in Eq. (5.20) for based conditions.

The charge density in the depletion layer, assuming it consists uniformly of donor ions, and the corresponding capacitance per unit area are obtained as

$$Q_n = qN_D W = \sqrt{2q\varepsilon N_D(V_{bi} - V_a)} \tag{11.6}$$

$$C_j = \left| \frac{dQ_n}{dV_a} \right| = \sqrt{(q\varepsilon N_D)/2(V_{bi} - V_a)} = \varepsilon/W \tag{11.7}$$

Equation (11.7) can be rewritten in the following form

$$(1/C_j^2) = 2(V_{bi} - V_a)/q\varepsilon N_D \tag{11.8}$$

Assuming a uniform N_D throughout the depletion layer, a plot of $1/C_j^2$ versus the applied voltage, V_a, is shown in Fig. 11.5. The intercept on the abcisse corresponds to the built-in voltage and the slope of the line is a measure of the doping density, as shown by Eq. (11.8).

EXAMPLE 11.1

The silicon in an aluminum-silicon Schottky barrier diode has $N_D = 10^{15} \text{cm}^{-3}$. Determine the built-in voltage, V_{bi}.

Solution We determine Φ_D from the density of states in the conduction band N_c, and the doping density N_D as

$$\Phi_D = (kT/q) \, \ell n \, (N_c/N_D)$$

$$\Phi_D = .0259 \, \ell n \, \frac{3.22 \times 10^{19}}{10^{15}} = 0.269 \text{V}$$

The built-in voltage is determined from Eq. (11.3) by using Table I.1 to obtain Φ_B.

$$V_{\text{bi}} = \Phi_B - \Phi_D = 0.72 - 0.269 = 0.451 \text{V}$$

Rectifying Metal-P Semiconductor Junction

The discussion so far has been limited to conditions at the contact between a metal and an N semiconductor. At that contact, the work function of the metal is greater than that of the semiconductor. We concluded that a barrier exists at this contact and is rectifying, allowing passage of current in only one direction. In a later section, we will show that when the work function of the metal is smaller than that of the N semiconductor, the contact is nonrectifying and is labeled ohmic. An ohmic contact provides no barrier to the flow of current and allows easy current flow in both directions—from metal to semiconductor and from semiconductor to metal.

We will now consider the contact between a metal and a P semiconductor, wherein $\Phi_m < \Phi_s$ and the work function of the metal is smaller than that of the semiconductor. Energy-band diagrams for the metal and semiconductor before contact are shown in Fig. 11.6(a).

Upon intimate contact, and because of the smaller work function of the metal, electrons move from the metal into the semiconductor and the Fermi levels are aligned.

The electrons that move into the semiconductor recombine with majority carrier holes, leaving the negatively charged uncompensated acceptor ions in the semiconductor to form a depletion layer whose width depends upon the doping of the semiconductor. On the metal side, a layer of positive charge is formed at the interface with the semiconductor. An electric field directed from the metal to the semiconductor and an energy barrier are formed. The downwards bending of the energy bands in the semiconductor corresponds to an upward bending of the potential and the built-in voltage, V_{bi}, in the semiconductor is equal to $(\Phi_s - \Phi_m)$. The contact is rectifying and the barrier from the metal to the semiconductor valence band, $q\phi'_B$, is seen from Fig. 11.6(b) to be $(q \, \chi_s + E_g - q \, \Phi_m)$. This is the energy separation between the aligned Fermi level and the surface of the valence band in silicon, in contact with the metal. Thus, the barrier, qV_{bi}, is the energy that holes in the bulk of the P region need to move into the metal.

Here, we remind the reader that the electrostatic potential barrier for hole motion is opposite to the direction of the barrier in the electron energy band diagram. Therefore, the highest energy level of a hole in the semiconductor is at the surface where the valence band curve meets the metal surface. While holes at that surface have the highest energy, they are not numerous because of the larger separation of this level from the Fermi level compared to the separation experienced by holes in the bulk of the semiconductor.

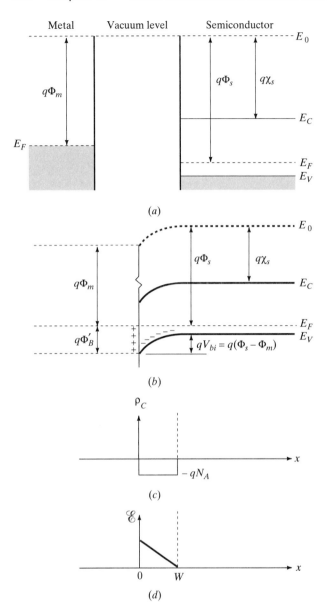

Figure 11.6 Metal P- semiconductor with $\Phi_m < \Phi_s$ (a) before contact, (b) thermal equilibrium after contact, (c) depletion layer charge density, and (d) electric field.

A comparison between the energy band diagrams for a rectifying metal-N semiconductor contact and a metal-P semiconductor contact is shown in Fig. 11.7.

When a voltage, V_a, is applied that makes the P semiconductor positive with respect to the metal (forward bias), the barrier to hole flow from the semiconductor to the metal is reduced. The reduction of barrier height is shown in Fig. 11.7. When the bias is reversed, making the metal positive, the barrier to hole flow from the

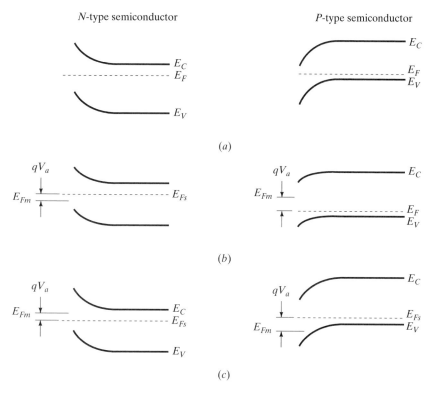

Figure 11.7 Energy band diagrams for metal N-semiconductor ($\Phi_m > \Phi_s$) and metal P semiconductor ($\Phi_m < \Phi_s$) at (a) equilibrium, (b) forward bias, and (c) reverse bias.

semiconductor to the metal is increased and the barrier in the metal at the interface is unchanged so that the current is negligibly small.

In conclusion, it is important to note that in both the metal-N rectifying contact and the metal-P rectifying contact, the forward current is due to the injection of majority carriers from the semiconductor to the metal. This is in contrast to the PN junction diode where the currents consist of the injection of minority carriers.

11.3 CURRENT-VOLTAGE CHARACTERISTICS OF METAL N-SEMICONDUCTOR SCHOTTKY DIODE

Simplifying Assumptions

Before proceeding with analytical relations for the current-voltage characteristics and in order to simplify the analysis, we make the following assumptions:

1. All electrons in the conduction band move with the same average thermal velocity and the motions are randomly distributed in direction.
2. All electrons in the semiconductor, which are incident on the barrier, cross into the metal.

3. At the semiconductor surface, the average electron velocity component directed toward the metal has been shown* to be approximately $v_{th}/4$, where v_{th} is the thermal velocity.

4. A certain number of electrons in the metal possess sufficient thermal energy to cross the barrier and move into the semiconductor.

Thermal Equilibrium Currents

The energies of the electrons at the surface of the conduction band of the semiconductor are separated from the Fermi level by $q\Phi_B$ so that by using Eq. (3.18) the density of surface electrons, n_s, at thermal equilibrium is given by

$$n_s = N_c \exp\left(\frac{-q\Phi_B}{kT}\right) \tag{11.9}$$

where N_c is the density of conduction band states that are assumed to be all located at the bottom of the conduction band.

At thermal equilibrium and from Fig. 11.2, the relation for $q\Phi_B$ is found to be

$$q\Phi_B = qV_{bi} + (E_C - E_F) \tag{11.10}$$

where E_C is the conduction band edge in the bulk of the semiconductor.

The electron density in the bulk of the semiconductor, away from the surface, which for normal doping is equal to the density of the donor atoms N_D, is found from

$$n = N_c \exp\left[-(E_C - E_F)/kT\right] \approx N_D \tag{11.11}$$

where the relation between n and N_c is defined by Eq. (3.18).

By substituting Eq. (11.10) in Eq. (11.9), we can write

$$n_s = N_c \left[\exp\left(-qV_{bi}\right)/kT\right]\left[\exp -(E_C - E_F)/kT\right] \tag{11.12}$$

By substituting the expression for N_D from Eq. (11.11) in Eq. (11.12), we write

$$n_s = N_D \exp\left(-qV_{bi}/kT\right) \tag{11.13}$$

Since all electrons at the surface of the semiconductor in the conduction band are assumed to cross into the metal with a mean velocity of magnitude $v_{th}/4$, the current of electrons that cross from the semiconductor to the metal (directed from metal to semiconductor), I_{ms}, at thermal equilibrium is given by

$$I_{ms} = qAn_s v_{th}/4 \tag{11.14}$$

where A is the cross sectional area of the junction and v_{th} is the thermal velocity of the electrons in the direction from the semiconductor to the metal.

---

*From D. Pulfrey and N. Tarr, *Introduction to Microelectronic Devices*, 1993, p. 271, Prentice Hall. Reprinted by permission of Prentice Hall, Upper Saddle River, New Jersey.

At thermal equilibrium, the diode current is zero, so that we can set the current from the semiconductor to the metal equal and opposite to the current from the metal to the semiconductor as

$$I_{sm} = -I_{ms} = -qAn_s v_{th}/4 \qquad (11.15)$$

The expression for the currents at thermal equilibrium can also be written in terms of the doping density N_D by using Eq. (11.13).

$$I_{sm} = -I_{ms} = -qA(v_{th}/4) N_D \exp(-qV_{bi}/kT) \qquad (11.16)$$

Currents With Bias Applied

We have indicated earlier that when a voltage is applied to the junction, the voltage appears totally in the semiconductor and the barrier to electron flow from the semiconductor to the metal is changed, whereas the barrier to electron flow from the metal to the semiconductor is unchanged at $q\Phi_B$.

Applying a forward bias to the diode by connecting the metal to the positive terminal of the voltage source increases the density of the electrons that can cross from the semiconductor, while keeping the density of electrons that can cross from the metal to the semiconductor at the thermal equilibrium value. Conversely, applying a reverse bias to the junction increases the height of the barrier that electrons at the semiconductor surface must surmount in order to cross into the metal. Again, this does not change the height of the barrier that electrons in the metal face. Therefore, the current with reverse bias, which makes the metal negative with respect to the semiconductor, is the current of electrons that cross from the metal to the semiconductor at thermal equilibrium.

We will consider the positive reference direction of current to be from the metal to the semiconductor across the junction. With reverse bias applied, the diode current is from semiconductor to metal and is determined by using Eq. (11.9) in Eq. (11.14) as

$$I_{ms} = -qA(v_{th}/4)N_c \exp(-q\Phi_B/kT) \qquad (11.17)$$

When a forward bias is applied to the diode and since all the applied voltage, V_a, appears in the semiconductor, the expression for the surface density of electrons given by Eq. (11.13) can be written as

$$n_s = N_D \exp[-q(V_{bi} - V_a)/kT]$$
$$= [N_D \exp(-qV_{bi}/kT)] [\exp qV_a/kT] \qquad (11.18)$$

where V_a is the applied voltage.

By replacing the first factor in the right hand side of Eq. (11.18) by its equivalence from Eqs. (11.9) and (11.13), we have

$$n_s = [N_c \exp(-q\Phi_B/kT)] [\exp qV_a/kT] \qquad (11.19)$$

In the expression for the current of electrons that cross from the semiconductor to the metal, given by Eq. (11.14), we replace n_s by its equivalence from Eq.

(11.19), so that we have the expression for the current from metal to semiconductor at forward bias given as

$$I_{ms} = qA(v_{th}N_C)/4 \exp\left(-q\Phi_B/kT\right) \exp\left(qV_a/kT\right) \tag{11.20}$$

The current of electrons from the metal to the semiconductor at thermal equilibrium, with reverse bias or with forward bias, is always the same and is given by Eq. (11.15). The total current is the sum of the two currents, the current of electrons that cross from the metal given by Eq. (11.15) with n_s replaced by its equivalence from Eq. (11.9) and the current of electrons that cross from the semiconductor given by Eq. (11.20) so that

$$I = I_{ms} + I_{sm} = qA(v_{th}/4)N_c \exp\left(-q\Phi_B/kT\right)\left[\exp\left(\frac{qV_a}{kT}\right) - 1\right] \tag{11.21}$$

The mean thermal velocity, defined by Eq. (4.1), is proportional to $T^{1/2}$ and N_c, from Eq. (3.20), is proportional to $T^{3/2}$, so that we can write

$$I = ART^2[\exp\left(-q\Phi_B/kT\right)]\,[\exp(qV_a/kT) - 1] \tag{11.22}$$

where R is $qv_{th}N_c/4$.

Equation (11.15) can be written as

$$I = I_S[\exp\left(qV_a/kT\right) - 1] \tag{11.23}$$

where for reverse bias $I = -I_S$ and I_S is given by

$$I_S = ART^2 \exp\left(-q\Phi_B/kT\right) \tag{11.24}$$

The term R, known as the Richardson constant, depends on the electron effective mass and other physical constants and has units of $A\mathrm{cm}^{-2}\mathrm{K}^{-2}$. The value of R depends on the particular semiconductor used and has an approximate value of $110 A\mathrm{cm}^{-2}\mathrm{K}^{-2}$ in N-type silicon and an approximate value of $8 A\mathrm{cm}^{-2}\mathrm{K}^{-2}$ for N-type GaAs.

11.4 COMPARISON OF SCHOTTKY DIODE WITH P⁺N DIODE

At this point, we highlight the advantages and disadvantages of the Schottky diode by comparing it with the P⁺N diode. As we indicated earlier, most Schottky barrier diodes used in integrated circuits use platinum silicide on N silicon so that the barrier height $q\Phi_B$ is about 0.9 eV.

We will use an example to illustrate the order of magnitudes of the currents and voltages for a Schottky barrier diode and a P⁺N diode.

EXAMPLE 11.2

A Schottky barrier is formed from platinum silicide on N-Si, which has a doping density of $10^{16}\mathrm{cm}^{-3}$ and an area of $10^{-3}\mathrm{cm}^2$. A PN diode has the same area and $N_A = 10^{19}\mathrm{cm}^{-3}$, $N_D = 10^{16}\mathrm{cm}^{-3}$, and $\tau_p = \tau_n = 1\mu s$.

a) Calculate the Schottky diode current at a forward bias of 0.4V at 300K.

b) Determine the value of the forward bias that has to be applied to the PN junction diode to deliver the same current as the Schottky diode.

Solution

a) From Table 11.1, the barrier is 0.85eV. Using 110Acm^{-2}K^{-2} for the Richardson constant, the reverse saturation current of the Schottky diode is found from Eq. (11.24) as

$$I_s = 10^{-3} \times 110 \times (300)^2 \exp{(-0.85/.0259)} = 5.53 \times 10^{-11}\text{A}$$

For a forward bias of 0.4V, Eq. (11.23) predicts a current of 0.28mA.

b) For the P$^+$N diode, μ_p is calculated from Eq. (4.16) and D_p is found to be 11.56 cm^2/s. Since the diode current is essentially the hole injection current from P to N, the saturation current is given by $qD_pp_{0n}A/L_p$, where $L_p = 3.4 \times 10^{-3}$cm and $p_{0n} = 10^4$cm^{-3}, so that $I_S = 5.44 \times 10^{-15}$A. To obtain a current of 0.28mA, the diode forward voltage has to be 0.64V.

Because the Schottky diode has a much higher current density than the P$^+$N diode, its turn-ON voltage is much lower. The current voltage characteristics are compared in Fig. 11.8. It is also important to note that PtSi forms a high barrier whose advantage is a smaller reverse saturation (leakage) current than that caused by aluminum or tungsten.

One of the important applications, as demonstrated from the above calculation, is the use of the Schottky diode in low-voltage, high-current rectifiers.

The lower cut-in voltage also makes the Schottky diode a valuable device used to clamp the collector to the base of a transistor at about 0.4V (PNP), making the the emitter to the collector voltage approximately 0.3V, and thus preventing the transistor from going into deep saturation. The use of the Schottky diode connected from collector to base considerably increases the switching speed of the BJT by up

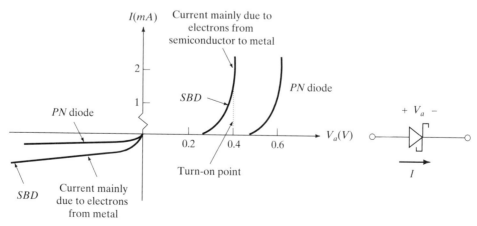

Figure 11.8   Current-voltage characteristics of PN and Schottky-barrier diodes and symbol for SBD.

to a factor of 10. As we have seen in discussing the switching of a BJT in saturation, there is a large injection of minority carriers into the base. This injection is responsible for slowing the switching. With the Schottky diode turning on at a very low forward voltage, the collector-base junction is clamped at about 0.4V and the Schottky diode provides a bypass (bypassing the base) for the excess minority carriers.

Another major advantage of the Schottky diode lies in the fact that the currents are majority carrier currents and hence the storage capacitance associated with minority carrier currents does not exist. Therefore, the transient response of the diode is determined only by the RC_j product, where C_j is the depletion layer capacitance and R is the diode series resistance. Consequently, when the diode is switched from a forward (ON) state to a reverse (OFF) state, the current tends to go to zero as soon as the driving voltage is switched to the reverse value. The reverse recovery time is therefore dependent upon RC. This product is usually about four orders of magnitude smaller than that of a corresponding P^+N junction diode.

The reverse recoveries of a P^+N and a Schottky diode are compared in Fig. 11.9.

The main disadvantage of the Schottky diode is a consequence of the property that permits a low cut-in voltage or that permits a high current for a low voltage. The

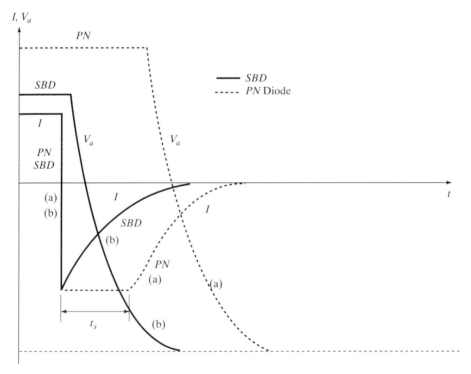

Figure 11.9 Reverse recovery transient for (a) P^+N junction diode and a (b) Schottky diode. Variations with time of currents and voltages.

disadvantage is the large current in the reverse direction, which is approximately four orders of magnitude greater than that of a P$^+$N junction diode.

REVIEW QUESTIONS

Q11-1   How is the work function of a metal determined?

Q11-2   Show, in equation form, how the work function of a P semiconductor varies with the doping.

Q11-3   The Fermi level of a metal does not change when electrons are added to the metal. Why?

Q11-4   Explain why a depletion layer is not formed in the metal at a metal-semiconductor contact.

Q11-5   Why is there no diffusion (storage) capacitance in a Schottky-barrier diode?

Q11-6   Why is the Schottky-barrier diode much faster, in switching, than the PN diode?

Q11-7   What requirement must be met to make tunnelling possible?

Q11-8   Identify the three uses to which metals are put in electronic circuits.

HIGHLIGHTS

- Metal-semiconductor contacts are either rectifying or ohmic. Rectifying contacts permit current in one direction only, whereas ohmic contacts allow easy current flow in both directions—from metal to semiconductor and from semiconductor to metal.

- Whether a contact is rectifying or ohmic is determined by the relations between certain energy levels in the semiconductor and in the metal.

- The energy levels of relevance are: The Fermi levels in both the metal and the semiconductor, and the affinity level in the semiconductor, which is the energy separation between E_0 and E_C.

- The work function is defined as the separation between E_0 and E_F. We note that for a metal, E_F is, for all practical purposes, located at E_C, whereas for a semiconductor, the work function varies with the doping because the energy level E_F is determined by the doping.

- For a metal-N semiconductor contact, wherein the metal work function is greater than that of the semiconductor, application of a bias that makes the metal positive with respect to the semiconductor causes electrons to move from the semiconductor to the metal and very few electrons to move in the opposite direction, so that a current results from metal to semiconductor. Reversing the bias does not change the height of the barrier that electrons in the metal face, so that the same number cross as at forward bias. However, very few electrons cross from the semiconductor to the metal and hence a small current results. The contact is thus rectifying and the device is a Schottky-barrier diode.

- Schottky-barrier diodes may also be formed from a metal and a P semiconductor when the metal work function is smaller than that of the semiconductor. For both the N and P semiconductors, the contacts with the metal are ohmic when the work function differences are opposite to those of the rectifying contacts.

- For the same applied voltage, the Schottky-barrier diode (SBD) conducts considerably larger forward and reverse currents than a PN junction diode.

- The PN junction diode is a minority carrier device, whereas the SBD is a majority carrier device.

EXERCISES

E11-1 An SBD formed on N-silicon is operating at 300K. Given $N_D = 2 \times 10^{15} \text{cm}^{-3}$, the affinity of silicon is 4.15eV, and the metal work function is 4.9eV. Determine: a) the built-in voltage, b) the barrier height and c) the width of the depletion layer with $V_G = 0$.

$$\text{Ans:} \quad \text{a) } V_{bi} = 0.5\text{V}$$

$$\text{b) barrier height} = 0.75\text{V}$$

$$\text{c) } W = 0.57\mu\text{m}$$

E11-2(a) Calculate the values of the two capacitances of the PN diode of Example 11.2 at $I = 5\text{mA}$. (b) Repeat (a) for the SBD.

$$\text{Ans:} \quad \text{a) } C_j = 68\text{pF}, C_s = 0.193\mu\text{F}$$

$$\text{b) } C_j = 72\text{pF}, C_s = 0$$

11.5 NON-RECTIFYING OHMIC CONTACTS

In Section 11.2, we showed that the Schottky barrier diode represents a rectifying contact just like the P$^+$N junction diodes. A rectifying metal-semiconductor junction permits easy current flow when the junction is forward-biased. A Schottky diode is formed when the metal work function is greater than that of the N semiconductor or when the metal work function is smaller than that of the P semiconductor. Such diodes, as indicated earlier, have special applications.

Another major application of metal-semiconductor junctions is in the formation of non-rectifying ohmic contacts.

In the fabrication of discrete devices and in the interconnections of integrated circuits, it is necessary to establish ohmic metallic contacts to the devices, to the connections between the semiconductor region and its external terminal, and to interconnect elements in an integrated circuit. Such contacts should not in any way interfere with the operation of the device or a circuit and therefore should exhibit negligible resistance to the flow of current into and out of the device. This resistance should be negligible compared to the resistance of the bulk regions of the device. The contact should consequently support a negligible voltage drop compared to the drop across the active region of the device. Such contacts are known as *ohmic contacts*.

Metal-Semiconductor Ohmic Contacts

An essential condition that is met by an ohmic contact is to permit easy and unopposed transfer of majority carriers between the metal and semiconductor.

An ohmic contact is formed when the metal work function is smaller than the N semiconductor work function with $\Phi_m < \Phi_s$. Energy band diagrams for the metal

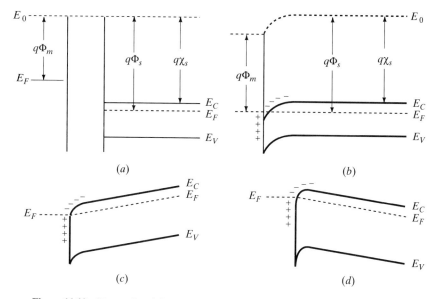

Figure 11.10   Energy band diagrams for metal N-semiconductor contact with $\Phi_m < \Phi_s$ (a) before contact, (b) after contact, (c) negative bias on the semiconductor, and (d) positive bias on the semiconductor.

and N semiconductor when isolated and when in contact are shown in Fig. 11.10(a) and Fig. 11.10(b).

Upon intimate contact and, because of the smaller work function of the metal, electrons flow from the metal to the semiconductor until thermal equilibrium is reached and the Fermi levels are aligned. Negative charges of electrons accumulate on the semiconductor surface adjacent to the metal and the metal surface becomes positively charged. After the alignment of the Fermi levels, a potential drop equal to $\Phi_m - \Phi_s$ is developed across the semiconductor and the energy bands in the semiconductor bend downwards at the surface.

The electrons that crossed from the metal to the semiconductor settled in the semiconductor at the junction. No depletion region exists in the semiconductor since there is no barrier to electron flow from the semiconductor to the metal nor in the opposite direction.

We observe an important difference between the ohmic contact and the Schottky contact, both made from a metal and an N semiconductor, in that the energy bands bend downward in the ohmic contact while they bend upward in the Schottky contact. The difference is manifested also in that the semiconductor charge consists of free electrons in the ohmic contact while positive donor ions accumulate at the surface of the rectifying contact. In the ohmic contact and because of the presence of excess electrons in the N semiconductor surface, the majority carrier concentration is enhanced near the surface compared to the bulk. Since no barrier is developed in the semiconductor, an applied bias appears across the neutral semiconductor, as shown in Fig. 11.10(c) and (d). As a follow up to the previous discussion, and for a metal-P semiconductor contact with $\Phi_m > \Phi_s$, there is no barrier to hole flow from the semiconductor to the metal and a negligible barrier to hole flow in the opposite direction. We conclude that such a contact is ohmic.

Tunnelling at a Metal-Semiconductor Contact

A more commonly used method for establishing an ohmic contact between a metal and a semiconductor is to make the contact favorable to the tunnelling of electrons in both directions. The contact is favorable when electrons are not required to climb a barrier, as they can cross directly from the metal to the conduction band of the semiconductor and also cross in the reverse direction.

At the end of Example 1.4 we considered an electron located in Region 1, having energy W separated from Region 3 by Region 2, which has a barrier of height E greater than W. We concluded that there is a finite probability for the electron to cross into Region 3 without going over the barrier, but rather by "tunnelling" through the barrier of Region 2. The probability of tunnelling depends on the width of the middle region.

The narrower the depletion layer at a junction, the higher is the probability of tunnelling. For two semiconductor regions in contact, the width of the depletion layer, given by Eq. (5.28), is determined mainly by the doping of the more weakly doped region. The higher the doping of both regions, the smaller the width of the depletion layer W. At a metal-N semiconductor contact, the depletion layer exists only in the semiconductor. To obtain a small W, the semiconductor is doped very highly and possibly to the degree of degeneracy at which the Fermi level is moved into the conduction band. For tunnelling to take place, it is necessary for the conduction band of one material to be located, energy-wise, opposite empty states in the conduction band of the other material.

At thermal equilibrium, the Fermi levels of the metal and semiconductor are aligned. With reverse bias applied, the Fermi level of the semiconductor is depressed so that some electrons in the conduction band of the metal are at the same energy level at, and across, the tunnel, from empty states in the conduction band of the semiconductor, as shown in Fig. 11.11(a). This results in the tunnelling of these electrons and easy current flow from the semiconductor to the metal.

When a forward bias is applied, making the metal positive with respect to the semiconductor, electrons in the conduction band of the semiconductor are opposite empty states of the conduction band of the metal, which makes it possible for electrons to tunnel across. The bending of the energy bands when forward bias is applied is shown in Fig. 11.11(b).

We conclude from the above discussion that a tunnelling ohmic contact is obtained if the semiconductor is heavily doped. Normally, the doping of the semiconductor is determined by other considerations, such as its function in a device. To form a tunnelling contact to a lightly doped semiconductor, a thin heavily doped layer of the same conductivity is formed over the lightly doped semiconductor. For example, an ohmic contact to a lightly doped N semiconductor can be formed by depositing a heavily doped N^+ layer over the N surface. The heavily doped layer is formed by ion implantation, or epitaxy. Such a contact generally possesses a linear current-voltage characteristic, as shown in Fig. 11.12(b).

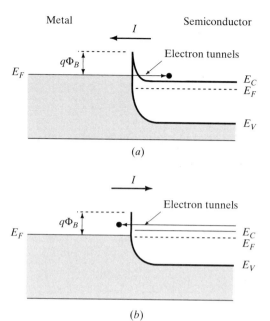

Figure 11.11   Tunnelling at a metal semiconductor junction with (a) reverse bias and (b) forward bias.

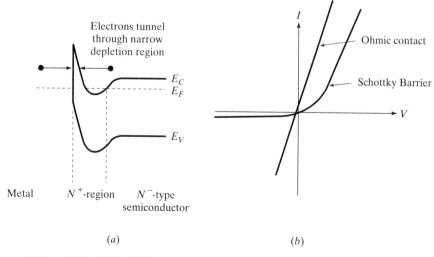

Figure 11.12   (a) Band diagram for a metal-N$^+$–N$^-$ ohmic contact and (b) current-voltage characteristics of a Schottky barrier diode and an ohmic contact.

The energy band diagram, showing tunnelling in both directions, for a metal-N^+–N^- contact is shown in Fig. 11.12(a).

11.6 THE MESFET

The basic structure and operation of the MESFET are identical to the JFET. While the main disadvantage of the JFET is its low gain-bandwidth product, two major differences in the fabrication of the MESFET make it more attractive. In the first place, the semiconductor gate of the JFET is replaced by a metal to form a Schottky barrier. This advantage is related to the simplicity of the Schottky barrier and since diffusion is not used, fabrication to close geometrical tolerances and smaller channel lengths is made possible. The second advantage is related to the use of N-GaAs for the channel in which the electron mobility is five times that of silicon (only at low fields of a few kV/cm). Both of these advantages serve to decrease the transit time of electrons from source to drain and provide a significant increase in the gain-bandwidth product of the MESFETs, thus leading to their use in monolithic microwave integrated circuits and in high speed digital circuits. For the same dimensions, GaAs MESFETs have a speed advantage over Si MESFETs of about a factor of 3. In addition, the use of semi-insulating GaAs provides excellent isolation between adjacent devices.

Fabrication of the MESFET

The starting material of the substrate is a GaAs wafer cut from an ingot produced by the Czochralski method and chromium doped, which places the Fermi level near the center of the bandgap, resulting in a material having a high resistivity of the order of 10^8ohm-cm. It is thus known as *semi-insulating gallium arsenide*. Devices and circuit-connections made on substrates of semi-insulating GaAs have lower capacitances, which lead to high speeds in integrated circuits.

Cross sections of a depletion MESFET and an adjacent Schottky barrier, formed on the same wafer, are shown in Fig. 11.13.

By using photolithography to define the regions, ion implantation is used to form the drain, source, and channel of the MESFET and the base of the Schottky

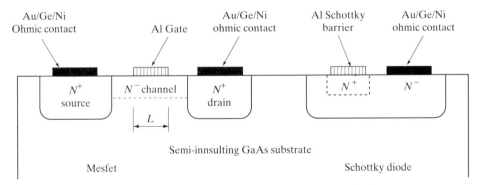

Figure 11.13   Cross section of a MESFET and a Schottky barrier diode.

diode. In these implants, silicon is the dopant of choice and it behaves as N-type in GaAs when introduced by ion implantation. A density of $10^{12}cm^{-3}$ is used for the channel and the diode base while densities of $10^{15}cm^{-3}$ are used in the drain, the source, and the N$^+$ region of the diode. The N$^+$ region of the diode is needed to make a good ohmic contact to the base.

Following the implantation and to cure the damage caused by the implantation, the wafer is annealed at a temperature of about 850°C. A problem arises in this process. Annealling GaAs at this temperature leads to its decomposition since arsenic evaporates at 600°C (GaAs melts at 1238°C). To prevent this, GaAs wafers are protected with a thin layer of silicon nitride prior to the implantation.

Once the implantation and annealing are completed, again by using photolithography, ohmic contacts are made to the drain, the source, and the diode base. A sequence of electron beam evaporation of Au/Ge/N$_i$ is used for the metal contacts, and the contacts are then annealed for several minutes at 400°C. Aluminum is used for the Schottky diode barrier using electron beam evaporation. The device has N$^+$ source and drain and N$^-$ channel.

A major advantage of MESFETs is their usefulness at very high frequencies because channel lengths of the order of 1μm and less can be fabricated. In order to make it possible to deposit ohmic metallic lines of this dimension, special methods must be used. One of the more common methods, known as the *lift-off method*, provides a powerful technique for defining a high resolution pattern for large scale integrated circuits. In this process, aluminum is used for the gate of the MESFET.

Lift-off is accomplished, as shown in Fig. 11.14, by first spreading a thin film of photoresist over the areas of the gate that *are not* to be covered by the metal. In this way, photoresist is used in a reverse role to that used in Chapter 6 in order to transfer patterns onto substrates. The photoresist is then developed as is normally done. The thin aluminum film is then deposited over the resist, as shown in the figure. Finally, the photoresist is removed by a solvent, which does not affect the aluminum, so that the metal covers only the gate.

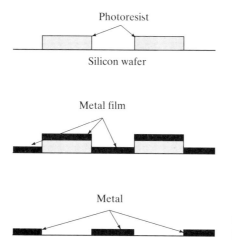

Figure 11.14   Illustrating the lift-off technique.

Modes of Operation

MESFETs are made in two types, known as the *enhancement* (normally-off) type and *depletion* (normally-on) type. The two modes are not possible in the same device. Thus, we are referring to two different types of devices whereby the type is characterized by the width of the channel that is formed under the gate. The width of the channel of the depletion-type device is about double that of the enhancement type.

In the MESFET, just as in the JFET, the gate voltage modulates the width of the depletion region in the semiconductor of the Schottky barrier diode (formed by the gate and the channel) and hence the resistance and therefore the resulting electron flow from source to drain in an N-channel device. In the enhancement device, the built-in voltage causes the channel to be completely closed at zero gate-source bias (assuming $V_D = 0$) and therefore operation of the device requires a positive V_G. In the depletion device, with $V_G = 0$ (and $V_D = 0$), the depletion region extends only partly across the channel, as shown in Fig. 11.15, and it becomes possible to modulate the width of the channel and decrease it by applying a negative voltage to the gate and by applying a positive voltage to the drain.

Threshold Voltage

We will identify two reference voltages for the MESFET by reproducing Eq. (11.4) for the width of the depletion layer at a Schottky barrier given as

$$W = \sqrt{\frac{2\varepsilon}{qN_D}(V_{bi} - V_a)} \tag{11.4}$$

In a MESFET, V_a represents the voltage that is applied from metal to semiconductor bulk and is a function of both V_G and V_D. For $V_D = 0$, $V_a = V_G$, while V_{bi} is the built-in voltage across the depletion layer. It is a positive quantity for both types of devices.

We determined from Eq. (10.3) that for a channel width a, the voltage that must exist across the depletion region to completely pinch the channel is defined as the pinchoff voltage and is given by

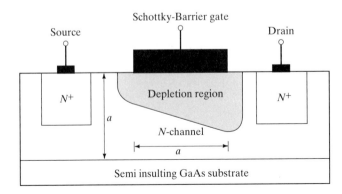

Source
Schottky-Barrier gate
Drain

N+
Depletion region
N+

a

N-channel

a

Semi insulting GaAs substrate

Figure 11.15 Depletion-type device showing the depletion region.

$$V_p = a^2 q N_D / 2\varepsilon \tag{11.25}$$

If (with $V_D = 0$) the built-in voltage causes the depletion layer to extend across only part of the channel width (as in the depletion type), and since the required voltage that will close the channel is V_p, then a negative voltage must be applied to the gate to close the channel between source and drain. This voltage is labeled the *threshold voltage*, V_T, and we can write

$$V_p = V_{bi} - V_T$$

$$\text{and } V_T = V_{bi} - V_p \tag{11.26}$$

However, when the built-in voltage is more than sufficient to close the channel, a positive voltage must be applied to the gate in order to reduce the width of the depletion layer so that the channel is just closed. For this condition, V_{bi} is greater than V_p and the threshold voltage is positive, given also by Eq. (11.26). This is what happens in an enhancement type device. To avoid confusion, a clear understanding of the relations derived in this section is obtained by assuming that $V_D = 0$. The relations in Eqs. (11.25)–(11.26) are valid, however, regardless of the value of V_D.

Thus, the threshold voltage, V_T, is positive for an enhancement device and negative for a depletion device, and the relations in Eq. (11.26) apply to both types. The conditions discussed above are illustrated by Fig. 11.16.

Depletion Device Characteristics

We observe from Fig. 11.15 that the depletion region becomes wider as we move from source to drain. This is due to the effect of the drain voltage, which increases the reverse bias across the Schottky junction and has its highest value at the drain end, causing the channel to have its smallest width at the drain.

At low values of V_D, and with an increase of V_D, the narrowing of the channel, at constant V_G, does not cause a sufficient increase of the resistance of the channel for that change of resistance to have an important effect on the drain current. Since it is assumed that for a constant resistance the current is directly proportional to V_D, the current increases linearly with an increase in V_D. At higher values of V_D, the narrowing of the channel due to an increase of V_D becomes important, and although the current continues to increase with an increase of V_D, the increase is slower and therefore less linear. These changes are illustrated in the characteristics shown in Fig. 11.17(a). A less negative value of V_G for a constant V_D decreases the width of the depletion layer and increases the width of the channel, causing an increase of its cross sectional area and a consequent decrease of resistance. Thus, a higher drain current results.

Enhancement Device Characteristics

In this device, the channel is pinched off (completely closed) by the built-in voltage at thermal equilibrium. To open up the channel and to reduce the depletion layer width, a positive voltage has to be applied to the metal gate at zero drain-source voltage. Because the applied voltage may cause the Schottky junction to be forward-biased, it becomes important to limit the forward current of that junction. With a

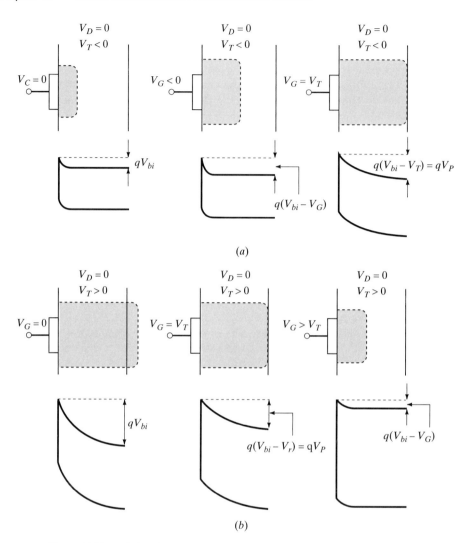

Figure 11.16 Relations between V_T, V_P, and V_{bi} for (a) a depletion device and (b) an enhancement device.

positive voltage at the drain $V_D > 0$, the largest forward bias and therefore the largest gate current occur at the gate-source end of the channel. A forward current at a forward voltage has the effect of reducing the input resistance of the device since a high input resistance is one of the advantages of these devices.

The input gate current can be limited by keeping the forward bias of the junction below the turn-ON voltage, identified in Fig. 11.8. The turn-ON voltage of GaAs Schottky diodes is of the order of 0.7V, compared to the 0.4V turn-ON voltage of silicon Schottky diodes.

The current-voltage characteristics of an enhancement type device are shown in Fig. 11.17(b).

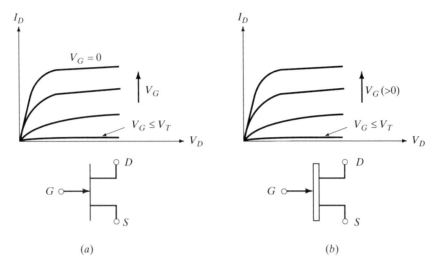

Figure 11.17   Current-voltage characteristics and symbol for (a) depletion device and (b) enhancement device.

Relations Between the Voltages

With respect to the current-voltage characteristics shown in Fig. 11.17, we will introduce the effect of the drain-source voltage into the relations for V_{bi}, V_p, V_G, and V_T. It is to be noted that both the gate voltage and the drain voltage, herein referred to as V_G and V_D, are measured with respect to the source, which is assumed to be grounded and connected to the substrate.

Both V_{bi} and V_p and therefore V_T are constants for a given device. The two other voltages that control the width of the depletion layer and hence the extent of constriction of the channel are V_D and V_G.

We first consider the relations for the *depletion-type device* where V_G is negative, so that the relation between the voltages that will cause pinchoff at the drain end of the channel is

$$V_D + V_{bi} - V_G = V_p \tag{11.27}$$

where a positive V_D and a negative V_G cause the depletion layer width to increase. The higher the value of V_D, the less is the magnitude of V_G that will pinch the channel at the drain.

For $V_G = 0$, the channel becomes pinched-off at the drain end when

$$V_D = V_{SAT} = V_p - V_{bi} \tag{11.28}$$

where V_{SAT} refers to the drain voltage at the point of saturation of the drain current. For a negative V_G, where $|V_G| < |V_T|$, pinchoff occurs at a smaller value of V_D and hence a smaller current, as shown by Fig. 11.16.

For $V_D = 0$, the channel is pinched-off throughout its length when $V_G = V_{bi} - V_p$, so that V_T is a particular value of V_G and V_T is negative and a constant for a given device. It is obvious that for $V_D = 0$, the channel is completely closed when $V_G = V_T$ and, consequently, the drain current is zero, as shown in Fig. 11.16.

In an *enhancement-type device*, the built-in voltage, V_{bi}, is equal to or greater than the pinchoff voltage, V_p. The drain current begins to flow when V_G exceeds V_T and $V_D > 0$. The relation for the threshold voltage is the same for both the depletion and enhancement-type devices, as given by Eq. (11.26) and repeated here as

$$V_T = V_{bi} - V_p \qquad (11.29)$$

Equation (11.27) applies to the enhancement-type device as well as indicating that, at values of $V_G > V_T$ and as V_G is increased ($V_G > V_T > 0$), the value of the drain voltage at pinchoff increases and thus the current is larger, as shown in Fig. 11.17.

The current-voltage relations that we derived in Chapter 10 for the JFET are equally applicable to the long-channel MESFET, provided that the expression for G_0 is one half the value. This is to account for the reduction of the channel width from $2a$ to a. Equation (10.14) for the linear region and Eq. (10.16) for saturation are repeated here as

$$I_D = G_0 \left[V_D - \frac{2V_p}{3}\left(\frac{V_{bi} + V_D - V_G}{V_p}\right)^{3/2} + \frac{2}{3}V_p\left(\frac{V_{bi} - V_G}{V_p}\right)^{3/2} \right] \qquad (11.30)$$

$$I_{SAT} = G_0 \left[V_p - V_{bi} + V_G - \frac{2V_p}{3} + \frac{2V_p}{3}\left(\frac{V_{bi} - V_G}{V_p}\right)^{3/2} \right] \qquad (11.31)$$

where, for the MESFET, $G_0 = q\mu_n Z a N_D / L$, V_G is negative for the depletion device and positive for the enhancement device. The channel has width a, length L, and depth Z.

We illustrate calculations for the MESFET in the following example.

EXAMPLE 11.3

An N channel GaAs MESFET has $\Phi_B = 0.88V$, $N_D = 10^{17}cm^{-3}$, $a = 0.25\mu m$, $L = 1\mu m$, and $Z = 10\mu m$. Given $\mu_n = 5000cm^2/V\text{-}s$ and $\varepsilon_r = 13.1$, determine:
a) The pinchoff voltage; b) The threshold voltage and the type of device c) The drain voltage at saturation for $V_G = 0$; and d) The drain current at saturation for $V_G = 0$.

Solution

a) $V_p = a^2 q N_D / 2\varepsilon = \dfrac{(0.25)^2 \times 10^{-8} \times 1.6 \times 10^{-19} \times 10^{17}}{2 \times 13.1 \times 8.85 \times 10^{-14}} = 4.31V$

b) $V_T = V_{bi} - V_p$
$qV_{bi} = q\Phi_B - q(E_C - E_F)$
$n = N_D = N_c \exp -(E_C - E_F)/kT$
From Table 3–3, N_c for GaAs is $4.21 \times 10^{17}cm^{-3}$,
So that $10^{17} = 4.21 \times 10^{17} -(E_C - E_F)/kT$
At $T = 300K$, $(E_C - E_F)/q = 0.37V$
From Eq. (11.10), $V_{bi} = 0.88 - 0.37V$
$V_{bi} = 0.842$ and $V_T = 0.842 - 4.31 = -3.467V$.
It is a depletion-mode device.

c) V_{SAT} for $V_G = 0$ is $V_p - V_{bi}$
$V_{SAT} = 4.31 - 0.842 = 3.468V$.
d) We calculate G_0 from $q\mu_n Z a N_D / L$ to be 2×10^{-2}s and from Eq. (11.31) I_{SAT} to be 16.8mA.

An expression for the saturation drain current, near threshold, for a long-channel MESFET is obtained* by taking the following steps: First, the equation for V_{bi} in Eq. (11.29) is substituted in Eq. (11.31) of the JFET. Second, the part of the expression for I_{SAT}, which includes the three-halves power, is expanded using the Taylor series while retaining terms up to the second order. Third, letting $(V_G - V_T)/V_p \ll 1$, the result is:

$$I_{SAT} \cong (\mu_n \varepsilon Z / 2aL)(V_G - V_T)^2 \qquad (11.32)$$

This equation is restricted to relatively low values of current since it is valid near the threshold where $V_G \approx V_T$.

Equation (11.30) is valid for both the depletion and enhancement mode devices, except that V_G and V_T are positive for the enhancement mode and they are negative for the depletion mode, as shown in the transfer characteristics of Fig. 11.18. Transfer characteristics are defined as those relating the output current to the input voltage.

While the elegance of the form of Eq. (11.32) cannot be underestimated, it is important to emphasize that it is only valid for long-channel devices that have $L \cong 2\mu$m. *In fact, current saturation in a MESFET takes place normally as a result of velocity saturation, caused by high electric fields.*

The advantages of the high electron mobility in the GaAs MESFET result from both the smaller channel lengths and the properties of the III-V compounds used. Basically, the gradual channel approximation is no longer valid for short-channels, and for very small channel lengths the one-dimensional analysis used in arriving at the characteristic equations is not sufficient.

We will use Eq. (11.32) to calculate the drain current at saturation for the device in Example 11.3.

EXAMPLE 11.4

Use Eq. (11.32) to calculate I_{SAT} at $V_G = 0$ for the device of Example 11.3 and comment on its validity.

Solution

$$I_{SAT} = (\mu_n \varepsilon Z / 2aL)(V_G - V_T)^2 = 13.93\text{mA}$$

We note that the current in Example 11.3 is 16.8mA. Thus, Eq. (11.32) is not valid for the conditions of Example 11.3 since for $V_D = 3.46V$, $V_G = 0$, $V_p = 4.3V$, and $V_T = 3.46V$, so that $(V_G - V_T)/V_p$ is not much smaller than unity (it being the condition for the validity of Eq. (11.32)).

---

*From S. M. Sze, *Semiconductor Devices, Physics and Technology*, p. 183, copyright © Wiley (1985). Reprinted by permission of John Wiley & Sons, Inc.

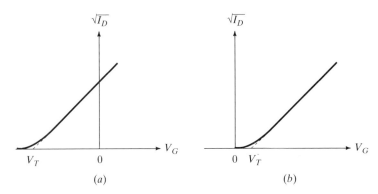

Figure 11.18 Transfer characteristics for (a) the depletion-type and (b) the enhancement type MESFET.

REVIEW QUESTIONS

Q11-9 Identify the major structural and material differences between the JFET and the MESFET.

Q11-10 Give one major reason why the GaAs MESFET has better high frequency response than other transistors.

Q11-11 Distinguish between an enhancement-mode and a depletion-mode MESFET, both in construction and in properties.

Q11-12 Define the threshold voltage for each of the modes of operation.

Q11-13 What are the signs of V_p, V_G, and V_T for each of the two modes?

HIGHLIGHTS

- The MESFET is a field-effect transistor similar to the JFET in construction and operation. The reverse-biased PN junction gate of the JFET is replaced by a metal-semiconductor, rectifying gate contact.

- MESFETs use a GaAs N channel implanted into a GaAs semi-insulating substrate. The SI GaAs substrate reduces parasitic capacitances and hence permits operation at microwave and millimeter-wave frequencies. Digital MESFETs can operate at higher speeds and consume less DC power than comparable ICs made from silicon.

- MESFETs may be of the depletion-mode or the enhancement-mode type. The channel in the enhancement device, also called the normally-off device, is completely pinched-off at zero gate bias by the built-in gate-channel voltage. A forward bias has to be applied to open the channel and allow current.

- The normally-on depletion-mode MESFET is turned-on at zero gate bias. The width of the channel is modified by a negative gate voltage.

- The voltages of importance in the MESFET are: the pinchoff voltage, the built-in voltage, and the threshold voltage. For an enhancement device, $V_{bi} > V_p$, so that V_T, given by $V_{bi} - V_p$, is positive. For a depletion device, $V_{bi} < V_p$ and V_T is negative.

- The expressions for the I-V relationships of a JFET apply to the MESFET provided $L \gg a$ and a constant mobility is assumed. Both conditions are not met by the short-channel GaAs MESFET. At the very small channel lengths ($L < 1\mu m$) that these devices are normally fabricated, the high electric field near the drain causes the mobility to decrease. Three models have been advanced to correct for the change in mobility.

EXERCISES

E11-3 A GaAs MESFET uses gold for the gate and the N channel doping is $5 \times 10^{16} \text{cm}^{-3}$. Given $a = 0.3\mu\text{m}$, $L = 3\mu\text{m}$, $Z = 10\mu\text{m}$, $\mu_n = 6500\text{cm}^2/\text{V} - \text{s}$, and $\varepsilon_r = 13.1$, determine V_T and identify the mode of operation.

Ans: $V_T = -2.4\text{V}$.

11.7 MODELS FOR SHORT-CHANNEL MESFETS

In Chapter 10, we derived current-voltage relations for the JFET based on the assumption of a long channel, $L \gg a$, and constant mobility. These expressions are equally valid for the MESFETs, subject to the assumptions made in the JFET derivation.

When we compare the current-voltage characteristics obtained by using the constant mobility assumption with experimentally determined characteristics for short-channel devices, serious discrepancies are observed in the values of the drain currents in saturation. The variations can be of the order of 50 percent. These discrepancies are attributed to the higher values of electric field intensity that occur in the short-channel devices.

With reference to Fig. 4.3, we note that for Si and GaAs, and at low values of electric field, the drift velocity increases linearly with the field so that the mobility is constant. As the field intensity increases, the velocity increases but at a smaller rate than that at low fields. Eventually, the drift velocity saturates at high fields. Therefore, it becomes necessary to include the changes in the mobility in determining the variation of drain current with an increase of the drain voltage. Three models have been advanced, each of which becomes necessary as the channel length is reduced.

Field-Dependent Mobility Model

To obtain a more accurate model than that of the constant mobility, a simple expression is assumed* for the drift velocity given by

$$v_d = \frac{\mu_n \mathscr{E}_x}{1 + \dfrac{\mu_n \mathscr{E}_x}{v_s}} \tag{11.33}$$

where $\mathscr{E}_x$ is the longitudinal electric field intensity in the channel, μ_n is the mobility at low values of electric field, and v_s is the saturation velocity.

When this relationship for the velocity is used, an expression for the drain current, for $V_D < V_{SAT}$, is determined that indicates that the drain current, compared to

---

*From S. M. Sze, *Physics of Semiconductor Devices*, p. 325, copyright © Wiley (1981). Reprinted by permission of John Wiley & Sons, Inc.

the constant mobility relation, is reduced by a factor of $[1 + (\mu_n V_D / v_s L)]$. This result yields reasonably acceptable values for silicon short-channel JFETs and MESFETs. However, because of the particular shape of the velocity-field characteristics of GaAs, the relationships resulting from the field-dependent mobility model are unsatisfactory for channel lengths less than $10\mu m$.

Two-Region Model

In the second model for the device in saturation, the GaAs channel is assumed to be divided into two regions: a constant mobility is assumed for a part of the channel near the source and, in the rest of the channel, it is assumed that the carriers are moving with a saturated drift velocity.

In carrying out calculations based on the results of the two-region model, it is found that the physical boundaries of the channel are nearly parallel to the gate surface. However, for large values of L/a, it appears that more than 90 percent of the channel is under the influence of the saturated velocity model because the electric field rises very rapidly as we move away from the source. This model provides reasonably close results to the observed short-channel characteristics.

Two-Dimensional Model

For very short channels, not only does the drift velocity take on its saturated value, but a two-dimensional solution of Poisson's equation becomes necessary. Such a model might produce results that agree fairly well with experimental measurements.

PROBLEMS

11.1 An N-type silicon Schottky barrier junction has $N_D = 5 \times 10^{15} cm^{-3}$ and $\Phi_m = 4.5V$. Using the affinity of silicon as 4.15V, determine, at $T = 300K$,

    a) the Schottky barrier height.

    b) the built-in voltage.

11.2 A metal with a work function of 4.3V is deposited on N-type silicon. Determine the doping density, which at zero bias results in the absence of a space-charge region at $T = 300K$.

11.3 A Schottky barrier diode is made by depositing tungsten on N-type silicon. Determine, at $T = 300K$, and for $N_D = 10^{15} cm^{-3}$,

    a) the built-in voltage.

    b) the depletion region width.

    c) the maximum electric field intensity.

11.4 A Schottky barrier contact is made from tungsten on P-type silicon having $N_A = 10^{16} cm^{-3}$. Determine, at $T = 300K$,

    a) the built-in voltage.

    b) the depletion layer width.

11.5 Gold is deposited on N-type GaAs to form a Schottky barrier diode. Determine, at $T = 300K$, the saturation current density using a value of $8 Acm^{-2}K^{-2}$ for Richardson's constant.

11.6 A Schottky barrier diode is made from aluminum deposited on N-type GaAs. The diameter of the junction is $100\mu m$. Plot the current-voltage characteristic for the diode starting at a reverse voltage of 1V and up to a forward voltage of 0.6V. Use $R = 8 Acm^{-2}K^{-2}$ and $T = 300K$.

11.7 A Schottky barrier diode is being compared to a P^+N junction diode. The Schottky diode is made from aluminum on N-type silicon. The PN diode has $N_A = 10^{18}cm^{-3}$, $N_D = 10^{16}cm^{-3}$, $\tau_p = \tau_n = 0.1\mu s$. Determine, at $T = 300K$, the voltages that need to be applied to each diode to obtain a current density of $1\ Acm^{-2}$.

11.8 A Schottky barrier diode is made from tungsten on silicon doped with antimony at a density of $5 \times 10^{15}cm^{-3}$. The junction area is $0.5mm^2$. Determine, at $T = 300K$,

    a) the current at a forward bias of 0.3V.

    b) the current, at $V_a = 0.3V$, in a P^+N silicon junction diode having $N_A = 10^{18}cm^{-3}$, $N_D = 10^{15}cm^{-3}$, and $\tau_p = \tau_n = 1\mu s$.

11.9 A Schottky barrier contact is made from gold on N-type silicon having $N_D = 10^{15}cm^{-3}$. The junction area is $1.0mm^2$. Determine, at $T = 300K$,

    a) the depletion layer capacitance at an applied voltage of $-3V$.

    b) Draw the $1/C_j^2$ versus V relationship identifying the values of the capacitance and the voltage at the respective intercepts.

11.10 A Schottky barrier diode is made from *PtSi* on silicon doped with $N_D = 10^{16}cm^{-3}$. The junction area is $10^{-5}cm^2$. Determine:

    a) the small-signal capacitance at zero DC bias.

    b) the reverse bias at which C is reduced by 30 percent.

11.11 A Schottky barrier diode is made on N silicon having doping $N_D = 1.5 \times 10^{15}cm^{-3}$. The metal work function is 5.1V. Determine:

    a) the barrier height.

    b) the built-in voltage.

    c) the depletion layer width at zero bias.

11.12 A Schottky barrier diode is made on N silicon with a junction diameter of $50\mu m$. The slope of the $1/C_j^2$ versus V_a plot is $-4 \times 10^{23}F^{-2}V^{-1}$. Determine the doping density in the silicon.

11.13 Tungsten is deposited on N-type silicon, with $N_D = 10^{15}cm^{-3}$, to form a Schottky barrier diode that has an area of $10^{-4}cm^2$. Determine:

    a) the diode current at $V_a = 0.4V$.

    b) the voltage that has to be applied to a P^+N junction diode, to obtain the same current, with the P^+N diode having the same area and $N_A = 10^{18}cm^{-3}$, $N_D = 10^{15}cm^{-3}$, and $\tau_p = \tau_n = 1\mu s$.

11.14 A GaAs MESFET uses gold as a Schottky barrier gate. The channel doping is $N_D = 10^{16}cm^{-3}$ and its width is $0.6\mu m$. Determine, at $T = 300K$,

    a) the pinchoff voltage.

    b) the built-in voltage.

    c) the threshold voltage.

11.15 To illustrate the idea that by closing the channel width, the device can be used as an enhancement or depletion mode transistor, determine, for the device of problem 11.14, the channel width for which the device is OFF when no gate bias is applied.

11.16 A GaAs enhancement-mode MESFET has Schottky barrier height of 0.8V, a channel doping of $10^{16}cm^{-3}$, and channel width of 0.25μm. Determine the gate bias needed to just open up the channel, as in Fig. 11.16(b).

11.17 A GaAs MESFET has barrier height of 0.76V, $a = 0.6$μm, $L = 15$μm, $Z = 20$μm, $N_D = 10^{16}cm^{-3}$, and $\mu_n = 7000cm^2V^{-1}s^{-1}$. Determine at $V_G = 0$, $V_G = -1.0$V, and $T = 300K$,

   a)   the saturation drain current.

   b)   the saturation transconductance.

11.18. A gold $-N$ channel GaAs MESFET has $N_D = 10^{16}cm^{-3}$, $a = 0.7$μm, $L = 1$μm, $Z = 10$μm, and $\mu_n = 5000cm^2V^{-1}s^{-1}$. Determine at $T = 300K$,

   a)   the pinchoff voltage.

   b)   the threshold voltage.

   c)   whether it is a depletion-or enhancement-mode device.

   d)   the saturation drain current at $V_G = 0$.

11.19 A tungsten on N channel GaAs MESFET has $N_D = 10^{16}cm^{-3}$, $a = 0.5$μm, $L = 1$μm, $Z = 100$μm, and $\mu_n = 5500cm^2V^{-1}s^{-1}$. Determine the channel resistance at $V_D = 0.1$V and,

   a) $V_G = -0.5$V

   b) $V_G = 0$

   c) $V_G = 0.5$V.

11.20 A metal N GaAs MESFET has a barrier height of 0.9V. The channel has $N_D = 5 \times 10^{15}cm^{-3}$ and $a = 0.8$μm.

   a)   Determine the pinchoff voltage.

   b)   Determine the threshold voltage.

   c)   Is the device depletion-mode or enhancement-mode? Explain.

11.21 Use the data of problem 11.19 to illustrate the control exerted by the channel doping on the threshold voltage and determine the threshold voltage for

   a) $N_D = 10^{15}cm^{-3}$

   b) $N_D = 10^{17}cm^{-3}$.

chapter 12

METAL-OXIDE-SILICON SYSTEMS

12.0 INTRODUCTION

A major application of metal-oxide-silicon (MOS) systems is the Metal-Oxide-Semiconductor Field-Effect Transistor (MOSFET). The use of MOSFETs in integrated circuits has been a major development in the implementation and permeation of fast and low-cost digital circuits. The MOSFET, also known as the Insulated Gate FET(IGFET), finds extensive use as a digital switching element and in semiconductor memories.

The advantages of MOSFETs as compared to BJTs are twofold: first, their dimensions, for executing a certain function, are smaller, thus assuring an increase in the component density; second, the manufacture of these devices requires fewer fabrication processes. They are therefore less expensive to manufacture than BJTs for an equivalent circuit function. A more comprehensive comparison between the two devices will be made at the end of the next chapter.

Another important application of the MOS system is in charge-coupled devices (CCD). These devices consist of arrays of precision capacitors and are used in the conversion of analog signals to digital representation. One application that has generated extensive interest is their use as solid-state imagers or video cameras for home and industrial application.

We will devote this chapter to the study of the properties of the metal-oxide-silicon systems. In the following chapter, we will consider the operation, characteristics, and properties of the MOSFET. Just as with metal-semiconductor junctions, a proper starting point for the study of MOS systems is the study of the energy band diagram.

12.1 ENERGY BAND DIAGRAMS

The basic device we consider first is the MOS capacitor or diode consisting of a thin layer of silicon dioxide sandwiched between a metal, such as aluminum or heavily doped polycrystalline silicon, and a silicon substrate, as shown in Fig. 12.1. In this chapter, aluminum will be the metal used in the illustrations and most of the examples. The advantages of using polysilicon will be discussed in the chapter on MOSFETs (Chapter 13).

The energy band diagram of solids in contact provides the necessary background information for the study of the properties of the system. We draw the separate energy level diagrams, shown in Fig. 12.2, for the metal, the oxide, and the silicon using the vacuum energy level as the common reference for all three solids. In this illustration, we have selected a P-type silicon substrate and aluminum as the metal.

We will make the following assumptions concerning the solids:

1. The insulator is ideal, its resistance is infinite, and hence no charges pass through it. For the ideal insulator, no charges exist on the surface of the oxide nor inside the solid.

2. The only charges that can exist in the system are in the semiconductor at the surface or near the surface of the oxide and in the metal surface adjacent to the oxide.

As the solids are brought into contact with $V_G = 0$ in Fig. 12.1, the Fermi levels of the metal and the silicon are immediately aligned.

The work function of the aluminum, $q\Phi_m$, is 4.1eV. Assuming that the work function of the substrate P silicon, $q\Phi_s$, is 5.05eV (for $N_A = 5 \times 10^{15} \text{cm}^{-3}$), so that the work function difference, defined as $(q\Phi_s - q\Phi_m)$, as calculated from Fig. 12.2, is 0.95eV.

With $V_G = 0$, thermal equilibrium, and because the work function of the metal is smaller than that of the silicon, electrons are transferred through the terminals of the device from the metal to the semiconductor. On the metal interface with the oxide, a thin sheet of surface positive charges is formed. On the semiconductor side,

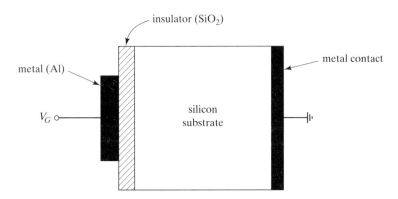

Figure 12.1   Basic composition of an MOS diode.

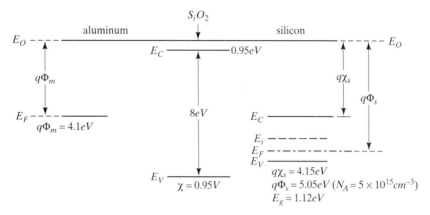

Figure 12.2 Energy level diagrams for aluminum, silicon dioxide, and P-silicon.

a voltage drop and an electric field are formed so that holes are driven away from the silicon adjacent to the surface of the oxide leaving an excess of negatively charged ions. The holes that left the silicon surface recombine at the ohmic contact of the substrate with the electrons that left the aluminum.

  With the alignment of the Fermi levels, the 0.95eV work function difference is absorbed by the semiconductor and the silicon dioxide since the metal is a perfect conductor and does not sustain a voltage drop. The voltage drop across the insulator is a consequence of the charges stored on either side. The energy band in thermal equilibrium is shown in Fig. 12.3. A voltage drop and an electric field occur in the semiconductor as the energy bands bend and a depletion layer is formed in the silicon adjacent to the oxide. Energy differences to the oxide are measured to the conduction band and the valence band of the oxide. The affinity of the oxide is 0.95eV.

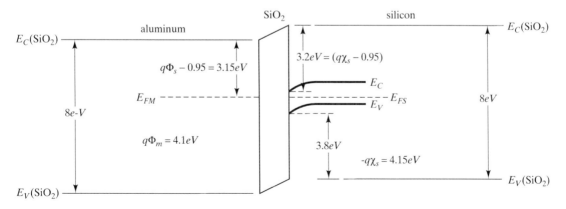

Figure 12.3 Energy level diagram after contact.

12.2 BAND BENDING AND THE EFFECT OF BIAS VOLTAGES

To simplify the presentation of the picture of energy band bending in semiconductors, we initially assume that the work function difference between the metal and the semiconductor is zero. As a result of this assumption, and for zero applied voltage, the Fermi levels of the metal and the silicon are aligned and there are no voltage drops in the oxide and in the silicon. Later on, and when it becomes necessary, we will take account of the work function difference together with other similar effects in our analytical relationships.

When a voltage is applied to the MOS diode of Fig. 12.1, the energy levels E_C, E_i, and E_V are bent, as shown by the drawings in Fig. 12.4, because some of the applied voltage is dropped in the semiconductor at the interface with the oxide. Since there is no current in the MOS diode, the position of the Fermi level in silicon does not change, but that of E_C, E_V, and E_i changes. When E_i is above E_F, the majority carriers are holes, and when E_i is below E_F, the majority carriers are electrons.

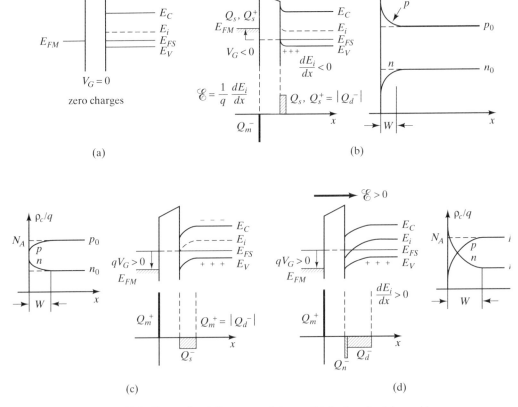

Figure 12.4 Energy band diagrams and charge distributions for (a) zero bias, (b) accumulation, (c) depletion, and (d) inversion.

The wider the separation, the greater is the density of majority carriers, and the smaller is the density of minority carriers.

When the levels E_C, E_i, and E_V in a P-type semiconductor are bent upwards, in the region near the oxide, E_i moves further away from E_F and an increase of the hole density takes place. When the bands bend downwards and E_i, at the surface of the semiconductor with the oxide, gets closer to E_F, the density of the holes decreases and the density of the electrons increases. When more downwards bending takes place and E_i at the surface is at E_F, an intrinsic condition occurs at the surface with both the electron density and the hole density equal to the intrinsic carrier density. Further downward bending moves E_i below E_F at the surface, causing the density of electrons to exceed n_i and the density of holes to be less than n_i. When additional bending causes E_i at the surface to be below E_F by the same energy difference that it is above E_F in the bulk of the semiconductor, the electron density at the surface is equal to the hole density in the bulk.

The Effects of Bias Voltage

The bending of the bands in silicon is accompanied by and results from a voltage appearing in the semiconductor, caused by a voltage applied to the metal gate. With the silicon P substrate held at ground potential and negative voltage applied to the gate, the metal acquires a negative charge on the surface with the oxide and, by capacitive action, additional holes accumulate at the semiconductor surface with the oxide. The hole density in the bulk of the silicon, away from the oxide interface, is equal to the density of acceptor atoms. The increase in the hole density at the surface is accompanied by an upwards bending of the energy bands in the silicon, as shown in Fig. 12.4(b). The intrinsic Fermi level has moved further away from E_F and an *accumulation* layer is said to be formed.

When a positive bias is applied to the metal gate, the bands in the silicon are bent downward. At low bias, the majority carrier holes in silicon are repelled away from the surface of the silicon and the energy level E_i at the surface has moved closer to E_F. This state of reduction of the majority carrier density near the surface is known as the *depletion* condition. As the bias is increased, the energy bands bend downward far enough for E_i to be equal to E_F at the surface. At this level, and as shown in Fig. 12.4(c), an intrinsic condition exists at the surface. With further increase in the bias voltage, E_i dips below E_F, the electron density at the surface exceeds the hole density, and an *inversion* condition results. The surface is inverted because an N region is beginning to form. At this level, the electron density at the surface, although much larger than the hole density, is still smaller than the hole density in the bulk. This is shown in Fig. 12.4(d).

The onset of *strong inversion* occurs when E_i, at the surface, has been bent so much that it is below E_F by the same amount of energy that it is above E_F in the bulk of the silicon. At strong inversion, the density of electrons at the silicon surface is equal to the density of holes in the bulk. Away from the silicon-oxide interface, a depletion layer is formed in the silicon, in which acceptor atoms are stripped of their holes. The ionized acceptor atoms form the depletion layer.

In the inversion conditions, the holes at the surface are depleted to values that are smaller than at equilibrium. Because of this, generation in this region exceeds recombination. An electric field and a voltage drop have been established. The electric field separates the generated electron-hole pairs, sweeping the holes towards the bulk of the silicon and moving the electrons to the surface with the oxide. The electrons accumulate at the surface and are held back by the huge potential barrier of the oxide.

We can conclude that at strong inversion, the semiconductor at the oxide surface has become N-type, and away from the surface, the bulk is P-type separated from the N silicon by a depletion layer of ionized acceptors. The applied bias has created a PN junction near the surface. It is noteworthy that we speak freely of an inversion N layer and a depletion layer, where, depending upon the substrate doping, the width of the depletion layer is of the order of 100Å or less and the inversion layer is only about a tenth of a micron.

The surface charge density of electrons at strong inversion is labeled Q_n and the depletion charge density is Q_d, with both having units of C/cm^2. In a P-type silicon substrate, both Q_n and Q_d are negative quantities and their sum is the charge density in silicon Q_s. The surface of the metal at the oxide acquires a positive charge density, Q_m, equal in magnitude to Q_s, as shown in Fig. 12.4.

12.3 ANALYTICAL RELATIONS FOR THE CHARGE DENSITIES

The importance of the electron charge density, Q_n, will become clear when we study the operation and characteristics of the MOSFET in the next chapter. The motion of the electrons within the very thin inversion region, which we later label the channel, results in the current of the MOSFET. In this section, we will determine analytical expressions for the charge densities in the inversion and depletion layers.

To describe the extent of the downward bending of the bands near the P semiconductor surface, we define a potential $\phi(x)$ as the potential or voltage anywhere in the semiconductor and $q\phi(x)$ as the amount of bending of the bands both measured from the equilibrium position of E_i in the bulk of the silicon. At $x = 0$, and as shown in Fig. 12.5, $\phi(x)$ becomes ϕ, known as the *surface potential*, so that $q\phi_s$ is the energy band bending at the surface. Since a decrease in potential energy as $E_i(x)$ is

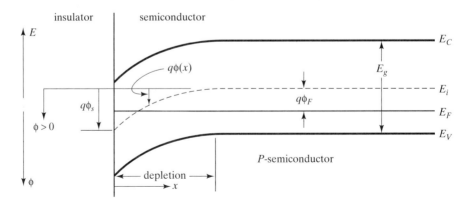

Figure 12.5 Energy band diagram at the surface and in the bulk of a P-type semiconductor when a positive bias is applied to the gate.

bent downward it corresponds to an increase in potential for an electron and $\phi(x)$ becomes positive when measured downwards from E_i in the bulk.

When a negative bias is applied to the metal gate, the bands are bent upward and $\phi_s < 0$. The bands are bent downward when a positive bias is applied to the gate, $\phi_s > 0$, and the semiconductor adjacent to the oxide is in a *state of depletion*. *Inversion* occurs when ϕ_s is positive and $q\phi_s$ has magnitude larger than the energy separation, $q\phi_F$, between E_i and E_F in the bulk. The location of ϕ_s and ϕ_F are shown in Fig. 12.5 and $q\phi_F$, in the bulk where $p_{0p} \cong N_A$, is determined by using Eq. (3.41), so that

$$p_{0p} = N_A = n_i \exp{(E_i - E_F)/kT} = n_i \exp{(q\phi_F/kT)} \tag{12.1}$$

and

$$q\phi_F = kT \ell n N_A / n_i \tag{12.2}$$

For N-type substrates

$$q\phi_F = -kT \ell n N_D / n_i \tag{12.3}$$

Inversion takes place when $\phi_s > \phi_F$ and the onset of *strong inversion* is assumed to occur at $\phi_s = 2\phi_F$.

With reference to Eq. (3.40) and Fig. 12.5, the electron concentration in the bulk of the P-type substrate (for E_i in the bulk) is found from the following relation

$$n_{0p} = n_i \exp{(E_F - E_i)/kT} = n_i \exp{(-q\phi_F/kT)} \tag{12.4}$$

At the surface, $x = 0$, of the silicon with the oxide, the density of electrons n_s becomes

$$n_s = n(x = 0) = n_i \exp{[E_F - E_i(0)]/kT} = n_i \exp{[E_F - E_i + E_i - E_i(0)]/kT} \tag{12.5}$$

$$= n_{0p} \exp{[E_i - E_i(0)]/kT} = n_{0p} \exp{[q\phi(0)/kT]}$$

$$n_s = n_{0p} \exp{q\phi_s/kT} \tag{12.6}$$

In Table 12.1, we show the various conditions in the P-type semiconductor as the applied voltage is changed.

As we indicated earlier, we are interested in the conditions occurring at strong inversion. At strong inversion, the charge Q_s in the silicon is made of an electron surface charge Q_n and a negative depletion layer charge Q_d. The capacitor formed by the metal and the semiconductor has a capacitance C_{ox} per unit area given as

$$C_{ox} = \varepsilon_{ox}/t_{ox} \tag{12.7}$$

where ε_{ox} is the permittivity of the oxide in F/cm, (for SiO_2, $\varepsilon_{ox} = 3.9 \times \varepsilon_o = 3.45 \times 10^{-13}$F/cm)$t_{ox}$ is the oxide thickness in cm, and C_{ox} is in F/cm², having surface area parallel to the oxide surface. The metal plate accumulates charge density Q_m^+ and the silicon accumulates charge density Q_s^- when a positive voltage V_G is applied to the metal gate with the substrate grounded and $|Q_s^-| = Q_m^+$. The voltage across the capacitor is V_{ox} and the voltage drop in the semiconductor at the surface at strong inversion is $\phi_s = 2\phi_F$, so that

$$V_G = V_{ox} + 2\phi_F \tag{12.8}$$

where $V_{ox} = |Q_s|/C_{ox}$.

TABLE 12.1   Effect of V_G on conditions at surface of P-substrate MOS capacitor

V_G	ϕ_s	Surface Condition	Surface Carrier Density
Negative	Negative $\phi_s < 0$	Accumulation	$p_s > N_A$
Positive	Positive $\phi_s < \phi_F$	Depletion	$n_s < p_s < N_A$
Positive large	Positive $\phi_s = \phi_F$	Instrinsic	$n_s = p_s = n_i$
Positive large	Positive $\phi_s > \phi_F$	Inversion	$p_s < n_s < N_A$
Positive larger	Positive $\phi_s = 2\phi_F$	Onset of strong inversion	$n_s = N_A$
	$\phi_s > 2\phi_F$	Strong inversion	$n_s > N_A$

We summarize the above by stating that a positive bias, applied between the metal and the substrate, causes (in a P substrate) a depletion layer of ionized acceptor atoms and an inversion layer of electrons occurring simultaneously with the bending of the bands in the semiconductor. The charges on either side of the oxide support the voltage across the oxide. A question of paramount importance is: How do Q_n and Q_d vary as V_G increases (while the system is in inversion)?

We have already shown that part of V_G causes ϕ_s, and we have also determined that the electron volume density, n_s, at the surface depends exponentially on ϕ_s. The value of n_s is obtained by dividing Q_n, having units of C/cm^2, by the product of the electron charge q and the thickness of the charge at the surface.

Depletion Region Thickness and Charge Density

Let us now investigate the relation between the depletion layer charge density Q_d and ϕ_s. We assume that the depletion layer thickness extends from $x = 0$ to $x = W$ and that, although a very thin layer of electrons (Q_n) occupies part of W, only negative acceptor ions populate the depletion layer. We can therefore write Poisson's equation in the depletion layer as

$$\frac{d^2\phi}{dx^2} = -\rho_c(x)/\varepsilon \tag{12.9}$$

where ε_s is the permittivity of the semiconductor and $\rho_c(x)$ is the total space charge density given by $\rho_c(x) = q(N_D^+ - N_A^- + p_{0p} - n_{0p}) = -qN_A^-$.

To simplify the mathematics, we assume that the region from $x = 0$ to $x = W$ is in depletion and consists only of ionized acceptors.

The band diagram together with the charge, field intensity, and potential distributions shown in Fig. 12.6, are for a device in inversion.

Integrating Eq. (12.9) once and assuming the electric field intensity $\mathcal{E}_x$ in the bulk is zero, at $x \geq W$, we obtain

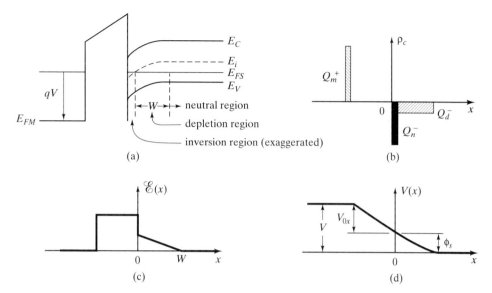

Figure 12.6 (a) Band diagram, (b) charge density distribution, (c) electric field distribution, and (d) potential distribution for a device in inversion.

$$\mathcal{E}_x = -d\phi/dx = (q/\varepsilon) N_A (W - x) \tag{12.10}$$

Our zero reference level for ϕ is at E_i in the bulk. Upon solving for ϕ and setting it equal to zero at $x = W$, we have

$$\phi(x) = \frac{q}{2\varepsilon} N_A W^2 \left(\frac{x}{W} - 1\right)^2 \tag{12.11}$$

Since $\phi = \phi_s$ at $x = 0$, we determine from Eq. (12.11) the following expression for the depletion layer width W, as

$$W = \sqrt{\frac{2\phi_s \varepsilon}{qN_A}} \tag{12.12}$$

Subject to our assumption that only negative acceptor ions are available in W, the charge density (C/cm^2) in the depletion layer, Q_d, is given by $-qN_A W$ and W depends on the square root of ϕ_s, as shown by Eq. (12.12). The electron surface density, n_s (cm^{-3}), however, depends exponentially on ϕ_s.

Once strong inversion is reached, $\phi_s = 2\phi_F$ and n_s, W and Q_d are given for a P substrate by

$$n_s = n_{op} \exp\frac{q}{kt}(2\phi_F) \tag{a}$$

$$W_m = \sqrt{\frac{2\varepsilon\, 2\phi_F}{qN_A}} \tag{b}$$

$$Q_{dm} = -qN_A W_m = -(4\varepsilon\, \phi_F\, qN_A)^{1/2} \quad (c) \tag{12.13}$$

where W_m and Q_{dm} represent the maximum values of W and Q_d respectively.

It is important to indicate that for a P substrate, ϕ_F is positive and Q_{dm} is negative, whereas for an N substrate, ϕ_F is negative and Q_{dm} is positive.

Beyond the onset of strong inversion, ϕ_s increases slightly above $2\phi_F$ and n_s increases very rapidly, whereas W_m and Q_{dm} are almost at the value of strong inversion since they are proportional to the square root of ϕ_s. An increase of V_G increases ϕ_s, which increases $|Q_n|$, the charge density in the inversion layer, $|Q_d|$, the charge density in the depletion layer, and hence, the total charge density and $|Q_s|$ in the semiconductor. The increase of $|Q_s|$ increases the voltage across the oxide. An increase in V_G is easily met by a very small increase in ϕ_s, which causes an appreciable increase in $|Q_n|$. This increase in $|Q_n|$ is sufficient to take care of the increase in V_G as $V_G = (|Q_s|/C_{ox}) + \phi_s$. Since at the onset of strong inversion $n_s = N_A$, an increase of ϕ_s of 60mV beyond the onset of strong inversion causes n_s to become $10N_A$, whereas this increase in ϕ_s results in a negligible increase in W and hence in the magnitude of Q_d. We therefore assume that, at strong inversion, W and the magnitude of Q_d have reached their maximum values, which we label Q_{dm} and W_m and are defined in Eqs. (12.13).

What physically happens is that the strong inversion charge near the oxide-semiconductor surface is sufficient to shield the bulk from any additional charge. A small increase in V_G has caused a very large increase in n_s, so that the charges required to terminate the fields established by V_G are provided by the inversion region and not by an extension of the depletion region. It is worthwhile repeating here that, whereas the maximum width of the depletion layer W_m is of the order of 100 angstroms, the width of the electron inversion layer, x_i, is of the order of a few atomic diameters.

A more rigorous analysis, assuming the volume charge density in the semiconductor, ρ, includes holes and electrons in each of the depletion and inversion regions, confirms the conclusions of the previous paragraph. The results are shown in Fig. 12.7, indicating a dramatic increase of the magnitude of Q_s with an increase of ϕ_s beyond strong inversion, whereas Q_d hardly changes.

12.4 THRESHOLD VOLTAGE

After having identified the processes and relations of the variables inside the diode, we need to relate the voltage that causes strong inversion to the physical characteristics of the oxide and the semiconductor.

We first define the quantity known as the *turn-on voltage*. This is the voltage that must be applied to the gate to cause strong inversion and it is an extremely important property of MOS devices.

At the onset of strong inversion, the voltage applied to the gate, V_G, is used up, as shown in Eq. (12.8), in bending the bands and at which point $|Q_s|$ in the semiconductor has reached its maximum value $|Q_{dm}|$. Although a small $|Q_n|$ exists, it is assumed to be zero. Thus, the turn-ON voltage is given by

$$V_T = V_{ox} + 2\phi_F = \frac{-Q_s}{C_{ox}} + 2\phi_F = \frac{-Q_{dm}}{C_{ox}} + 2\phi_F \tag{12.14}$$

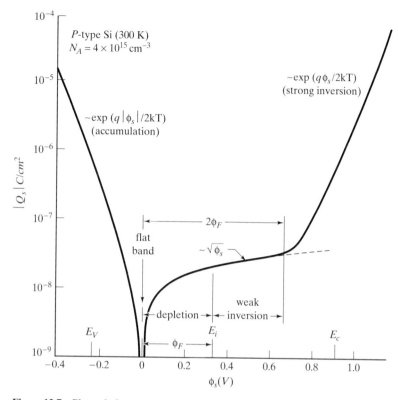

Figure 12.7 Plots of $|Q_s|$ versus ϕ_s. From S. M. Sze, *Physics of Semiconductor Devices*, p. 369, copyright © Wiley (1981). Reprinted by permission of John Wiley & Sons, Inc.

The turn-ON voltage, also known as the *threshold voltage*, is determined for a P-type substrate by substituting for Q_{dm}, in Eq. (12.14), its expression from Eq. (12.13c), so that

$$V_T = 2\phi_F + \frac{1}{C_{ox}} \sqrt{4\varepsilon q N_A |\phi_F|} \tag{12.15}$$

It does seem surprising that the threshold voltage depends on properties of the oxide and the semiconductor but not on the metal. In the following section, we will include a property of the metal as well.

We will now consider an example that illustrates the order of magnitude of the quantities involved.

REVIEW QUESTIONS

Q12-1 Identify the three conditions that take place in the P-substrate of an NMOS capacitor as the gate voltage is increased from zero.

Q12-2 What is the significance of the negative sign in Eq. (12.3)?

Q12-3 Clearly define ϕ_s and ϕ_F, and indicate under what condition they are equal.

Q12-4 What happens to the holes at the oxide surface of a P substrate when the capacitor is driven into strong inversion?

Q12-5 What are the signs of ϕ_F and Q_{dm} in an N substrate capacitor?

Q12-6 Distinguish between n_s and Q_n in a P substrate capacitor.

Q12-7 What is the basis of the assumption that W and Q_d have reached their maximum values at strong inversion?

HIGHLIGHTS

- A MOSFET basically consists of an MOS capacitor to which a source and a drain are attached. The MOS capacitor has a metallic plate labeled the gate and a silicon substrate forms the other plate with the gate and substrate separated by a dielectric, such as silicon dioxide.

- By the application of a bias to the metal gate, part of the voltage appears across the insulator and the rest is across the silicon substrate. The voltage that is dropped in the silicon causes the energy bands to bend with the largest bending occurring at the surface of the semiconductor in contact with the insulator.

- For a P substrate, a negative voltage applied to the gate causes the energy bands in silicon to bend upward. When a positive voltage is applied, they bend downward. A negative voltage at the gate increases the hole density at the surface of the semiconductor and a state of accumulation is said to take place. At the same time, the intrinsic Fermi level moves farther away (upward) from E_F, indicating an increase of hole density at the surface accompanied by a depletion region of positively ionized acceptors whose electrons are repelled from the surface.

- The application of a positive bias to the metal gate of a P-device causes E_i to move toward (downward) E_F at the surface of silicon. Further increase of positive bias causes E_i to be at E_F at the surface and an intrinsic condition results at the surface.

- An increase of the gate bias beyond the generation of an intrinsic condition causes the accumulation of electrons at the surface followed by a narrow depletion layer consisting of negatively charged acceptors. At the surface, an N region begins to form and inversion is said to occur.

- At a sufficiently high positive gate bias, E_i at the surface has moved below E_F by the same energy separation that it is above E_F in the bulk of the silicon (far away from the surface), indicating the onset of strong inversion. This N region at the surface becomes the conduction channel of the MOSFET. The total positive charge on the metal gate is balanced, at all times, by the negative charges of the electrons and the depletion layer.

- At gate voltages greater than the one that causes strong inversion, increase of the gate voltage is accompanied mainly by an increase of the density of electrons in the inversion layer and an increase of the positive charge on the metal gate.

EXERCISES

E12-1 A silicon P substrate is doped with $N_A = 10^{17} \text{cm}^{-3}$. Determine the surface potential that causes strong inversion.

$$\text{Ans: } \phi_s = 0.834\text{V}.$$

E12-2 A silicon N substrate is doped with $N_D = 5 \times 10^{15} \text{cm}^{-3}$. Determine the surface potential needed to make the surface a) intrinsic and b) strong inversion.

Ans: a) $\phi_s = -0.34\text{V}$,   b) $\phi_s = -0.68\text{V}$.

EXAMPLE 12.1

For a silicon P substrate doping $N_A = 10^{15}\text{cm}^{-3}$ and at the onset of strong inversion, calculate:
a) The width of the depletion layer.
b) The charge density in the depletion layer.
c) The electron density n_s at the surface.
d) The threshold voltage.
  Given for silicon: $\varepsilon = 11.8 \times 8.854 \times 10^{-14}$ F/cm, and $n_i(300\text{K}) = 1 \times 10^{10}\text{cm}^{-3}$.
  Given for the oxide: $\varepsilon_{ox} = 3.9 \times 8.854 \times 10^{-14}$F/cm and $t_{ox} = 90$ Angstroms.

Solution

$$\phi_F = \frac{kT}{q} \ell n \frac{N_A}{n_i} = 0.298\text{V}$$

$$C_{ox} = \varepsilon_{ox}/t_{ox} = (3.9 \times 8.854 \times 10^{-14})/90 \times 10^{-8} = 3.83 \times 10^{-7}\text{F/cm}^2$$

a) $W_m = (4\varepsilon\phi_F/qN_A)^{0.5} = 8.82 \times 10^{-5}\text{cm} = 0.882\mu\text{m}$ from Eq. (12.13b)
b) $Q_{dm} = -qN_A W_m = -1.41 \times 10^{-8}\text{C/cm}^2$
c) $n_s = n_{0p} \exp\left(\dfrac{2q\phi_F}{kT}\right) = n_{0p} \times 9.996 \times 10^9$

$$n_{0p} \cong \frac{n_i^2}{N_A} = 10^5\text{cm}^{-3}$$

$$n_s = 9.996 \times 10^{14}\text{cm}^{-3}$$

Precise calculations yield $n_s = 10^{15}\text{cm}^{-3}$.
d) $V_T = 2\phi_F - Q_{dm}/C_{ox} = 0.596 + (1.41 \times 10^{-8}/3.83 \times 10^{-7}) = 0.596 + 0.0368 = 0.6328\text{V}$

12.5 CAPACITANCE-VOLTAGE MEASUREMENTS

Measurements leading to the capacitance-voltage characteristics of an MOS capacitor provide valuable information, which can be used to identify the deviations from the ideal in both the oxide and the semiconductor.

These measurements are made by using an electronic bridge. The bridge circuits supply a slowly varying DC gate voltage superimposed by an AC voltage having an effective value of about 10mV at a frequency of 1MHz. Plots of capacitance versus gate voltage are displayed on a recorder and they have the shape shown in Fig. 12.8. The capacitance is determined from the AC voltage and current.

Since, in the measuring bridge, an AC voltage is superimposed on the DC gate bias, it is important to briefly explain the effect of the frequency of the AC signal on the majority and minority carriers.

Majority carriers respond to changes in the electric field with a time known as the *dielectric relaxation time*. This is the time within which majority carriers are

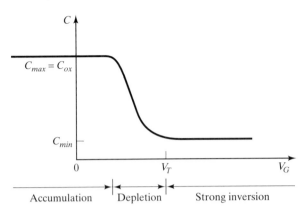

Figure 12.8   Capacitance-voltage characteristics of a P substrate MOS capacitor.

redistributed in response to an electric disturbance. It is given by the ratio of the dielectric constant to the conductivity ε/σ. For semiconductor samples, such as silicon and gallium arsenide, having a resistivity of 1 ohm-cm, this time is of the order of 10^{-12}s. Therefore, at the frequency of the AC signal voltage, corresponding to a period of 1μs, majority carriers will have no trouble responding to the variation in the AC signal.

Those minority carriers that, in the MOS capacitor, form an inversion layer at the surface of the oxide and originate by generation in the bulk, take a fraction of a second to form the inversion layer. Therefore, at the frequency of 1 MHz and a period of 1μs of the AC signal voltage, it will not be possible for the inversion layer to acquire ΔQ minority carriers in response to the AC signal voltage, as we will show later.

The characteristics shown in Fig. 12.8 are obtained by slowly varying the DC gate voltage. At every value of the gate voltage, the AC signal causes a varying charge such that the capacitance in F/cm^2 is given by

$$C = dQ_m/dV_G \tag{12.16}$$

where Q_m in C/cm^2 is the charge density on the gate and the gate voltage is the sum of the voltage drops in the oxide and across the semiconductor, which is equal to the surface potential ϕ_s, so that V_G becomes

$$V_G = V_{ox} + \phi_s \tag{12.17}$$

The charge density on the gate is balanced by the charge density in the semiconductor Q_s, made of the charges in the inversion and depletion layers.

$$Q_m = -Q_s = -(Q_n + Q_d) \tag{12.18}$$

By using Eqs. (12.17) and (12.18) in Eq. (12.16), we have

$$C = dQ_m/(dV_{ox} + d\phi_s) = \{[1/(dQ_m/dV_{ox})] + [1/(dQ_m/d\phi_s)]\}^{-1}$$
$$= 1/[1/C_{ox} + 1/C_s] \tag{12.19}$$

where C_{ox} and C_s represent the capacitances of the oxide and the semiconductor respectively. On the semiconductor side, C_s consists of the depletion layer capacitance and the inversion layer capacitance.

Let us now consider the variations of the charge in the semiconductor that accompanies the applied AC signal as the DC bias is changed from accumulation to depletion and then to inversion. The MOS capacitor used has a P substrate.

Accumulation

A negative bias applied to the gate causes an accumulation of majority carrier holes at the interface of the oxide and the semiconductor, and a corresponding surface sheet of electrons is formed at the metal interface with the oxide. At an AC signal frequency of 1 MHz and the majority carrier dielectric relaxation time of 1 ps, the state of the system is changed very rapidly. The device follows the signal with the small signal causing the addition or subtraction of a charge ΔQ on the two sides of the oxide, as shown in Fig. 12.9(a).

With reference to Eq. (12.19), we note that the total capacitance is made up of an oxide capacitance and a semiconductor capacitance caused by accumulation. A portion of ΔQ is attributed to V_{ox} and the rest to ϕ_s. However, the surface hole density, as shown by Eq. (12.6), is proportional to $\exp(K\phi_s)$, so that a small change in ϕ_s results in a large change in Q_s and hence a very large capacitance C_s. Effectively,

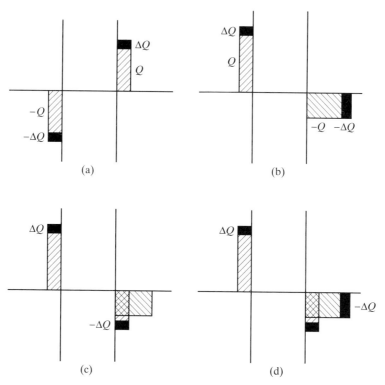

Figure 12.9   Charge changes in the P-type MOS capacitor in response to the AC signal and the DC bias (a) accumulation, (b) depletion, (c) inversion low-frequency signal, and (d) inversion high-frequency signal.

since the two capacitances are in series, one can safely state that the total capacitance at accumulation is the oxide capacitance C_{ox}, as shown in Fig. 12.8, and given by Eq. (12.20) where t_{ox} is in cm and ε_{ox} is in F/cm

$$C_{ox} \,(\text{F/cm}^2) = \varepsilon_{ox}/t_{ox} \qquad (12.20)$$

where t_{ox} is the oxide thickness and ε_{ox} is the dielectric constant of the oxide.

Depletion

As the gate voltage is increased and made positive, holes are driven away from the semiconductor interface and a depletion region develops in the semiconductor surface. This region is made up of negatively charged acceptor atoms. By using Eq. (12.12) for the width of the depletion layer, we determine the charge in it and its capacitance as

$$Q_s = Q_d = -qN_AW = -\sqrt{2\varepsilon q\phi_s N_A} \qquad (12.21)$$

$$C_s = -dQ_s/d\phi_s = \sqrt{\varepsilon q N_A/2\phi_s} = \varepsilon/W \qquad (12.22)$$

Because only majority carriers are involved, the system reaches equilibrium very rapidly. The AC signal causes, as shown in Fig. 12.9(b), small-signal variations in the charge, ΔQ, in the depletion layer. The overall capacitance, consisting of the series connection of the oxide capacitance, C_{ox}, and the capacitance, C_s, across the depletion layer is given as

$$1/C = (1/C_s) + (1/C_{ox}) = (W/\varepsilon) + (1/C_{ox}) \qquad (12.23)$$

A further increase of V_G increases ϕ_s, resulting in a wider depletion layer and a smaller depletion capacitance. The total capacitance decreases as the gate voltage increases, as shown in Fig. 12.8.

By using Eqs. (12.21) and (12.22), both the voltage across the oxide and the surface potential can be expressed in terms of W and their sum is equal to the gate voltage V_G. One can then use Eq. (12.23) to obtain the total capacitance as a function of C_{ox} and V_G as

$$C = \frac{C_{ox}}{[1 + (2V_G C_{ox}^2/q\,N_A\,\varepsilon)]^{1/2}} \qquad (12.24)$$

Inversion

An additional increase of the gate voltage takes the capacitor to inversion: electrons tend to pile up at the semiconductor surface and as ϕ_s increases, more energy band bending takes place at the surface. The width of the depletion layer reaches its maximum value, W_m. The density of surface electrons is proportional to $\exp(K\phi_s)$, so that a small increase in ϕ_s should be accompanied by a large change in Q_s and a proportionately large capacitance if the AC signal changes very slowly. Under these conditions, the resulting capacitance is that of the oxide capacitance in a series with the semiconductor capacitance, which is extremely large. The charge fluctuation in the semiconductor will take place in the inversion layer, as shown in Fig. 12.9(c), and the capacitor reacts as an ordinary capacitor and $C(\text{inv}) \cong C_{ox}$.

When the measurement frequency is very high, the slow generation-recombination process, which supplies electrons and retrieves them from the surface (at the rate of $10^{-5} - 10^3$s) is not able to respond to the AC signal whose period is 1μs. Therefore, the density of electrons at the surface is fixed at the DC value and the depletion layer width will increase and decrease about its DC value, as shown in Fig. 12.9(d). The measured capacitance approaches a value corresponding to the series connection of the oxide capacitance and the depletion layer capacitance (at its maximum width and smallest C). The overall capacitance is given by C_{min}, as

$$1/C_{min} = 1/C_{ox} + W_m/\varepsilon \qquad (12.25)$$

We note that the onset of strong inversion occurs at the threshold voltage, V_T, corresponding to the capacitance C_{min} where W_m has settled at its largest value.

The maximum width is given by Eq. (12.13b), so that the capacitance C_{min} becomes

$$C_{min} = \frac{C_{ox}}{1 + C_{ox}\,[4\phi_F/q\,N_A\,\varepsilon]^{1/2}} \qquad (12.26)$$

We summarize the calculations of capacitance by the following:

1. For a P substrate MOS structure, a negative gate voltage, greater in magnitude than V_T, indicates operation in the accumulation region. The dynamic capacitance is given by the oxide capacitance, C_{ox}.
2. Depletion region operation occurs when the gate voltage is positive but smaller than V_T. The capacitance depends upon the gate voltage and is given by Eq. (12.24).
3. At values of gate voltage greater than V_T, operation is in the inversion region and the capacitance is given by Eq. (12.26).

From the C-V measurements, it becomes possible to determine certain important properties of the MOS capacitor. The value of V_T is determined as the gate voltage V_G at which the capacitance first reaches its minimum value. By using the value of the oxide capacitance at accumulation, the oxide thickness is determined. The maximum depletion layer width, W_m, at strong inversion, is calculated by using Eq. (12.13b). The doping density, N_A, is calculated by using iteration in the expressions of Eqs. (12.2) and (12.13b). In these calculations, the magnitude of ϕ_F is also found.

It is important to mention here that the C-V characteristics have been obtained for the ideal case. It is ideal since we are not including the effects of work function difference between the metal and semiconductor and the effects of positively charged particles in the oxide and in the interface of the Si-SiO$_2$. These will be considered in the following sections. For an aluminum gate and a P substrate, both of these effects tend to reduce the value of V_T and therefore translate the C-V characteristics to the left of those shown in Fig. 12.8.

12.6 OXIDE CHARGES IN MOS CAPACITORS

The study of MOS capacitors was profoundly influenced in its early stages by the instabilities that were observed in the devices. One of the earliest problems was the observed shift in the capacitance-voltage characteristics by tens of volts. The shifts occurred when the devices were subjected to reliability-testing procedures. On certain occasions, the characteristics were observed to change with time as a result of a DC bias that was applied for a certain time duration.

After thorough and careful studies, it was initially concluded that these instabilities were caused by the presence of positively charged mobile alkali ions in silicon dioxide. While these ions seemed to be the only cause of instabilities, it was later concluded that, in addition to the mobile ions, there were other sources of charges in the oxide or in the interface between the silicon and the silicon dioxide. These other charges consist of: fixed oxide charges, oxide trapped charges, and interface trapped charges.

We will briefly consider each of these charge-categories separately. Their locations, in the silicon dioxide and in the interface of the Si-SiO$_2$, are shown in Fig. 12.10.

Mobile Ions (Q_M)

It was concluded that alkali ions, in particular sodium ions (N_a^+), can move through crystalline SiO$_2$, at temperatures below 250°C. The effect of these ions is to shift the capacitance-voltage characteristics always in a direction opposite to that of the applied gate voltage. These ions tend to drift when an electric field is sustained across the oxide, are particularly mobile in SiO$_2$ at high temperatures, and cause a shift of several volts in the threshold voltage when a bias is applied to the gate. To reduce their influence on the threshold voltage, a negative bias has to be applied to the gate.

Alkali ions, and sodium ions more specifically, are abundant in the processing atmosphere. They are located in chemical reagents, in glass apparatus, in the hands of laboratory personnel, and even in the quartz tube.

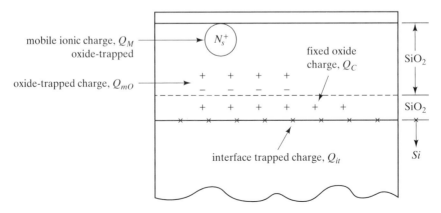

Figure 12.10 Types and locations of charges in oxidized silicon.

The first step that is taken to eliminate these contaminants is to maintain a strictly clean processing environment. In addition, stabilization procedures are used to reduce their effects. One of the methods used is to introduce a small quantity of chlorine in the form of gaseous hydrochloric acid into the oxidation furnace during the growth of the SiO_2 layer.

Fixed Oxide Charge (Q_f)

These are positive charges located within the oxide at a distance of about 30 Angstroms from the silicon-silicon dioxide interface. It has been postulated that these charges are formed during the oxidation process and are due to excess ionic silicon, which break away from silicon and form a thin layer when the oxidation process is terminated. Their presence is attributed to the oxidizing environment and to the furnace temperature.

Their effect, while not as severe as that of the mobile ions, is to cause a translation of several volts in the C-V characteristics. Because these charges are positive, they cause a translation in V_G of about $-Q_f / C_{ox}$.

These contaminants can be effectively reduced by annealing the wafer in an inert atmosphere containing argon or nitrogen.

Interface Trapped Charges (Q_{it})

These charges, also known as *surface states*, refer to both electrons or holes trapped at the interface of the silicon and silicon dioxide. They are assumed to be produced by the presence of excess oxygen and impurities.

The surface states tend to produce, throughout the silicon and in the forbidden energy band, energy levels in which electrons are physically trapped in the vicinity of the surface. It is acknowledged that while these traps represent standard defects caused by oxidation, they arise from what are known as "dangling bonds" at the surface of the semiconductor. They are "dangling" since when the periodic crystal lattice of silicon is abruptly terminated, one of the four covalent bonds of each of the atoms at the surface is left unattached and left dangling, while some of the bonds are tied up by the SiO_2 layer. It has been observed that a small density of these charges can have a significant effect on device characteristics, as they commonly tend to distort the shape of the C-V characteristics.

The effect of these traps can be considerably minimized by annealing the oxidized silicon wafers in a hydrogen ambient at a relatively low temperature of about 450°C.

Oxide Trapped Charges (Q_{mo})

These are positively charged holes trapped in the silicon dioxide. These charges are photogenerated during the processes of etching and deposition of thin films while the wafers are being exposed to low-energy x-rays. Since the band gap of SiO_2 is approximately 8eV, a photon that is incident on the SiO_2 layer and has energy greater than 8eV generates electron-hole pairs. The electrons, having greater mobilities, are swept away by any electric field present, while the holes remain trapped in the silicon dioxide.

While they do not produce serious disruptions in the C-V characteristics, the excess positive charges tend to cause a shift in the threshold voltage.

These charges can be completely eliminated by annealing the wafers, subsequent to the exposure to the radiation that produces them, at a temperature of about 400°C.

12.7 EFFECTS OF WORK FUNCTIONS AND OXIDE CHARGES

Early on, we assumed, in the ideal device and with the applied voltage set to zero, that the Fermi levels are aligned and that there was no band bending. We label the device ideal since we neglected the work function difference between the metal and the semiconductor and we also assumed an oxide that did not contain charges that influence the threshold voltage.

Work Function Difference

Both the 0.95eV work function difference in Fig. 12.2 and the presence of positive charges in the oxide cause the bands to bend with $V_G = 0$. The work function difference depends upon the type of metal used and on the density of the dopant atoms in the silicon. With reference to Fig. 12.2, the expression for the work function difference, $q\Phi_{ms}$, between aluminum and a P-type semiconductor is given by

$$q\Phi_{ms} = q(\Phi_m - \Phi_s) = q\left[\Phi_m - (\chi_s + E_g/2q + \phi_F)\right] \tag{12.27}$$

where $q\Phi_m$ is the metal work function. For the semiconductor, $q\chi$ is the affinity, $q\phi_F$ is $(E_i - E_F)$, and E_g is the band gap energy.

In Fig. 12.2, Φ_m for aluminum is 4.1V, $E_g = 1.12$eV, and $\chi_s = 4.15$V for silicon. For a dopant density of 4×10^{15}cm^{-3}, ϕ_F is 0.33V, so that Φ_{ms} is calculated to be approximately -0.95V, indicating that the metal work function is less than that of silicon by 0.95V. The lower work function causes electrons to transfer from the metal to the semiconductor. This effect is identical to that caused by the application of a positive V_G to the metal. The bands in the silicon bend downwards by $\phi_s = 0.95$V at thermal equilibrium. A plot of work function difference versus substrate doping, using Al on Si, is shown in Fig. 12.11.

Another widely known material used as a gate electrode in MOS devices is heavily doped polycrystalline silicon, commonly known as polysilicon, which when N^+ doped has a work function of 3.95eV.

Oxide Charges

The oxide charges are positive and their presence attracts electrons to the silicon surface. This causes a voltage to appear in the silicon and is accompanied by downward bending of the bands. These charges having effective charge surface density Q_i cause a voltage in the silicon, as if they were placed on the metal-silicon dioxide interface. The voltage has magnitude Q_i/C_{ox} and adds to the work function difference, causing an additional downward bend in the energy bands in the silicon. To

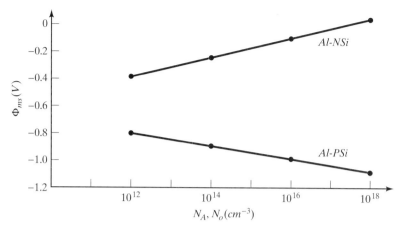

Figure 12.11   Work function differences as a function of substrate doping for Al on Si.

cancel their effect, it is necessary to apply a negative voltage to the gate in order to flatten the energy bands in the silicon.

Flatband Voltage

In order to compensate for the voltages of the work function difference and the oxide charges, a negative voltage has to be applied between the metal and the silicon to achieve flattening of the bands. We must point out that a diode in which the bands are flattened by the application of a voltage to the metal is no longer in thermal equilibrium. This voltage is known as *the flatband voltage*, V_{FB}, and is given by

$$V_{FB} = \Phi_{ms} - Q_i / C_{ox} \tag{12.28}$$

where Φ_{ms} is a negative quantity for aluminum on silicon and Q_i is a positive quantity.

With the flatband voltage so defined, the threshold voltage, defined by Eq. (12.14), is modified to include the flatband voltage.

$$V_T = V_{FB} - \frac{Q_{dm}}{C_{ox}} + 2\phi_F$$

$$= \Phi_{ms} - \frac{Q_i}{C_{ox}} - \frac{Q_{dm}}{C_{ox}} + 2\phi_F \tag{12.29}$$

For a device using aluminum as the metal and P-type silicon substrate biased to the onset of strong inversion, Q_{dm} is a negative quantity, $2\phi_F$ is positive, Q_i is positive, and Φ_{ms} is negative. Both the work function difference and the oxide charges serve to reduce V_T. They both cause a translation of the C-V characteristics to the left of the ideal ones shown in Fig. 12.8.

Tabulated below are the signs of the quantities that make up the threshold voltage in Eq. (12.29).

TABLE 12.2

	Φ_{ms}*	$-Q_i/C_{ox}$	$-Q_{dm}/C_{ox}$	$2\phi_F$	V_T
N channel	<0	<0	>0	>0	positive or negative
P channel	<0	<0	<0	<0	negative

*The sign of Φ_{ms} depends upon the metal and the semiconductor doping. It is negative for aluminum on silicon.

Why the Label "Flatband"?

The inquiring reader will wonder why is it that we need to "flatten the bands" when we are trying (in a P substrate) to bend the bands downward? Both the work-function difference and the oxide charges cause the bands to bend downward and seem to work in our favor.

In fact, we do not flatten the bands. The label "flatband" may convey the wrong connotation and it may be a misnomer. The flatband voltage helps by bending the bands downward and the answer to our earlier question lies in the definition of the threshold voltage.

The threshold voltage has been defined as the voltage that is applied to the metal, with the source grounded, that causes the onset of strong inversion and makes the magnitude of Q_n incrementally greater than zero. An increase of the metal voltage above V_T increases the magnitude of Q_n that is achieved at strong inversion. If it were not for the flatband voltage, one would need to apply a larger voltage to the metal, hence a larger V_T, to achieve the onset of strong inversion.

In the majority of applications of MOSFETs (NMOS P-substrate devices) in digital circuits, a positive V_T is required. Because a large magnitude of V_{FB} (which is negative) may result in a negative V_T, additional methods are used to control V_T and make it positive. These methods, which include ion implantation, are discussed in the following chapter on MOSFETs.

The following examples illustrate the calculations of the flatband and threshold voltages.

EXAMPLE 12.2

Determine the flatband voltage and the threshold voltage for an MOS device that has a silicon P substrate with $N_A = 10^{16} cm^3$ and uses aluminum for the gate. Given $Q_i = 5 \times 10^{10} qC/cm^2$, $\varepsilon_{ox} = 3.9 \times 8.85 \times 10^{-14} F/cm$, $t_{ox} = 200$ Angstroms, and $\varepsilon = 11.8 \times 8.85 \times 10^{-14}$ F/cm, use $n_i = 1 \times 10^{10} cm^{-3}$, $E_g = 1.12 eV$, and $\chi_s = 4.15 V$.

Solution

$$\phi_F = (kT/q) \ell n (N_A/n_i) = .0259 \ \ell n \frac{10^{16}}{10^{10}} = 0.357V$$

$$\Phi_{ms} = \Phi_m - (\chi_s + E_g/2q + \phi_F)$$

$$= 4.1 - (4.15 + 0.56 + 0.357) = -0.967V$$

$$Q_{dm} = -(4\varepsilon \ \phi_F \ qN_A)^{0.5} = -4.88 \times 10^{-8} C/cm^2$$

$$Q_i = 5 \times 10^{10} \times 1.6 \times 10^{-19} = 0.8 \times 10^{-8} C/cm^2$$

$$t_{ox} = 200 \times 10^{-8} = 2 \times 10^{-6} cm$$

$$C_{ox} = \varepsilon_{ox}/t_{ox} = \frac{3.9 \times 8.85}{2} \times 10^{-8} = 1.725 \times 10^{-7} F/cm^2$$

We use Eq. (12.29) to calculate V_T as

$$V_T = \Phi_{ms} - Q_i/C_{ox} - Q_{dm}C_{ox} + 2\phi_F$$

$$V_T = -0.967 - \frac{0.8 \times 10^{-8}}{1.725 \times 10^{-7}} - \frac{-4.88 \times 10^{-8}}{1.725 \times 10^{-7}} + 0.714$$

The flatband voltage is made up of the first two terms in V_T and is $V_{FB} = -1.0133V$.

$$V_T = -0.967 - 0.282 + 0.473 + 0.714 = -0.016V$$

EXAMPLE 12.3

Repeat Example 12.2 but use $N_A = 10^{17} cm^{-3}$.

$$\phi_F = 0.417V, \Phi_{ms} = 4.1 - (4.15 + 0.56 + 0.417) = -1.027V$$

$$Q_{dm} = -16.66 \times 10^{-8} C/cm^2$$

All other terms are the same as in Example 12.2.

$$V_T = \Phi_{ms} - Q_i/C_{ox} - Q_{dm}/C_{ox} + 2\phi_F$$

$$V_T = -1.027 - 0.046 + 0.965V + 0.834 = 0.726V$$

The flatband voltage has the same value as in Example 12.2.

We observe a substantial increase in the threshold voltage resulting from an increase in the doping.

We will now carry out calculations using an N substrate.

EXAMPLE 12.4

Determine the flatband voltage and the threshold voltage for a PMOS device that has silicon N substrate with $N_D = 10^{15} cm^{-3}$ and uses an aluminum gate. Use the values of the constants given in Example 12.2.

Solution

$$\phi_F = -kT/q \, \ell n \, N_D/n_i = -0.298V$$

We use Eq. (12.27) to calculate the work function difference Φ_{ms}.

$$\Phi_m = 4.1V, \chi_s = 4.15V, E_g/2q = 0.56V, \text{ and } \phi_F = -0.298$$

$$\Phi_{ms} = -0.312V$$

$$C_{ox} = 1.725 \times 10^{-7} \text{F/cm}^2, Q_i/C_{ox} = -0.046\text{V}$$

$$Q_{dm} = (4\varepsilon \phi_F q N_D)^{0.5} = 1.41 \times 10^{-8} \text{C/cm}^2$$

The flatband voltage is given by

$$V_{FB} = \Phi_{ms} - Q_i/C_{ox} = -0.312 - 0.046 = -0.358\text{V}$$

$$V_T = \Phi_{ms} - Q_i/C_{ox} - Q_{dm}/C_{ox} + 2\phi_F$$

$$V_T = -0.312 - 0.046 - 0.0817 - 0.596 = -1.035\text{V}$$

EXAMPLE 12.5

Use the data and results of Example 12.3, but increase t_{ox} to 300 Angstroms in order to calculate the threshold voltage.

Solution   From Example 12.3, $\phi_F = 0.417\text{V}$, $Q_{dm} = -16.66 \times 10^{-8} \text{C/cm}^2$, $Q_i = 0.8 \times 10^{-8} \text{C/cm}^2$, and $\Phi_{ms} = -1.027\text{V}$.

$$C_{ox} = 3.9 \times 8.85 \times 10^{-14}/300 \times 10^{-8} = 11.5 \times 10^{-8} \text{F/cm}^2$$

$$V_T = \Phi_{ms} - Q_i/C_{ox} - Q_{dm}/C_{ox} + 2\phi_F$$

$$V_T = -1.027 - (5 \times 10^{10} \times 1.6 \times 10^{-19}/11.5 \times 10^{-8}) + 16.66 \times 10^{-8}/11.5 \times 10^{-8} + 0.834$$

$$V_T = -1.027 - 0.069 + 1.44 + 0.834$$

$$V_T = 1.18\text{V}$$

Upon comparing the data of Examples 12.2 and 12.5, we note that the substrate doping and the oxide thickness have increased. The effect of these changes is to increase the threshold voltage from 0.174V to 2.066V. We will use this technique in the fabrication of the MOSFET to avoid any interaction between adjacent transistors on the same chip. The threshold voltage will be increased to values that are higher than the highest voltage supplied to the chip.

REVIEW QUESTIONS

Q12-8   Briefly define the flatband voltage.

Q12-9   Define the threshold voltage.

Q12-10   Qualitatively, explain how the work function affects the threshold voltage.

Q12-11   For an N-substrate capacitor, identify the signs of the various terms in the equation for the threshold voltage.

Q12-12   In achieving a positive, V_T, for a P-substrate capacitor, is it more favorable to have a larger or smaller work function difference between the metal and the semiconductor? Explain your answer.

Q12-13   Qualitatively explain how the oxide charges affect the threshold voltage. In aiming for a positive V_T for a P substrate and a negative V_T for an N substrate, is the larger or the smaller oxide charge more favorable?

HIGHLIGHTS

- In our early definition of threshold voltage, two terms were included in the expression for V_T: the potential at the surface at strong inversion and the potential due to the charges in the depletion layer.

- Two additional phenomena contribute to the threshold voltage. One is the work function difference between the metal and the semiconductor. For example, aluminum has a higher work function than P silicon, so that at thermal equilibrium, band bending takes place.

- A second defect that influences the magnitude of the threshold voltage is the presence of positively charged ions in the oxide insulator. The presence of these ions tends to reduce the threshold voltage for a P substrate.

- The total contribution of the work function difference and the ions in the oxide makes up the flatband voltage.

- For reasons related mainly to the improvement of frequency response, heavily doped polycrystalline silicon is used for gate material. It is known as polysilicon.

EXERCISES

E12-3 Determine the threshold voltage for a P-silicon substrate MOS capacitor with an aluminum gate that has $N_A = 3 \times 10^{15} \mathrm{cm}^{-3}$, oxide thickness of 1000Å, and $Q_i = 4 \times 10^{10}q$ cm^{-2}, given $\Phi_{ms} = -0.9$V, $\varepsilon_r(\mathrm{Si}) = 11.8$, and $\varepsilon_r(\mathrm{SiO_2}) = 3.9$.

Ans: $V_T = 0.309$V.

E12-4 An Au gate MOS capacitor is formed on an N-silicon substrate having $N_D = 10^{15} \mathrm{cm}^{-3}$. The thickness of the gate oxide is 300Å and the oxide charge density Q_i is $3 \times 10^{11}q$ cm^{-2}. Use the plots in Fig. 12.12 and determine: a) the flatband voltage and b) the turn-on voltage.

Ans: a) $V_{FB} = 0.14$V,   b) $V_T = -0.64$V.

Polysilicon Gate Work Function

The most common gate materials are: aluminum and heavily doped polycrystalline silicon (polysilicon). The reasons for the pervasive use of polysilicon in MOSFETs will be discussed in the following chapter.

Polysilicon gates are deposited by the CVD (chemical vapor deposition) process. They are then heavily doped by either diffusion or ion implantation. For an NMOS device, the N^+ doping is usually carried out simultaneously with the doping of the drain and the source. The work function for the polysilicon is given by

$$q\,\Phi_m = q[\chi_s + (E_C - E_F)_{\text{poly-si}}] \tag{12.30}$$

where χ_s is the affinity of silicon.

For an N^+ polysilicon gate, the Fermi level is at the bottom of the conduction band, so that the gate work function is the silicon affinity (4.15eV). On the other

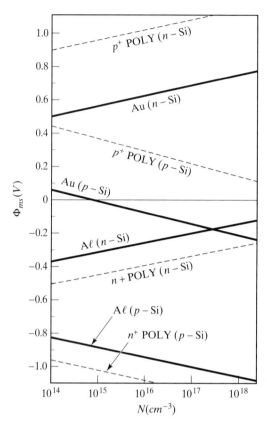

Figure 12.12 Dependence of the work function on silicon doping density. *Source*: M. S. Tyagi, *Introduction to Semiconductor Materials and Devices*, p. 481, copyright © Wiley (1991). Reprinted by permission of John Wiley & Sons, Inc.

hand, for a P^+ polysilicon gate, the Fermi level is at the top of the valence band and the work function is (5.27eV).

The work function difference between an N^+ polysilicon gate and a silicon P substrate becomes

$$\Phi_{ms} = \frac{1}{q}\left[(E_C - E_F)\text{poly-si} - (E_C - E_F)_{\text{substrate}}\right] \qquad (12.31)$$

We have exhibited in Fig. 12.12 the work function difference as a function of silicon substrate doping for Al, Au, P^+, and N^+ polysilicon.

We now consider an example using polysilicon gate.

EXAMPLE 12.6

Determine the flatband voltage and the threshold voltage for an NMOS silicon device that has $N_A = 4 \times 10^{16}\text{cm}^{-3}$ and uses SiO_2 and an N^+ polysilicon gate, given $Q_i = 5 \times 10^{10}\text{C/cm}^2$, $t_{\text{ox}} = 200$ Angstroms.

Solution

$$\phi_F = kT/q \ \ln N_A/n_i = 0.393\text{V}$$

$$Q_{dm} = -1.63 \times 10^{-7}\text{C/cm}^2, Q_i = 8 \times 10^{-9}\text{C/cm}^2$$

$$C_{ox} = 1.725 \times 10^{-2}\text{F/cm}^2$$

$$\Phi_{ms} = \Phi_{\text{poly-si}} - (\chi_s + E_g/2q + \phi_F)$$

$$\Phi_{ms} = 4.15 - (4.15 + 0.56 + 0.393) = -0.953\text{V}$$

$$V_{FB} = \Phi_{ms} + Q_i/C_{ox} = -0.953 - 0.046 = -1.0\text{V}$$

$$V_T = V_{FB} - Q_{dm}/C_{ox} + 2\phi_F$$

$$V_T = -1 + 0.945 + 0.786 = 0.731\text{V}$$

In the next chapter, we will apply the concepts developed in this chapter to the study of the MOSFET.

PROBLEMS

12.1 For an MOS capacitor, the silicon substrate is doped by $N_D = 5 \times 10^{16}\text{cm}^{-3}$. Calculate the surface potential required to make the semiconductor surface

    (a)   intrinsic.

    (b)   at strong inversion.

12.2 Repeat Problem 12.1 for a substrate doped by $N_A = 10^{17}\text{cm}^{-13}$.

12.3 An MOS capacitor uses a silicon substrate with doping of $N_A = 10^{14}\text{cm}^{-3}$ and has an oxide thickness of 0.1μm. Calculate for each of $V_G = 0.4$V and $V_G = 1$V,

    (a)   the surface potential.

    (b)   the voltage across the insulator.

12.4 Reproduce Table 12.1 in the text for an N-substrate MOS capacitor.

12.5 An MOS capacitor has silicon substrate doping $N_A = 10^{16}\text{cm}^3$ and oxide thickness of 300 Angstroms. Calculate for i) intrinsic semiconductor interface, and ii) onset of strong inversion

    (a)   the applied voltage.

    (b)   the electric field intensity at the interface.

12.6 An MOS capacitor has silicon substrate doping $N_A = 10^{16}\text{cm}^{-3}$. Calculate the maximum depletion layer width.

12.7 A silicon MOS capacitor substrate is doped with $N_A = 5 \times 10^{16}\text{cm}^{-3}$ and has an oxide thickness of 100Å. Calculate for i) intrinsic semiconductor surface, and ii) strong inversion:

    (a)   the gate voltage.

    (b)   the depletion layer width.

12.8 By calculating the surface charge density, n_s, and the charge in the depletion layer, verify by calculation the following statement in the text: "an increase of ϕ_s of 60mV, beyond the onset of strong inversion, causes n_s to become $10N_A$, whereas this increase in ϕ_s results in a negligible increase in W and hence in the magnitude of Q_d." Use $N_A = 10^{15}\text{cm}^{-3}$.

12.9 Calculate the threshold voltage for an N-channel MOS with oxide thickness of 80 Angstroms for substrate doping of: (a) $5 \times 10^{14}\text{cm}^{-3}$ and (b) $5 \times 10^{16}\text{cm}^{-3}$. Neglect oxide charge and work function difference.

12.10 Repeat Problem 12.9 for a P-channel MOS having doping of $5 \times 10^{16} \text{cm}^{-3}$ and oxide thickness of (a) 80 and (b) 300 Angstroms.

12.11 An Al-SiO$_2$-Si capacitor has substrate doping $N_D = 5 \times 10^{15} \text{cm}^{-3}$ and an oxide thickness of 200 Angstroms. The charge density at the silicon-silicon dioxide interface is $5 \times 10^{10} q/\text{cm}^2$. Calculate:

    (a)   the flatband voltage.

    (b)   the threshold voltage.

12.12 Derive Eq. (12.24).

12.13 Calculate the capacitance per unit area of a silicon MOS capacitor having $N_A = 4 \times 10^{15} \text{cm}^{-13}$, $t_{ox} = 100\text{Å}$, and $\phi_m = \phi_s$ at an applied gate voltage of:

    (a) -1V

    (b) 0.5V

    (c) 1V

    Where relevant, calculate the capacitance at (i) low frequency and (ii) high frequency.

12.14 For an ideal MOS diode having doping $N_A = 5 \times 10^{16} \text{cm}^{-3}$ and an oxide thickness of 120 Angstroms, calculate the value of the minimum capacitance on the C-V curve.

12.15 For an Al-SiO$_2$-Si capacitor, the doping of the substrate is $N_A = 10^{14} \text{cm}^{-3}$ and the oxide thickness is 80 Angstroms. Given that the flatband voltage, V_{FB}, is -0.4V, calculate at strong inversion

    (a)   the surface potential.

    (b)   the maximum depletion layer width.

    (c)   the gate voltage.

12.16 A polysilicon-SiO$_2$-Si diode has substrate doping of $N_A = 5 \times 10^{14} \text{cm}^{-3}$, an oxide thickness of 100 Angstroms, and an oxide charge of $5 \times 10^{11} q$ C/cm^2. Calculate:

    (a)   the flatband voltage.

    (b)   the threshold voltage.

12.17 A silicon substrate is doped with $N_A = 4 \times 10^{16} \text{cm}^{-3}$. The thickness of the silicon dioxide is 120 Angstroms. To form an MOS capacitor, a polysilicon gate is deposited. Calculate the threshold voltage, V_T, for

    (a)   zero oxide charge.

    (b)   oxide charge of $10^{11} q$ C/cm^2.

    The work function of polysilicon is $\Phi_m = 3.95\text{eV}$.

12.18 An MOS capacitor using Al-SiO$_2$ has substrate doping $N_D = 10^{17} \text{cm}^{-3}$, gate oxide thickness of 100 Angstroms, and oxide charge of $5 \times 10^{11} q$ C/cm^2. Calculate:

    (a)   the flatband voltage.

    (b)   the threshold voltage.

12.19 Show that the number of electrons on a MOS gate that can cause an oxide to break down depends on the gate area and the oxide permittivity. (Assume the breakdown electric field intensity is known).

12.20   **(a)**   For an MOS that has length of 6μm and depth of 20μm, which is charged by static electricity, how many electrons could be transferred to the gate before it breaks down? $\mathscr{E}_{br} = 8 \times 10^6 \text{V/cm}$.

    (b)   How long would it take to deliver these electrons to the gate if an average current of 1pA flowed because of the transfer of static electricity.

    (c)   For a one volt drop across the oxide, determine the minimum thickness of the oxide to avoid breakdown.

chapter 13

METAL-OXIDE-SEMICONDUCTOR FIELD-EFFECT TRANSISTOR

13.0 INTRODUCTION

Metal-oxide-semiconductor field-effect transistors have played a major role in the development of complex large scale integrated circuits. In particular, they have provided the basis for most large scale integrated digital circuits. Acronyms for the MOSFET are the IGFET (*insulated-gate field-effect transistors*) and the MISFET (*metal-insulator-semiconductor field-effect transistors*). The major reason for this pervasive use is that they are significantly less expensive to manufacture than bipolar circuits of equivalent functions.

The MOSFET operation depends on the conductance modulation of a channel of carriers that are induced by an applied gate voltage. Modulation is achieved by the variation in the carrier surface density. In the JFET on the other hand, conductance modulation is accomplished by the variation of the cross-sectional area of a channel formed from a PN junction. FETs are also known as unipolar devices because conduction is by means of one type of carrier only.

One very important application of MOSFETS is in the arrangement known as a complementary metal oxide semiconductor system (CMOS). The CMOS forms, at the present time, the mainstream of high density digital system design technology.

At present, most MOS circuits operate at a supply voltage of 3.3 volts. The very small size of MOS circuits raises problems of heat removal. To alleviate this problem, it is expected that a 3-volt supply will become the standard for MOS circuits.

13.1 CONSTRUCTION AND BASIC OPERATION

In Fig. 13.1, we reproduce Fig. 12.1 transposed by 90° and modified by the addition of two metal contacts located at the two ends of the inversion layer in order to form the basic components of a MOSFET.

The functions of the contacts that are made to the inversion layer are there so as to form the source and drain between which an electric field is established by an applied voltage, V_D, from drain to source. The electric field accelerates the electrons in the inversion layer from source to drain. The inversion layer is formed by capacitor action of the voltage across the insulator between gate and substrate.

The device is therefore composed of a metal gate and an insulator separating the gate from the semiconductor region where conduction is to take place. We assume a P-type semiconductor, N^+ drain, and source contacts, so that conduction is carried out by the drift of electrons in the inversion layer from source to drain. The minority carrier density of electrons in the semiconductor at thermal equilibrium is very small, so that it becomes necessary to provide, by other means, a high density of electrons in the channel. In this particular model, a channel is not physically constructed; rather, it is the accumulation of the electrons in a thin layer in the semiconductor adjacent to the oxide that becomes the channel.

By applying a positive voltage to the metal gate, greater than a threshold voltage V_T, and assuming the bulk or substrate is grounded, a very thin layer of electrons is induced by capacitor action in the surface between the semiconductor and the insulator. These electrons become the current carriers. The region in which they move is the *channel*. In fact, a thin N region has been formed on the surface of the substrate that is in contact with the insulator.

In Fig. 13.2, we show a perspective view of the MOSFET together with a cross section of a P-substrate, N-channel device.

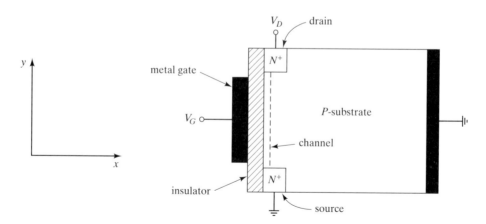

Figure 13.1 A crude layout of a P-substrate MOSFET showing basic components.

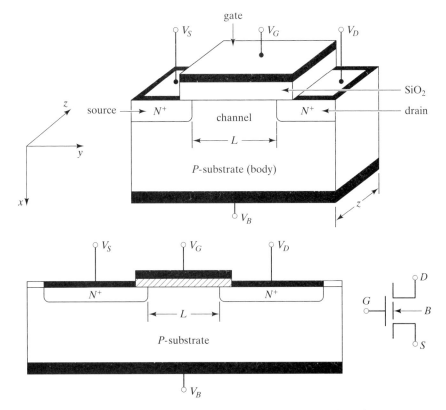

Figure 13.2   Perspective view, cross-section, and symbol of an N-channel MOSFET (NMOS).

13.2 FABRICATION OF N-TYPE MOSFET (NMOS) ON AN INTEGRATED CIRCUIT CHIP

Two major processes are peculiar to MOSFET fabrication. First, there is the scheme for isolation between two adjacent transistors and, second, the use of polysilicon for the gate, which permits the almost perfect self-alignment of the source and drain with respect to the gate.

Isolation Process

With BJTs, isolation between adjacent transistors on a chip was provided by a reverse-biased PN junction. A major advantage of MOSFETs, and one that results in considerable area saving compared to BJTs, is provided by their inherent self-isolation. Isolation in a BJT occupies a large percentage of the area of the transistor. Mainly because of self-isolation, the packing density of MOSFETs is about 20 times that of BJTs.

We have seen earlier that the drain-source current consists of the motion of carriers that are induced in the substrate. These carriers are formed by the capacitor action of the gate-insulator-substrate capacitor. In a P-substrate device, the application of a positive voltage to the gate (with respect to the substrate) that is equal to, or exceeds, a threshold voltage, V_T, causes the formation of an inversion layer of electrons at the substrate-oxide interface, which becomes the channel.

Two transistors normally separated by a P region covered by an insulator and a metal may have the drain of one connected to the adjacent source of the other. A "parasitic" transistor may be formed by the two adjacent electrodes since the intervening metal-insulator-substrate causes a "parasitic" channel. The formation of a "parasitic" channel and transistor is avoided if the threshold voltage of the "parasitic" device is greater than the highest supply or signal voltage on the chip. As shown in Eq. (13.7), this is accomplished in two ways: first, by increasing the doping underneath the oxide separating the devices, and, second, by making this oxide, known as the *field oxide*, much thicker than the gate oxide, as we shall discuss in Sec. 13.7.

The doping level in the P substrate of the field region is increased by implanting boron (for a P-substrate NMOS) to make it a P⁺ region, as will be illustrated in the section titled basic steps in fabrication. The thicker oxide and the higher doping serve to increase the threshold voltage of the parasitic devices.

The process of local oxidation, LOCOS, is used to form both the thick oxide region and thin oxide of the gate. An additional requirement of LOCOS is that the transition from the thick to the thin oxide should be a smooth and tapered step, rather than a steep wall, so that a smooth path for the overlaid metal is provided. This is shown in Fig. 13.3.

In local oxidation, a thin layer, 1000nm thick, of oxide is grown over the whole wafer. A thicker layer of silicon nitride is grown over the silicon dioxide. The silicon nitride is subsequently removed, with a masking step, from those areas of the field

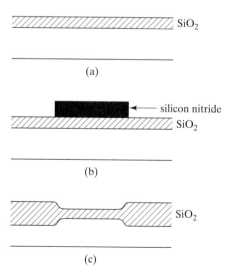

Figure 13.3   Local oxidation (a) SiO₂ covers wafer, (b) silicon nitride covers SiO₂ and, (c) after oxidation and nitride removal.

regions where a thick oxide is to be grown. Silicon nitride has the property that it blocks oxygen diffusion, which might cause additional oxidation. Following this, a high temperature oxidation is carried out in which a thick oxide is grown in the regions not covered by the nitride and no additional oxidation takes place under the nitride.

Polysilicon

A major element that limits the high-frequency gain of MOS transistors is the capacitance formed by the overlap between the gate and drain, as seen in Fig. 13.2. Some overlap is necessary between the gate and each source and drain, so as to identify the boundaries of the channel as the channel is formed by the junction edges in the y-direction. When a metal, such as aluminum, is used for the gate electrode, the source and drain are formed before the gate is overlaid. As a result, additional overlap, source, and drain diffusion depths (2μm) are required to accommodate the tolerances in masking alignment during fabrication. This overlap results in the relatively large capacitance that limits the high-frequency gain of MOS transistors.

Polycrystalline silicon, or polysilicon, is an almost pure semiconductor and is the form that silicon takes prior to the growth of the very pure silicon in the Czochralski process. It is made of small crystallites with roughly submicrometer dimensions. A major advantage of using polysilicon, in contrast to aluminum, for the gate stems from its property of being resistant to high temperatures and corrosion by chemicals. It can withstand the high temperature required in the diffusion of the drain and source, so that polysilicon electrodes do not need to be placed after all other operations have been carried out, as is the case for metal electrodes. By being in place, ahead of the drain and source, the gate can serve as a self-aligning mask. The use of polysilicon ensures that there will always be some overlap between gate and drain due to the lateral diffusion of the dopant atoms.

Furthermore, very shallow depths (0.2 μm) can be used for source and drain diffusions concurrent with the reduced overlap. Thus, lower parasitic capacitances and the advantage of smaller area devices decrease the cost and improve device performance.

Next, we briefly discuss the process of deposition of the three types of films discussed in this section.

The Deposition of Silicon Dioxide, Silicon Nitride and Polysilicon

The deposition of these films is carried out by a process of chemical vapor deposition (CVD), as explained in the epitaxial growth section of Chapter 6. The films are deposited out of a gaseous mixture onto the surface of the wafer.

Silicon dioxide is deposited by heating the wafers at a temperature of 400°C, using the reaction of silane (SiH_4) with oxygen.

Silicon nitride (Si_3N_4) is deposited at a temperature of 700°C using a mixture of dichlorosilane and ammonia.

Polysilicon is deposited by heating the wafers to a temperature of 625°C and exposing them to an atmosphere of silane, which decomposes into silicon and hydrogen.

In the CVD process, all elements forming the deposited layer are introduced into the quartz enclosure; nothing comes from the silicon wafer. CVD polycrystalline films are, as indicated earlier, made of many submicrometer crystallites and form uniform layers.

Basic Steps in Fabrication

In the next section, there will be frequent references to masks. We remind the reader that our label of the mask is made up of numerous processing steps. A mask contains a pattern or drawing that is transferred to the silicon wafer using photolithography. In this process, the drawing on the mask is transferred from the mask to a photoresist. Etching is then used to transfer the pattern from the photoresist to the surface of the wafer. In Fig. 13.4, we illustrate the fabrication steps, including the deposition of the metal contacts.

A P-type silicon substrate having a resistivity of about 5 ohm-cm is the starting wafer. The doping density is determined by the drain-substrate breakdown voltage of 20–30V [see Eq. (7.11)]. A thin (100nm) layer of silicon dioxide is formed over the substrate to provide stress relief to the wafer. A thicker silicon nitride (Si_3N_4) layer is deposited by the CVD process on top of the silicon dioxide. The silicon nitride permits selective oxidation so that a thick oxide (1μm) can be formed in the field region. These operations result in Fig. 13.4(a).

Mask No. 1 defines the transistor areas, so that the SiO_2 and the silicon nitride are chemically etched out except where the transistor is formed.

In Fig. 13.4(b), boron has been ion-implanted in the field regions to form P^+ islands in the substrate so as to help increase the threshold voltage, V_T, and prevent the formation of "parasitic" transistors between adjacent devices on the wafer. Following this, and also shown in Fig. 13.4(b), a field oxide layer, about 1μm thick is formed over the P^+ implanted region. This also helps to increase V_T.

As shown in Fig. 13.4(c), the remaining silicon nitride and silicon dioxide are etched away and a thin (about a hundred angstroms) layer of SiO_2 is grown over the transistor area (not over the field). This forms the gate oxide of the transistor.

A layer of polysilicon is next deposited over the entire wafer surface. This is shown in Fig. 13.4(d).

Mask No. 2 defines the gate region, whereupon the polysilicon is etched away except over the gate, as shown in Fig. 13.4(e).

By ion-implantation, and using the polysilicon gate and the field oxide as the mask, the source and drain N^+ regions are formed. The N^+ layers laterally diffuse a sufficient distance to insure proper alignment so that the channel length is well-defined. The dopants do not penetrate the field oxide. A thin layer of SiO_2 is then grown, by the CVD process, over the wafer.

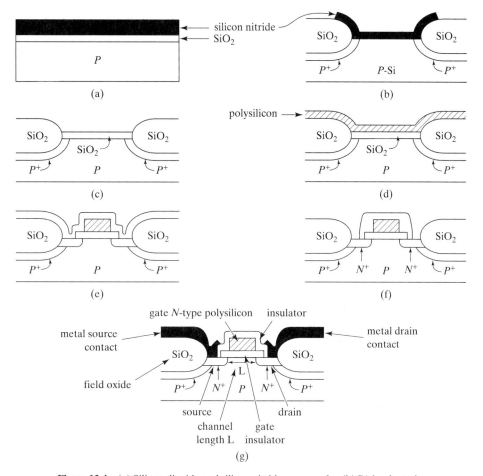

Figure 13.4   (a) Silicon dioxide and silicon nitride cover wafer, (b) P$^+$ implantation and field oxide, (c) gate oxide grown, (d) polysilicon layer, (e) polysilicon etched except over gate, (f) N$^+$ source and drain implanted and contact windows opened, and (g) metal deposition and patterning.

Mask No. 3 is used to open windows for the metal contacts to the transistor regions, as shown in Fig. 13.4(f).

The thin layer of SiO$_2$ is etched away and aluminum is deposited over the surface of the wafer by evaporation or by sputtering. *Mask No. 4* defines the interconnection pattern that is etched in the aluminum, as shown in Fig. 13.4(g). This is followed by the deposition of a protective passivation layer of phosphosilicate glass over the entire surface of the wafer.

Just as with the BJT, a final *Mask No. 5* is used to open windows, so that bonding wires can be connected to the pads on the IC chip.

13.3 REGIONS OF OPERATION

In this section, we will explain and identify, on the current-voltage characteristics, the three regions of operation of the MOSFET. The three regions are *cutoff, linear,* and *saturation* labeled on the graphical characteristics of Fig. 13.5. In this and the following sections, the transistor is an NMOS device in which the current-carrying channel consists of electrons only in contrast to the PMOS device whose channel is made of holes.

We consider a device made up of a P substrate, an insulator, a metal gate, and an N⁺ drain and source. We assume that the source and substrate are connected to ground.

Cutoff Region

For V_G equal to, or less than, the threshold voltage, which is defined by Eq. (12.29), no inversion layer is formed and the drain current is zero. This segment of the characteristics that coincides with the V_D axis is known as cutoff. In cutoff, with no inversion layer, the structure between the N⁺ drain and source consists of two N⁺P junction diodes back to back, so that the current for any voltage V_D applied between drain and source is, for all practical purposes, zero.

Linear Region—$V_D = V_{D1}$

For the linear and saturation regions, the voltage applied to the gate, V_G, is greater than the threshold voltage, V_T, and an inversion layer of electrons (N layer) is formed in the silicon at the silicon-oxide interface by capacitor action between the

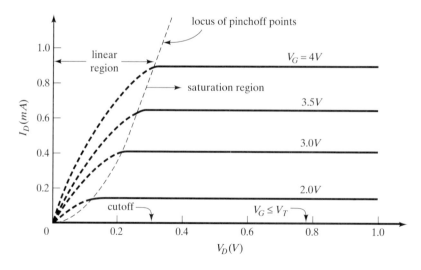

Figure 13.5   Characteristic curves of MOSFET showing the regions of operation.

metal gate and the semiconductor. A depletion layer is also formed between the N channel and the P substrate as happens at a PN junction.

With a fixed V_G, a small increase in V_D causes the channel to be biased positively with respect to the substrate, which is at ground potential. This reverse bias widens the depletion layer. For values of $V_D = V_{D1}$ of several hundred millivolts, the resistance of the channel is practically unchanged, although the voltage between the gate and the channel has been slightly reduced, causing a negligible change in the channel electron density. For the range of values of V_G, the channel acts as a resistance with the drain current, I_D (electrons from source to drain), being linearly related to V_D, as shown in the characteristics of Fig. 13.5.

Linear Region—$V_D = V_{D2} > V_{D1}$

As V_D is increased beyond the linear segment of the characteristics (which is very close to the origin), the depletion layer becomes wider and the inversion layer charge density, $|Q_n|$, decreases with the largest decrease occurring at the drain end, as shown in Fig. 13.6(a), where the voltage across the oxide, $(V_G - V_D)$, has its smallest value. The drain voltage thus negates some of the effects of the gate voltage.

The decrease in the channel charge density decreases its conductivity and increases its resistance. An increase of V_{D2} above V_{D1} causes a smaller increase in the drain current from I_{D1} to I_{D2}, resulting in a smaller slope of the characteristic curves, as shown in Fig. 13.5.

This process continues as V_D is increased, causing an increase of I_D but with a smaller slope, dI_D/dV_D, until the point is reached where the increase of V_D culminates in pinchoff at the drain end where the electron density in the channel has become zero. The effect of the gate voltage that exceeds V_T has been negated by the increase in V_D, resulting in the termination of the linear region, as shown in Figs. 13.5 and 13.6.

Saturation Region—$V_{D3} \geqslant V_{SAT}$ and $I_D = I_{SAT}$

The inversion charge density at the drain end of the channel is determined by the voltage across the oxide, $(V_G - V_D)$. When that becomes equal to or less than V_T, $|Q_n|$ at the drain end becomes zero and $V_{D3} = V_G - V_T$. The drain current reaches a constant value at $V_D \geqslant V_{D3}$. The region where the current is constant for $V_D \geqslant (V_{SAT} = V_{D3})$ is labeled the saturation region.

13.4 CURRENT-VOLTAGE ANALYTICAL RELATIONS

In the derivation of the $I_D - V_D$ analytical relationships, for the structure shown in Fig. 13.7, the following assumptions are made:

1. The source and the substrate are grounded ($V_B = 0$ in Fig. 13.7).
2. The surface mobility is assumed to be constant throughout the channel.

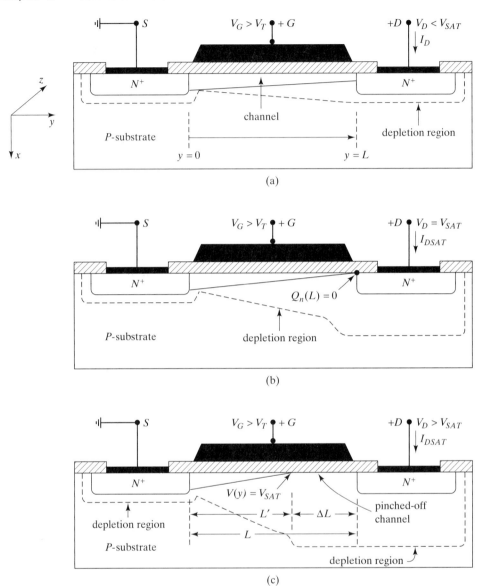

Figure 13.6 Conditions in an NMOS in the channel at a fixed V_G and as V_D increases, corresponding to (a) the linear region, (b) pinchoff, and (c) beyond pinchoff.

3. The threshold voltage is defined as the gate voltage that must be applied for an inversion layer to form at $V_D = 0$.

4. The potential gradient from drain to source is much smaller than that from gate to channel. Accordingly, it is the gate-to-channel potential that determines the electron density in the channel. This is known as the *gradual channel approximation* and has been defined in Sec. 10.2.

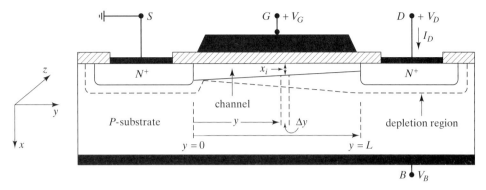

Figure 13.7 Idealized NMOS showing applied bias voltages.

The procedure that will be followed to determine an expression for the drain current, in terms of V_G, V_D, and the transistor constants, is to first express the channel electron density at a certain point y as a function of V_G, and the voltage at y with respect to the source, $V(y)$, including the device constants.

The voltage applied from gate to substrate (ground) is dropped partly across the oxide, partly in the semiconductor, and the rest flattens the bands in accordance with Fig. 12.6(d), so that we write

$$V_G = V_{ox} + \phi_s(y) + V_{FB} \tag{13.1}$$

where V_{ox} is the voltage across the oxide given by $|Q_s|/C_{ox}$, and Q_s is the negative charge density in the semiconductor given by $(Q_n + Q_d)$, V_{FB} is the flatband voltage, and $\phi_s(y)$ is the surface potential in the semiconductor and at strong inversion has a value of $2\phi_F$ at the source.

At strong inversion, the voltage at the surface of the oxide with respect to the substrate is given in terms of the channel-bulk voltage, which, for the bulk grounded, is $V(y)$, as

$$\phi_s(y) = 2\phi_F + V(y) \tag{13.2}$$

The reader may wonder how in Eq. (13.2) it appears that the surface potential exceeds $2\phi_F$. It is important to realize that the $V(y)$ represents a voltage drop V_{CB} from channel to bulk (ground here). In fact, the Fermi level is at the E_F of the substrate in the bulk and, because of the voltage drop, E_F in the channel moves down as we approach the surface to become E_{Fn} (see Fig. 4.2). The condition for strong inversion becomes $(2\phi_F + V_{CB})$, where $V_{CB} = V_{CS} + V_{SB}$, V_{CS} is channel to source, V_{SB} is source to bulk, and V_{CB} is the channel to bulk voltage. Now ϕ_F, which is $(E_i - E_F)$ in the bulk, is also $(E_{Fn} - E_i)$ at the surface. Equation (13.1) becomes

$$V_G = (-Q_s/C_{ox}) + 2\phi_F + V_{FB} + V(y) \tag{13.3}$$

where $|Q_s|$ is the charge density in the semiconductor. Replacing Q_s by $(Q_n + Q_{dm})$ in Eq. (13.3) and solving for Q_n, we have

$$Q_n(y) = -C_{ox}[V_G - V_{FB} - 2\phi_F - V(y)] - Q_{dm}(y) \tag{13.4}$$

As we move in the channel from source to drain and for a fixed V_G, the voltage across the oxide is $[V_G - V_{FB} - 2\phi_F - V(y)]$ and this voltage results in a smaller magnitude of total charge $|Q_s|(y)$. In addition to this, the positive voltage $V(y)$ between the N channel at y and the substrate (assumed grounded) causes an increase in the charge of the depletion layer of the NP reverse-biased diode. The decrease of $|Q_s|$ and the increase of the depletion layer charge, $|Q_{dm}|$, serves to reduce the magnitude of Q_n as we move from source to drain.

Therefore, two phenomena contribute to the decrease of $|Q_n(y)|$ with an increase of y. First, the total semiconductor charge is less and, second, the depletion layer charge is larger. The expression for $Q_{dm}(y)$ at strong inversion, from Eq. (12.13c) modified by $V(y)$ is now written as

$$Q_{dm}(y) = -\sqrt{2\varepsilon qN_A[2\phi_F + V(y)]} \tag{13.5}$$

After having determined the expression for the electron charge density, we will now determine two solutions for the current-voltage characteristics. In the first simple model, labeled the *square-law model*, we will assume Q_{dm} does not change with $V(y)$. In the second model, labeled the *bulk-charge model*, we will include the effect of V_D on Q_{dm}.

Square-Law Model

For this model, Eq. (13.4) can be written as

$$Q_n(y) = -C_{ox}[V_G - V(y) - V_T] \tag{13.6}$$

where V_T, the threshold voltage and a constant for a particular device, has been defined by Eq. (12.29) as

$$V_T = V_{FB} + 2\phi_F - Q_{dm}/C_{ox} = V_{FB} + 2\phi_F + (4\varepsilon qN_A\phi_F)^{0.5}/C_{ox} \tag{13.7}$$

and Q_{dm}, the depletion charge density, is described by Eq. (12.13c) and is assumed not to be a function of $V(y)$.

From Fig. 13.7, the resistance dR of a length dy of the channel is given as

$$dR = \frac{dy}{[q\,\bar{\mu}_n\,n(y)]\,[Z\,x_i(y)]} \tag{13.8}$$

where $(q\,\bar{\mu}_n\,n(y))$ is the conductivity of the channel at any y, $\bar{\mu}_n$ is the effective mobility,* and $Z\,x_i(y)$ is the cross sectional area of the channel, with x_i being the width of the channel (the inversion layer) at any y and Z being the thickness of the device perpendicular to the plane of the paper. The total charge, $Q_N(y)$, in coulombs contained within the volume $Z\,x_i(y)dy$ is

$$Q_N(y) = -q\,n(y)\,Z\,x_i(y)\,dy \tag{13.9}$$

The surface charge density $Q_n(y)(C/cm^2)$ is determined by dividing $Q_N(y)$ by the incremental surface area of the capacitor, $Z\,dy$, as

$$Q_n(y) = Q_N(y)/A = \frac{-qn(y)Zx_i(y)dy}{Zdy} = -qn(y)x_i(y) \tag{13.10}$$

By using Eq. (13.10) in Eq. (13.8), we have

---

*Effective mobility is discussed in Sec. 13.5.

$$dR = \frac{-dy}{\mu_n Z \, Q_n(y)} \qquad (13.11)$$

Since the current is constant throughout the channel, the voltage drop, $dV(y)$, along a length of the channel dy, for the references of Fig. 13.7, is

$$dV(y) = I_D dR = \frac{-I_D}{Z \, \overline{\mu}_n \, Q_n(y)} \, dy \qquad (13.12)$$

By using the expression for $Q_n(y)$ from Eq. (13.6) in Eq. (13.12) and rearranging terms, we have

$$I_D \, dy = Z \, \overline{\mu}_n \, C_{\text{ox}} \, [V_G - V(y) - V_T] \, dV(y) \qquad (13.13)$$

Integrating the left-hand side of Eq. (13.13) from $y = 0$ to $y = L$ and the right-hand side from $V = 0$ to $V = V_D$, we obtain the expression for the drain current, labeled I_{D1} for this model:

$$I_{D1} = \overline{\mu}_n \, C_{\text{ox}} \frac{Z}{L} \left[(V_G - V_T) \, V_D - \frac{V_D^2}{2} \right] \qquad (13.14)$$

Equation (13.14) is subject to two limitations. In the first place, we have neglected the change in the depletion layer charge as we move from source to drain. As we indicated earlier, the depletion layer charge increases with $V(y)$, causing $|Q_n(y)|$ to decrease and dR to increase, so that the actual drain current will be smaller than that given by Eq. (13.14). Second, the expression is valid as long as there is a Q_n throughout the channel. Referring to Eq. (13.6), Q_n at the drain becomes zero when $V(y)$ at the drain, V_D, labeled V_{SAT1}, is given by

$$V_{\text{SAT1}} = V_D = V_G - V_T \qquad (13.15)$$

For $V_D \leqslant V_G - V_T$, Eq. (13.14) is valid. At $V_D = V_G - V_T$, pinchoff at the drain has occurred, $Q_n = 0$ at the drain end of the channel, and the drain current is found from Eq. (13.14) at $V_D = V_{\text{SAT1}}$. Using Eq. (13.15) in Eq. (13.14), we determine the saturation drain current at a specific V_G as

$$I_{D1S} = \frac{\overline{\mu}_n \, C_{\text{ox}} \, Z}{2L} (V_G - V_T)^2 \qquad (13.16)$$

At $V_D = V_{\text{SAT1}}$, the current has reached the largest value at that V_G. We point out that the value of V_{SAT1} is also found by setting the derivative dI_D/dV_D of Eq. (13.14) equal to zero (zero slope) in order to determine the relationship required for maximum drain current at $V_{\text{SAT1}} = V_G - V_T$.

We define a *device parameter*, k_p, a measurable quantity, as discussed in Section 13.8, as

$$k_p = \overline{\mu}_n \, C_{\text{ox}} \, Z/L \qquad (13.17)$$

By using the device parameter and recalling that the region of the characteristics preceding the pinchoff points is known as the linear region, we write Eq. (13.14) together with the regions of its validity as

$$I_{D1} = k_p \, [(V_G - V_T) \, V_D - V_D^2/2] \text{ for } V_G \geqslant V_T \text{ and } V_D \leqslant V_G - V_T \quad (13.18)$$

Similarly, Eq. (13.16) is only valid in the saturation region, beyond pinchoff, and using the device parameter it is written together with the regions of its validity as

$$I_{DIS} = k_p/2 \, (V_G - V_T)^2 \text{ for } V_G \geq V_T \text{ and } V_D = \geq (V_G - V_T) \qquad (13.19)$$

In cutoff, $I_D = 0$ and $V_G \leq V_T$. Equations (13.18) and (13.19) describe the square-law model of the MOSFET in the linear and saturation regions respectively. These characteristics are displayed in Fig. 13.5.

EXAMPLE 13.1

Consider a silicon N-channel MOSFET that has an N^+ polysilicon gate and has the following constants: $N_A = 5 \times 10^{16} \text{cm}^{-3}$, $Q_i = 5 \times 10^{10} q \text{Ccm}^{-2}$, $t_{ox} = 300 \text{Å}$, $\bar{\mu}_n$ (effective) $= 500 \text{cm}^2 \text{V}^{-1} \text{s}^{-1}$, $Z = 50 \mu\text{m}$, and $L = 5 \mu\text{m}$, given $n_i = 10^{10} \text{cm}^{-3}$, $\varepsilon_{ox} = 3.9 \varepsilon_0$, and $qE_g = 1.12 \text{eV}$.
Assume E_F for polysilicon is at E_C.
Use the square-law model to calculate the drain current at

a) $V_G = 2\text{V}$ and $V_D = 1\text{V}$
b) $V_G = 3\text{V}$ and $V_D = 4\text{V}$

Solution We first determine the threshold voltage

$$V_T = \Phi_{ms} - Q_i/C_{ox} - Q_{dm}/C_{ox} + 2\phi_F$$

$$\phi_F = kT/q \, \ell n \, N_A/n_i = 0.4\text{V}$$

$$\Phi_{ms} = (E_C - E_F)_{\text{poly-si}} - (E_C - E_F)_{\text{subs}} = -0.96\text{V}$$

$$\Phi_{ms} = -0.96\text{V}$$

$$C_{ox} = \varepsilon_{ox}/t_{ox} = 3.9 \times 8.85 \times 10^{-14}/300 \times 10^{-8} = 11.50 \times 10^{-8} \text{Fcm}^{-2}$$

$$V_{FB} = \Phi_{ms} - Q_i/C_{ox} = -0.96 - (5 \times 10^{10} \times 1.6 \times 10^{-19})/C_{ox} = -1.02\text{V}$$

$$Q_{dm} = -\sqrt{4\varepsilon\phi_F qN_A} = -11.53 \times 10^{-8} \text{Ccm}^{-2}$$

$$V_T = -0.96 - 0.8/11.5 + 11.5/11.5 + 0.8 = 0.78\text{V}$$

a) We determine the drain voltage that separates the linear region from saturation by

$$V_{SAT1} = V_G - V_T = 1.22\text{V}$$

Since $V_D < V_{SAT1}$, this point is in the linear region, so that the drain current is

$$I_{D1} = k_p \left[(V_G - V_T) V_D - V_D^2/2 \right] \text{ where } k_p = \bar{\mu}_n C_{ox} Z/L = 5.75 \times 10^{-4} \text{ A V}^{-2}.$$

At $V_G = 2\text{V}$, $V_D = 1\text{V}$ and $V_T = 0.78\text{V}$

$$I_{D1} = 0.414\text{mA}.$$

b) At $V_G = 3\text{V}$ and $V_D = 4\text{V}$, $V_{SAT1} = 1.22\text{V}$. Since $V_D > V_{SAT1}$, operation is in the saturation region, so that

$$I_{DIS} = k_p/2 \, (V_G - V_T)^2 = 1.41\text{mA}.$$

The following symbols will refer to the square-law model: I_{D1} in the linear region, I_{D1S} and V_{SAT1} in saturation.

Bulk Charge Model

In the bulk charge model of the N-channel MOSFET, account is taken of the increased charge in the depletion layer as a result of the positive voltage drop along the channel from source to drain. This is done by using the expression for $Q_{dm}(y)$ from Eq. (13.5) in Eq. (13.4), so that $Q_n(y)$ is given, for the source and substrate grounded, by

$$Q_n(y) = -C_{ox}[V_G - V_{FB} - 2\phi_F - V(y)] + \sqrt{2\varepsilon q\, N_A[2\phi_F + V(y)]} \quad (13.20)$$

We replace the expression for $Q_n(y)$ from Eq. (13.20) in Eq. (13.12) and integrate the new expression between the limits $y = 0$, $V = 0$ and $y = L$, $V = V_D$ in order to obtain

$$I_{D2} = \overline{\mu}_n C_{ox} Z / L \left\{ \left(V_G - V_{FB} - 2\phi_F - \frac{V_D}{2} \right) V_D \right. \quad (13.21)$$

$$\left. \frac{-2\sqrt{2\varepsilon q\, N_A}}{3\, C_{ox}} [(V_D + 2\phi_F)^{3/2} - (2\phi_F)^{3/2}] \right\}$$

where V_D and V_G are measured with respect to the source and the substrate, which are assumed to be grounded. Equation (13.21) describes the drain current for the bulk charge model of the MOSFET in the linear region. We have labeled the current I_{D2}. The expression for V_T is found by setting $Q_n = 0$ in Eq. (13.20) for $V_D = 0$, hence $V(y) = 0$, as

$$V_T = V_{FB} + 2\phi_F + (4\varepsilon q\, N_A \phi_F)^{0.5}/C_{ox}$$

It can be shown that the expression for I_{D2} in Eq. (13.21) reduces to Eq. (13.14) if it is assumed that $V_D \ll 2\phi_F$. This is obtained if the last part of Eq. (13.21), enclosed within the brackets, is expanded using the binomial expansion and allowing $V_D \ll 2\phi_F$. This assumption implies that the surface potential, ϕ_s, has a value of $2\phi_F$ throughout the channel instead of being $(2\phi_F + V(y))$. This signifies that the depletion layer has the same charge density Q_{dm} throughout the device and hence the excess voltage, $(V_G - V_T)$, is consumed by a higher total channel charge density, resulting in a higher current in the square-law model.

The expression in Eq. (13.21) is valid as long as $Q_n(y)$ exists. When the drain voltage is increased to the point where $Q_n(y)$, at $y = L$, becomes zero, pinchoff takes place. To determine the condition for pinchoff, we set $Q_n(y)$ at $y = L$, in Eq. (13.20), equal to zero in order to obtain V_{SAT2} as

$$V_{SAT2} = V_G - V_{FB} - 2\phi_F + \frac{\varepsilon q\, N_A}{C_{ox}^2} \left[1 - \sqrt{1 + \frac{2C_{ox}^2}{\varepsilon q\, N_A}(V_G - V_{FB})} \right] \quad (13.22)$$

The saturation current, I_{D2S}, can be obtained by substituting Eq. (13.22) in Eq. (13.21). We note that to every V_G, there corresponds a V_{SAT2}.

EXAMPLE 13.2

For the MOSFET of Example 13.1, use the bulk charge model to determine the drain current at saturation for $V_G = 3$V.

Solution   To determine the current at saturation, we need to determine V_{SAT2} for $V_G = 3$V from Eq. (13.22).

$$V_G = 3\text{V}, V_{FB} = -1.02\text{V}, \phi_F = 0.4\text{V}, \varepsilon q N_A / C_{ox}^2 = 0.63\text{V}$$

$$V_{SAT2} = 2.25\text{V}.$$

We then calculate the drain current from Eq. (13.21) using $\bar{\mu}_n C_{ox} Z / L = 5.74 \times 10^{-4} \text{ AV}^{-2}$ and $V_{SAT2} = 2.25$V, so that $I_{D2S} = 0.72$mA.

Calculations using the square law model yield a saturation current of 1.42mA at $V_{SAT1} = 2.22$V, which is about double the value calculated in this example.

The following symbols will refer to the bulk charge model: I_{D2} in the linear region, I_{D2S} and V_{SAT2} in saturation.

Comparing the Square-Law and the Bulk-Charge Models

As we indicated earlier, the square-law model neglects the increase of the depletion layer charge brought about by the drain voltage. It thus underestimates the width of the depletion layer and overestimates the magnitude of the inversion charge density. By overestimating Q_n, a larger drain current results.

Calculations were carried out using both models and the results are shown in the plots in Fig. 13.8.

We observe in Fig. 13.8 a discrepancy of between 30–40 percent in the values of the currents in saturation with the higher currents corresponding to the square-law model. The discrepancy is quite a bit less in the linear region, as expected, since V_D has a smaller effect on the depletion layer charge.

The simplicity of the square-law model and the clear relevance of the various transistor parameters to the current makes it a very useful tool for pencil and paper calculations. However, it is still very difficult to actually predict the current-voltage characteristics from a knowledge of device dimensions and the physical properties of the transistor. In order to obtain satisfactory results, it is necessary to obtain some of the constants from measured data taken on a sample of MOSFETs. The data that are obtained are used in the square-law model and fairly satisfactory results are obtained. The principal parameter that is measured is the device parameter, which depends directly on the mobility. No exact measurement of the mobility, however, is possible, as we will see in the next section.

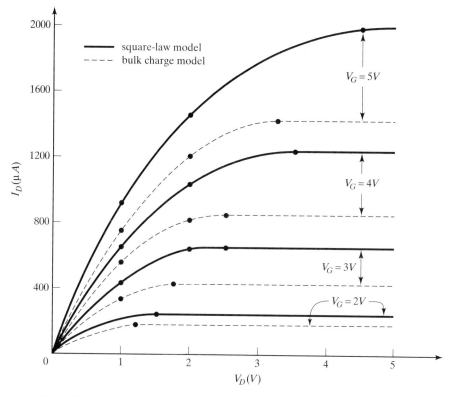

Figure 13.8   Plots of the characteristics for the square-law model given by Eqs. (13.14) and (13.16) and the bulk-charge model given by Eq. (13.21) combined with Eq. (13.22).

REVIEW QUESTIONS

Q13-1 Compare the isolation processes used in the fabrication of the BJT and the MOSFET.

Q13-2 For an NMOS device, give applicable voltage relationships for the three regions of operation.

Q13-3 Why and how does a depletion layer form when the NMOS is operating at strong inversion?

Q13-4 The depletion layer forms at the edge of the inversion layer. Identify a point on the energy band diagram that indicates the termination of the depletion layer.

Q13-5 Why is it that for a voltage applied to the gate of a MOSFET, no portion of this voltage appears in the metal?

Q13-6 What voltage components make up V_G when the MOSFET is beyond strong inversion?

Q13-7 What mechanism accounts for the continuity of the drain current beyond pinchoff?

HIGHLIGHTS

- The MOSFET consists of an MOS capacitor to which are added a source and a drain. An N-channel MOSFET (NMOS) uses a P-type silicon substrate and N^+ silicon for the source and drain. A sufficient positive voltage applied to the gate causes strong inversion and the formation of an N-channel. The electrons in the channel are caused to drift from source to drain by the application of a voltage, V_D. The source and substrate are assumed to be at ground potential.

- At strong inversion and for $V_D = 0$, the surface potential in the semiconductor, ϕ_s, is equal to $2\phi_F$, where $q\phi_F$ is the energy separation between E_i and E_F in the bulk of the semiconductor, far away from the surface. The gate voltage is dropped partly in the oxide as V_{ox}, partly to flatten the bands, and partly in the surface potential. The oxide voltage, $V_{ox} = |Q|/C_{ox}$, and Q is Q_m (at the metal gate) or $|Q_s^-|$ in the semiconductor. Q_s is the sum of the depletion layer and inversion layer charge densities.

- By increasing the gate voltage to a value higher than that which causes strong in version, a higher V_{ox} results and thus a higher $|Q_s^-|$, which implies higher $|Q_n|$ and a higher $|Q_d|$. The higher $|Q_n|$ requires a higher ϕ_s, but a very small change in ϕ_s is needed to increase $|Q_n|$ and hence ϕ_s may be assumed to remain constant at $2\phi_F$.

- A drain voltage applied between drain and source causes the following to happen: First, a drift of electrons from source to drain, second, an increase of the surface potential along the channel so that $\phi_s(y) = 2\phi_F + V(y)$, where y is the direction from source to drain and $V(y) = V_D$ at the drain end.

- As a result of positive $V(y)$ and for a fixed V_G, V_{ox} is smaller, causing a smaller $|Q_s^-|$. However, the increase of the voltage across the reverse-biased NP junction (channel to substrate) causes a wider depletion layer and a larger depletion layer charge density, $|Q_{dm}^-|$. The higher depletion charge density coupled with the decrease of $|Q_s^-|$ causes a gradual decrease in $|Q_n^-|$ with an increase of y since $Q_n + Q_{dm} = Q_s$.

- Two analytical models for the current-voltage characteristics are derived. In the square-law model, the increase of depletion charge density along the length of the channel, due to V_D, is neglected. The bulk charge model, on the other hand, takes account of the increased depletion layer charge density.

- The effect of the drain voltage on the channel charge density is that as we approach the drain end the charge density in the inversion layer decreases. For a fixed V_G and at a high enough V_D, the channel charge density at the drain becomes zero and the channel is said to be pinched-off.

EXERCISES

E13-1 An N-channel silicon MOSFET has the following properties: $V_{FB} = 0$, $N_A = 10^{15}\text{cm}^{-3}$, $t_{ox} = 0.1\mu\text{m}$, $Z/L = 10$, and $\mu_n = 700\text{cm}^2/\text{V-s}$. Given $\varepsilon_r(\text{silicon}) = 11.8$ and $\varepsilon_r(\text{SiO}_2) = 3.9$, determine: a) V_{SAT} for the square law model and b) V_{SAT} for the bulk charge model at $V_G = 4\text{V}$.

Ans: b) $V_{SAT} = 2.48\text{V}$.

E13-2 A silicon PMOS is designed to have a saturation current of 4mA when $V_G = -5\text{V}$. Given $V_T = -0.8\text{V}$, $t_{ox} = 400\text{Å}$, and $\mu_p = 300\text{cm}^2/\text{V} - \text{s}$, use the square-law model to determine the required Z/L.

Ans: $Z/L = 9$.

Why would this ratio be smaller for an NMOS having the same oxide thickness, $V_G = 5V$ and $V_T = 0.8V$?

13.5 IMPORTANT SECONDARY EFFECTS

Effective Mobility

We have defined mobility of a carrier as the ratio of the average drift velocity imparted to the carrier to the electric field that accelerates the carrier. In other words, mobility refers to the ease of motion of a carrier.

In general, as was shown in Chapter 4, mobility in the bulk of the semiconductor is dependent on two types of carrier scatterings: lattice, or temperature scattering, and impurity scattering.

In a MOSFET, the motion of the carriers is restricted to an extremely thin region in the inversion layer, sandwiched between the bulk of the semiconductor and the oxide, where the motion is strongly influenced by the electric field generated by the gate voltage. The carriers are therefore made to bounce between the surfaces of the semiconductor and the oxide. Therefore, in addition to the two types of scatterings that a carrier is subjected to in the bulk, the carrier in the inversion layer is also under the influence of surface scattering. The mobility of the carrier is therefore considerably reduced.

As a result of the uncertainty of conditions at the surface of the oxide and the inversion layer, it is very difficult to predict the values of the mobility.

The mobility of the carriers in a MOSFET is known as the effective mobility $\bar{\mu}_n$ for electrons. Depending on the gate voltage, the effective mobility varies between 30 to 50 percent of the bulk mobility. Its value is best determined by a measurement of the device parameter k, using a large sample of devices.

Effect of Substrate Bias

It is possible to change the value of the threshold voltage by applying a voltage to the substrate. So far in our derivations, we have assumed that the substrate is connected to the source and grounded. The substrate voltage is applied in a direction to reverse-bias the substrate-channel junction, so that the substrate voltage, also known as the *body voltage*, V_B, is a negative quantity with respect to the source in a P-type substrate.

With the body at a negative voltage, more band bending is required in order to cause the onset of strong inversion at the silicon surface and therefore more charge stored in the depletion layer. The larger voltage drop in the silicon depletion layer reduces the voltage across the oxide for a fixed gate voltage. The lower oxide voltage means less total charge, both in the metal and in the silicon, thus reducing the magnitude of Q_n while the magnitude of Q_d is increased. Consequently, a larger negative bias at the body reduces the drain current for a given V_D and V_G and lowers the value of V_{SAT}. The net effect of a source-substrate reverse bias, as will be shown in the following derivation, is that the threshold voltage becomes more positive for an N-channel transistor and more negative for a P-channel transistor.

At the onset of strong inversion, the voltage drop at the silicon surface with respect to the body and using the subscripts C for channel, S for source, and B for body for the voltage is

$$\phi_s(y) = 2\phi_F + V_{CS} + V_{SB} = 2\phi_F + V(y) + V_{SB} \tag{13.23}$$

where V_{CS} is the channel to source voltage and V_{SB} is the source to body voltage. Upon rewriting Eq. (13.1), replacing V_G by V_{GB} and using Eq. (13.23), we have

$$V_{GB} = V_{ox} + \phi_s(y) + V_{FB} = V_{ox} + 2\phi_F + V(y) + V_{SB} + V_{FB}$$

$$V_{GB} = -Q_s(y)/C_{ox} + 2\phi_F + V(y) + V_{SB} + V_{FB} \tag{13.24}$$

We replace Q_S by $(Q_n + Q_{dm})$, V_{GB} by $(V_{GS} + V_{SB})$, and we solve Eq. (13.24) for Q_n, assuming the source is grounded, to have

$$Q_n(y) = -C_{ox}[V_{GS} + V_{SB} - V_{FB} - 2\phi_F - V_{SB} - V(y)] - Q_{dm}(y)$$

$$= -C_{ox}[V_{GS} - V_{FB} - 2\phi_F - V(y)] + \sqrt{2\varepsilon q N_A[2\phi_F + V(y) + V_{SB}]} \tag{13.25}$$

By using the expression for $Q_n(y)$ in Eq. (13.2) and integrating from $y = 0$, $V(y) = 0$ to $y = L$, $V(y) = V_{DS}$, and letting $V_{SB} = -V_{BS} = -V_B$, we obtain

$$I_{D3} = \bar{\mu}_n C_{ox} Z/L \left\{ \left(V_{GS} - V_{FB} - 2\phi_F - \frac{V_{DS}}{2} \right) V_{DS} - (2/3C_{ox})\sqrt{2\varepsilon q N_A} \right.$$

$$\left. [(V_{DS} + 2\phi_F - V_B)^{3/2} - (2\phi_F - V_B)^{3/2}] \right\} \tag{13.26}$$

The threshold voltage, defined as the gate-source voltage at which $Q_n = 0$ for $V_{DS} = 0$ is found from Eq. (13.25) as

$$V'_T = V_{FB} + 2\phi_F + \sqrt{2\varepsilon q N_A(2\phi_F - V_B)}/C_{ox} = V_T \text{ (of Eq. 13.7)} + \Delta V_T \tag{13.27}$$

so that

$$\Delta V_T = \gamma(\sqrt{2\phi_F - V_B} - \sqrt{2\phi_F})$$

and γ, labeled the *body factor*, is given by $\sqrt{2\varepsilon q N_A}/C_{ox}$

EXAMPLE 13.3

Use the device of Example 13.1 to determine the required substrate bias that will increase the threshold voltage from 0.78V to a) 1.0V and b) 1.2V. In Example 13.1, use $\varepsilon = 11.8$, $N_A = 5 \times 10^{16} \text{cm}^{-3}$, $\phi_F = 0.4\text{V}$, and $C_{ox} = 11.5 \times 10^{-8} \text{Fcm}^2$.

Solution   We use Eq. (13.26) to calculate V_B as

$$\Delta V_T = \gamma(\sqrt{2\phi_F - V_B} - \sqrt{2\phi_F}) = 1 - 0.78 = 0.22\text{V}$$

$$\gamma = (1/C_{ox})\sqrt{2\varepsilon q N_A} = 1.12\text{V}^{0.5}$$

a) $\Delta V_T = 1.12(\sqrt{0.8 - V_B} - \sqrt{0.8}) = 0.22\text{V}$
     Solving for V_B, we obtain $V_B = -0.39\text{V}$.
b) For a ΔV_T of 0.42V, V_B is calculated to be $V_B = -0.81\text{V}$.

Channel-Length Modulation

In the derivation of the current-voltage characteristics, we have assumed that at $V_D = V_{SAT}$, corresponding to pinchoff at the drain end of the channel, the drain current reaches a saturation value I_{SAT} and further increase of V_D above V_{SAT} maintains the current at I_{SAT}. In fact, the drain current increases slightly as V_D is increased. For $V_D > V_{SAT}$, the pinched-off region of the channel becomes wider, spreading out from just a point at the drain into a depleted section, as shown in Fig. 13.9(a). The depleted section is ΔL and absorbs the drain voltage that is in excess of V_{SAT}, while V_{SAT} is dropped from the new pinchoff point to the source.

For long-channel devices, ΔL is much smaller than L, so that a drop of V_{SAT} across $(L - \Delta L)$ results in a small increase in the current. However, in modern short-channel devices, where ΔL is no longer negligible, an increase of V_D above V_{SAT} causes an increase in the drain current. This is verified by noting that, in the expression for the drain current, I_D is inversely proportional to L so that a decrease in the effective channel length causes an increase in the current. This phenomenon is similar to base-width modulation in bipolar junction transistors. The effect of channel-length modulation on the characteristics is shown in Fig. 13.9(b).

This effect is introduced into the square-law model equation, so that Eq. (13.19) becomes

$$I_{D1S} = (k_p/2)(V_G - V_T)^2 (1 + \lambda_p V_D) \tag{13.28}$$

where λ_p is known as the ***channel-length-modulation parameter***. Its value ranges from 0.01 to 0.02V^{-1}. An empirical value for the reciprocal of λ_p can be obtained by finding the intercept on the V_D axis of the extension of the characteristics, as shown in Fig. 13.9(b).

EXAMPLE 13.4

Use the device of Example 13.1 in the square-law model to determine the current in saturation at $V_G = 4$V and

a) $V_D = 3.6$V   and   **b)** $V_D = 10$V
Given $\lambda_p = 0.02$V^{-1} and from Example 13.1, $k_p = 575 \times 10^{-4}$AV^{-2}.

Solution

a) With $V_T = 0.78$V and $V_{SAT1} = V_G - V_T = 3.22$V, the device is in saturation because $V_D > V_{SAT1}$. The current is given by

$$I_{D1S} = k_p/2 \, (V_G - V_T)^2 (1 + \lambda_p V_D)$$

For $V_D = 3.6$V

$$I_{D1S} = 3.2\text{mA}$$

If channel length modulation is not included, I_D becomes 2.98mA.

b) For $V_D = 10$V

$$I_{D1S} = 3.2\text{mA, since the device is in saturation.}$$

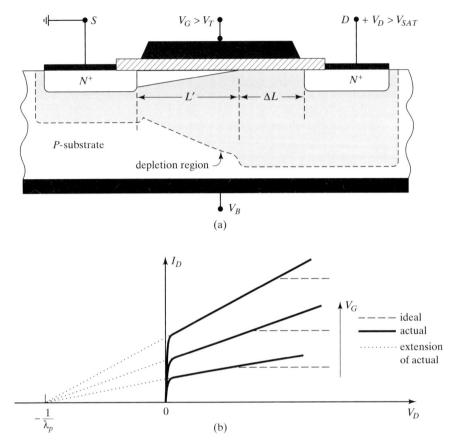

Figure 13.9   (a) Widening of the pinched-off region. (b) Increase of drain current with V_D in the saturation region.

Temperature Dependence

The current-voltage characteristics of a MOSFET are influenced by temperature variations through two mechanisms. First, an increase of temperature increases the lattice scattering, thus reducing the mobility and the drain current. Second, the bulk Fermi potential, ϕ_F, is affected by temperature variations both by a change of the intrinsic carrier density and through the thermal voltage kT/q. Changes in the bulk Fermi potential, ϕ_F, cause changes in the threshold voltage, which changes the drain current. In general, an increase of temperature causes a decrease of the threshold voltage.

It has been shown that at low currents the effect of the increase of temperature on the threshold voltage is dominant and an increase of temperature increases the drain current. At higher drain currents, the reduction of mobility that accompa-

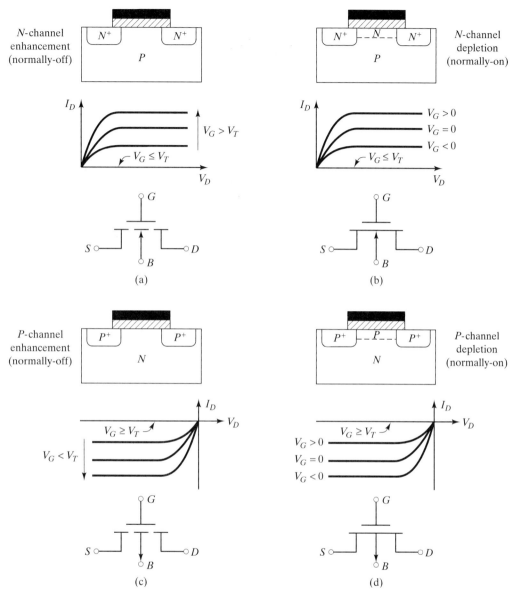

Figure 13.10 Types of MOSFETs: (a) N-channel enhancement, (b) N-channel depletion, (c) P-channel enhancement, and (d) P-channel depletion.

nies an increase of temperature is dominant and leads to a decrease in the drain current. The value of the drain current that separates the two regions has to be determined experimentally.

In general, the effect of temperature changes on MOSFET characteristics is much less significant than their effect on BJT characteristics.

Isolation in MOSFET Integrated Circuits

Unlike bipolar junction transistors, the prevention of interaction between adjacent MOSFET transistors on a chip does not require the diffusion of isolation regions. In order for interaction to take place, a channel must exist between two neighboring transistors. To prevent this interaction in MOSFETS, the regions separating the transistors are made with a large V_T. The threshold voltage is made large enough that the highest voltage in the circuit will not cause the formation of a channel between adjacent devices. This is accomplished in two ways: first, by depositing a thick field oxide outside of the region occupied by the device, as shown in Fig. 13.4, and, second, by increasing the doping density below this field oxide. The thick oxide is covered by the metallic interconnections between the elements. The field oxide is made much thicker than the gate oxide so as to reduce the oxide capacitance and thus increase V_T. Similarly, the increased doping serves to increase the threshold voltage of the region separating transistors, thus preventing the formation of conducting channels between transistors.

Destructive Handling of a MOSFET

The gate of a MOSFET uses silicon dioxide, which has a thickness of several tens of angstroms. The use of a thin oxide possesses certain advantages and disadvantages. One of the shortcomings is the low breakdown voltage of the capacitor, C_{ox}.

The electric field intensity at which silicon dioxide breaks down*, at $t_{ox} = 50\text{Å}$, is about $2 \times 10^7 \text{V/cm}$. The highest transverse electric field occurs at the source so that the electric field is the ratio of the gate-source voltage to the oxide thickness. For an oxide thickness of 100 Angstroms, the oxide will break down at a V_G of 20 volts.

Caution must be exercised in handling MOSFETs and unless internal protective devices are included, it is quite possible that the static voltage developed on the body of a person touching the MOSFET can destroy the device immediately. A person walking on a carpet or the motion of an arm can help build voltages in excess of thousands of volts on the person. Touching the gate of a MOSFET placed on a bench can cause a static voltage discharge and an immediate destruction of the MOSFET. Before handling a MOSFET, it is necessary to touch the bench and discharge the accumulated static electricity.

Devices are now being fabricated that incorporate diodes at the input that serve as protective devices against this important phenomenon known as ESD (electrostatic discharge).

13.6 TYPES OF MOSFETS

So far, we have classified MOSFETs based on the type of substrate, as N channel (P substrate) and P channel (N substrate). Within each of the two classifications, devices may be *enhancement-mode* or *depletion mode* MOSFETs. If for zero gate

---

*From Sze *Physics of Semiconductor Devices* p. 406, Copyright © Wiley 1985. Reprinted by permission of John Wiley & Sons, Inc.

bias the channel conductance is negligibly small, a positive voltage must be applied to the gate to form an N channel. This is normally an OFF (OFF at $V_G = 0V$) device known as an enhancement-mode transistor. When an N channel exists at $V_G = 0$, a negative gate voltage has to be applied in order to deplete the electron density in the channel and reduce the channel conductance to zero. This depletion mode is normally an ON (ON at $V_G = 0V$) transistor. We conclude that an N channel device having a positive threshold voltage is an enhancement-mode, whereas one having a negative threshold voltage is a depletion-mode device.

In the depletion-mode device, a channel exists at thermal equilibrium. This is done by conventional doping of the region under the gate oxide using ion implantation techniques described in the following section. A negative voltage applied to the gate causes a restriction of current through channel depletion. A positive gate voltage attracts additional electrons to the channel and the current increases.

Similarly, P-channel MOSFETs may be of the enhancement-mode (normally OFF) and of the depletion mode (normally ON).

It is important to point out that both N-channel and P-channel devices are commonly used. Because the electron mobility in silicon is about three times that of holes, N-channel devices are generally preferred.

The enhancement MOSFET is commonly used as a switch or a driver in digital logic gates, whereas the depletion mode is usually the load to the enhancement driver.

Cross sections, symbols, and output characteristics of the four types are shown in Fig. 13.10.

13.7 CONTROL OF THE THRESHOLD VOLTAGE

Control of the threshold voltage is of major importance in the fabrication of MOSFETs since the manufacturing spread in the value of V_T, in general, may be relatively large. This is especially so when compared to the spread of the value of V_{BE} in bipolar transistors. It is thus important to adjust or restrict the value of V_T so that it may conform to the required specifications.

The general expression for the threshold voltage of a MOSFET, assuming that the source and body are grounded, is given by Eq. (13.7) as

$$V_T = V_{FB} - Q_{dm}/C_{ox} + 2\phi_F \qquad (13.29)$$

Let us look into the various components of V_T and consider what changes may be made to attain the desired value. In a previous section, we have determined the effect of the applications of a body bias on V_T.

The flat band voltage, V_{FB}, is negative for both P-channel and N-channel devices. For P-channel devices, Q_{dm} is positive and ϕ_F is negative, so that V_T is negative as required by PMOS devices. For N-channel devices, both Q_{dm} and ϕ_F contribute positive quantities, so that V_T may be positive or negative. Most commonly, N-channel devices are needed in digital circuits that operate in the enhancement mode, so that V_T must be positive. To guarantee a positive V_T, the sum of the magnitudes of $2\phi_F$ and Q_{dm}/C_{ox} must be greater than the magnitude of V_{FB}.

In order to increase the contributions of Q_{dm}/C_{ox} and $2\phi_F$, two changes may be made. The doping density N_A and the oxide thickness can be increased. Increasing the impurity density increases the depletion charge density, Q_{dm}, as seen from Eq. (13.7), while causing a small increase in $q\phi_F$ that results in a more positive value for V_T. An upper limit is set on the doping by the junction breakdown voltage and by the high substrate capacitance. Another disadvantage of using a large concentration of impurities is the decrease of mobility as the doping density increases. The advantage of the higher speed N-channel devices that is obtained, because of the higher mobility of electrons when compared to holes, may be lost when the impurity density is increased to high values.

The second contribution that makes V_T positive is a decrease in the oxide capacitance made possible by an increase in the oxide thickness. The price to be paid for a smaller oxide capacitance is a reduction in the drain current, as seen from Eqs. (13.16) and (13.21), which in turn results in a decrease of the transconductance in the saturation region, which is translated into a smaller voltage gain.

In spite of the strong dependence of the threshold voltage on the substrate doping, on the oxide capacitance, and on the ability to cause some improvements in V_T for N-channel devices, the values of the substrate doping and the oxide capacitance are determined mainly by other design considerations. Furthermore, calculations using Eq. (13.26) for the threshold voltage do not yield exact results in practical devices because of variations of oxide thickness, mobility, and oxide positive charges.

The most common method presently employed to improve the threshold voltage of MOSFETs is *ion implantation*. This method allows not only for a fairly precise control of V_T but also makes it possible to tailor different threshold voltages for transistors on the same wafer, which may include N-channel and P-channel devices. This method has been discussed in detail in Chapter 6. We will review here the basic steps relevant to this section.

The technique of ion implantation is the process of introducing ionized dopants by direct bombardment of a surface. First, impurity atoms are ionized and are accelerated in a strong electric field so they acquire energies in the range of 20 to 200keV. The ions are generated when an arc discharge is caused to occur in a gas containing the dopant. These ions are made to strike the oxide before the metallic gate electrode is deposited. The depth of penetration of the ions is precisely adjusted by the accelerating voltage and the density of the ions deposited is determined by the current of the ions and the duration of bombardment. While the implantation is carried out at room temperature, the damage to the lattice caused by this method is removed by annealing at temperatures of 600°C to 1000°C.

For N-channel devices, boron ions are used that are made to rest at the silicon-silicon dioxide interface. When the MOS is biased to strong inversion, the implanted ions act to oppose the effect of the positive ions present at the oxide surface. The result is a positive shift in the threshold voltage. The implantation of phosphorus ions causes a negative shift in the threshold voltage. The shift in V_T is found from $\Delta V_T = -qD_I/C_{ox}$, where D_I is the ion density per square centimeter, which is negative for acceptor ions and positive for donor ions.

EXAMPLE 13.5

Use the device of Example 13.1 to determine the new value of V_T if boron ions of density $D_I = 2 \times 10^{11}q$ Ccm^{-2} are implanted and the substrate is shorted to the source. The original V_T in Example 13.1 is 0.78V and $C_{ox} = 11.5 \times 10^{-8}$Fcm^{-2}.

Solution

$$\Delta V_T = \frac{2 \times 10^{11} \times 1.6 \times 10^{-19}}{11.5 \times 10^{-8}} = 0.27V$$

$$V_T' = 0.27 + 0.78 = 1.05V$$

13.8 MEASUREMENT OF MOS TRANSISTOR PARAMETERS

An experimental arrangement for the measurement of the major transistor parameters, namely k_p and V_T, is shown in Fig. 13.11. The gate is connected to the drain so that $V_D = V_G$ and hence $V_D > V_G - V_T$, placing the device in the saturation region. A plot of $\sqrt{2I_{D1S}}$ versus V_D by using Eq. (13.19) yields a straight-line relationship.

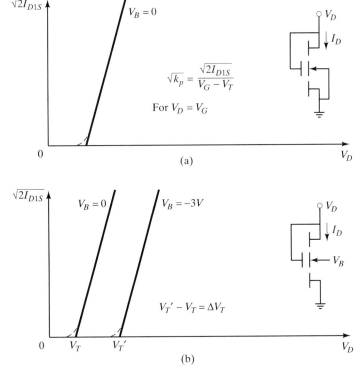

Figure 13.11   Plots of $\sqrt{2I_{D1S}}$ versus V_D to determine V_T and V_T' for (a) $V_B = 0$ and (b) $V_B < 0$.

For $V_B = 0$, in Fig. 13.11, and from Eq. (13.19), the intercept of the extrapolation of the straight line on the V_D axis is the threshold voltage. The slope of the line is used to calculate the value of k_p. The effect of the bulk voltage on the threshold voltage, V'_T, as given by Eq. (13.26), is determined from Fig. 13.11(b).

13.9 MOSFET SMALL-SIGNAL EQUIVALENT CIRCUITS

Low-Frequency Circuit

Since the gate terminal is connected to the device by means of an insulator precluding any gate current, there is only one current in the MOSFET and that is the drain-to-source current, I_D. To determine the elements of the low-frequency equivalent circuit, we look for the answer to the following question: How does the drain current vary as changes are made in the drain and gate voltages? We are assuming the source is grounded and neglecting the effect, on the current, of changes in the substrate bias.

The quiescent operating point drain current, I_D, is a function of V_D and V_G, so we can write

$$I_D = I_D(V_G, V_D) \tag{13.30}$$

Although the procedure we used for determining the low-frequency equivalent circuit for the BJT can be used here, we will use a much less elaborate method. We assume incremental changes v_d and v_g in the drain and gate voltage respectively and we determine an expression for the incremental change in the current. The expression for the total current made up of the quiescent value and the incremental change is written as

$$i_D = I_D + \Delta I_D = I_D(V_G + \Delta V_G, V_D + \Delta V_D) \tag{13.31}$$

We use small signal quantities to replace the delta changes in the voltages and the current as

$$i_D = I_D + i_d = I_D(V_G + v_{gs}, V_D + v_{ds}) \tag{13.32}$$

The small-signal current is given by

$$i_d = i_D[\text{Eq. (13.32)}] - I_D[\text{Eq. (13.30)}] \tag{13.33}$$

The terms on the right-hand side of Eq. (13.32) can be expanded using the Taylor series expansion. We will include in the expansion the first two terms of the series only by assuming that $v_{gs} \ll V_G$ and $v_{ds} \ll V_D$. The result is written as

$$i_D = I_D + i_d = I_D + \frac{\partial I_D}{\partial V_G}\bigg|_{V_D} v_{gs} + \frac{\partial I_D}{\partial V_D}\bigg|_{V_G} v_{ds} \tag{13.34}$$

The small-signal component of the drain current is obtained from Eq. (13.34) and is written as

$$i_d = g_m v_g + g_d v_d \tag{13.35}$$

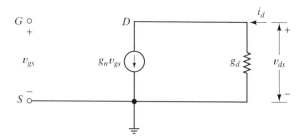

Figure 13.12   Small-signal low-frequency equivalent circuit of the MOSFET.

where g_m is the partial derivative of I_D with respect to V_G at constant V_D and g_d is the partial derivative of I_D with respect to V_D at constant V_G.

The equivalent circuit of the MOSFET in the common-source connection as constructed from the quantities in Eq. (13.35) is shown in Fig. 13.12.

The symbol g_m represents the transconductance of the device and the prefix "trans" refers to the control that the input gate voltage has over the drain current.

The conductance, g_d, represents the control that the output voltage has over the drain current.

The application of MOSFETs in analog circuits is mainly for amplification. When used as an amplifier, the device is operated in the saturation region. The expressions for g_m and g_d in saturation are obtained for the square-law model from the relation for the drain current in saturation, Eq. (13.28), as

$$g_m \equiv \left. \frac{\partial I_{D1S}}{\partial V_G} \right|_{V_D} = k_p(V_G - V_T)(1 + \lambda_p V_D) \tag{13.36}$$

$$g_d \equiv \left. \frac{\partial I_{D1S}}{\partial V_D} \right|_{V_G} = \frac{k_p}{2}(V_G - V_T)^2 \lambda_p \tag{13.37}$$

The expressions in Eqs. (13.36) and (13.37) may be written in terms of the quiescent point drain current by using Eq. (13.28). After introducing the expression for I_{D1S} from Eq. (13.28), we assume $\lambda_p V_D \ll 1$, so that

$$g_m = \sqrt{2k_p I_{D1S}} \tag{13.38}$$

$$g_d = \lambda_p I_{D1S} \tag{13.39}$$

EXAMPLE 13.6

Use the device of Example 13.1 and $\lambda = 0$ to determine

a) The transconductance g_m at $V_G = 3V$ in saturation.

b) The resistance in the linear region at $V_D = 0$ and $I_D = 0$ for

i) $V_G = 1V$ and ii) $V_G = 3V$.

Solution   For the device in Example 13.1, $k_p = 575.5 \times 10^{-6} \mathrm{AV}^{-2}$ and $V_T = 0.78\mathrm{V}$, so that at $V_G = 3\mathrm{V}$:

a) $g_m = k_p(V_G - V_T) = 1.27 \times 10^{-3}\mathrm{S}$

b) The expression for the drain current in the linear region is given by Eq. (13.18) as

$$I_D = k_p \left[(V_G - V_T) V_D - V_D^2/2 \right]$$

The drain conductance in the linear region becomes

$$g_d = dI_D/dV_D = k_p(V_G - V_T - V_D)$$

The drain resistance at $V_D = 0$ is

$$r_d = 1/g_d = 1/k_p(V_G - V_T)$$

We observe that the drain resistance depends on the variable V_G only. The device acts as a variable resistor controlled by V_G. For $V_T = 0.78\mathrm{V}$, the resistances at $V_D = 0$ and $I_D = 0$ are calculated to be

$$\text{At } V_G = 1\mathrm{V}, r_d = 7.9\,\mathrm{k}\Omega$$

$$\text{At } V_G = 3\mathrm{V}, r_d = 783\,\Omega$$

High-Frequency Circuit

The high-frequency circuit is obtained by adding, to the low-frequency circuit, the capacitances inherent to the operation of the device plus parasitic capacitances. The effect of the capacitances is to reduce the voltage gain at high frequencies, with the oxide capacitance as the major contributor to the decline of the gain. This capacitance is distributed throughout the length of the channel and a reasonable approximation is to assume that this capacitance is made up of a gate-to-source capacitance, C_{gs}, and a drain-to-gate capacitance, C_{gd}. In the saturation region, C_{gs} can be assumed to form a large part of $C_{ox}ZL$ and C_{gd} is approximately zero. This is because in saturation the channel near the drain is no longer made of Q_n and the drain exerts very little influence on the channel and the gate charge.

The complete high-frequency equivalent circuit is shown in Fig. 13.13(a).

The capacitances, C_{os} and C_{od}, are parasitic elements between the gate and source and between the gate and drain respectively. They are the overlap capacitances of the gate over the source and over the drain introduced in the fabrication process to allow the gate to extend beyond the channel on both sides. This is so because of the difficulty of precisely aligning the gate oxide deposition with the channel, hence the gate oxide is extended slightly to guarantee full control by the gate. The locations of the capacitances are shown in Fig. 13.13(b).

Two additional capacitances, C_{ts} and C_{td}, represent the depletion layer capacitances between the source and substrate and drain and substrate respectively because reverse-biased PN junctions exist at both ends. In the derivation of the equivalent circuit, we have assumed that the substrate is connected to the source, hence the reason for not showing C_{ts} in Fig. 13.13(a).

A major innovation that is used to extend the upper limit of the operating frequency of MOSFETs was undertaken with the utilization of polycrystalline silicon

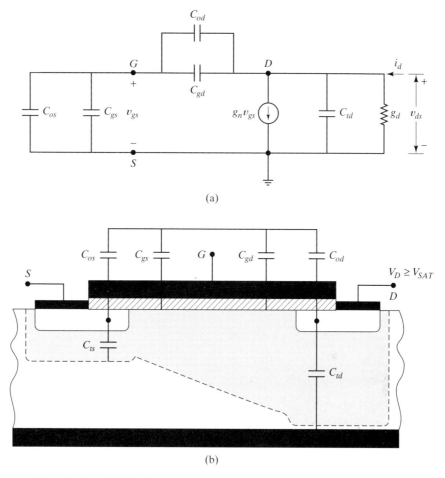

Figure 13.13   (a) High-frequency equivalent circuit of MOSFET, (b) Section
showing sources of capacitances.

(polysilicon) to replace the aluminum gate electrode. The use of silicon permits the
fabrication of the gate prior to the diffusion of the source and the drain, thus drasti-
cally improving the alignment of the gate over the channel. This has the advantage
of practically cancelling the overlap capacitances and improving the high frequency
performance of the transistor. It must be emphasized that polysilicon is used for the
gate electrode and all other interconnections between elements on a wafer. In addition to polysilicon, tungsten and thalium are also used.

13.10 HIGH-FREQUENCY PERFORMANCE

The f_T of the MOSFET

The measure of the high-frequency performance of a transistor is the *figure of merit*
or *cutoff frequency, f_T.* The cutoff frequency is defined as the frequency at which the
magnitude of the short-circuit current gain of the device is unity. Although the input

DC current of a field-effect transistor is practically zero, an AC current exists as a result of the capacitances of the device.

The expression for the cutoff frequency is derived in the following manner: Place a short circuit across the output, apply a sinusoidal current having an effective value I_i to the input, and determine the expression for the current I_o in the short-circuit.

The new equivalent circuit is shown in Fig. 13.14 in which the overlap capacitance from gate to drain has been lumped with the gate-drain capacitance C_{gd}. The symbol V_{gs} and I_o are the effective values of the sinusoidal gate voltage and short-circuit drain current. We will assume, as an approximation, that the total capacitance at the input is equal to $C_{ox}ZL$. The input current becomes

$$I_i = j\omega(C_{gs} + C_{gd})V_{gs} \tag{13.40}$$

The short-circuit output current is given by

$$I_o = g_m V_{gs} \tag{13.41}$$

The magnitude of the current gain becomes

$$|I_o/I_i| = g_m/\omega\,(C_{gs} + C_{gd}) \tag{13.42}$$

The frequency at which the magnitude of the current gain is unity is therefore found to be at the frequency f_T as

$$f_T = g_m/2\pi\,(C_{gs} + C_{gd}) \tag{13.43}$$

We define $(C_{gs} + C_{gd})$ to be $C_{ox}ZL$, so that f_T becomes

$$f_T = \frac{g_m}{2\pi C_{ox}ZL} \tag{13.44}$$

We now replace g_m by using Eq. (13.36) and making $V_D \ll 1$, so that

$$f_T = \frac{k_p\,(V_G - V_T)}{2\pi\,ZL\,C_{ox}} = \frac{\bar{\mu}_n\,(V_G - V_T)}{2\pi\,L^2} \tag{13.45}$$

where k_p has been replaced by $\bar{\mu}_n\,C_{ox}Z/L$.

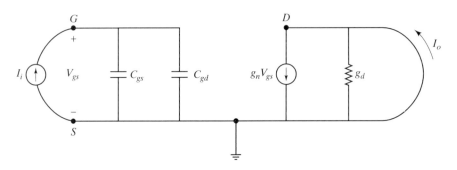

Figure 13.14 Circuit for calculating f_T.

EXAMPLE 13.7

For the MOSFET of Example 13.1, calculate an approximate value for f_T at $V_G = 3$V.

Solution   The value of the effective mobility is $500 \text{cm}^2 \text{V}^{-1} \text{s}^{-1}$, the channel length is $5 \mu \text{m}$, and V_T is 0.78V, so that by using Eq. (13.45), f_T is calculated to be

$$f_T = \frac{500 \times (3 - 0.78)}{2\pi \times 25 \times 10^{-8}} = 0.709 \text{GHz}$$

13.11 COMPARING THE MOSFET AND THE BJT

One of the advantages that MOS transistors have over bipolar transistors is that, in general, they are less costly to fabricate. The reason for this is that they are self-isolating. By referring to the sections on fabrication in Chapter 8 for the BJT and in this chapter for the MOSFET, one important difference stands out: Bipolar transistors require tubs to isolate devices from each other on one integrated circuit. Isolation in MOS transistors is provided by heavy doping and a thick oxide in the regions between adjacent devices. Because of this, the MOS transistors are smaller and less costly to fabricate. Furthermore, in general, they require fewer number of processing steps.

We will now identify the advantages of each device as they relate to their application in digital and analog circuits.

In digital circuits, MOSFETs possess two distinct advantages: First, they are operated by a binary threshold gate voltage. Below this voltage, the device is completely nonconducting. Nonconduction in a BJT is a gradual mechanism. Second, MOS transistors consume less DC power.

For analog applications, there are two possible criteria for comparison: transconductance, g_m, or, preferably, the ratio of g_m to the current at the operating point, and the cutoff frequency. Transconductance can be looked at as a measure of amplification of a circuit, it being the ratio of a change in the output current to the incremental change in the input voltage. For proper comparison, it becomes necessary to find g_m/I.

The ratio of the transconductance to the collector current g_m/I_C of a BJT is given by Eq. (9.62) to be q/kT, which at room temperature is 38.6V^{-1}. The expression for g_m/I_{D1S} of a MOSFET, by assuming $\lambda V_D \ll 1$, is found from Eqs. (13.36) and (13.19) to be $2/(V_G - V_T)$. For the same current and since $2/(V_G - V_T)$ for normal operation is much smaller than 38.6V^{-1}, we conclude that in this respect, the BJT is superior to the MOSFET.

The limiting cutoff frequency will be calculated from the transit time of an electron across the basewidth, W_B, of an NPN-BJT and across the channel length, L, of an N-channel MOSFET, both of which are assumed to use silicon.

The transit time, τ_{tr}, for an NPN is given by Eq. (9.30) to be $W_B^2/2D_n$, where D_n is the diffusion constant for electrons. For a MOSFET, it is the ratio of the channel length to the saturation velocity of an electron in silicon, L/v_s, where v_s is assumed to be 10^7cm/s. While D_n and v_s are known quantities, it is the possible values of W_B and L that determine the cutoff frequency that is approximately $1/2\pi\tau_{tr}$. Using these calculations for the cutoff frequency, one might be led to conclude that for the same channel length and basewidth, the MOSFET is superior to the BJT. Of course, one would need to confirm the possibility of fabricating such dimensions for both devices.

REVIEW QUESTIONS

Q13-8 How does substrate biasing affect the threshold voltage?

Q13-9 Identify and explain the two modes of operation of the MOSFET.

Q13-10 One method that can be used to increase the threshold voltage for an NMOS is to increase the doping of the substrate. What factor sets the limit on the doping?

Q13-11 Identify another mechanism, available in the expression for V_T, that may be used to increase the threshold voltage. Indicate the price that is paid for this increase.

Q13-12 Explain how ion implantation is used to improve V_T.

Q13-13 Briefly explain the significance of g_m and g_d.

Q13-14 Why is there no dynamic input resistance in the equivalent circuit of the MOSFET (compared to the BJT)?

Q13-15 Which capacitances of the MOSFET impose the larger limitation on the frequency response?

HIGHLIGHTS

- The mobility of carriers in the channel of a MOSFET, known as the surface mobility, is considerably smaller than the mobility in the bulk of the semiconductor. This is because carriers in the channel are restricted to a very thin region at the surface of the channel and hence are subjected to frequent collisions with the irregular surface of the oxide.

- The threshold voltage of an NMOS can be made more positive by applying a negative bias to the body. This has the effect of increasing the magnitude of the negative depletion layer charge. Application of a positive bias to the body of a PMOS makes V_T more negative.

- Control of the threshold voltage of an NMOS can be achieved through two other mechanisms, as shown by the expression for V_T. Increasing the substrate doping and using a thicker oxide make V_T more positive. Furthermore, implantation of the proper type of ions in the semiconductor at the oxide interface provides precise control of V_T.

- Depending upon whether a channel is pre-formed or induced by capacitor action, MOSFETs may be either of the enhancement or depletion mode. In the enhancement mode, normally OFF device, a positive gate voltage (for the NMOS) exceeding V_T must be applied to form an N channel. In the depletion mode, where a channel may have been formed by ion implantation, a gate voltage equal to $-V_T$ must be applied to turn the device OFF. The depletion mode device may be operated with both negative and positive gate voltages.

- In digital applications, MOSFETs are preferred over BJTs for two reasons: First, they operate on a binary basis, with V_T representing the dividing boundary, and, second, they consume less DC power. For analog applications, BJTs have a higher gain factor, defined by the ratio of the transconductance to the operating current.

- Isolation of BJTs in microcircuits results in devices that consume more chip area than MOSFETs and are therefore more costly to manufacture. The high-frequency response of MOSFETs is improved by the use of polysilicon for the gate metal, which results in a considerable reduction of the overlap capacitances.

EXERCISES

E13-3 Two transistors, a BJT and MOSFET, are operating at collector and drain current respectively of 1mA. Calculate the ratio of the transconductance to the current for each of the devices. Use the square-law model and let $V_G = 3$V and $V_T = 0.7$V.

Ans: For the BJT: $g_m/I_C = 38.6$.

For the MOSFET: $g_m/I_D = 0.87$.

13.12 THE MOSFET SWITCH AND THE CMOS INVERTER

The use of the MOSFET as a switching element is a major application of the device. Compared to the BJT, it occupies a smaller area on a chip and consequently a larger number can be placed within a given area. The MOSFET has found widespread use in digital gates and in memory circuits. The use of the complementary MOSFET (CMOS) as the principal device in very large-scale integrated circuits (VLSI) has made possible the latest state-of-the-art microprocessors. A unique advantage of the CMOS device lies in the small amount of power that it dissipates.

A switch in the open state can be modeled by a very high resistance resulting from low-current, high-voltage operation of a device. When a switch is in the closed position, it may be represented by high-current low-voltage operation of an element. Such states of operation can occur in a transistor that is made to switch from cutoff to a high current point.

An electronic switch using a transistor is formed by an *inverter* circuit, which is a major building block of switching and memory circuits.

The Inverter

An inverter is made up of a transistor, known as the driver, connected in series with a load element. The load may be a resistor or it may be another transistor. The input signal to a MOSFET inverter is applied between gate and source and the output is taken from the drain to the source of the driver transistor. The basic property of an inverter is the inversion of the input signal. When the input signal is low (L), the output is high (H). When the input is high (H), the output is low (L).

Digital signals are made of pulses having a nominal voltage of 3 volts (H) separated by lows of approximately zero volts.

Resistor Inverter

The circuit of an inverter using an enhancement-mode N-channel MOSFET for the driver and a resistor load is shown in Fig. 13.15(a). The current-voltage characteristics of the transistor and the load are shown in Fig. 13.15(b) and the composite characteristics of the inverter are in Fig. 13.15(c). The composite characteristics are obtained by plotting the load characteristics on the MOSFET characteristics since

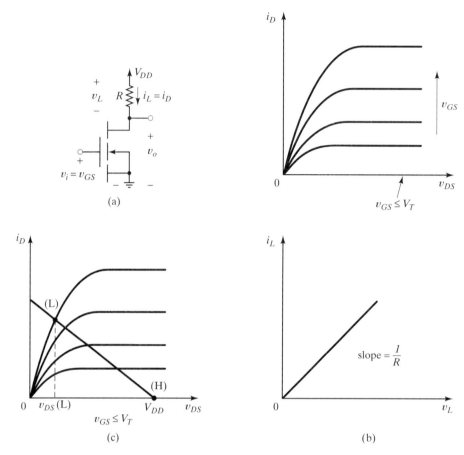

Figure 13.15   Resistor-load inverter: (a) Circuit, (b) current-voltage characteristics of NMOS and load, and (c) composite characteristics of inverter.

$$i_L = (V_{DD} - v_{DS})/R = (V_{DD} - v_o)/R \tag{13.46}$$

The threshold voltage of the NMOS is V_T, so that for $v_{GS} = v_i \leq V_T$, the current is zero and operation is along the v_{DS} axis at $v_o = V_{DD} = 5V$. Thus, for a low (L) input the output is at high (H). When v_i is increased to a high ($H = 5V$), the transistor is ON, drawing a high current and operation at a point of low (L) voltage.

Therefore, a low amplitude pulse, less than V_T, causes a high to appear at the output (high-voltage, zero current) and a high input pulse causes v_o to become low and the current to be high.

Enhancement-Load Inverter

An all-transistor inverter, shown in Fig. 13.16(a), consists of enhancement-mode NMOS driver and load. The driver is also known as the *pull-down transistor* and the load is the *pull-up transistor*. So that the load is made to operate as a two-terminal

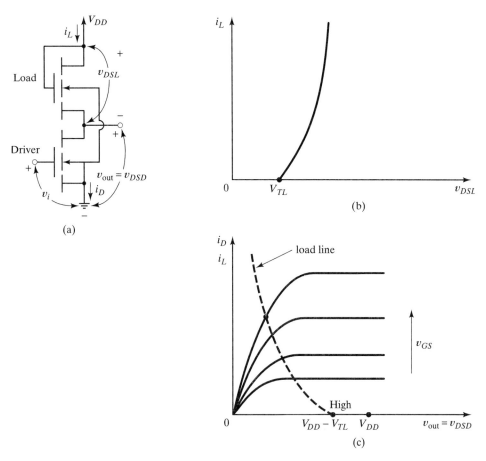

Figure 13.16   MOSFET-load inverter: (a) Circuit, (b) characteristics of load and,
(c) composite characteristics of inverter.

device, its gate is connected to its drain. The input terminals, at the driver, are the
gate and the source and the output is taken from drain to source.

By connecting the gate of the load to its drain, the current-voltage characteristics of the load are converted from a family of curves, as shown in Fig. 13.15(b), to
one curve, as shown in Fig. 13.16(b). We will now briefly explain how this curve is
obtained.

A MOSFET is said to be operating in saturation when $v_{DS} \geq v_{GS} - V_T$. By
shorting the gate to the drain, $v_{DS} = v_{GS}$, so that the inequality required for operation in saturation is satisfied. The load characteristic equation in saturation is given
by

$$i_L = k_{pL}(v_{GSL} - V_{TL})^2 = k_{pL}(v_{DSL} - V_{TL})^2 \tag{13.47}$$

where quantities with the subscript L refer to the load and $k_{pL} = Z\bar{\mu}_n C_{ox}/2L$. Figure
13.16(b) is a plot of Eq. (13.47). The drain-source voltage of the load is given by

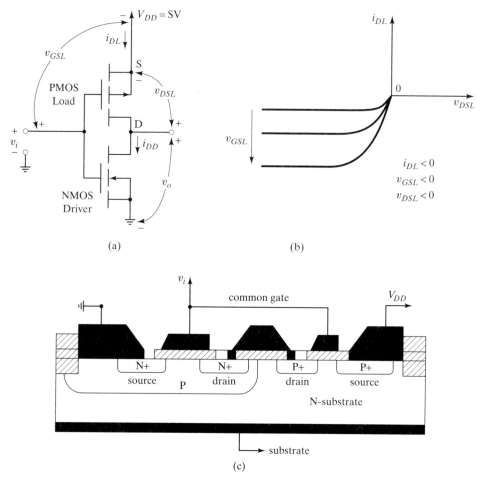

Figure 13.17 CMOS inverter: (a) Circuit, (b) PMOS characteristics, (c) cross section of the CMOS inverter.

$$v_{\text{DSL}} = V_{\text{DD}} - v_{\text{DSD}} = V_{\text{DD}} - v_o \qquad (13.48)$$

where v_{DSD} refers to the drain-source voltage of the driver.

Since the load current is equal to the driver current, an expression for the current of the load in terms of the axes of the driver characteristics is obtained by using Eq. (13.48) in Eq. (13.47)

$$i_D = i_L = k_{pL}(V_{\text{DD}} - v_o - V_{\text{TL}})^2 \qquad (13.49)$$

Equation (13.49) is plotted on the output characteristics of the driver to obtain the composite characteristics of Fig. 13.16c, where all operation is along the dotted load line.

For low $v_i (< V_{\text{TD}})$, the current is zero and the load curve intersects the driver characteristics at the point H where, from Eq. (13.49), $v_o = V_{\text{DD}} - V_{\text{TL}}$. For v_i having a high value (5V), the intersection of the load curve with the appropriate NMOS curve, for which $v_i = v_{\text{GS}}(H)$, occurs at a low value of v_o.

For both the resistor and transistor load inverters, there is no current in the circuit when the input is low. However, when the driver is in the state where the input is high, the load and the transistor carry current corresponding to the low point on the characteristics. Hence, both inverters dissipate power in the static low (L) state.

The CMOS Inverter

The CMOS inverter consists of an enhancement-mode NMOS as the driver and an enhancement-mode PMOS as the load. The two gates are connected and the input signal is applied between the gates and the driver source. As shown in Fig. 13.17(a), the two drains are connected and the output is taken from the common drain. The characteristics of the PMOS is shown in Fig. 13.17(b) and a cross-section of the CMOS device is shown in Fig. 13.17(c). Assume the input is 5V.

The characteristics of the two devices are related by

$$i_{\text{DL}} = -i_{\text{DD}}$$

$$v_{\text{DSL}} = -V_{\text{DD}} + v_{\text{DSD}} = -V_{\text{DD}} + v_o \qquad (13.50)$$

where, as shown in the PMOS characteristics, i_{DL} and v_{DSL} are negative quantities.

Let us determine the composite characteristics of the inverter for a high input and then for a low input. For the sake of simplicity, we assume $V_{\text{TL}} = -V_{\text{TD}}$, where the subscript D refers to the NMOS driver and L refers to the load.

For a high input signal, v_{GS} of the driver is 5V and v_{GSL} is zero. The driver is said to be ON but the load device is OFF. For the load to be ON, the gate-source voltage of the load must be negative and have a magnitude greater than V_{TD}. The current in the circuit is zero and the characteristics of the driver and load are shown in Fig. 13.18a. The two characteristics intersect at the origin and, v_o is zero (L).

When input is low ($v_i = 0$), the driver is OFF but the load transistor is ON because $v_{\text{GSL}} = -5\text{V}$. The composite characteristics are shown in Fig. 13.18(a). The

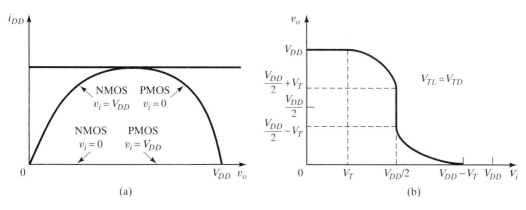

Figure 13.18   (a) Composite characteristics of the CMOS inverter for high input and low input. (b) Voltage transfer characteristics.

load characteristics are drawn on the driver axes by using Eqs. (13.50). The intersection of the characteristics of the two devices occurs at zero current and $v_o = V_{DD}$.

The very important conclusion to be drawn from the above analysis is that in both static states, high output and low output, the current in the devices is zero, so that there is no power dissipation in the circuit. This is in contrast to the two earlier inverter circuits in which power is dissipated when the input is high. The zero power dissipation of the CMOS inverter is the principal reason for its prevalent use. To appreciate this, one must realize that a VLSI chip contains a great number of inverters and excessive power dissipation becomes a serious problem. To alleviate this problem, either the number of inverters in a given area has to be reduced or a means of cooling the circuits must be provided. A cooling fan requires a power source and additional space. The heat generated is a serious shortcoming, particularly in laptop computers, where not only is the size small but battery drain that supplies the energy must be considered.

Switching of Inverters

In switching applications, the output of a master driver is connected to the inputs of several inverters. Both the driver and driven inverters have parasitic capacitances that are parallel and appear as one capacitance at the output of the driver, as shown in Fig. 13.19(a). Because of this capacitance, switching of an inverter does not take place in zero time since this capacitor is charged and discharged in the process of switching, as shown in Fig. 13.19(b). When the input to a driver inverter is changed from a low to a high, the low at the output does not appear instantaneously and, during this portion of the switching cycle, currents flow in the load and the driver and power are dissipated in the devices. The CMOS inverter does not dissipate power in either of the two states and power is dissipated in CMOS inverters only when switching takes place. This is known as dynamic power. All other inverters dissipate both static and dynamic power, where static power is dissipated when the inverter is operating in one of its two states.

PROBLEMS

Unless otherwise indicated, use the square-law model of the MOSFET.

13.1. An NMOS device has $V_T = 3V$, $L = 2\mu m$, $Z = 14\mu m$, $C_{ox} = 12 \times 10^{-8} Fcm^{-2}$, and $\bar{\mu}_n = 500 cm^2/V\text{-s}$. Calculate the drain current for

    (a) $V_G = 5V$ and $V_D = 8V$

    (b) $V_G = 2V$ and $V_D = 8V$

    (c) $V_G = 6V$ and $V_D = 2V$

13.2 An NMOS device has the gate connected to the drain.

    (a)   Determine the region of operation of the device.

    (b)   Sketch I_D versus V_D, clearly labeling all critical points.

13.3 A MOSFET is made using Al-SiO$_2$-Si. Given $N_A = 10^{17} cm^{-3}$, $L = 1\mu m$, $Z = 10\mu m$, oxide thickness $= 200\text{Å}$, $\bar{\mu}_n = 400 cm^2/V\text{-s}$, and $Q_i = 5 \times 10^{11} q$ C/cm^2, determine:

    (a) V_T

    (b) I_D for $V_G = 2V$ and $V_D = 6V$

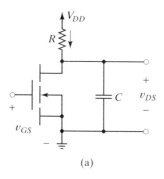

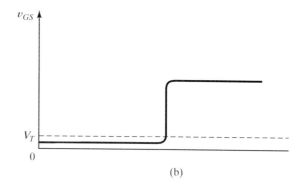

(a)

(b)

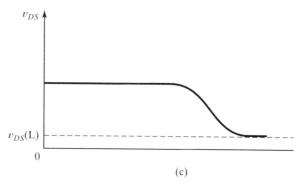

(c)

Figure 13.19 MOSFET inverter with resistive load: (a) Circuit diagram showing parasitic capacitance, (b) input waveform, and (c) output waveform showing switching delay due to capacitor discharge.

(c) I_D for $V_G = 2V$ and $V_D = 1V$.

13.4 An NMOS silicon device has $\Phi_{ms} = 0$, $Q_i = 0$, and is designed to have $\phi_F = 0.21V$ and $V_T = 1.42V$ at an insulator thickness of 200Å. During fabrication, a mistake is made so that the insulator thickness is 100Å while everything else is unchanged. For $V_G = 3V$ and $V_D = 6V$, determine the percentage change in the drain current as a result of the mistake.

13.5 An NMOS silicon transistor has $N_A = 4 \times 10^{14}cm^{-3}$, oxide thickness 100Å, $L = 2\mu m$, $Z = 10\mu m$, and $\bar{\mu}_n = 700cm^2/V\text{-s}$. Given also that $V_{SB} = 0$ and $V_{FB} = -0.2V$, at a gate voltage of 5V and a drain voltage of 4V, determine the drain current using:

   (a)   the square-law model
   (b)   the bulk-charge model.

13.6 An NMOS has $Z/L = 10$, $V_{FB} = -0.3V$, $N_A = 10^{18}cm^{-3}$, $\bar{\mu}_n = 600cm^2/V\text{-s}$, and oxide thickness = 200Å. For $V_G = 5V$ determine:

   (a)   the drain current at saturation
   (b)   the saturation drain voltage
   (c)   repeat parts (a) and (b) using the bulk-charge model.

13.7 An NMOS transistor has $Z/L = 12$, $V_T = 0.5V$, $\bar{\mu}_n = 500cm^2/V\text{-s}$, and oxide thickness of 200Å. At $V_G = 5V$, calculate:

- **(a)** I_D, g_m, and g_d at $V_D = 6V$
- **(b)** I_D, g_m, and g_d at $V_D = 3V$

13.8 An NMOS silicon transistor has $Z/L = 15$, oxide thickness of 150Å, and $\bar{\mu}_n = 500 \text{cm}^2/\text{V-s}$. It is to be used as a controlled resistor. Determine:

- **(a)** the free electron density in the channel required to present a 3KΩ resistance between drain and source at low values of V_D.
- **(b)** the value of $(V_G - V_T)$ that will produce such a resistance.

13.9 An MOS capacitor uses N silicon with $N_D = 10^{15} \text{cm}^{-3}$ and oxide thickness of 200Å. Determine the surface charge density, Q_n, in the inversion layer for $V_G = -5V$. Let $\Phi_{ms} = 0$ and $Q_i = 0$.

13.10 A P-channel silicon MOSFET uses an Al gate. Given $N_D = 10^{15} \text{cm}^{-3}$, $t_{ox} = 80$Å, $Z/L = 10$, $\bar{\mu}_p = 450 \text{cm}^2/\text{V-s}$, and $Q_i = 5 \times 10^{11} q$ C/cm^2, determine:

- **(a)** the turn-on voltage
- **(b)** the value of V_{SAT} at $V_G = -5V$
- **(c)** the value of V_{SAT} at $V_G = -7V$
- **(d)** I_{SAT} for $V_G = -5V$

13.11 An Al-SiO$_2$-Si NMOS has $N_A = 3 \times 10^{16} \text{cm}^{-3}$, $t_{ox} = 90$Å, and $Q_i = 8 \times 10^{11} q$ C/cm^2. Calculate the threshold voltage for:

- **(a)** $V_{SB} = 0$
- **(b)** $V_{SB} = 5V$

13.12 **a)** Modify Eq. (13.27) so that it applies to a P-channel device.
  - **b)** A silicon PMOS has $t_{ox} = 500$Å and $N_D = 10^{16} \text{cm}^{-3}$. Calculate the body-source voltage that shifts the threshold voltage from the $V_{BS} = 0$ value by $-1V$. Assume $\Phi_{ms} = 0$ and $Q_i = 0$.

13.13 A silicon NMOS device has $N_A = 2 \times 10^{16} \text{cm}^{-3}$ and oxide thickness of 400Å. Calculate the shift in the threshold voltage when a source-body voltage of 1V is applied.

13.14 An NMOS has $V_T = 1.42V$ and $\bar{\mu} Z C_{ox}/L = 10^{-4} \text{AV}^{-2}$. It is used in the circuit shown below. Neglect the effects of work function difference and oxide charges. Determine the range of gate voltages during which the transistor operates in the saturated region.

13.15 A MOSFET has $\bar{\mu}_n C_{ox} Z/L = 2 \times 10^{-4}$ AV^{-2}. We wish to compare its g_m to that of a BJT, both operating at a quiescent current of 1mA.

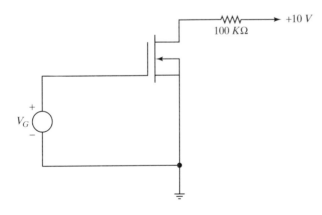

(a)   Find the ratio of the g_m of the BJT to that of the MOSFET.

(b)   Explain the reason for the large differences between the two.

13.16 An NMOS device has $Z = 40\mu m$, $L = 4\mu m$, $t_{ox} = 80\text{Å}$, $\bar{\mu}_n = 550\text{cm}^2/\text{V-s}$, and $V_T = 1.1\text{V}$. Calculate:

(a) I_{SAT} at $V_G = 3\text{V}$ using the square-law model

(b)   repeat (a) for the bulk charge model

(c)   the power dissipated in the device for condition (a).

(d) g_d at $V_G = 3\text{V}$ and $V_D = 0$

13.17 Plot on the same set of axes, V_T versus N for both P-substrate and N-substrate Al-SiO$_2$-Si devices. Allow N to vary from 10^{13} to 10^{18}cm^{-3}. Let $Q_i = 5 \times 10^{10}q$ and $t_{ox} = 100\text{Å}$.

13.18 A silicon NMOS uses an aluminum gate and has $N_A = 10^{15}\text{cm}^{-3}$, oxide thickness = 100Å, $Z/L = 10$, and $\bar{\mu}_p = 800\text{cm}^2/\text{V-s}$. It is to be used as a variable resistor. Determine the value of the resistance at $V_D = 0.5\text{V}$ and

(a) $V_G = 5\text{V}$

(b) $V_G = 7\text{V}$

13.19 An NMOS and a PMOS each has a current $I_{SAT} = 5\text{mA}$ when $V_G = 5\text{V}$ for N channel and $V_G = -5\text{V}$ for the P-channel device. Given $t_{ox} = 80\text{Å}$, $\bar{\mu}_n = 500\text{cm}^2/\text{V-s}$, $\bar{\mu}_p = 300\text{cm}^2/\text{V-s}$, $V_T = 0.7\text{V}$ for NMOS and $V_T = -0.7\text{V}$ for PMOS, determine Z/L for both.

13.20 An NMOS has $Z = 10\mu m$, $L = 2\mu m$, and $C_{ox} = 10^{-7}\text{F/cm}^2$. At $V_D = 0.1\text{V}$, the drain current is given by

$$\text{at } V_G = 1.6\text{V}, I_D = 40\mu A$$

$$\text{at } V_G = 2.6\text{V}, I_D = 90\mu A$$

Calculate:

(a)   the electron mobility.

(b)   the threshold voltage.

chapter 14

OPTOELECTRONICS

14.0 INTRODUCTION

Optoelectronics is the science that deals with the interaction between light and matter. The consequences of the interaction may be either the production of devices that are sensitive to light or of devices that generate light. Some devices respond to light by the generation of current and other devices radiate light when activated by a source of electrical energy. *Photodetectors* convert light energy into current and *photovoltaci cells* convert optical radiation into electrical energy. On the other hand, electro-luminescent devices such as *light-emitting diodes* (LEDs) and *lasers* emit radiation when energized by a current.

Light-emitting diodes generally serve as indicator lamps while lasers are sources of light characterized by high-intensity directional radiation and are used in medical and industrial applications, and in particular as sources of light in fiber optic communications. Optical fibers consist of thin cylindrical waveguides made of silica or silica-based glass through which information-carrying light is made to travel between physical locations.

Photodetectors are used in a variety of industrial applications such as in counters and readers of light, that is reflected from bar-codes, as well as serving as the elements that convert information-carrying light, at the receiving end of a fiber optic network, into electrical signals. The use of photovoltaic cells has made possible extensive space exploration and satellite communication by providing to space-crafts electrical energy converted from the sun's rays. They are constantly explored as commercial sources of electrical energy, especially at locations that are distant from conventional sources of electrical power.

This chapter is divided into five sections. In the first section, we study the principles of light emission from a semiconductor diode and discuss the operation, properties, and construction of the LED. The laser, another light-emitter with a sharply focused beam, is the subject of the second section. Following the LED and the laser sections, we consider the operation of photodetectors. In section four, the photovoltaic cell is covered together with the use of solar radiation in the generation of electrical energy. The very fascinating application of light-emitting devices, optical fibers, and detectors, to optical communication systems is the subject of the last section.

14.1 LIGHT-EMISSION AND LIGHT-EMITTING DEVICES

Our interest here is in devices made from semiconductors that act as emitters of light. These devices are: *light-emitting diodes* and *lasers*. Before proceeding, we acknowledge the following devices that act as sources of light and are not semiconductor-based: the incandescent tungsten-filament heated lamp, the television receiver and the computer screen, in which the emission of light is based on the phosphorescence of a screen when impacted by an electron beam, display applications based on gas-discharges, such as neon lamps and other forms of lasers that do not employ semiconductors.

Light-emitting semiconductor diodes generally serve as linkages between electronic instruments and the human eye. They are used in the combination of light emitter and photodetector, serving as bar code reader. The bar code consists of several bars having different widths and varying separations that reflect the LED light to a detector, resulting in a series of numerals that are transmitted to a computer that identifies the product. LEDs are also used as displays, in LED printers, as digital disk readers in audio systems, and occasionally as optical signal generators in fiber optic communications (these systems more commonly use lasers, which are considered in a later section).

The early LEDs were very inefficient providing an output of 0.15 lumens per watt of electrical energy compared to the 3.5 lumens per watt of the red-filtered incandescent lamps. It is now possible to fabricate LEDs that exceed the efficiency of incandescent lamps and that are sufficiently bright enough and efficient enough to replace incandescent lamps in applications such as automatic braking lights, which are placed in the center of the rear window of automobiles. In this application, the faster response time of LEDs compared to incandescent lamps is of particular importance. LEDs are inexpensive, consume very little power, have a very long life, and are easily adaptable to many applications.

In the various applications of LEDs, we are particularly interested in devices that emit light in two regions, each identified by the wavelength of the emitted radiation. These two regions are: the visible region extending from 0.4 to 0.7μm and the infrared region covering the 0.7 to 15μm range as shown in Fig. 14.1. As a means of comparison, we have displayed in Fig. 14.2 spatial distributions of the responses of the eye, the radiation from a tungsten lamp, a light emitter, and a light detector. The responses at various wavelengths are indicated with respect to the peak response and hence are labeled relative responses.

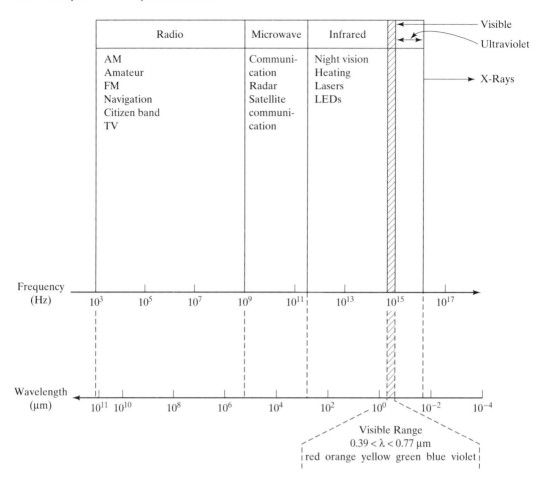

Figure 14.1 Electromagnetic radiation spectrum indicating the ranges from radio waves to x-ray frequencies.

With respect to emitters of light, the question that needs to be addressed is: How is radiation from semiconductors generated and emitted?

Radiative and Non-Radiative Transitions

Electromagnetic radiation from a semiconductor takes place, most commonly, when excess electrons that have been injected into the conduction band (of the P region) of a semiconductor fall *directly* into the valence band releasing the energy difference as photons. A photon is a zero-mass particle having energy and traveling at the speed of light. The band gap energy, E_g, the difference between the energies of the bottom of the conduction band and the top of the valence band, is converted to a quantum of radiation and the wavelength of the radiation is given by

$$\lambda = \frac{hc}{E_g} = \left(\frac{1.24}{E_g(\text{in eV})} \right) \mu m \tag{14.1}$$

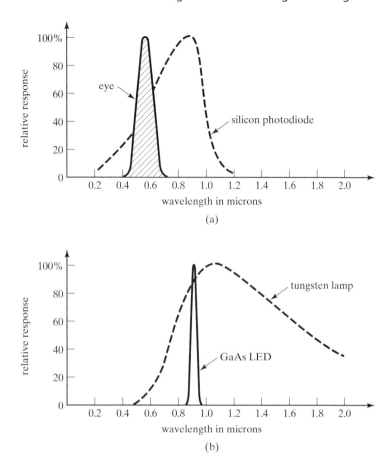

Figure 14.2   Spectral distributions: (a) light-sensitive devices and (b) light-emitting devices.

where h is Planck's constant in J-s, c is the velocity of light in m/s, E_g is in J, and λ is in m. The product of E_g in eV and λ in μm is 1.24.

There are thus two basic requirements for the release of photons of radiation. First, a means is required whereby excess electrons are injected into the conduction band and, second, that the transition of these excited electrons to the valence band is accompanied by the generation electromagnetic radiation.

Excitation of a semiconductor is produced by the passage of a current, as occurs by the injection of minority carriers in a PN junction diode that is forward-biased. The excess electrons in P and the excess holes in N recombine with holes and electrons respectively, causing radiative recombination. This process is illustrated in Fig. 14.3.

Radiative recombination occurs in what are known as *direct bandgap semiconductors*, discussed in Section 4.10, in which the recombination takes place when an electron from the conduction band drops directly into the valence band. GaAs is

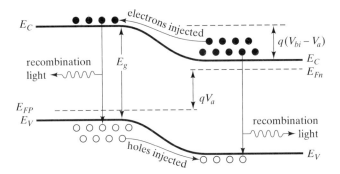

Figure 14.3 Interband recombination in a forward-biased PN junction diode results in radiation.

an example of a direct band gap semiconductor. In indirect band gap semiconductors, such as Ge and Si, recombination takes place, most frequently, via a recombination center in the forbidden band. There is a very low probability of radiation when recombination takes place in indirect band gap semiconductors.

In general, when recombination occurs via recombination centers in the forbidden band (such as happens in Ge and Si), energy is released in the form of heat. On the other hand, when recombination takes place in the valence band (such as happens in GaAs) energy is released in the form of light. In most semiconductors, however, both types of recombinations take place to a certain extent. The efficiency of the process in emitting radiation depends on the *semiconductor material*, the *doping level* (hence, the injection level), and *temperature*. A direct band gap semiconductor is much more likely to produce an efficient electroluminescent device.

It may be possible to produce radiation from certain indirect semiconductors by purifying the material so that non-radiative recombination becomes negligible. Another workable method is by adding certain dopants to indirect III-V compound semiconductors.

Direct semiconductors are always compound semiconductors, such as GaAs and InP. The difference between direct and indirect semiconductors is exhibited in the relative locations on an energy-momentum plot of the conduction band minimum and the valence band maximum, as shown in Fig. 14.4, for two semiconductors.

For an electron having energy E_1 to drop from the conduction band to the valence band energy E_2, two requirements must be met. They are: the conservation of energy and the conservation of momentum. The conservation of energy requires that a photon with energy $h\nu$ is emitted such that

$$E_1 - E_2 = h\nu \qquad (14.2)$$

The conservation of momentum, defined in Sec. 1.6 as $h\mathbf{k}/2\pi$ requires that

$(h\mathbf{k}/2\pi)$ electron in the valence band + $[h\mathbf{k}/2\pi]$ photon = $[h\mathbf{k}/2\pi]$ electron in the conduction band. $\qquad (14.3)$

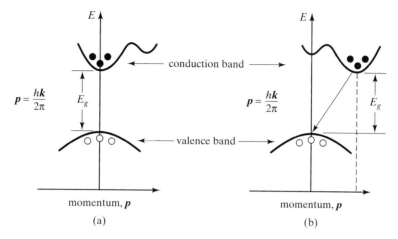

Figure 14.4   Energy versus momentum for (a) direct and (b) indirect band gap semiconductors.

The momentum of a photon is negligibly small. Compared to that of an electron, it is about 1000 times smaller. Hence, the recombination requirement is realized if the initial and final momentum of the electron is the same. In a direct semiconductor, the bottom of the conduction band and the top of the valence band occur at the same momentum; thus, radiative recombination takes place.

The conduction band electrons in an indirect semiconductor have a different momentum than the holes in the top of the valence band, so that radiative recombination does not take place. However, recombination takes place usually at an impurity level in the forbidden band.

Quantum and Overall Efficiency

The internal *quantum efficiency* of an LED is defined as the ratio of rate of photon emission to the rate of carrier injection. This efficiency is determined by both the number of electrons injected and the number of radiative recombinations that take place. If all the injected electrons recombine and all the recombinations are radiative, the quantum efficiency is 100 percent. It is quite possible that all injected electrons eventually recombine but it is not necessarily true that all recombinations are radiative. This may happen because of crystal imperfections where recombination does not result in a photon but in the generation of heat that is dissipated in the crystal.

The *overall or total efficiency* is a measure of the quantum efficiency as well as the efficiency with which the photons become useful by being extricated from the device.

For *electroluminescent devices* to be useful and efficient, four properties are required: First, the material of the device must be a direct semiconductor; second, it must be possible to excite the material by injection across a PN junction; third, if the detector is the eye, the band gap of the material must be such that the emitted radiation is in the visible part of the spectrum having an E_g between 1.8 and 2.8eV, according to the relation in Eq. (14.1); and, fourth, the device must have a sufficiently high total efficiency so as to emit a large fraction of the generated light.

Let us consider how the various semiconductors meet these requirements.

Column IV Semiconductors

Both Ge and Si are indirect semiconductors, hence their quantum efficiency is very low. Silicon Carbide, another indirect semiconductor, exists in various combinations of silicon and carbon with one form having a band gap of 3eV. The total efficiency, defined as the ratio of the quantity of light emitted, in lumens, to the power input, in watts, of such an LED is very low (0.04 lumens/watt). In spite of this, blue LEDs have been fabricated by the addition of certain impurities that cause various radiative transitions.

Compound Semiconductors III-V

There are three III-V compounds that are most important and most efficient for the generation of optical radiation in the visible and infrared parts of the electromagnetic spectrum. These are: gallium arsenide, gallium phosphide, and the alloy gallium arsenide phosphide.

Gallium arsenide devices ($E_g = 1.4eV$) emit radiation in the infrared region ($\lambda = 860nm$) and are very efficient. Spectral responses of GaAs diodes* are shown in Fig. 14.5, illustrating the effect of an increase of the current and a change of temperature.

For emission in the green region of the visible spectrum, gallium phosphide (GaP, $E_g = 2.24eV$, $\lambda = 549nm$) is used. Because it is an indirect semiconductor, the probability of radiative recombination is low. The radiation is enhanced by the addition of nitrogen, an impurity that helps conserve momentum during transition, which is introduced in both the P and N regions of the diode. A GaP:N LED has a total efficiency of 2.6 lumens/watt at a wavelength of 565nm (yellow-green).

Neither GaAs nor GaP strictly meet all the requirements of an LED because GaAs emission is not in the visible part of the spectrum and GaP is indirect, requiring the addition of impurities to enhance its response.

To obtain radiation that is close to the ultraviolet region, an increase of the bandgap of the materials is needed. This is achieved in alloys of III-V compounds, mainly $GaAs_{1-x}P_x$, by distributing randomly the atoms of one group V element, such as phosphorus, so as to reside in arsenic lattice sites. The subscript x, in the symbol, represents the ratio of the number of P atoms to the total number of group V atoms (As + P) in the alloy. The band gap of this compound can be adjusted by varying x so that at $x = 0$, $E_g = 1.42eV$ ($\lambda = 873nm$) and at $x = 1$, $E_g = 2.26eV$ ($\lambda = 548nm$). This covers the optical range from near infrared to green. For alloys wherein $0 < x < 0.44$, the alloy is direct and high luminous output is obtained. For $x > 0.44$, an indirect semiconductor results and the efficiency of radiative recombination is considerably reduced. An excellent alloy that emits light that appears very

---

*C. H. Gooch, *Injection Electroluminescent Devices*, p. 82, Wiley (1973).

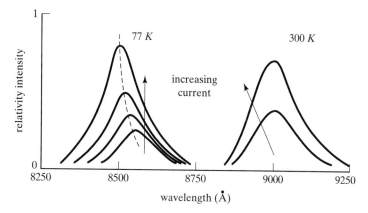

Figure 14.5   Spectral response of GaAs diodes at two temperatures showing the increase of light intensity with increased current. *Source*: C. H. Gooch, *Injection Electroluminescent Devices*, p. 82, Wiley (1973). Copyright John Wiley & Sons Limited. Reproduced with permission.

bright to the eye is formed at $x = 0.4$. For an increase of x from 0.28 to 0.44, the overall efficiency of the LED decreases and the wavelength decreases but, with the reduction of the wavelength, the response of the eye increases considerably more than offsetting the decrease of the efficiency. A value of $x = 0.4$ seems to provide an optimum value satisfying both the eye and the efficiency and hence is used for the production of commercial devices.

The spectral responses of selected compounds are shown in Fig. 14.6.

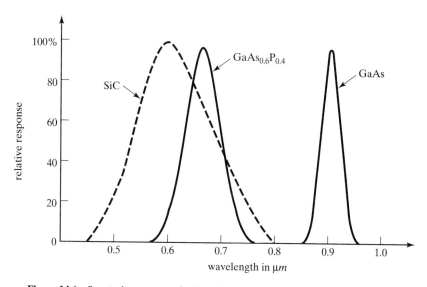

Figure 14.6   Spectral responses of selected compounds.

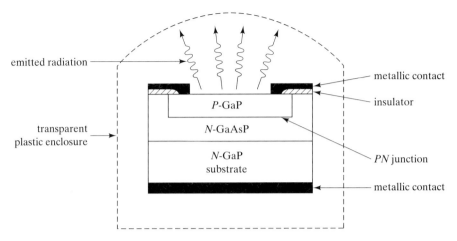

Figure 14.7 Schematic of LED diode.

LED Construction

Figure 14.7 shows the construction of a GaAsP LED. The LED consists of a GaAsP substrate upon which a $GaAs_{0.6}P_{0.4}$ layer is directly deposited, which forms the N region of the diode. The P region is formed by diffusing zinc into the N material. When the diode is forward-biased, photons are generated by the recombination of electrons, injected from N, with holes in the P side of the junction. Holes injected into the N region also generate radiative recombinations. However, because of the location of the N region, as shown in Fig. 14.7, the radiation is absorbed within the diode as it is generated and does not contribute to the light output.

Therefore, the efficiency of light production is improved if the current crossing the junction is predominantly due to electrons. The ratio of electron to hole current across the junction, obtained from Eq. (5.44), indicates that higher electron current is obtained for $\mu_n > \mu_p$ and by making $N_D \gg N_A$ since

$$\frac{J_n}{J_p} = \frac{D_n n_{op} L_p}{D_p p_{on} L_n} = \frac{\mu_n N_D L_p}{\mu_p N_A L_n} \tag{14.4}$$

Not all the photons generated at the junction, however, reach the eye of the observer. There are two sources of losses. First some light is absorbed by the N region and some is lost in the P region. Since the light leaves the diode through the P region, the narrower the P region, the less light is absorbed. Second, as the light travels from the semiconductor to the outside, some of it is reflected back into the semiconductor because of the different indices of refraction of the semiconductor and the outside medium. The index of refraction of a material is defined as the ratio of the velocity of light in vacuum to the velocity of light in the material. Many compound semiconductors have high indices of refraction, so that light from these semiconductors strikes the semiconductor-air interface at angles greater than the critical angle, resulting in complete reflection. For instance, the critical angle for GaAs is 16 degrees.

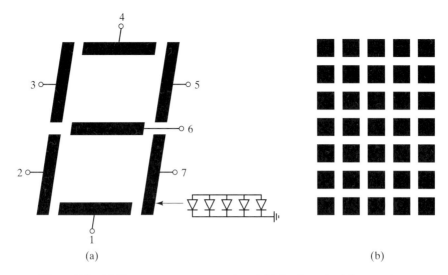

Figure 14.8   (a) The seven segment numeric and (b) 5×7 matrix alphanumeric display.

Increased extraction of light from a diode is obtained by placing the diode in a dome-shaped plastic enclosure that has a high refractive index. The plastic lens is usually colored to enhance contrast.

When used in displays, the LEDs are arranged to be of the numeric type (showing a number only) or of the alphanumeric type (showing letters and numbers). Shown in Fig. 14.8 are the seven segment numeric and the 5×7 dot matrix alphanumeric type. Several LEDs connected in parallel may be used to display numbers and letters. Each of the seven segments of Fig. 14.8(a) may consist of several diodes in parallel as shown. A digital control systems is used to direct current to the appropriate LEDs. The number 3 in Fig. 14.8(a) is displayed by causing the system to direct current to the array of diodes in segments 1, 4, 5, 6 and 7.

14.2 SEMICONDUCTOR LASERS

The word *laser* is an acronym for: *light amplification by the stimulated emission of radiation*. While the mechanisms that cause light emission in the LED and the laser are very similar, the difference between the two is that light output in the LED is spontaneous whereas it is stimulated in the laser. Spontaneous emission occurs when an electron falls from the conduction to the valence band and no further interaction occurs between the photon and the electron. Emission is stimulated when other factors in the device cause a generated photon to be reflected inside the device and is used to induce an electron to fall to the valence band and emit another photon. In fact, the light is amplified as it is fed back to generate additional photons. This process is repeated until a steady state is reached when the losses in the diode are balanced by the additional photons generated.

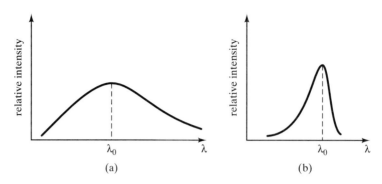

Figure 14.9   Spectral response of (a) LED diode and (b) laser diode.

Because light is stimulated, the laser light has unique properties. It is said to be *highly directional* and *monochromatic*. A light is monochromatic when it is principally emitted at one wavelength with a very narrow bandwidth, as illustrated in Fig. 14.9. If we assume that the peak of a response has amplitude R, the bandwidth is defined as the separation in terms of wavelength (or frequency) between the two points on the response curve where the magnitude is $0.707R$. The bandwidth of the spectral response may be of the order of tens of nanometers for an LED but is a fraction of a nm for a laser.

The light emitted by an LED is said to be incoherent, which means that the intensity of light that is to be emitted at a future time cannot be predicted from its value at the present time. Light from a laser is coherent (in space and in time), so that the intensity of the light emitted in a certain direction at a certain time can be predicted from what it was at a previous time. Expressed in equation form, the coherent light source electric field is given by $\mathscr{E}(t) = A \sin(2\pi f t + \phi)$, where A and ϕ are constants. In an incoherent source, A and ϕ are random functions of time.

Lasers have found applications in both industrial and material working applications, in information processing, such as recording and reading compact audio discs (CDs), optical computer disc systems, laser printers, medical equipment, and last, but not least, optical communication systems where the laser beam acts as the carrier on which information is loaded.

Absorption and Emission of Radiation

The interaction between electromagnetic radiation and matter may result in absorption or in the emission of electromagnetic radiation. We consider a two-energy level system, shown in Fig. 14.10, where E_1 is the lower ground state and E_2 is the upper excited state, and identify the processes of light absorption and emission. We assume that the material is in the ground state. In the presence of radiation, it becomes possible for an electron to absorb a photon with energy $h\nu$ and be lifted to the upper state. This processs, labeled *absorption*, is shown in Fig. 14.10(a), where one electron is raised to the upper state and one photon is missing from the radiation.

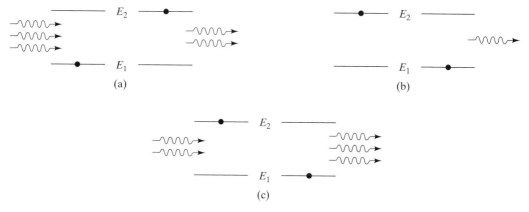

Figure 14.10   An electron interacts with a photon; (a) an electron absorbs photon and moves to the higher energy level; (b) an electron in the upper energy level, without excitation, spontaneously drops to the lower level emitting photon; (c) photons are incident on a system, move an electron to the lower level and an additional photon is generated, causing stimulated emission.

Spontaneous emission takes place when an electron at the excited level, E_2, drops to level E_1 and emits a photon with energy $h\nu$. This is shown in Fig. 14.10(b). This type of emission does not require the presence of any radiation or electromagnetic energy.

Now we assume that electrons are in the upper excited states and photons of the proper energy are incident on the system. This radiation causes a transition of an electron from the upper state to the lower vacant state, with the electron giving up energy in the form of a photon to the incident radiation. Compared to the incident photon, the emitted photon has the same energy or wavelength, is in the same phase, and has the same direction. This is *stimulated emission* and is shown in Fig. 14.10(c).

For stimulated emission to be possible, electrons must be located at higher energy levels than the location of vacant states into which they fall. This process of accumulation of electrons at higher energy levels, such as E_2, and accumulation of vacant states at E_1 is known as *population inversion*.

Population Inversion

To form population inversion, whereby excess electrons are located at the bottom of the conduction band while vacant states are located at the top of the valence band, three requirements must be satisfied. First, the semiconductor must be of the direct type; second, a PN junction is formed that is forward-biased and in which the P and N regions are degeneratively doped; and, third, the applied voltage must be at least as large as the band gap voltage. The semiconductor must be of the direct type to enhance radiative recombination. By heavily doping both the N and P regions, the Fermi levels move into the conduction band and valence band of the N and P regions respectively. To cause accumulation of electrons in the conduction band and holes in the valence band, the applied voltage should be greater than the band gap voltages, as we shall see later.

A typical compound semiconductor that is direct is GaAs. Let us now investigate the processes that may take place in a degenerate semiconductor when the Fermi levels move into the allowed bands.

Degenerate doping of both P and N causes the Fermi levels to move into the bands; the conduction band of N and the valence band of P. Consequently, and within each band, all states above the Fermi level are empty and all states below it are occupied. Degeneracy is illustrated in Fig. 14.11.

Assume light of frequency ν is propagating through the N and P semiconductors. In Fig. 14.11, vacant states are available in both the valence and conduction bands above E_{Fp} and E_{Fn} respectively and filled states exist below the Fermi levels, with electrons filling the states below E_{Fn} and holes filling the states above E_{Fp}. We assume that the electrons with the highest energy are ΔE_C above the band edge. The holes with the highest energy are located ΔE_V below E_V. Light waves that have frequency $\nu > (E_g + \Delta E_C + \Delta E_V)/h$ are absorbed and lift an electron from the valence band to the conduction band. However, if the frequency of the light is such that $E_g < h\nu < E_g + \Delta E_C + \Delta E_V$, then the incident wave can cause only downward transition of electrons from the conduction band to the valence band, resulting in the electrons giving up their energy in the form of light having the same frequency, phase, and direction as the incident light, thus causing light amplification.

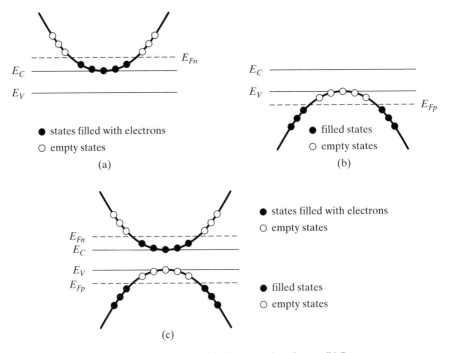

Figure 14.11 Degenerate junctions; (a) N-type semiconductor, (b) P-type semiconductor, and (c) P and N degenerate junctions.

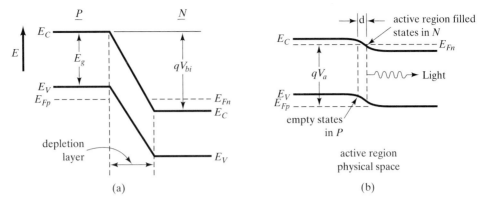

Figure 14.12   Laser diode energy band diagram (a) at thermal equilibrium and (b) with forward bias $V_a \geq E_g/q$ applied.

Diode Laser

In a PN junction laser, both P and N are degenerately doped. The condition at thermal equilibrium is shown in Fig. 14.12(a). The Fermi levels are aligned. When a forward bias, V_a, is applied to the diode with $V_a \geq E_g/q$, the N region moves up with respect to the P region, causing a significant lowering of the barrier voltage, as shown in Fig. 14.12(b). We note the built-in voltage at thermal equilibrium, in Fig. 14.12(a), exceeds the band gap voltage. The lowering of the barrier sends a large current across the junction, causing the formation of a small region near the junction that has excess electrons in the conduction band in N and excess holes in P. This region, in the same volume of spaces in which population inversion has taken place, is known as the *active region*, identified as the region d. As the voltage is built up, spontaneous emission of light with frequency v takes place at the active region as an electron falls into the valence band and recombines with a hole. This radiation, whose frequency is related by $E_g < hv < (E_{Fn} - E_{Fp})$, causes additional recombination as it travels in the material.

The wave of light-emission is spontaneous at low values of applied voltage. This emission is reflected inside the diode. As the voltage is increased, the emission becomes stimulated, as shown in Fig. 14.13. The emitted photons have a frequency $v = E_g/h$.

The stimulated emission is amplified by positive (reinforcing) feedback, resulting in reflections from the physical boundaries inside the diode in what is known as the cavity, which is the subject of the next section. It is important to note that in a GaAs laser, the active region, d, is about one micron thick, whereas the cavity is 2 to 3 times thicker. A section and a perspective view of the laser are shown in Fig. 14.14.

The Laser Cavity

When a current is passed through the diode by the application of a forward bias, the junction emits radiation. The optical cavity, described below, provides the necessary feedback so that laser action can take place. Laser light will be continuously emitted

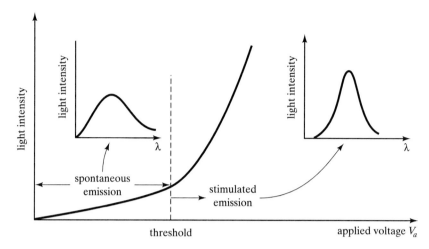

Figure 14.13 Stimulated emission is preceded by spontaneous emission.

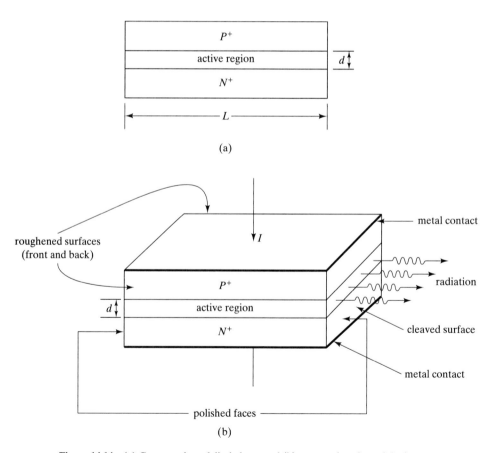

Figure 14.14 (a) Cross section of diode laser and (b) perspective view of the laser and its cavity.

at a frequency E_g/h when a sufficiently large current, known as the *threshold current*, identified in Fig. 14.13, is reached. The threshold current is that below which mostly spontaneous emission takes place and radiation is incoherent. Above the threshold current, stimulated emission is predominant. The threshold may be defined also as the point at which the amplification due to stimulated emission is just equal to the losses due to absorption by the cavity walls and scattering due to the interaction between the waves.

The cavity is in the shape of a parallelepiped, as shown in Fig. 14.14. Metallic contacts are made to the top and bottom of the diode. The front and back are roughened so that no emission takes place from them. Two side surfaces perpendicular to the plane of the junction are polished and cleaved so they become semi-reflecting. The smooth reflecting surfaces are essential to the lasing action (amplification).

The first spontaneously emitted radiation wave travels down the cavity and is reflected back from the cleaved surface in order to stimulate the production of additional photons. The spontaneously emitted photons reach the cleaved surface and, because of the high refraction index of this surface, about one third of the photons are reflected back into the active region. This process is repeated many times as the photons undergo many reflections. The light intensity reaches a steady-state value when the light amplification by stimulated emission is balanced by the loss mechanisms in the diode. These losses occur by the absorption of the light in the walls of the cavity that make up the bulk of the semiconductor away from the junction, the nonradiative recombinations that normally occur, and the relatively large portion of the light that passes through the mirror surface.

The radiation exits from one of the cleaved surfaces on one end of the cavity. To prevent radiation from the opposite surface, the reflectivity is increased to 100 percent by coating the surface with a layer of insulator and then covering it with a good reflector, such as aluminum. It is the nature of such a cavity to reinforce any reflections that are an integral number of wavelengths along the axis of the cavity and to dampen out all others.

Light can also resonate across the transverse direction, normal to the direction of emission of laser light.

A typical semiconductor laser will emit light at a central wavelength surrounded by a number of spectral lines separated by several nanometers from the central one. Emission spectra near threshold are shown in Fig. 14.15.

Heterojunction Lasers

Diodes that have the same semiconductors, such as GaAs, on both sides of the junctions, are called homojunction lasers. Such diodes require large values of threshold current density at room temperature and dissipate large amounts of power when continuously operated, thus requiring heat sinks. They may be operated, intermittently, as in a pulsed mode. Lasers used in fiber optic communication systems are normally of the heterojunction type, which contain junctions between two or more different semiconductors. On one laser, a thin layer of P-type GaAs is grown on an N-GaAs substrate followed by a P-type AℓGaAs layer. Such a layer has substantially lower threshold current than the homojunction type.

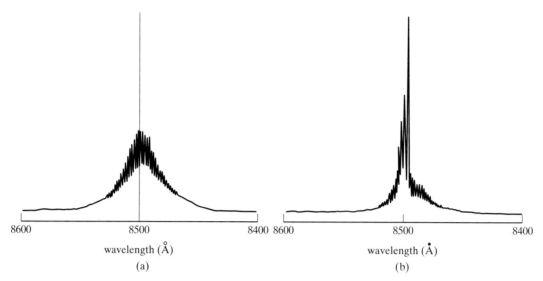

Figure 14.15   Emission spectra from gallium arsenide laser near threshold: (a) below threshold, and (b) above threshold. C. H. Gooch, *Injection Electroluminescent Devices*, p. 140, Wiley (1973). Copyright John Wiley & Sons Limited. Reproduced with permission.

Laser Applications

The high frequency of optical radiation, which is of the order of 10^{14}Hz, makes possible the use of large bandwidths in optical communication systems. Even if only one percent of this frequency is used for carrying information, a bandwidth of 10^{12}Hz is very significant. Considering that a commercial television channel requires 6MHz of bandwidth while a radio channel requires about 15kHz and a telephone conversation needs 3kHz, the applications of laser technology to communications are only limited by advances in allied systems, such as electronic components.

Lasers are compact and highly efficient and their applications to linking of high-speed computers make them critical components in the high-speed transmission of data. It is anticipated that when it becomes possible to produce lasers operating at the blue wavelength, the data density that could be read from compact discs would be four times larger and compact disc sizes could be significantly reduced.

REVIEW QUESTIONS

Q14-1   Why is it not likely that radiative recombination can occur in indirect semiconductors?

Q14-2   What two conditions must be met for recombination to be labeled radiative?

Q14-3   Since light is emitted from an LED by spontaneous emission, why is the light continuous and not spontaneous?

Q14-4   Define the term *population inversion*.

Q14-5   What conditions must be met for population inversion to occur?

Q14-6 Why does the temperature of a semiconductor affect the wavelength of the emitted radiation?

Q14-7 Define the term *threshold current*.

Q14-8 What factors set the limit to the growth of stimulated emission in a laser?

Q14-9 Explain what is meant by coherent and monochromatic light.

HIGHLIGHTS

- Semiconductors, in general, are classified as direct band gap and indirect band gap types. Gallium arsenide and some other III-V compounds are of the direct type, whereas silicon and germanium are indirect semiconductors.

- Direct semiconductors are those in which the bottom of the conduction band and the top of the valence band are located at the same momentum on an energy-momentum plot. In indirect semiconductors they are not.

- Light emission from a direct semiconductor occurs in three processes. The first is the generation of electron-hole pairs by excitation of the semiconductor. The second is the recombination of electron-hole pairs and the third is the extraction of photons from the semiconductor as a result of the recombination.

- For a radiative recombination process to take place, both energy and momentum must be conserved. In direct semiconductors, electrons and holes are at the same momentum and energy is conserved when, as a result of recombination, optical energy is radiated with $h\nu = E_g$.

- LEDs and lasers are the two semiconductor devices that emit light upon excitation. The laser light is highly directional and of almost a single frequency.

- The recombination that results in the emission of radiation occurs in a PN junction diode. In LEDs, recombination results in spontaneous emission of photons. In lasers, the initial emission is spontaneous but the structure of the device causes the spontaneously generated photons to be reflected and to cause recombination to occur and further emission of photons. This stimulated emission continues to build up until the losses inside the structure balance the photon emission. Part of the light that is labeled lost is directed out as useful light through one of the faces of the structure.

- LEDs are used in visual displays as indicators and sometimes as the source of light in optical communication systems. Lasers are used in optical communication systems, in medicine for diagnosis and surgery, in compact discs recording and reading, and in a variety of industrial applications.

EXERCISES

E14-1 A certain semiconductor has energy gap of 2.48eV. Determine (a) the range of wavelengths that are absorbed and (b) the range transmitted by the semiconductor.

Ans.: b) $\lambda > 500$nm

E14-2 In a semiconductor ($E_g = 1.78$eV) laser, the bandwidth of the emission spectrum is approximately 0.01eV. Determine the frequency bandwidth of the light.

Ans.: $BW = 24 \times 10^{12}$Hz

14.3 PHOTODETECTORS

A photodetector is a device that is sensitive to optical radiation and converts this radiation into an electric current. This device has a variety of applications such as in optical fiber communication systems where it is used, at the receiving end, to convert optical signals carrying information into an electric current. It is used in this application as a demodulator. Photodetectors are also used to measure light intensity, operate a relay, or control the lens exposure of a camera.

The absorption of light creates electron-hole pairs that result in major increases in the conductivity of a semiconductor. Optical energy is absorbed if its wavelength is less than a critical value, λ_{max}. For radiation that has energy $h\nu$ greater than the band gap energy of the semiconductor, each photon releases sufficient energy to create an electron-hole pair. The energy of a photon is related to its wavelength by

$$E = h\nu = \frac{1.24}{\lambda(\mu m)}\, eV \tag{14.5}$$

The critical frequency and wavelength for absorption are given by

$$\nu_{min} \geq E_g/h,\ \lambda_{max} = hc/E_g = \frac{1.24}{E_g(eV)}\, \mu m \tag{14.6}$$

where ν_{min} is the lowest frequency below which no absorption occurs and λ_{max} is the maximum critical wavelength given by $(1.24/E_g)(\mu m)$, where E_g is the bandgap energy in eV. Radiation having wavelength greater than λ_{max} is not absorbed by the semiconductor, which is then said to be transparent to that wavelength radiation. If the wavelength of the radiation is smaller than λ_{max}, implying that the radiant energy is greater than the band gap, each photon produces an electron-hole pair with the balance of the energy absorbed as heat.

The simplest type of photodetector, the photoconductive detector, shown in Fig. 14.16, is formed from a slab of semiconductor that has a large area exposed to radiation and to which a voltage is applied.

Prior to exposing the slab to light, an equilibrium condition of recombination and thermal generation exists. After exposure to light, an equilibrium condition is reached at which the carrier density is higher. We will determine an expression for the photocurrent in the circuit as follows: If the total optical power absorbed by the detector is P_{ph} (J/s) and $h\nu$ is the energy (J) of each photon, the incident number of photons per second is $P_{ph}/h\nu$. If we define quantum efficiency, η, as the ratio of the number of electron-hole pairs generated by each incident photon, then the rate of generation of carriers per unit volume equals the rate of recombination as

$$(\eta\, P_{ph}/h\nu)\frac{l}{w\ell d} = \frac{n}{\tau} \tag{14.7}$$

where n is the carrier density, τ is the lifetime of carriers, and $w\ell d$ is the volume of the slab shown in Fig. 14.16. The carriers thus generated drift under the action of the electric field created by the applied voltage, so that the expression for the photogenerated current can be written as

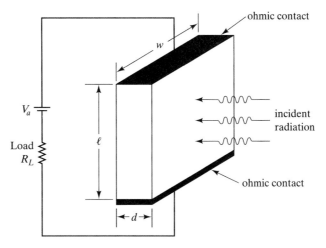

Figure 14.16   Schematic of a photodetector.

$$I_{ph} = (q\mu_n n\mathscr{E})wd \tag{14.8}$$

where $\mathscr{E}$ is the electric field intensity, assumed to be uniform and given by V/d.

By substituting for n from Eq. (14.7) into Eq. (14.8), we have

$$I_{ph} = (q\mu_n \tau V/\ell^2)(\eta P_{ph}/h\nu) \tag{14.9}$$

With this background, let us consider the operation and properties of the PN photodiode, upon which detectors in optical communication systems are based.

The Photodiode

The photodiode consists of a PN junction diode, shown in Fig. 14.17, and specially fabricated with a window at the top to permit incident radiation to impact the P region.

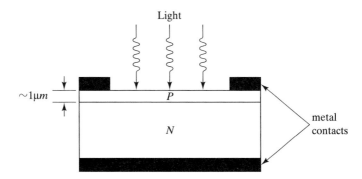

Figure 14.17   Cross section of a photodiode.

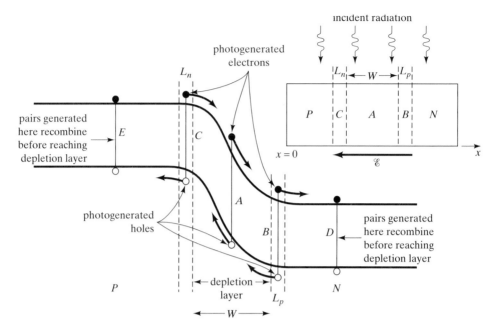

Figure 14.18 Four possible locations where a photon is absorbed. Most productive location is A; D is the least productive (widths of depletion layer and diffusion length in the diode section are exaggerated).

The diode is assumed to be reverse-biased. An incident photon having $\lambda <$ λ_{max} generates an electron-hole pair. These carriers cause a photocurrent in the external circuit, which is added to the thermally generated minority carrier currents.

To illustrate the process of photogeneration of carriers, we consider the schematic of a diode shown in Fig. 14.18. Radiation is assumed to be incident on the whole diode (depletion region width is exaggerated).

We identify four possible locations in which electron-hole pairs are generated, as illustrated in Fig. 14.18.

First, if the photon is absorbed in the depletion layer, where it generates an electron-hole pair, as at A in Fig. 14.18, both the electron and the hole generated drift under the influence of the electric field with the electron going into the N region and the hole going into the P region, both contributing to the photogenerated current in the external circuit. The second and third locations where electron-hole pairs are photogenerated lie in the bulk within a distance of one diffusion length from the edge of the depletion layer, as shown in B and C. Since the diffusion length of a minority carrier is the average distance, it can travel in a "sea" of majority carriers before recombining, then those carriers generated at B and C have a high probability of surviving recombination, diffusing to the edges of the depletion layer, and drifting across it to contribute to the photogenerated current.

Electron-hole pairs are also photogenerated in the bulk of the P and N regions on the average more than a diffusion length distant from the edge of the depletion layer, as at D and E, but these carriers recombine before they reach the depletion layer and do not contribute to the current.

The total current in the external circuit thus consists of carriers photogenerated in or very near the depletion layer and thermally generated carriers. To emphasize the role of the depletion layer, we restrict the discussion to the carriers generated in it.

The total diode current becomes

$$I = I_s \left[\exp\left(qV_a/kT\right) - 1\right] - I_{ph} \tag{14.10}$$

where I_{ph}, the photogenerated current, is $qg_{ph}A(W + L_n + L_p)$ and g_{ph} is the number of photogenerated electron-hole pairs per second per unit volume, where A is the area, L_n and L_p are the diffusion length of electrons and holes respectively, and W is the depletion layer width.

When the diode is forward biased, thermally generated electrons cross from N to P and holes cross from P to N, while the photogenerated carriers drift under the influence of the electric field in the opposite direction. When the junction is reverse-biased, both the thermally generated and photogenerated electrons cross from P to N and the holes cross from N to P—hence, the reason for the sign of I_{ph}. The generation rate g_{ph} increases with light intensity and the resulting photogenerated current adds to the thermally generated dark current, I_S, so that the resulting characteristics become as shown in Fig. 14.19(a). A circuit model for the photodiode and its symbol are shown in Fig. 14.19(b) and (c).

In the first quadrant of the characteristics, both current and voltage are positive. Because the photogenerated current is opposite in direction to the normal current, the total current is smaller than the normal diode current. The characteristics in the third quadrant exhibit operation in the reverse direction, while the sensitivity to light for $g_{ph} > 0$ increases the reverse current, as shown by the characteristics in Fig. 14.19(a). In this region, the diode is used in light-sensitive switching circuits and in light meters. In the fourth quadrant, a positive voltage appears across the device, resulting only from the light falling on the device. The device is therefore operated as a photovoltaic cell in this fourth quadrant.

Overall Efficiency

In its applications, we are particularly interested in two properties of the photodiode: the efficiency and the speed of response of the device. We define the overall efficiency as the ratio of the power absorbed in the depletion region to the total power incident on the detector. We consider the realistic case of the diode, shown in Fig. 14.17, and use the diode section of Fig. 14.18 with the direction of radiation rotated 90° counterclockwise, so that the light is incident normal to the P region. A fraction of the light falling on the P region is absorbed in that region as determined by the ratio of the width of the P region to the width of the depletion region. Furthermore, a fraction of the light reaching the depletion region is absorbed while some of the light is reflected from the whole device. These factors cause the efficiency of this device to be very low.

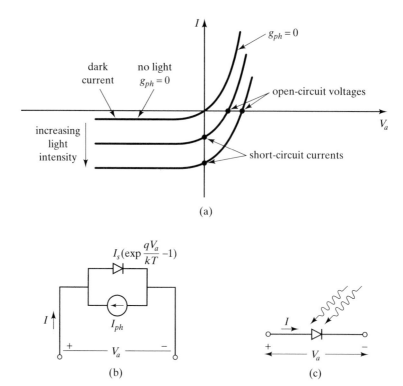

Figure 14.19 Photodiode (a) characteristics, (b) circuit model, and (c) symbol.

We define two factors that determine the portion of the light that is absorbed by the depletion region, namely: the Fresnel reflection factor, R_f, and the attenuation factor alpha, (α).

The Fresnel factor represents the fraction of the light reflected from the P region and is given by

$$R_f = \left(\frac{n_2 - n_2}{n_2 + n_1}\right)^2 \tag{14.11}$$

where n_2 and n_1 represent the indices of refraction of the P region and the external medium respectively. Hence, the fraction of the light entering the P region is $(1 - R_f)$. The amount of light leaving the P region is determined by the attenuation factor, which is determined as follows:

The amount of optical power absorbed in an incremental length Δx along the path of the radiation is given by

$$\Delta P(x) = P(x + \Delta x) - P(x) = P(x)\alpha\Delta x \tag{14.12}$$

where $P(x)$ is the power at point x and α is the attenuation factor. The attenuation factor is a function of the material and the wavelength of the incident light. Assuming that Δx is very small, the equation can be written in differential form as

$$dP(x)/dx = P(x)\alpha \tag{14.13}$$

The solution to this equation, subject to the condition that the power entering the P region at $x = 0$ is $P_0(1 - R_F)$ (P_0 is incident on P), becomes

$$P(x) = P(0)e^{-\alpha x} \tag{14.14}$$

where $P(x)$ is the power in the P region at any x and the fraction of power remaining at the end of the P region is $e^{-w_p\alpha}$, where w_p is the width of the P region. Hence, the power, P_d, entering the depletion region is

$$P_d = P_0(1 - R_f)\, e^{-\alpha w_p} \tag{14.15}$$

where P_0 is the power incident on the P region and w_p is the width of the P region.

The depletion region will absorb some of the power, reducing the entering power by $e^{-\alpha W}$, so that the power absorbed by the depletion region becomes $P_d(1 - e^{-\alpha W})$. The fraction of power absorbed by the depletion region labeled the overall efficiency, η_0, becomes

$$\eta_0 = (1 - R_f)e^{-\alpha w_p}(1 - e^{-\alpha W}) \tag{14.16}$$

where w_p is the width of the P region and W is the width of the depletion region.

A smaller P region and a wider depletion region increase the quantum efficiency.

EXAMPLE 14.1

An 0.3μm single crystal of silicon is subjected to incident light normal to the surface having power of 20mW. Given the attenuation factor for silicon at the frequency of the light is 5×10^4/cm and the index of refraction of silicon is 3.5, determine the power absorbed by the crystal.

Solution   The fraction of light reflected from the silicon is

$$R_f = \left(\frac{3.5 - 1}{3.5 + 1}\right)^2 = 0.308$$

The fraction of the power in the sample that reaches the end of the sample is

$$\exp(-5 \times 10^4 \times 0.3 \times 10^{-4}) = 0.223$$

The power absorbed by the silicon becomes

$$20 \times (1 - R_f)\,(1 - 0.223) = 10.74\text{mW}$$

From this example, we conclude that the longer the sample of the silicon the smaller is the exponential factor and the larger the power absorbed. It follows that to increase the absorption of light in the depletion region of a PN junction diode, we need to make the P region as narrow as possible and the depletion region very wide.

The speed of response of the photodiode is limited by both the time it takes to collect the carriers and by the capacitance of the depletion layer. Making the depletion region too wide increases the drift time of the carriers and reduces the frequency response. On the other hand, making it too thin results in a large depletion capacitance, which together with the load resistance, R_L, results in a large time constant, $R_L C$.

To improve the efficiency and reduce the capacitance, one needs to make the depletion region wider. There are two possible solutions for that: one is to apply a large reverse bias to the diode, which by Eq. (5.30) increases the width of the depletion layer and reduces the capacitance in accordance with Eq. (7.16). Both the width of the depletion layer and its capacitance vary inversely as $(V_{bi} - V_a)^{1/2}$, where V_a is negative. This therefore requires a relatively high voltage that might cause breakdown.

The second solution is to use a photodiode with an intrinsic region separating P and N. This is the PIN photodiode.

The PIN Photodiode

The efficiency of the photodiode is considerably improved by making the intrinsic region wide enough to absorb most of the radiation. A cross section of the PIN photodiode is shown in Fig. 14.20.

The PIN diode is by far the most commonly used photodiode because the width of the depletion layer (the intrinsic region) can be controlled to satisfy the requirements of high emission efficiency and improved response speed.

In this diode, a wide region formed from an intrinsic semiconductor is sandwiched between the P and N regions. The intrinsic region has a relatively low conductivity, thus a high resistivity, when compared to the P and N regions. Therefore, most of the reverse-applied voltage appears across the intrinsic layer. This region is

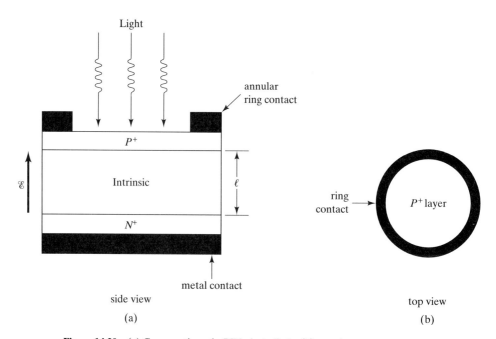

Figure 14.20   (a) Cross section of a PIN photodiode; (b) top view.

made long enough so that most of the incident radiation on the diode is absorbed within the intrinsic region. Although there will be a sacrifice of speed, the efficiency in the production of current is considerably higher than that of the photodiode.

Additional increase of efficiency of photocurrent generation is obtained if the PIN photodiode is biased at a reverse voltage that is close to, but not quite equal to, the avalanche breakdown voltage. Because of the high reverse voltage, avalanche multiplication of carriers takes place in the intrinsic region, thereby resulting in internal current gain and increased sensitivity of the device. As a by-product of the high voltage, the electric field causes high acceleration of the carriers so that faster switching times are achieved.

Noise in PIN Diode

One of the problems associated with the use of intrinsic semiconductors as optical detectors operated at room temperature is that thermally generated electron-hole pairs become a source of noise. This noise is superimposed on the information that is contained in the optical signal. This problem is particularly serious for optical energy that is in the infrared and higher wavelength regions. In these regions, the semiconductor to be used has a relatively narrow band gap, so that the intrinsic carrier density is very high. The longer the wavelength of the radiation, the more serious the problem. For radiation having wavelengths smaller than about 10-15 microns, semiconductors are operated at liquid nitrogen temperature (77K) or sometimes at liquid helium temperature (4K), so as to increase the band gap and reduce the densities of the thermally generated carriers.

As an illustration, InSb has a bandgap of 0.18eV at room temperature and its intrinsic carrier density is metal-like. Large improvements in signal to noise ratio are obtained in a detector using InSb when it is operated at 77K. At this temperature, the bandgap is increased to 0.23eV with a consequent reduction of the intrinsic carrier density by six orders of magnitude. At 77K the critical wavelength becomes 5.5μm, making the material sensitive to infrared radiation.

Another example is the use of Ge (λ_{max} = 1.4μm, E_g = 0.66eV) in infrared detectors. This is accomplished by doping Ge with copper, which is an acceptor with an ionization energy of 0.04eV. When the resulting semiconductor is operated at 77K, thermal generation produces a negligible density of carriers, whereas optical radiation having λ = 32μm (infrared) ionizes the copper atoms and lifts electrons from the copper impurity level to the conduction band.

The Phototransistor

Amplification of current and hence increased photosensitivity is obtained in the phototransistor detector. Light is directed through a window at the collector-base junction, which is reverse biased and at which additional minority carriers are photogenerated. The transistor operates with the base open-circuited and hence has no external base lead. The model and symbol of an NPN phototransistor are shown in Fig. 14.21.

Light energy focused on the collector-base junction causes electron-holes pairs to be generated. With a positive voltage applied between the N collector and

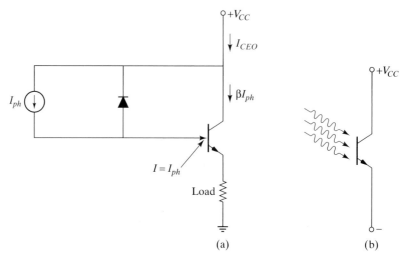

Figure 14.21 Phototransistor (a) model and (b) symbol.

the N emitter, the emitter-base junction becomes forward-biased and the collector-base junction becomes reverse-biased. We assume the number of thermally generated carriers at the collector-base junction to be negligible compared to those generated by photon absorption. The electrons generated at the reverse-biased junction travel to the collector and cause a collector current, I_{ph}, across the junction, whereas the holes drift to the base. The resulting excess density of holes in the base draws electrons from the emitter to the base in order to form a base current. This base current contributes an additional $\beta_F I_{ph}$ current of electrons to the collector, so that the total collector current is given by

$$I_{CEO} = (\beta_F + 1) I_{ph} \tag{14.17}$$

where I_{CEO} is the collector to emitter current with the base open. The dark current (in the absence of light) component of I_{CEO} is of the order of nanoamps, whereas I_{ph} may be of the order of microamps and I_{CEO} may reach milliamps.

An obvious advantage of the phototransistor, when compared to the photodiode, is the current gain obtained. The higher gain is made possible at the expense of a much slower device because of the increased transit time of the electrons that originate in the emitter and diffuse across the base to reach the collector. While a photodiode has a typical response time of 1ns, that of a phototransistor is in microseconds.

The phototransistor is commonly used in an opto-isolator package constructed as one unit and consisting of an LED and the phototransistor.

14.4 THE PHOTOVOLTAIC EFFECT AND SOLAR CELLS

In Section 14.3, we derived relations for the current in a photodetector as a function of the voltage, the photogeneration rate, and other diode constants. The graphical characteristics of the photodiode are reproduced in Fig. 14.22.

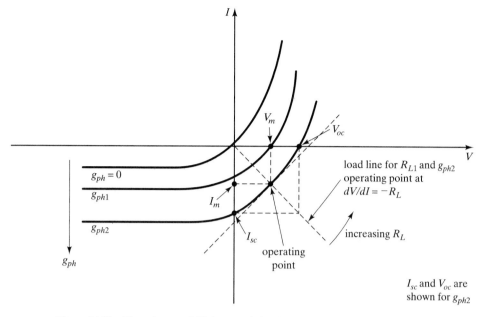

Figure 14.22 Photodetector I-V characteristics showing region of operation of photocell.

In the fourth quadrant of the current-voltage characteristics, the voltage is positive but the current is negative, indicating that the diode is operating as a source of energy. More specifically and for open-circuit, $I = 0$, a voltage given by V_{oc} appears across the diode. The diode that is exposed to natural light is operating as a photocell.

At thermal equilibrium, a depletion layer and a barrier voltage, V_{bi}, exist across the junction. When the cell is exposed to light, electron-hole pairs are generated near the junction with electrons drifting to the N region and holes drifting to the P region. The photogenerated carriers neutralize ionized donors in the N region and ionized acceptors in the P region. This causes a reduction in the barrier voltage (without a load connected) to $(V_{bi} - V_{oc})$, resulting in a forward-biased junction having a voltage V_{oc} across its terminals.

For an infinite load, R_L, the current is zero and the cell delivers a maximum voltage V_{oc}. As the load changes, both the voltage (positive) of the cell and the current change. The respective values of the voltage and current are determined at the intersection of the load line with the graphical characteristics.

An expression for the open-circuit voltage, V_{oc}, of a cell is determined by setting the current to zero in Eq. (14.10), to obtain

$$I_s \left(e^{qV_{oc}/kT} - 1 \right) = I_{ph} \tag{14.18}$$

so that V_{oc} is determined to be

$$V_{oc} = kT/q \, \ell n \left[(I_{ph}/I_s) + 1 \right] \cong \frac{kT}{q} \, \ell n \left(I_{ph}/I_s \right) \tag{14.19}$$

where I_s is the thermally generated reverse current and I_{ph} is the photogenerated current. Typical values of V_{oc} are of the order of one-half the band gap voltage E_g/q.

We note from Eq. (14.10) that I_{ph} represents the magnitude of the short-circuit current, identified in Fig. 14.22, at the intersection of the characteristic and the ordinate for $V = 0$.

Power and Efficiency

Upon referring to Fig. 14.22, we note that the power delivered by a photocell, when connected to a load, is given by VI, where V and I are measured at the operating point that occurs at the intersection of the relevant g_{ph} curve and the load line. The question becomes: What value of load resistance will cause the photocell to deliver maximum power to the load? Since $P = VI$, power is maximized when

$$dP/dI = V + I\,dV/dI = 0 \tag{14.20}$$

and

$$dV/dI = -V/I = -R_L \tag{14.21}$$

Maximum power is therefore delivered when the load resistance is equal to the negative of the slope of the cell characteristics at the operating point, corresponding to the point defined by I_m and V_m, as shown in Fig. 14.22. This maximum power, given by $V_m I_m$ is less than the product $V_{oc} I_{sc}$, as can be seen by comparing the areas of the two rectangles. We therefore define a fill factor, FF, as

$$FF = \frac{V_m I_m}{V_{oc} I_{sc}} \tag{14.22}$$

The efficiency of power generation of a photocell, defined as the ratio of the maximum electrical power to the incident solar power, becomes

$$\eta_p = \frac{V_m I_m}{P_{ph}} = \frac{(V_{oc} I_{sc})FF}{P_{ph}} \tag{14.23}$$

Typical values of efficiency range from 10 to 15 percent.

The structure of a typical photocell is shown in Fig. 14.23. An increase of the power delivered is enhanced by the following measures:

1. Increase the area of the diode at the surface so as to absorb larger amounts of sunlight.
2. Coat the surface with antireflecting material.
3. Make the width of the N region smaller than the diffusion length, L_p, of holes, so that the holes can diffuse to the metal contact before recombining.
4. Increase the lifetime by using a semiconductor, such as silicon, which has an indirect band gap.
5. Reduce the series resistances of the diode by increasing the area of the N region and by including finger contacts to the N region, as shown in Fig. 14.23.

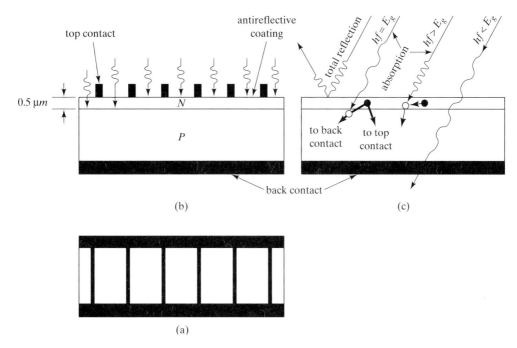

Figure 14.23   Solar cell (a) cross section, (b) top view, and (c) effects of photons having different amounts of energy.

A typical solar cell is 10cm long and 2.5 cm wide. Assuming that the incident solar radiation at the equator at noon (known as air mass 1-AM1) is 1000W/m^2, this device, operating at an efficiency of 15%, delivers less than half a watt of power.

Solar Energy

So that solar energy production can be seriously considered for the generation of large amounts of electrical energy, two obstacles have to be overcome: First, the conversion of solar energy to electrical energy must be carried out economically and, second, the energy must be available at all times and in all places. Not only must these obstacles be overcome, but the cost of solar energy must be competitive with other means of power generation.

The incentive for the development of solar energy has been led by the space program where solar energy provides a continuously available source.

Some general information about solar radiation and conversion is helpful.

On a clear day, it is estimated that solar energy reaches the earth at an approximate rate of one kilowatt per square meter and, in the least sunny part of the United States, enough energy falls on an area of 80 square meters to meet the needs of an average family.

By using silicon to form solar cells, and to excite electrons from the valence band to the conduction band, an energy of 1.12eV at room temperature is needed.

This corresponds to a wavelength of radiation of $\lambda = 1.1\mu m$. However, more than half of the radiation received from the sun is in the infrared region at $\lambda > 1.1\mu m$. Radiation having a wavelength greater than this critical value does not have sufficient energy to lift electrons from the valence band to the conduction band in silicon. Fundamental physics laws limit the efficiency of an ideal photovoltaic cell to less than 25 percent. Furthermore, the contacts provided on the front of the cell prevent sunlight from reaching the crystal at the location of the contacts. There are also other losses in the semiconductor that must be reckoned with.

In general, individual cells are connected in series to obtain higher voltages and connected in parallel to provide larger amounts of current.

REVIEW QUESTIONS

Q14-10 When the wavelength of the radiation incident on a semiconductor is smaller than λ_{max}, are electron-hole pairs produced?

Q14-11 What are the consequences if the incident optical energy is greater than the band gap energy of a semiconductor?

Q14-12 When optical radiation is incident on the neutral region of a semiconductor PN diode, are electron-holes generated? Do these pairs contribute to the photocurrent? Explain.

Q14-13 Explain, in a sketch, the effect of the absorption coefficient on the light intensity incident on a semiconductor as x increases.

Q14-14 What are the advantages of using a PIN photodiode?

Q14-15 What is the physical significance of the short-circuit current of a photocell?

Q14-16 On the current-voltage characteristic of a solar cell, show why the power $V_m I_m$ is smaller than $V_{oc} I_{sc}$.

HIGHLIGHTS

- Radiation of optical energy having wavelength greater than $(1.24/E_g(eV))\mu m$ is not absorbed by a semiconductor having bandgap energy E_g.

- Optical radiation incident on a photodetector at the appropriate wavelength generates electron-hole pairs at a rate given by $\eta(P_{ph}/hv)(1/wld)$, where η is the quantum efficiency, P_{ph} is the incident optical power, and wld is the volume of the detector. The ratio P_{ph}/hv represents the number of photons per unit time.

- Quantum efficiency, η, represents the number of electron-hole pairs generated by each incident photon.

- In a PN photodiode, the carriers that contribute to the photocurrent are those that are generated within the depletion layer and those generated within a diffusion length of the depletion layer.

- Optical radiation incident on the P region of a photodiode undergoes partial reflection from the P region, attenuation by partial absorption in the P region, and the absorption by the depletion layer.

- For optimum photocurrent in a PN junction diode, it becomes necessary to make the depletion layer wide. The increase in the width, however, reduces the speed of a photodiode.

- To form a diode in which the depletion region is wide enough to absorb a large fraction of the radiation incident on the diode, it becomes necessary to form a PIN diode.
- In a semiconductor diode that is not biased and subjected to optical radiation, a voltage is developed across the terminals of the diode. The diode becomes a solar cell that can act as a source of energy to a load connected across its terminals.

EXERCISES

E14-3 A photodetector slab is illuminated by a 0.6μm light having incident power density of 2mW/cm^2. The semiconductor has E_g = 2eV, μ_n = 500cm^2/V-s, and lifetime = 10^{-4}s. The slab dimensions are l = 0.5cm, w = 0.24cm, and d = 0.02cm. Determine the photocurrent when 6V is applied to the device.

E14-4 In solar cell applications, and to obtain the required voltage and current, cells are connected in series and blocks of these are connected in parallel. Consider two cells connected in parallel. One cell has V_{oc} = 0.6V and I_{sc} = 1.2A, and the other cell has V_{oc} = 0.72V and I_{sc} = 0.9A. Determine:

  a) the resulting V_{oc}
  b) the resulting I_{sc}

Ans.: a) 0.646V

b) 2.1A

14.5 FIBER OPTIC COMMUNICATIONS

Communication Systems

Before discussing the use of light-energy and light systems to carry information, we will briefly explain the principles of transmission of information through the atmosphere.

To send information, such as audio for telephones and radio, television pictures, and computer data, to distant locations, it is necessary that these signals be superimposed on high-frequency signals through a process known as modulation. These high-frequency waves that carry the information can be transmitted over short distances by copper wires while for long distances they are radiated into space by transmitting antennas with the size and shape of the radiating element determined by the frequency of the wave. For transmitting and receiving high-frequency electromagnetic waves efficiently, antennas' dimensions are usually of the order of half a wavelength of the relevant wave.

The process of modulation consists, essentially, in having the information-carrying signal, such as audio or video, ride on top of a much higher frequency known as the carrier frequency. Carrier frequencies for AM radio signals range from 500 kHz to approximately 30 MHz; FM carrier frequencies range from 88-108MHz, and television carrier frequencies may range up to several hundreds of megahertz. Microwave signals using radiating dish antennas operate at thousands of megahertz (see Fig. 14.1).

The audio information for telephone communication covers a bandwidth range of frequencies of about 3kHz that for AM radio requires about 15kHz, whereas FM signals need about 150kHz. We note the increase in bandwidth improves the quality of the received audio by including the high frequencies required for faithful reproduction of music. Processing a TV signal requires a bandwidth of 6MHz. Every telephone conversation requires about 3kHz; every AM radio station needs about 15kHz; and every TV station needs 6MHz. The electromagnetic spectrum is being covered by all sorts of information being processed. The question becomes, how is this information transmitted?

Low-frequency telephone sound uses copper wires in local areas. AM radio signals are radiated to the lower atmosphere and are reflected back to earth. Depending on the power of the station, the signal may cover a radius of 50 miles. Short-wave radio signals, from 15 to 30 MHz, are radiated into the ionosphere and reflected back to distances exceeding thousands of miles. For carrier frequencies exceeding 30MHz, such as TV, electromagnetic signals follow the curvature of the earth and it is necessary that the transmitting antenna and the receiving antennas be within horizon sight of each other. Hence, the distance for useful reception is small.

It is quite obvious that more bandwidth is available at higher carrier frequencies, recalling that bandwidth covers a certain fraction of the carrier frequency. With all the information that is being processed and dispatched nowadays, the frequency spectrum is becoming overcrowded. Thus, more space is required in the electromagnetic spectrum and that space is available at the frequencies of light waves.

There is no reason why one cannot operate at frequencies of visible and infrared radiation, modulate these frequencies, and radiate them into space. After all, light has been used for centuries as a means of communication. The most common and ideal form of generating this light is by the use of a laser. But a laser beam radiated into space has problems. First, it is distorted in free space. Furthermore, it is attenuated by fog, smog, rain, and snow. It is attenuated so much that even over short distances, the signal becomes so weak that it cannot be distinguished from noise. Light beams are also reflected from structures since they travel in a line of sight direction. In fact, it is said that it is easier to transmit light to the moon than to send it across a metropolitan city.

While there is no question as to the availability of large bandwidths at the very high frequencies of light, means had to be discovered to convey these waves to distant locations. This led to the discovery of small cylindrical glass waveguides through which light waves travel, guided by the structure of the waveguide. These are known as optical fibers. These optical fibers, used as waveguides, are formed from a continuous lens that confines the light rays, refocuses these rays, and allows them to travel around bends. Compared to other means of communication, such as copper wire, coaxial cable, and microwave links, optical fibers have the widest bandwidth, lower attenuation, immunization from electromagnetic interference and monitoring, and are cost effective.

At the present, an optical fiber system that had been laid under the Atlantic Ocean several years ago is being reinforced and extended throughout Europe to reach Asia.

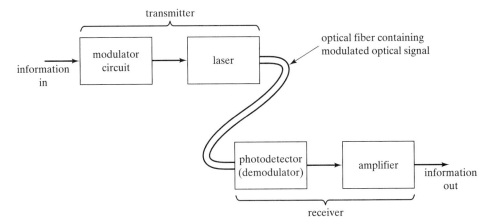

Figure 14.24   Block diagram of an optical communication system.

Fiber Optic System

A fiber optic communication system consists of the following processes, shown in Fig. 14.24:

- Generation of a light beam, such as by laser, that possesses a very sharp spectral response at a single frequency and that acts as the carrier.
- Modulation of the light beam by the information that is to be communicated.
- Transmission of the light through a glass fiber.
- When the light beam is to be transmitted over long distances and because it is attenuated, it is amplified along the way by what are known as repeaters.
- Reception of the light beam, which is then converted (demondulated) to an electrical signal, by means of a photodetector. The demodulated signal contains the original information, which modulated the light wave at the sending end.

As indicated earlier, the light source is commonly a laser. Compared to the LED, the laser has a much smaller area, much greater brightness, is highly efficient, and is ideally of a single frequency.

The optical fiber, which has a total diameter of a few hundredths of a millimeter, consists of a continuous long waveguide made of fused silica, which includes what are known as the core and cladding of the fiber, as shown in Fig. 14.25. Typically, the core has a diameter of about 12.5μm and the cladding and core have a total diameter of about 100μm. A fiber cable is formed from several fibers that are covered by a polyethylene jacket and protected by additional steel wires covering the jacket. The plastic layer protects the glass from scratches and abrasions, which could cause irregular reflections and may lead to fracture of the glass.

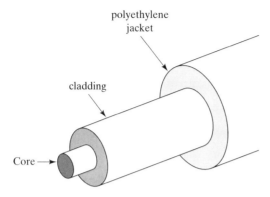

Figure 14.25 Optical fiber cross section.

The speed with which light travels through a material is represented by its index of refraction—the index of refraction being the ratio of the speed of light in vacuum to the speed of light in a material. Because the core has a higher index than the cladding, light travels at a slower speed in the core. Light rays that enter the core at a shallow angle to the axis are reflected back into the core, while if they are at a large angle, they are likely to escape through the cladding. The light rays that are reflected back continue to bounce back and forth in the core as they travel through. Provided there are no sharp bends in the fiber, all waves are confined to the core.

The real measure of the quality of any communication system is the *signal to noise ratio* at the receiving end detector. An effective means of increasing the signal to noise ratio is to encode the signal into digital form before transmission. The digital system of pulses has a greater tolerance to noise since minor distortion in, or weakening of the pulses, does not alter the intelligence being conveyed. In analogue transmission, a signal to noise ratio of 60dB (10^6 times) is required at the detector, whereas a ratio of 20dB (100 times) is satisfactory with digital encoding. There is, however, a major limitation to the bandwidth that can be transmitted using digital signals. That limitation is the spreading of a signal in time known as *dispersion*. As the binary input spreads in time, pulses overlap and distortion occurs.

The satisfactory range of a light-wave communication system is dependent on (a) the power of the source, (b) the per-unit-length attenuation of the fiber, and (c) the noise level tolerated by the detector. Attenuation, as explained next, is the main determining factor.

The attenuation of a signal results from, (a) intrinsic material absorption, (b) absorption due to impurities, and (c) bending and scattering. The total power loss in a fiber is represented by an attenuation coefficient defined by decibels per kilometer. For an attenuation of one decibel per kilometer, P watts of optical power radiated into a fiber one kilometer long emerges as 0.79P watts. Compared to ordinary glass, which has an attenuation of thousands of decibels per kilometer, optical communication systems require that losses be held to a level of one decibel per kilometer and less. It is interesting to note that early fibers were made of very pure and transparent silicon dioxide. It is estimated that if sea water was as transparent, one could see the bottom of the sea. The transparency is a measure of reduction of impurities.

Although we have restricted our discussion of optical fibers to their application in communication systems, it is essential to note that the use of optical fibers has also made major impacts on the medical profession in imaging, diagnosis, and surgery.

PROBLEMS

14.1 Light is incident on a semiconductor surface whose index of refraction at the wavelength of the radiation is 3.6. Determine:

   (a)   The fraction of light reflected.

   (b)   Repeat part (a) for glass that has an index of refraction of 1.5.

14.2 A slab of semiconductor is subjected to incident radiation normal to the surface. The absorption coefficient of the semiconductor at the wavelength of the radiation is $5 \times 10^4 \text{cm}^{-1}$. Determine:

   (a)   The thickness that will absorb 90 percent of the radiation.

   (b)   Repeat part (a) if 99 percent of the radiation is absorbed.

14.3 A GaAs N^+P diode has the following properties: $A = 10^{-2}\text{cm}^2$, $N_A = 10^{16}\text{cm}^{-3}$, $N_D = 5 \times 10^{17}\text{cm}^{-3}$, $\mu_n = 6{,}000\text{cm}^2/\text{V-s}$, $\mu_p = 200\text{cm}^2/\text{V-s}$, $n_i = 2.5 \times 10^6\text{cm}^{-3}$, and $\tau_p = \tau_n = 10^{-8}$. It is connected to a forward bias of 1V. Determine:

   (a)   The number of photons generated per second.

   (b)   The optical power that is generated.

14.4 Light having energy $h\nu$ of 2.5eV illuminates a 0.4μm thick silicon crystal. The crystal has an absorption coefficient, at the wavelength of the radiation, equal to $2 \times 10^4\text{cm}^{-1}$. The incident power is 15mW. Determine:

   (a)   The energy absorbed each second by the crystal.

   (b)   The rate of dissipation of heat energy in the lattice.

   (c)   The number of photons per second emitted from recombination.

14.5 A silicon solar cell has a short-circuit current of 4.4A and a saturation current of 0.4nA. It is subjected to light at a power of 10 watts. Determine:

   (a)   The open-circuit voltage.

   (b)   The energy-conversion efficiency assuming a fill factor of unity.

   (c)   Repeat part (b) for a fill factor of 0.3.

14.6 A GaAs photodetector is subjected to optical radiation of 12W/m^2 at a wavelength of 0.7μm. The absorption coefficient is 10^4cm^{-1}. Determine:

   (a)   The rate at which electron-hole pairs are generated.

   (b)   The carrier density if the lifetime is 10^{-8}s.

14.7 A silicon PN junction photodiode has the following properties: $A = 10^{-4}\text{cm}^2$, $N_A = 10^{17}\text{cm}^{-3}$, $N_D = 10^{16}\text{cm}^{-3}$, $\mu_n = 800\text{cm}^2/\text{V-s}$, $\mu_p = 400\text{cm}^2/\text{V-s}$, and $\tau_p = \tau_n = 10^{-8}$s. It is subjected to a reverse bias of 3V. The rate of electron-hole pairs generated by light is $10^{23}\text{cm}^{-3}\text{s}^{-1}$. Determine:

   (a)   The photocurrent in the diode, assuming that $W \ll L_n, L_p$.

   (b)   Repeat part (a) but include the width of the depletion layer.

14.8 **a)** Derive an expression for the short-circuit current of a solar cell.

b) A silicon cell has a photocurrent of 4A and a saturation current of 0.4mA. Determine the open-circuit voltage.

14.9 A certain solar cell has a saturation current of 5nA and when it is subjected to optical radiation it has a short-circuit current I_{sc} = 40mA.

a) Determine the open-circuit voltage.

b) Derive an expression for V_m at which maximum power is delivered.

c) Determine the maximum power available.

14.10 A solar cell has a I_{ph} = 35mA, I_S = 5 × 10^{-11}A, and area = 2 cm^2. Determine:

a) The open-circuit voltage.

b) The short-circuit current.

c) The power delivered by the cell if the fill factor is 80%.

14.11 A solar cell has an open-circuit voltage of 0.6V and a short-circuit current = 35mA. It is to be used in a circuit that requires a maximum of 20W at 12V. Determine:

a) If the fill factor is 0.81, the maximum power delivered by each cell.

b) The number of cells required in series, assuming $V_m = (FF)^{0.5}V_{oc}$.

c) The number of cells required in parallel, assuming $I_m = (FF)^{0.5}I_{sc}$.

d) The total number of cells needed.

APPENDICES

A: UNITS

FUNDAMENTAL QUANTITIES

Quantity	Unit	Symbol
Capacitance (C)	Farad	F
Conductance (G)	Siemens	S
Current (I)	Ampere	A
Electric Charge (Q, q)	Coulomb	C
Energy (W, E)	Joule	J
Force (F)	Newton	N
Frequency (f)	Hertz	Hz
Length (L)	Meter	m
Mass (m)	Kilogram	kg
Potential (V, ϕ)	Volt	V
Power (P)	Watt	W
Resistance (R)	Ohm	Ω
Temperature (T)	Degrees Kelvin	K
Time (t)	Second	s

DERIVED QUANTITIES

Quantity	Unit	Symbol
Angular Frequency (ω)	Radian/sec	rad/s
Conductivity (σ)	Siemens/meter	S/m
Dielectric constant (ε_r) (Relative Permittivity)	Dimensionless	
Permittivity ($\varepsilon, \varepsilon_{ox}$)	Farad/meter	F/m
Resistivity (ρ)	Ohm-m	Ω-m
Velocity (v)	Meters/second	m/s
Wavelength (λ)	Meters	m

B: PHYSICAL CONSTANTS AND FACTORS

B.1. PHYSICAL CONSTANTS

Quantity	Symbol	Value
Avogadro's Number	N	$6.022 \times 10^{23}/m$
Boltzmann Constant	k	$1.38066 \times 10^{-23} J/K$
Boltzmann Constant	k/q	$8.61738 \times 10^{-5} eV/K$
Magnitude of Electronic Charge	q	$1.602 \times 10^{-19} C$
Electronvolt	eV	$1.602 \times 10^{-19} J$
Electron Rest Mass	m_0	$9.109 \times 10^{-31} kg$
Permittivity in Vacuum	ε_0	$8.854 \times 10^{-12} F/m$
Planck's Constant	h	$6.626 \times 10^{-34} J\text{-}s$
Speed of Light in Vacuum	c	$2.998 \times 10^8 m/s$
Thermal Voltage	$V_t = kT/q$	$0.02586\ V(T = 300K, 27°C)$

B.2 PREFIXES

$$k = kilo, M = mega = 10^6, G = giga = 10^9, m = milli = 10^{-3},$$
$$\mu = micron = 10^{-6}, n = nano = 10^{-9}, p = pico = 10^{-12}$$

B.3 CONVERSION FACTORS

$$1\text{Å} = 10^{-8} cm = 10^{-10} m$$

$$1nm = 10^{-9} m = 10\text{Å} = 10^{-7} cm$$

$$1\mu m = 10^{-4} cm = 10^{-6} m$$

$$1eV = 1.602 \times 10^{-19} J$$

C: THE DENSITY OF STATES N(E)

By assuming the free electron model for a metal, the energy states for electrons will be the same as for electrons in a three-dimensional potential well where the potential is V = 0 inside the well and infinite elsewhere.

Schrodinger's equation in rectangular coordinates inside the well for electrons having constant total energy E becomes

$$\frac{\partial^2 \Psi}{\partial x^2} + \frac{\partial^2 \Psi}{\partial y^2} + \frac{\partial^2 \Psi}{\partial z^2} + \frac{8\pi^2 mE}{h^2} \Psi = 0 \tag{C.1}$$

This equation is solved by the method of separation of variables by assuming

$$\Psi = g_x(x)g_y(y)g_z(z) \tag{C.2}$$

By substituting this solution in Eq. (C.1) we have

$$\frac{1}{g_x}\frac{\partial^2 g_x}{\partial x^2} + \frac{1}{g_y}\frac{\partial^2 g_y}{\partial y^2} + \frac{1}{g_z}\frac{\partial^2 g_z}{\partial z^2} = -\frac{8\pi^2 mE}{h^2} \tag{C.3}$$

Since the right-hand side of Eq. (C.3) is a constant, it follows that each of the terms is a constant so that

$$\frac{\partial^2 g_x}{dx^2} = G_1^2 g_x, \frac{d^2 g_y}{dy^2} = G_2^2 g_y, \frac{d^2 g_z}{dz^2} = G_3^2 g_z \tag{C.4}$$

We assume for the sake of simplicity that each of the sides of the cubic well has length a. The boundary conditions for ψ are

$$\psi = 0 \text{ at } x = y = z = 0$$
$$x = y = z = a \tag{C.5}$$

The solution to each of the equations in (C.4) contains sine and cosine terms. Possible solutions, wherein the cosine terms drop out and, subject to the boundary conditions in Eq. (C.5), become

$$g_x = A \sin \frac{n_x \pi x}{a} \qquad \text{(a)}$$

$$g_y = B \sin \frac{n_y \pi y}{a} \qquad \text{(b)} \tag{C.6}$$

$$g_z = C \sin \frac{n_z \pi z}{a} \qquad \text{(c)}$$

where n_x, n_y and n_z are quantum numbers and A, B, and C are new constants.

Using the above expressions in Eq. (C.3) we have

$$n_x^2 + n_y^2 + n_z^2 = \frac{8mEa^2}{h^2} \tag{C.7}$$

In the three coordinate system that contains n_x, n_y, and n_z, we assume that their maximum values are N_x, N_y, and N_z. The reason for this is to determine the number of states that have energies less than some value E given by

$$E \geqslant \frac{h^2(N_x^2 + N_y^2 + N_z^2)}{8\,ma^2} \tag{C.8}$$

We are actually interested in a section of the sphere of the solid that is made by joining the three maximum values for which the radius of the sphere R is $\sqrt{N_x^2 + N_y^2 + N_z^2}$. That section is one for which the values of n_x, n_y, n_z are all positive. This section forms one eighth of the sphere, shown in Fig. CF.1, having volume $\pi R^3/6$ or

$$\text{Volume} = \frac{\pi}{6}(N_x^2 + N_y^2 + N_z^2)^{3/2} \tag{C.9}$$

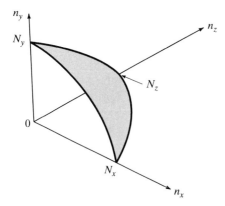

Figure C.F.1 Section of sphere having n_x, n_y, and n_z as coordinates.

By substituting Eq. (C.8) in Eq. (C.9), we obtain

$$\text{Volume} = \frac{\pi}{6}\left(\frac{8\,ma^2E}{h^2}\right)^{3/2} \tag{C.10}$$

Each point inside the section of Fig. C.F.1, corresponding to integer values of n_x, n_y and n_z represents an energy state that contains a pair of electrons. Thus the total volume inside this section represents the total number of states. Since each quantum state, because of the quantum spin, can accommodate up to two electrons, at the absolute zero of temperature the volume represents one half the number of electrons N. We can then write

$$N = \frac{\pi}{3}\left(\frac{8\,ma^2E}{h^2}\right)^{3/2} \tag{C.11}$$

The question is, what is the distribution of the density of the electrons as a function of energy? We first introduce the following terms defined as

$N(E) = $ volume density of states per unit energy at energy E

$N(E)dE = $ volume density of states with energies from E to $E + dE$

$N_S = $ volume density of occupied states where

$$N_S = \frac{1}{2} \times \frac{\text{Number of electrons}}{\text{Volume}} = \frac{1}{2}\frac{N}{a^3}.$$

The term N_s can be written as

$$N_s = \frac{\pi}{6}\left(\frac{8mE}{h^2}\right)^{3/2} \tag{C.12}$$

where N_s represents the total density of states from zero energy to E.

The rate of change of the density of states becomes

$$\frac{dN_s}{dE} = \frac{\pi}{4}\left(\frac{8m}{h^2}\right)^{3/2} E^{1/2} \tag{C.13}$$

Thus Eq. (C.13) represents the increment in the density of states in an increment dE at energy E. This is what we defined as $N(E)$ so that

$$N(E) = \frac{\pi}{4} \left(\frac{8m}{h^2}\right)^{3/2} E^{1/2} \tag{C.14}$$

At absolute zero, electrons occupy the lowest levels possible. We assume that all states up to $x = y = z = a$ are occupied and for $x > a$ there are no electrons at all. Thus one can determine the maximum energy that electrons can have at absolute zero from Eq. (C.11). We label that energy the Fermi energy E_F. Once E_F is fixed, the highest quantum states at absolute zero are fixed and E_F is defined as the maximum energy that an electron in a certain material can have at absolute zero.

We determine E_F from Eq. (C.11) as

$$E_F = \frac{h^2}{8m} \left(\frac{3N}{a^3\pi}\right)^{2/3} \tag{C.15}$$

As an example, assume that the density of free electrons in silver is $5.9 \times 10^{28} \text{m}^{-3} = N/a^3$. We solve for E_F to be $5.51 eV$.

D: EINSTEIN RELATION

By considering the expressions for the drift and diffusion current densities given by Eqs. (4.15) and (4.16), it seems reasonable to expect that, other than the particle charge, there must be some other commonality between these 2 equations since both refer to the motion of carriers in the same solid. This commonality exists between the mobility and the diffusion constant.

To determine the relation between the mobility and the diffusion constant, we consider a slab of semiconductor that has been doped to a nonuniform distribution of electrons in a region, as shown in D.F1a. For x < 0 the material is intrinsic, for 0 < x ≤ d the distribution of electrons is increasing while for x > d the density of electrons is constant.

At thermal equilibrium electrons diffuse from the region around x = d to that at x = 0 leaving behind positively ionized atoms. An electric field is thus established between x = d and x = 0 directed from x = d to x = 0. Electrons continue to diffuse from x = d to x = 0, and since at thermal equilibrium the electron current is zero, the electric field returns the same number of electrons from x = 0 to x = d. In accordance with Eq. (4.15) the electron drift current is directed from right to left and in the direction of the electric field while the electron diffusion current is from left to right. For the total electron current to be zero we have

$$q\mu_n n \mathscr{E} = \frac{-q D_n dn}{dx} \tag{D.1}$$

We show, in Fig. D.F1b the potential distribution in the slab, and in Fig. D.F1c the energy band diagram. The electron density in Fig. D.F1 can be expressed as

$$n(x) = n_i \exp\left[\frac{E_F - E_i(x)}{kT}\right] = n_i \exp\left[\frac{E_i(0) - E_i(x)}{kT}\right] \tag{D.2}$$

The electric field intensity and the potential energy of an electron are related to the potential by

$$\mathscr{E}(x) = -\frac{d\phi(x)}{dx} \qquad \text{(a)}$$

$$E_i(x) = -q\phi(x) \qquad \text{(b)} \tag{D.3}$$

Since $\phi(0) = 0$ and hence $E_i(0) = 0$ we can write Eq. (D.2) as

$$n(x) = n_i \exp\left[\frac{q\phi(x)}{kT}\right] \text{ for } 0 \leqslant x \leqslant d \tag{D.4}$$

By using Eqs. (D.3) and (D.4) in Eq. (D.1) we have

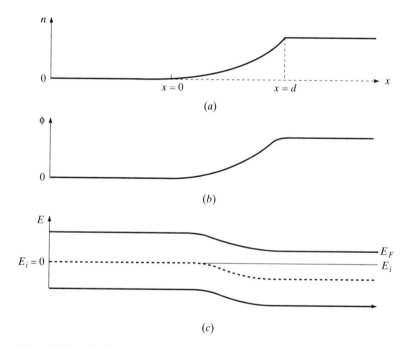

Figure D.F1 Distribution of electrons, potential and energy versus distance.

$$\mu_n\left(\frac{-d\phi(x)}{dx}\right) n_i \exp\frac{q\phi(x)}{kT} = -q\, D_n \frac{n_i}{kT}\frac{d\phi(x)}{dx}\exp\frac{q\phi(x)}{kT} \tag{D.5}$$

so that

$$D_n = \left(\frac{qT}{q}\right)\mu_n \tag{D.6}$$

The expression in Eq. (D.6) is known as the Einstein relation.

E: FREQUENTLY USED SYMBOLS

a	=	acceleration (cm^2/s), metallurgical channel halfwidth for JFET (cm)
A	=	area (cm^2)
B, E, C	=	base, emitter, collector of a BJT
BV_{CBO}	=	BJT breakdown voltage collector to base with emitter open
BV_{CEO}	=	BJT breakdown voltage collector to emitter with base open
c	=	speed of light (cm/s)
C_j	=	diode junction capacitance, depletion capacitance, transition capacitance (F)
C_s	=	diode charge storage capacitance, diffusion capacitance (F)
C_{je}	=	BJT emitter-base junction (depletion region) capacitance (F)
C_b	=	BJT base storage (diffusion) capacitance
C_u	=	BJT collector-base capacitance
C_{gd}, C_{gs}	=	JFET and MOSFET gate-drain capacitance, gate source capacitance
D_n, D_p	=	diffusion constant for electrons, holes (cm^2/s)
D_{nE}, D_{nC}	=	diffusion constant for electrons in the emitter, in the collector (cm^2/s)
D, G, S	=	drain, gate, source of an FET
$\mathscr{E}$	=	electric field intensity (V/cm)
E	=	energy (J, eV)
E_A, E_D	=	acceptor, donor energy level (J, eV)
E_C, E_V	=	energy level of bottom of the conduction band, top of the valence band
E_F	=	equilibrium Fermi level
E_g	=	bandgap or forbidden gap (J, eV)
E_i	=	intrinsic Fermi level
E_0	=	zero vacuum level of energy
F	=	Force (N)
f	=	frequency (Hz)
$f_c(E)$	=	Fermi function for conduction band
$f(E)$	=	Fermi–Dirac distribution function
f_T	=	unity current-gain frequency
g_d	=	small-signal drain conductance
g_m	=	small-signal transconductance
g_{ph}	=	number of photogenerated electron-hole pairs per second per unit volume

G_o	=	channel conductance of JFET, MESFET in the absence of a depletion region
h	=	Planck's constant (J-s, eV-s)
hν	=	photon energy
I	=	current (A)
I_B, I_C, I_E	=	BJT DC base, collector, emitter current
i_b, i_c, i_e	=	BJT varying component of base, collector, emitter current
i_C	=	BJT total collector current (AC + DC)
i_c	=	BJT varying component of collector current
I_{CBO}	=	BJT DC collector to base current with emitter open
I_{CEO}	=	BJT DC collector to emitter current with base open
I_{CS}, I_{ES}	=	BJT reverse saturation current with collector, emitter shorted to base
I_D	=	FET drain current
I_{D1}, I_{D1S}	=	MOSFET DC drain current, saturation drain current for square-law model
I_{D2}, I_{D2S}	=	MOSFET DC drain current, saturation drain current for bulk-charge model
i_d	=	FET varying component of drain current
i_D	=	FET total drain current (AC + DC)
I_{ph}	=	photogenerated current
I_{SAT}	=	FET saturation drain current
I_R	=	Diode recombination current in depletion layer
I_S	=	PN junction diode reverse saturation current
J_n, J_p	=	current density electrons, holes
k	=	Boltzman constant
k	=	wave vector
k_p	=	MOSFET device parameter
L	=	dimension of lattice cube, FET channel length
L'	=	FET effective channel length (reduced by modulation)
L_n, L_p	=	diffusion length of electrons in P, of holes in N
L_{nE}, L_{pB}	=	diffusion length of electrons in emitter, of holes in base
M	=	carrier multiplication factor
m	=	mass
m_0	=	electron rest mass
$m*_n, m*_p$	=	effective mass of electron, of hole
$n, \mathbf{n}$	=	density of electrons (cm^{-3}), quantum number
n'	=	excess density of electrons over equilibrium
n_0	=	thermal equilibrium electron density
n_{0C}	=	thermal equilibrium electron density in collector
n_{0E}	=	thermal equilibrium electron density in emitter
n_{0p}	=	thermal equilibrium electron density in P region

n_{0n}	=	thermal equilibrium electron density in N region
$n(0)$	=	electron density at origin ($x = 0$)
n_i	=	intrinsic density of electrons and holes
n_s	=	surface density of electrons
N, P	=	semiconductor doped with donors, with acceptors
N^+, P^+	=	heavily doped N, P
N_A, N_D	=	density of acceptors, donors
$N_{AE}, N_{DB},$ N_{AC}	=	density of acceptors in emitter, donors in base, acceptors in collector
$N(E)$	=	density of states function per volume per unit energy
N_c, N_v	=	effective density of states per unit energy at E_C, at E_V
p	=	hole density
p_{0n}	=	thermal equilibrium hole density in N region
p'	=	excess hole density over equilibrium
p_{0p}	=	thermal equilibrium hole density in P region
$p(0), p_n(0),$ $p_p(0)$	=	hole density at $x = 0$, in N region, in P region
q	=	magnitude of electron charge
P_{ph}	=	total power absorbed by optical detector
Q	=	total charge
Q_B	=	excess minority carrier charge stored in BJT base
Q_d	=	MOSFET charge density in depletion layer
Q_{dm}	=	maximum value of Q_d
Q_n	=	MOSFET electron charge density in channel
Q_s, Q_m	=	MOSFET charge density in semiconductor, in metal
Q_i	=	MOSFET oxide charge density
R_o	=	lattice spacing
r, R	=	resistance
r_d	=	diode small-signal resistance (AC)
r_o	=	BJT output resistance (AC)
r_u	=	BJT AC resistance from collector to base
r_x	=	BJT AC input resistance from base to emitter
r_b, r_e, r_c	=	BJT ohmic resistance; base, emitter, collector
t_s	=	storage time
t_{ON}	=	time to turn device ON
t_{OFF}	=	time to turn device OFF
t_{ox}	=	thickness of insulator in MOSFET
T	=	depress Kelvin
V	=	potential

V_A	=	BJT Early voltage
V_a	=	applied voltage
V_B	=	MOSFET body voltage
V_{bi}	=	built-in voltage
V_j	=	junction voltage $= V_a - V_{bi}$
V_{br}	=	breakdown voltage
v_{ce}, v_{be}	=	BJT varying component of voltage, collector to emitter, base to emitter
v_{ds}, v_{gs}	=	FET varying component of voltage, drain to source, gate to source
V_{EB}, V_{CB}, V_{EC}	=	BJT DC voltage; emitter to base, collector to base, emitter to collector
V_{CC}, V_{BB}	=	DC supply voltage to BJT; collector circuit, base circuit
$V_G (V_{GS}), V_D (V_{DS})$	=	FET DC voltage; gate to source, drain to source
V_{SB}, V_{CS}	=	MOSFET DC voltage; source to bulk, channel to source
V_P	=	FET DC pinchoff voltage
V_{FB}	=	MOSFET DC flatband voltage
V_T	=	FET DC threshold (turn-on) voltage
V_t	=	thermal voltage kT/q
V_{SAT}	=	FET DC voltage at saturation
v_d	=	drift velocity
v_s	=	saturation velocity
v_{th}	=	thermal velocity
W	=	depletion layer width, potential energy
W_B	=	BJT base width
W_p, W_n	=	width of P region, N region
Z	=	FET depth

Greek

α	=	BJT: DC ratio of collector to emitter current
α_F	=	BJT: DC ratio of collector to emitter current with collector shorted to base in active region
α_R	=	BJT: DC ratio of collector to emitter current with emitter shorted to base in inverse active region
β	=	BJT: DC ratio of collector to base current
β_F	=	BJT: DC ratio of collector to base current with collector shorted to base in active region
β_R	=	BJT: DC ratio of collector to base current with emitter shorted to base in inverse action region
β_0	=	BJT: AC ratio of collector to base current with collector shorted to base

γ = BJT: emitter injection efficiency

δ = BJT: base transport factor

ε = permittivity or dielectric constant

ε_r = relative permittivity

ε_o = permittivity of free space

ε_{ox} = permittivity of oxide

λ = wavelength

λ_p = MOSFET channel length modulation factor

μ = mobility

μ_n, μ_p = mobility of electron, holes

$\overline{\mu}_n, \overline{\mu}_p$ = MOSFET effective mobility of electrons, holes

ν = frequency of light

ρ = resistivity

ρ_c = volume charge density

σ = conductivity

σ_n, σ_p = conductivity of N region, of P region

τ_B = carrier base transit time

τ_p, τ_n = lifetime of holes in N region, of electrons in P region

ϕ, ϕ_s = potential, surface potential

ϕ_F = potential from E_i to E_F in bulk of MOSFET

Φ_m = metal work function voltage

Φ_s = semiconductor work function voltage

Φ_{ms} = work function difference from metal to semiconductor, $\Phi_m - \Phi_s$

χ_s = affinity of semiconductor, $E_0 - E_C$

INDEX